Computations and Computing Devices in Mathematics
Education Before the Advent of Electronic Calculators

MATHEMATICS EDUCATION IN THE DIGITAL ERA
Volume 11

Alexei Volkov · Viktor Freiman
Editors

Computations and Computing Devices in Mathematics Education Before the Advent of Electronic Calculators

 Springer

Editors
Alexei Volkov
Center for General Education
 and Institute of History
National Tsing Hua University
Hsinchu
Taiwan

Viktor Freiman
Université de Moncton
Moncton, NB
Canada

ISSN 2211-8136   ISSN 2211-8144   (electronic)
Mathematics Education in the Digital Era
ISBN 978-3-319-73394-4   ISBN 978-3-319-73396-8   (eBook)
https://doi.org/10.1007/978-3-319-73396-8

Library of Congress Control Number: 2017962537

This Springer imprint is published by the registered company Springer Nature Switzerland AG
The registered company address is: Gewerbestrasse 11, 6330 Cham, Switzerland

Contents

Editors and Contributors

About the Editors

Alexei Volkov is a professor of the Center for General Education and of the Graduate Institute of History of the National Tsing Hua University (Hsinchu, Taiwan). His research focuses on the history of mathematics and mathematics education in East and Southeast Asia. He has published a number of papers and book chapters on these topics, including "Didactical dimensions of mathematical problems: 'weighted distribution' in a Vietnamese mathematical treatise", in Alain Bernard and Christine Proust (eds.), *Scientific Sources and Teaching Contexts Throughout History: Problems and Perspectives*, Dordrecht etc: Springer, 2014, pp. 247–272 and "Argumentation for state examinations: demonstration in traditional Chinese and Vietnamese mathematics", in Karine Chemla (ed.), *The History of Mathematical Proof in Ancient Traditions*, Cambridge University Press, 2012, pp. 509–551.

Viktor Freiman is a Full Professor of Mathematics Education at the Université de Moncton, Canada. His work, besides the history of mathematics education, focuses on the use of digital technology in teaching and learning, interdisciplinary learning, computational thinking, design thinking, creativity, mathematical problem-solving, and giftedness. Since 2014, he is director of CompeTI.CA (Compétences en TIC en Atlantique) partnership network, studying a lifelong development of digital competences. He is co-editor of the Springer Book Series Mathematics Education in the Digital Era.

Contributors

Yifu Chen Independent Scholar, Kaohsiung, Taiwan

Jean Christianidis Department of History and Philosophy of Science, National and Kapodistrian University of Athens, Athens, Greece; Centre Alexandre Koyré, Paris, France

Marion Cousin Institut d'Asie Orientale, École normale supérieure de Lyon, Lyon, France

Viktor Freiman Université de Moncton, Moncton, NB, Canada

Jens Høyrup Roskilde University, Roskilde, Denmark; Institute for the History of Natural Sciences, Chinese Academy of Sciences, Beijing, China; Max-Planck-Institut Für Wissenschaftsgeschichte, Berlin, Germany

Alexander Karp Teachers College, Columbia University, New York, NY, USA

Takanori Kusuba History and Philosophy of Science, Osaka University of Economics, Osaka, Japan

Jean-François Maheux Université du Québec à Montréal, Montréal, QC, Canada

Dragana Martinovic University of Windsor, Windsor, ON, Canada

Athanasia Megremi Department of History and Philosophy of Science, National and Kapodistrian University of Athens, Athens, Greece

Duncan J. Melville St. Lawrence University, Canton, NY, USA

Young Sook Oh Program of History and Philosophy of Science, Seoul National University, Seoul, South Korea

Charlotte-V. Pollet Center for General Education, National Chiao-Tung University, Hsinchu, Taiwan

Xavier Robichaud Université de Moncton, Shippagan, NB, Canada

Gert Schubring Universidade Federal do Rio de Janeiro, Rio de Janeiro, Brazil

Alexei Volkov National Tsing-Hua University, Hsinchu, Taiwan

Part I
Introduction

History of Computations, Computing Devices, and Mathematics Education from the Teaching and Learning Perspective: Looking for New Paths of Investigation

Viktor Freiman and Alexei Volkov

Abstract In this introductory chapter, the editors of the volume discuss the main directions of research on the history of counting instruments and computing devices and of their role in the transmission of mathematics knowledge from antiquity to the mid-twentieth century. They briefly present the issues discussed in the book, as well as in each contribution collected in the volume, and suggest that they can be organized according to the five following axes of research on the history of computing devices: (1) history of development of computational instruments per se and of the related computational practices; (2) transmission of expertise related to the counting skills and counting devices; (3) arithmetical operations performed with the instruments and their optimization; (4) mathematical writing and its interrelationship with computations; (5) role of the instruments in mathematics instruction and possible influence of the teaching practices on development of counting instruments.

Keywords History of counting instruments · History of computing devices
Arithmetical operations and their implementation on computing devices
History of mathematics education

It might appear somewhat surprising to find in the series of monographs devoted to *mathematics education in the digital era* a volume on computing devices and computing techniques of the mathematical traditions that antedated the massive advent of electronic calculators and computers in the late twentieth century. However, it can be argued that the development of mathematics, as well as

---

V. Freiman (✉)
Université de Moncton, Moncton, Canada
e-mail: viktor.freiman@umoncton.ca

A. Volkov
National Tsing-Hua University, Hsinchu, Taiwan
e-mail: alexei.volkov@gmail.com

mathematics education, over the human history was intimately interconnected with the evolution of some tangible and visual tools and devices and decline of the others; such tools as pebbles, rods, tables and diagrams, rulers, *abaci*, blocs, and more recently, calculators, computers, and other electronic devices were systematically used as support for teaching and learning. Accordingly, a number of ideas, methods, and approaches discussed by modern educators already existed, albeit in somewhat different form, long before the beginning of the so-called digital era, when learning mathematics became more visual, dynamic, multimodal, and cyber-oriented (Martinovic et al. 2013). However, the multidimensional connection between the mathematical concepts and procedures, didactical approaches, and early computing devices has not been sufficiently studied and arguably merits much more attention from educators, historians of mathematics, and historians of mathematics education.

While the role of computations is still a "hot" topic of mathematics curriculum development worldwide, several issues related to the questions of "what to teach" and "how to teach" in mathematics classroom remain open to debates among mathematics educators. Numerous reforms of mathematics curricula and teaching methods have been undertaken in the twentieth century and continue to take place in the twenty-first century. Their key axes are student-centered *versus* teacher-centered classroom, learning with understanding *versus* rote memorization, dialectical interplay between conceptual and procedural knowledge, students' and teachers' creative use of algorithms including their development by the learners in the context of teaching arithmetic *versus* their automatic execution. Their didactical rationale includes, among other elements, a continuous call for the use of a great variety of computing tools and devices seen not only as aids for calculations but also as important didactical tools at all educational levels, from early schooling up to the university.

Several questions can be raised at this point. In the era of electronic computers and Internet, does everyone need to memorize the so-called number facts? If yes, what kind of "facts" should be learned, and how are they supposed to be memorized? Do the students need to master computational procedures with integer numbers, and if they do, what kind of procedures they should focus upon? With what purpose? In which context? Another series of questions is related to the role of *physical* and *virtual* manipulatives as learning tools:[1] What are they supposed to be, what is their *presumed role* in the instruction, and what is their *actual* role? Whatever the answers to these and many other questions related to teaching and learning calculations may be, historical arguments are often evoked to appeal to the use of various kinds of counters, pebbles, rods, *abaci*, and other devices to be introduced in the classroom, presumably, to enhance the learners' understanding.[2]

---

[1]For an in-depth investigation of virtual manipulatives, we would refer the reader to the Volume 7 of the Series edited by Moyer-Packenham (2016). This new book is devoted to the structure, and the use of physical devices analyzed from the historical and didactical perspectives.

[2]See, for example, Boggan et al. (2010).

It is clear that computing devices and related algorithms used during the long period stretching from the third millennium BC to the twentieth century AD, in East and West, underwent considerable changes. Yet, the present volume provides some insights concerning the history of use of instruments in educational practices that can be, at least to some extent, transposed to be easily recognizable by the present-day educators. The book is written by historians of mathematics and mathematics educators working on historical development of teaching and learning mathematics; it seeks to cater, in the first hand, to a larger community of historians of mathematics, historians of mathematics education, and mathematics educators, while providing them with information about the history of the development of computational techniques, computing devices, and their use for educational ends. We hope that it will be also useful for professional mathematicians interested in didactical aspects of computations and computing devices seen in the historical perspective.

Historically, the development of computing devices and related methods of calculations involved a complex dynamic interaction with mathematical concepts and, in particular, with contemporaneous mathematical practices. The development of mathematical concepts, especially of the concept of number, was different in different sociocultural contexts; accordingly, the foundational principles of the instruments were not identical. The same can be said about the didactical practices in general: They were arguably different as well. The resulting combination of mathematical and didactical backgrounds defining the environment in which a given instrument was used, and of the impact that the instrument, in turn, made upon mathematical and didactical practices, was markedly different in different societies starting from the very beginning of mathematics education up to the twentieth century when a substantial global transformation occurred with the arrival of electronic calculators, computers, and the Internet (Monaghan and Trouche 2016).

The development of methods of numerical solution of computational problems led to algorithms supposed to be performed with the help of the tools that were readily available to the calculator (such as human hand, set of pebbles or sticks, among others) or were especially constructed to perform the operations (various forms of *abaci*, tables, mechanical and, lately, electronic calculators and computers); special attention was often paid to finding the simplest and most efficient algorithms. The algorithms themselves can thus be viewed as particular "virtual instruments," and one can conjecture that their transmission from generation to generation was conducted via especially designed didactical activities and methods. These activities, together with the methods related to the operations with the instruments, thus constituted the essential part of the mathematics instruction in numerous traditions. The didactical methods related to algorithms were, in turn, constantly adjusted to a particular culture and context, and sometimes redefined in the process of instruction.

Traders used computing instruments for commercial transactions; engineers and scientists (in particular, astronomers and mathematicians) used them to perform their (sometimes rather complex) computations while executing a variety of tasks, and professional mathematics educators taught operations with the instruments in their classrooms or used them for educational purposes (e.g., as visual support);

finally, there were textbooks presenting information related to arithmetical operations which in some (but not all!) cases were supposed to be performed with the instruments. Computing devices and related algorithms thus belonged to two interrelated realms: On the one hand, they were part of the domain of computational mathematics; on the other, they constituted the medium of interaction between the learners, instructors (teachers), and mathematical knowledge. In other words, they might have generated, in a process of *instrumental genesis,* "co-emergence of schemes and techniques for using the artefact" (Drijvers et al. 2010, p. 213). Each application of an instrument imposed particular requirements and limitations, and when the use of the instrument in one of these two realms became obsolete, the requirements of the other defined the transformation of the instrument and of the modes of operations with it. For instance, the beads abacus, originally designed as a computational instrument, gradually became a "didactical device" (in particular educational contexts, such as primary school mathematics); in the process of this transformation, its "software" and "hardware" were adjusted accordingly. One can even say that once the instruments and related techniques of computations entered classrooms and textbooks, they *changed their nature:* From computing devices per se they became pedagogical artifacts or tools, and their affiliation, at least in classroom, shifted from the field of (applied and abstract) mathematics to the field of didactics.

This was, for instance, the case of arithmetic teaching in several parts of the world in the nineteenth century, when several new didactical trends emerged; they continue influencing mathematics classes in schools even now. One of these trends was related to the introduction of physical manipulatives along with other kinds of visual support for representation of numbers and for performing calculations. Among the tools that could be used for this end there were concrete objects well known to the learners from everyday life (e.g., sticks or matches), objects especially designed for representation of numbers (such as tables and number figures), and instruments adopted for performing arithmetical operations (in particular, various kinds of *abaci*).[3] While the history of introduction of the beads abacus in schools in the nineteenth and twentieth century is still a relatively underexplored topic, at least two directions of further investigation could be mentioned: (1) the transmission of Russian abacus to Europe, often credited to the French mathematician J.-V. Poncelet (1788–1867; see Gouzévitch, & Gouzévitch 1998) and its further adaptation as didactical tool, and (2) the influence of Pestalozzian school of pedagogical thought, especially in the North American education (Menninger 1969; Roberts et al. 2012; Roberts 2014).

By all means, this tradition seems to be alive and thriving in today's schools: for instance, the most recent curriculum (2016) in New Brunswick, Canada, puts

---

[3]Like three tables with marks, introduced in Pestalozzi's school, number figures (points) or Tillich's arithmetic box or "brickbox" (see, for example, Barnard 1906; Bjarnadóttir 2014; Szendrei 1996). Different models of numeral frames (bead *abaci*) were used in infant schools (called in France *salles d'asile*), like Wilderspin's (1840) device (Kidwell et al. 2008), or the "boulier" designed by Marie Pape-Carpantier (1815–1878), see (Pape-Carpantier 1878, 1901; Regnier 2003).

emphasis on the use of concrete objects, dot plates ("assiettes à points" in French), ten frames, number line, and the twenty-bead abacus *Rekenrek* ("calculating frame" in Dutch) to represent numbers and operations in primary school grades (MEDPENB 2016, p. 61). Moreover, the capacity of doing calculations with the use of the *Rekenrek* becomes a specific learning outcome.

The present-day growing interest in introducing such seemingly archaic tools and devices in the arithmetic teaching in the era of iPads, laptops, and Internet can be even more intriguing if we take into consideration the expansion of the use of the Chinese abacus by today's students worldwide. For instance, a 2013 study of the impact of the abacus learning on cognitive development of 8–12-year-old students in India reports the improvement in the following seven abilities: concentration, problem solving, associative memory, working memory, concept formation, creativity, and ability to create mental images and to perform operations with them ("spatial ability") (Vasuki 2013). Another three-year long study of 204 children (5–7 year old) on the learning mathematics in a visual/spatial format using so-called Mental Abacus technique was conducted in India (Khudson 2016); it confirmed that an experimental group outperformed a control group on arithmetic tasks and suggested that Mental Abacus expertise can be achieved by children in standard classrooms (Barner et al. 2016). Recently, a Turkish study on the use of the Soroban abacus (Japanese version of Chinese abacus) by 14 teachers working on mental arithmetic with students of 7–12 year old seems to confirm a positive effect of such training on their problem-solving abilities, creativity, understanding of the concepts, and attraction toward mathematics lesson (Altiparmak 2016).

Not only physical manipulatives, as well as other computing devices, but also the procedures themselves remain important items of educational agenda with the recent shift from the (almost automatic) execution of technical steps of algorithms to the quest for deeper understanding of *how* the algorithms work, why they produce the expected results, what are the other possible ways to perform the same operations or obtain the same results, and what way is the most efficient in a given context. For example, Varol and Farran (2007) discuss the benefits of mental calculations where students create different approaches for better understanding of how computational procedures work. While the devices used in today's schools, as well as the implicit goals of introducing them to the students (such as increasing students' capacity to reason about quantities and deepening their understanding of numbers and operations), can be different from those formulated by Pestalozzi and his followers in the nineteenth century, the main idea of providing students with visual, intuitive, and hands-on didactical support is still an important educational issue in the twenty-first-century classrooms. Moreover, from the computational practices prior to the nineteenth century, mathematics educators can find remarkably ingenious and timely examples of dealing with tasks that required performing complex calculations, finding a way to simplify them, and often producing new tools as aids to calculations. In this creative process, the task of producing tools and methods of calculations led to the problem of translating knowledge from one person to another, from one group to another, from one context to another, etc., and eventually to some important transformation of both tools and methods, which is, to our understanding, an important part of purely didactical work, a process that certainly deserves a deeper investigation.

In this respect, we found it meaningful to initiate a discussion of the ways in which different cultures in different historical periods were dealing with computations, what types of counting devices were used, and how particular features of these devices may have influenced the procedures designed for them. From the outset of the project, the main intention was to find what may have been the didactical repercussions imposed by the devices upon the didactical practices and, conversely, to identify the ways in which the devices were interpreted and used for didactical purposes. Certainly, it would not be realistic to expect that the authors who contributed to the present volume could provide answers to all these questions; instead, every chapter of the book is treating only some of them—while some authors focused mainly on the history of particular devices, others dealt with their use in mathematics instruction or discussed the algorithms used for solutions of problems that may or may not have been designed for computational tools while being certainly related to operations with numbers or even with symbolic expressions.

Going back to historical roots of mathematical practice, the expertise of historians of science is particularly valuable since it could provide us with deeper insights into a number of questions which certainly need to be answered. How were the computing devices and related algorithms conceived? How were the algorithms described, applied, and improved? How were they taught? Did the algorithms and computing devices used to perform them influence each other, and if they did, how? How the use of the computing devices and algorithms influenced didactical practices and design of mathematics textbooks? Did these didactical practices, in turn, influence computing devices and algorithms and/or the ways in which they were used? The primary sources that may be used for an attempt to answer these and related questions are often scarce, if not missing, especially in early periods of the human history, and only sometimes can we use historical or even archeological artifacts to make relatively informed guesses concerning the educational practices related to these devices.

The book is subdivided into **four sections**, three of which are dealing with particular geographical zones and specific chronological periods (with possible overlaps). The *first section* is devoted to the computational practices in ancient Mesopotamia, Byzantine Empire, and medieval India; it opens with the contributions of **Duncan J. Melville** "Computation in Early Mesopotamia" and of **Jens Høyrup** "Computational Techniques and Computational Aids in Ancient Mesopotamia." The authors focus on slightly overlapping periods in the history of Mesopotamian mathematics: While Melville's contribution is devoted especially to the third millennium BC, Høyrup's chapter deals with materials dated from the mid-third millennium BC to the Seleucid period (third–second centuries BC). Both authors treat in great detail the computational procedures employed in the respective periods and suggest that a counting board was used in Mesopotamia. The first author who advanced such a conjecture was Adolf Leo Oppenheim (1904–1974); in his paper of 1959, he wrote:

We know of the use of stone counters (Latin *calculi*) on counting boards from Classical and later sources, and it is not excluded that similar devices were used in the ancient Near East for addition, subtraction, etc. [...] although no indications concerning the use of this technique can be found in cuneiform mathematical texts or elsewhere. (p. 124)

The particularities of the grid drawn or carved on the surface of this hypothetical device (and especially the number of positions) remained unknown. Later, while providing a meticulous analysis of arithmetical errors found in the tablet N 3958 from Nippur that had been published and studied by Sachs (1947, pp. 228–229),[4] Proust (2000, p. 302) asked whether the author of the table found in this tablet used a computational device with only five sexagesimal positions. In his contribution to the present volume, J. Høyrup accepts Proust's hypothesis concerning this partic- ular computing instrument and offers its tentative reconstruction: According to him, the instrument may have been designed as a wooden tablet with one or two rows of square or rectangular cells; the number of cells in each row was equal to five.

Even more important for the present discussion is the very nature of Mesopotamian mathematical texts essentially based on the concept of algorithm. This particularity was already discussed by Donald Knuth in his pioneering paper (1972) published in a journal primarily addressed to the community of computer scientists; this, at least to some extent, may explain why it remained virtually unknown to historians of mathematics.[5] More recently, Jim Ritter (2010) analyzed several problems found on Old Babylonian cuneiform tablets and demonstrated that the Babylonian mathematical texts resembled computer programs with embedded sub-algorithms; he also stressed that "the structural reuse of early algorithms in later ones" is related to "pedagogical design" of the tablets (p. 359).

The chapter "Interpreting Tables of the *Arithmetical Introduction* of Nicomachus Through Pachymeres' Treatment of Arithmetic: Preliminary Observations" is written by **Athanasia Megremi** and **Jean Christianidis**. The authors explore computational and procedural (or algorithmic) tradition present in the Greek mathematical literature; they focus on the work of Nicomachus of Gerasa *Arithmetical introduction* (first half of the second century AD) that has been conventionally seen as purely theoretical. The authors argue that this text was an important contribution to a certain arithmetical practice that used arithmetical tables as a device for problem solving. The authors also offer important insights into the arithmetical parts of the *Quadrivium* compiled by Georgios Pachymeres (1242–ca. 1310) and conjecture that Pachymeres may have used the table of Nichomachus. It could be interesting to compare these mathematical tables with those used by the Armenian mathematician Anania Shirakatsi (also known as Ananias of Shirak/Širak, 610–685) whose work was strongly influenced by Byzantine mathematical tradition.[6]

---

[4]In Sachs (1947) the order of pages 228–229 containing the discussion of N 3958 is reversed.

[5]Ritter (2010, p. 350, n. 1).

[6]On mathematical work of Shirakatsi, see Orbeli (1918), Hewson (1968), Eganyan and Maistrov (1982) and Greenwood (2011). We would like to thank Professor Timothy W. Greenwood for kindly sending us a copy of his paper (2011).

In his chapter "The So-Called "Dust Computations" in the *Līlāvatī*," **Takanori Kusuba** deals with the Indian arithmetical treatise *Līlāvatī* written in 1150 by the outstanding medieval Indian mathematician Bhāskara II (1114–?) also known as Bhāskarācārya. In the opening part of his arithmetical treatise, Bhāskarācārya provides a systematic presentation of the elementary arithmetical operations presumably performed on a dustboard, i.e., a (white) wooden board covered with a thin layer of colored dust or sand. The chapter provides a very detailed description of the procedures (algorithms) used to perform the basic arithmetical operations. Kusuba's well-researched analysis is based on the explanations and examples provided in commentaries of medieval Indian mathematicians on Bhāskarācārya's treatise, but it is certain that the arithmetical operations presented in the twelfth century work of Bhāskara had a long history going back to the first millennium AD.

In her chapter "Reading Algorithms in Sanskrit: How to Relate Rule of Three, Choice of Unknown, and Linear Equation?" **Charlotte-V. Pollet** discusses the methods of solution of algebraic equations presented in the *Bījagaṇitāvataṃsa* (*Garland of Seed-Mathematics*) by Nārāyaṇa (or Nārāyaṇa Paṇḍita, fl. mid-fourteenth century) that remained underexplored until now. Pollet provides a meticulous analysis of the method of solution of linear equations used by Nārāyaṇa and, one more time, draws attention to the question concerning the material support used to perform calculations. Her analysis of solutions of equations highlights the role of tabular settings which may indicate the use of some writing medium such as sand or a dustboard allowing for particular organization and modification of data in the process of computations. The functioning of the algorithm which can be itself seen as a kind of "computing device" is discussed here. The text studied by Pollet offers a new insight into the operations with numerical arrays representing linear equations. Representations of this kind used earlier by Brahmagupta (b. ca 598–d. after 665) in a treatise of 628 to work with quadratic equations are briefly mentioned in a footnote of Kusuba's chapter, but in the chapter of Pollet their application to the solution of linear equations is discussed in great detail, and her interpretation helps us better understand not only a certain "automatization" of the procedure, but also approaches to its verification and justification, which brings the question of meaning of mathematical procedure up front to the reflection and discussion by mathematics educators.

The chapters of the *second section* of the book focus on the use of counting instruments in East Asia (China, Japan, and Korea) from the very beginning of their use in China in the first millennium BC, to the early twentieth century. The section begins with the paper by **Alexei Volkov** "Chinese Counting Rods: Their History, Arithmetic Operations, and Didactic Repercussions," which deals with a tradition established in mathematical practice in China in the first millennium BC where computations were performed with the help of so-called counting rods made of wood, bamboo, ivory, or other materials. The author emphasizes the versatility of the instrument which combined simplicity, from the technological point of view, and complexity of mathematical objects that could be realized with it, starting from mere operations with integer and fractional numbers (positive as well as negative) and including algebraic operations involving the solution of simultaneous linear equations with unlimited number of unknowns, polynomial equations of higher degrees, and simultaneous polynomial equations in up to four unknowns. This variety and complexity of the mathematical objects represented and operated with this relatively simple instrument was directly related to the well-designed system of representation of the mathematical objects as well as to a large number of algorithms offering efficient and adaptable tools for solution of different classes of problems. Mathematical manuals styled as collections of problems to be solved with the counting rods were used for mathematics instruction in state-run and private mathematical schools, which made this computing instrument very popular for a long period of time in China and across Chinese border, in Korea, Japan, and Vietnam.

In her second contribution to the volume titled "Interpreting Algorithms Written in Chinese and Attempting the Reconstitution of Tabular Setting: Some Elements of Comparative History," **Charlotte-V. Pollet** focuses on the methods of solution of algebraic equations discussed in the mathematical treatise *Yigu yanduan* 益古演段 (The Development of Pieces [of Area according to the Collection] Augmenting the Ancient [Knowledge]) compiled in 1259 by one of the most outstanding medieval Chinese mathematicians, Li Ye 李冶 (1192–1279). Pollet's meticulous analysis of the treatise shows, in great detail, how the text problems were transformed into quadratic equations, how these equations were represented with counting rods, and how they were solved. In the concluding part of her chapter, Pollet offers valuable insights into the similarities and differences between Chinese methods of operations with quadratic equations with those practiced by contemporaneous Indian mathematicians whose work the author discussed in the first section of the book. She concludes that in both traditions, while being essentially different, the studied methods shared some common points, namely the operation of division was perceived as the foundation of the methods used for solving equations, and tabular settings played equally important role when a special counting device was used (China) or the tables were represented on a dustboard (India).

The educational, or more precisely, didactical aspects of evolution of computations and computational tools add complexity to their analysis in terms of variability of social and cognitive practices and traditions which developed with the time and travelled through different cultures and manifested substantial differences even within one country or tradition; this complexity is exemplified by **Young Sook Oh** in her chapter "Same Rods, Same Calculation? Contextualizing Computations in Early Eighteenth-Century Korea." By analyzing computational practices presented in two mathematical texts authored by representatives of two different social groups, the *yangban* 兩班 *literati* having high hereditary social status and the technical experts *chungin* 中人 placed lower on the social ladder, the author analyzes two specific ways of teaching and learning matters related to counting rods, one based on physical manipulations with them conducted for mathematical calculations, and the other, based on commentarial tradition accompanying classical Chinese texts and aiming at metaphysical considerations. These differences are not only remarkable in terms of rivalry between two obviously different didactical traditions but also in terms of the influence of socio-ideological biases of two societal backgrounds on the mathematics curriculum.

The last two chapters of the section, "The Education of Abacus Addition in China and Japan Prior to the Early 20th Century," authored by **Yifu Chen**, and "Teaching Computation in 19th-Century Japan: The Transition from Individual Coaching on Traditional Devices at the End of the Edo Period (1600–1868) to Lectures on Western Mathematics During the Meiji Period (1868–1912)," by **Marion Cousin**, are devoted to the history of beads abacus, a device that historians of informatics often mention as precursor of today's electronic computers, but whose origin still remains obscure. This instrument started being increasingly popular in China in the early second millennium AD and apparently made a great impact on mathematics and mathematics education in China, Korea, Japan, Vietnam, and, more recently, worldwide. In his contribution, **Yifu Chen** analyzes how addition, the most elementary arithmetic operation performed with the beads abacus, was taught over the five centuries (15th–20th), first in China and later in Japan, while paying special attention to the related didactical approaches used to teach the most efficient techniques of computations, in particular, the positions and motions of the fingers of operator, eventually leading to the further automatization of calculations. In her turn, **Marion Cousin** focuses on teaching computations in the nineteenth-century Japan (1868–1912), still under a strong influence of the Chinese traditions of computations with abacus, but at the same time undergoing important shift to the Western ways of doing calculations with pen and paper. The apparent tension between two distinctive philosophies of teaching arithmetic and corresponding practices eventually led to a form of coexistence of both methods with emphasis on the importance of beads abacus as part of a Japanese cultural heritage, and an efficient teaching tool for elementary schools.

The third section of the book is dealing with the counting devices used in Europe from the seventeenth century to the early twentieth century. It includes a brief historical overview of calculating machines, a chapter entitled "A Short History of Computing Devices from Schickard to de Colmar: Emergence and Evolution of Ingenious Ideas and Technologies as Precursors of Modern Computer Technology," written by **Viktor Freiman** and **Xavier Robichaud**. The authors bring to the discussion issues of increasing automatization of calculations in Europe which was a result of the attempts to handle complexity and difficulty of operations with numbers as part of the development of science, commerce, and management of the *époque* coupled with unique combination of trends in philosophical thought, technological creativity, and mathematical culture, which led to the invention of calculating machines. It appears that the desire to introduce mechanical calculations increased the need of standardizing procedures through uniform technological steps, including particular way of representing numbers and operations with them using machines, coordinating different parts of the process, keeping track of the intermediate results, to put it briefly, all what constitutes today parts of electronic computing, including processing data, organization of input-output, as well as of memory. In these circumstances, the programming language becomes *the* key element in communication between humans and the machine, and between different parts of the machine (and, as we would add today, communication between the machines, referring to the Internet) and even Internet of Things, representing an increasing role of artificial intelligence in the everyday life.

Although the production and commercialization of calculating machines had remarkable success in the nineteenth century, it did not lead to substantial use of them as educational devices until the era of electronic calculators in 1970–80s. Instead, there were rather early inventions that attracted attention of mathematics educators, such as logarithmic tables and slide rules analyzed in the chapter "Computation Devices in Nineteenth-Century Mathematics Instruction in Europe," by **Gert Schubring**. The author investigates the development of mathematical instruction in several European states, and the role of computations and computing devices as didactical tools. Namely, France is mentioned as having played an important role in introducing logarithmic tables into secondary education as tool for simplifying calculations as well as professional tool for further professional practice; the emphasis was put on exactness and rigor in computations. Overall, the chapter presents a complex portrait of the use of tables across Europe related to cultural and political differences and shows that, as result, there was a visible disparity in educational systems and, consequently, in mathematics curricula. For instance, it appears that in England, where stress was laid on classical education, the tables were not in use in schools until the first half of the twentieth century, although it is plausible to conjecture that the tables may have been used in some specific fields of professional training, such as sailing or military training. Regarding the use of the slide rules, whereas England was the country where the instrument originated from and where it was extensively used by practitioners at the end of the eighteenth century, it was France again who took the lead in its further development and implementation through the nineteenth century, in practice, as well as in teaching mathematics in schools. Again, continent-wide, the use of the slide rules in mathematics education shows a great disparity; for example, the chapter informs us that the Netherlands seem to introduce it in the curriculum as late as 1968. This disparity, according to the author, deserves further study, by historians of mathematics and historians of education.

Two remaining chapters of the section, "Counting Devices in Russia" by **Alexei Volkov** and "Toward a History of the Teaching of Calculation in Russia" by **Alexander Karp**, focus on Russia, a country at the crossroad between Europe and Asia which found itself under important influence of both neighbors while developing its own authentic ways and traditions to do mathematics and to teach it. A particularly intriguing question about the origins of the Russian abacus called "schyoty" widely used in the country from the seventeenth to the later twentieth centuries leads Volkov in his chapter to a discussion about the history of the instruments and of its use. This interest is apparently nurtured by the complexity of the processes of transmission of mathematical ideas in general and in the case of the Russian abacus in particular. Prior to the mid-twentieth century, a number of authors claimed that the Russian "schyoty" originated from the Chinese abacus *suanpan* 算盤; this theory was rejected by I. G. Spasskiĭ who in his book-length article (1952) suggested that the Russian instrument was invented independently from the Chinese one. Later, Simonov (2001b) provided additional arguments in support of Spasskiĭ's theory; more specifically, Simonov (2001c) claimed that an authentic Russian abacus operating with beads of two kinds existed as early as the eleventh century. In his chapter, Volkov critically evaluates the arguments of Spasskiĭ (1952) and suggests that the early mentions of the *schyoty* may indicate that the instrument was brought to Russia in the fourteenth century from Golden Horde, the state located in the southwestern part of present-day Russia, while the hypothesis concerning its transmission from China still lacks supporting evidence. In the second part of his chapter, Volkov provides an analysis of operations with common fractions performed with a version of the instrument used in Russia in the seventeenth century, in particular, of lists of identities involving common fractions found in arithmetical manuals of that time. His concluding remarks briefly discuss the role of the instrument in educational context in Russia and Soviet Union in the nineteenth and twentieth centuries.

While Volkov's chapter is devoted to the Russian computing instruments, their development, and their use throughout a quite long period of history, the contribution of **Alexander Karp**, also dealing with Russian mathematics education, focuses upon teaching calculations in a variety of forms and contexts and includes an in-depth analysis of didactical traditions related to teaching arithmetic, from the early second millennium AD up to the seventeenth century, and their substantial transformation in the eighteenth century beginning with the publication of the famous *Arithmetic* of L. Magnitskiĭ (1703).[7] While this development features written calculations as the dominant method of computation, there was an important part of people's activities, in their everyday life, that required complex computations and where the use of some kinds of computing devices, including *schyoty*, was rather common, and involved adaptation of the tools to specific problems that people solved in everyday life and also to the socioeconomic development of country at this particular historical period. When analyzing didactical traditions of different arithmetical textbooks of the eighteenth and nineteenth centuries, the author emphasizes the complexity of evolution of methodological ideas going beyond teaching elementary operations; his analysis provides interesting insights into the dialectic of apparently opposite approaches, namely the "practical, or material" one, and the other, more of developmental nature, arguably inspired by the ideas of Pestalozzi. Didactical debates that accompanied attempts for bringing forth innovative methods seem to lead to a rather eclectic profile of the educational system in the country where different educational methods (and different vectors of influences corresponding to the underlying ideas) coexisted, thus bringing the didactical debates into the twentieth century. Besides the methods of (written) arithmetic, the chapter provides interesting details about the exercises in "mental calculations" practiced by some educators, which constituted important part of Russian didactical traditions of giving students numerous oral exercises and training them not only in mere technical skills but also in strategies best suitable for particular type of calculations.

---

[7]On Magnitskiĭ's textbook, see Freiman and Volkov (2015).

The concluding section contains two chapters, "The Unsettling Playfulness of Computing," by **Jean-François Maheux** and "Calculating Aids in Mathematics Education Before the Advent of Electronic Calculators: Didactical and Technological Prospects" by **Dragana Martinovic**, which offer insights into the use of counting devices in mathematics classroom from the late twentieth century to present, while attempting to bridge the present and the past. More specifically, Maheux offers an interesting discussion of the nature of computing and computations, particularly focusing on the role of algorithms in mathematics and tools that facilitate their execution. The author underlies the aesthetical value of the algorithms (sometimes hidden from the learner) and of the elements of play with numbers; he focuses upon appreciation of beauty of arithmetic that brings its teaching and learning beyond a mere memorization of computational routines. The author claims that a variety of classroom activities, including, among others, the use of play, can eventually transform mathematical learning into exploration of and experimenting with the meaning of numbers and operations, thus offering mathematically richer and conceptually more meaningful experiences.

In her turn, Martinovic provides important details of the development of computing devices and corresponding educational practices of the twentieth century mathematics classroom, up to the appearance of modern electronic devices and Internet. While analyzing main pedagogical and technological ideas, with the focus on their interaction with the practices of teaching and learning constituting both cyclic and transformational development in mathematics education, the author underlies the importance of looking into the past in an attempt to better understand the recent development and, what is even more important, future trends. The author is also making connections between psychology, computer science, and education. In the twentieth century, cognitive science emerged, connecting artificial intelligence, linguistics, anthropology, psychology, neuroscience, philosophy, and education. Mathematics and mathematics education, in addition to language, were always at the forefront of the interest of cognitive science, as the biggest marvel of what human mind is capable of.

While reflecting on the scope of the collection of contributions as a whole, we realized that it would be possible to identify five main axes of research on development, transmission, and application of computations and computing devices presented in the book. They include:

A1. History of development of computational instruments per se and of the related computational practices. This axis includes descriptions of the numerical systems that were used as conceptual foundations of the respective instruments, the particularities of construction of the instruments, and description of computations performed with their help. These topics were discussed in the chapters devoted to Chinese counting rods (Pollet and Volkov), Chinese abacus (Chen), arithmetical tables (Megremi and Christianidis), Russian *schyoty* (Volkov), calculating machines (Freiman and Robichaud), logarithmic tables and slide rules (Schubring).

A2. Transmission of knowledge. There are two types of transmission of expertise related to the counting skills and counting devices: educational activities focused on the learners' ability to use the devices, and transmission of the counting devices and related methods from one sociocultural context to another. As our authors suggest, an essential part of educational activities is reflected in production of textbooks and descriptions destined to learners. In the present collection, this aspect of the transmission of knowledge is discussed in the contributions of Cousin and Karp. The latter process, i.e., the transmission of the devices and related expertise, historically had three different forms: (1) both contexts belonged to the same or close cultural traditions (these were the cases of transmission of Chinese counting devices and related practices to Korea and Japan described in the contributions of Oh and Chen); (2) the contexts were markedly different (this is the case of the Russian *schyoty* or Chinese *suanpan* being adopted as a didactical device in Western schools, as briefly mentioned above); and (3) different contexts were embedded into a continuous process of creation, modification, and description of computing devices and their potential or actual use (as it is the case of calculating machines in the seventeenth to nineteenth centuries; see the chapter authored by Freiman and Robichaud), and more recent development of teaching devices in the twentieth century that took place prior to electronic era (see the chapter authored by Martinovic).

A3. Arithmetical operations performed with the instruments and their optimization. This topic is crucial for understanding the historical evolution of the instruments and of the operations performed with or without instruments. Contributions dealing with this theme include the chapters authored by Melville, Høyrup, Kusuba, and Chen who provided very detailed descriptions of arithmetical operations. This axis is closely related to the following axis.

A4. Writing and its interrelationship with computations. The role of writing in mathematical and didactical work can hardly be overestimated; this is why a number of contributions (Melville, Høyrup, Kusuba, and Pollet, among others) focus on the practices of written representations of operations, arithmetical as well as algebraic ones (see the contribution of Pollet devoted to Indian mathematics). One can distinguish the topic of "writing *and* tools" from the topic "writing *as* tool": While the former one is related to the interplay between the operations performed with an actual instrument and their records on a writing medium (see the contributions of Volkov and Pollet on Chinese mathematics), the latter deals with the cases in which the writing medium itself became a tool of special kind to be used not only to record the results, but also to produce them. Here, once again, we deal with the particular formats of representations of data (numerical or even symbolic) mentioned or discussed in the chapters on Mesopotamian, Indian, Byzantine, Chinese, and Korean mathematics.

A5. The last but, arguably, one of the most important axes discussed by the authors is the role of the instruments in mathematics instruction and, conversely, the influence of the teaching practices on the ways in which the instruments were designed and used. This axis is directly related to the phenomenon of appearance of a large variety of counting instruments (including traditional ones, their modifications, and a newly created instruments) adapted for mathematics instruction. This new movement began in Europe and in the North America in the early nineteenth century and continues until the present day. This phenomenon probably can be explained if we take into consideration not only the traditional didactical practices but also the theories of mathematics instruction mentioned earlier in this introduction. The contributions devoted to this phenomenon include, the chapters authored by Schubring, Karp, Maheux, and Martinovic.

The diversity of contributions to the volume does reflect a complexity of interaction between mathematics, technology, and didactics that seems to be rather increasing over the time and across the cultures. The five main axes identified above are related to the development and the use of computations and computing devices, to their role in transmission of mathematical knowledge, and to the influence of didactical practices on the design and use of the instruments. Below, we would like to list five categories of problems that emerge from the contributions constituting the present collection; these categories may eventually define an agenda for future research. They include (1) problems arising from counting objects of different kinds which historically led to a great diversity of systems of representations of data (e.g., number systems, measurement systems, and systems of monetary units), as well as their interaction and development within particular contexts, and in response to specific needs, tasks, and approaches; (2) problems related to performing arithmetic operations with these categories of data which bring forth the issues of dealing with different representations, complexity of calculations, as well as their accuracy in order to solve particular or general tasks (in the case of general tasks, the designed procedures become identical with generally applicable algorithms); (3) problems of keeping tracks of the performed operations as well as of the obtained results (including intermediate ones); (4) problems of creating, sharing, and translating knowledge from one task-context-culture-*époque* to the other; in particular, the ways to explain to the learners or to the readers how the procedures work and why they yield the expected results; (5) problem of understanding the general trends, tensions, reforms along with the complexity and multidimensional nature of undergoing transformations not only from the technological and mathematical viewpoint, but also from the cultural, political, ideological, and pedagogical (didactical) perspectives.

To summarize, one can identify two types of didactical approaches related to the counting devices: The first of them was the family of teaching practices used for the transmission of the operational skills necessary for using a given computational tool, be it a numerical table, a dustboard, a set of counting tokens, a beads abacus, or even a mechanical or electronic calculator; the second one comprised the didactical practices related to the arithmetical (or sometimes even algebraic) concepts closely related to the structure of the used counting devices. Their complex interrelationship reflects the ambivalent nature of the counting devices: While being originally devised for some particular mathematical practices, they themselves, in turn, considerably influenced the imagery and conceptualization of arithmetical (or broader, mathematical) objects. Any change of one of these two elements led to changes of the other; accordingly, the didactical approaches of the first and second type mutually influenced each other and thus cannot be studied separately. This dual influence was mirrored in the respective educational practices. The contents of the present volume reflect this complex interplay of the two aspects of the counting devices: Some contributions deal with the history of the devices per se, while some others discuss computational practices, including operations not only with numbers but with symbols as well.

The present volume was conceived as an attempt to explore the aforementioned intuition about the elusive connection between the computational instruments, related computational methods, and didactical practices. We did not mean to apply any sort of restrictive interpretation to the term "instrument": every device, be it a set of tokens, a beads abacus, a human hand, a table, or even a standard routine of written computations was considered worth of study. The mentioned initial intuition that led us to the research agenda set forth in this volume was based on the understanding of the existence of an interrelationship between the history of counting devices and the history of didactical practices featured in the contemporaneous mathematical traditions. The relationship between them, as the contributions collected in this volume may suggest, was extremely complex: Historically, not only the didactical practices depended on the structure and design of counting devices, but a number of devices, in turn, were constructed *for* particular educational purposes, and thus reflected, at least to some extent, the visions of mathematics instruction shared by their creators.

It is obvious that the chronological and geographical scope of the invited contributions cannot suffice to present all the computational practices that historically existed, nor can it account for the complexity of didactical approaches related to the selected computing devices. The editors, however, hope that the selection presented in this collection provides sufficient illustration for the theses advanced in the opening part of this introduction.

Learning from our authors, as well as from the reviewers of the book to whom we express our deepest gratitude, we argue that each chapter, while being devoted to a description of a particular computing device (including computational procedures or algorithms), as well as to an account of its creation, development and, in some case, teaching and learning activities into which it was involved, can be further analyzed within each of the five above-mentioned categories of problems; however, we leave this task to the reader for now, while hoping that the collection of contributions captured her or his attention and triggered interest that eventually will result in future fruitful explorations into the history of counting tools and their didactical functions.

References

Altiparmak, Kemal. 2016. The teachers views on Soroban abacus training. *International Journal of Research in Education and Science (IJRES)* 2 (1): 172–178.

Barnard, Henry. 1906. *Pestalozzi and his educational system*. Syracuse: Bardeen.

Barner, David, George Alvarez, Jessica Sullivan, Neon Brooks, Mahesh Srinivasan, and Michael C. Frank. 2016. Learning mathematics in a visuospatial format: A randomized, controlled trial of mental abacus instruction. *Child Development* 87 (4): 1146–1158.

Bjarnadóttir, Kristín. 2014. History of teaching arithmetic. In *Handbook on the history of mathematics education*, eds. Alexander Karp and Gert Schubring, 431–458. New York, NY: Springer.

Boggan, Matthew, Sallie Harper, and Anna Whitmire. 2010. Using manipulatives to teach elementary mathematics. *Journal of Instructional Pedagogies* 3 (1): 1–6. Online version retrieved from http://www.aabri.com/manuscripts/10451.pdf on August 7, 2017.

Drijvers, Paul, Michiel Doorman, Peter Boon, Helen Reed, Koeno Gravemeijer. 2010. The teacher and the tool: Instrumental orchestrations in the technology-rich mathematics classroom. *Educational Studies in Mathematics* 75 (2): 213–234.

Eganyan, Aleksandr M. [Еганян, Александр Меликсетович], and Leonid E. Maĭstrov [Леонид Ефимович Майстров]. 1982. О математических таблицах Анания Ширакаци [O matematicheskikh tablitsakh Ananiya Shirakatsi] (On mathematical tables of Anania Shirakatsi, in Russian). *Istoriko-Matematicheskie Issledovaniya* 26: 189–197.

Freiman, Viktor, and Alexei Volkov. 2015. Didactical innovations of L.F. Magnitskiĭ: Setting up a research agenda. *International Journal for the History of Mathematics Education* 10 (1): 1–23.

Gouzévitch, Irina, and Dmitri Gouzévitch. 1998. La guerre, la captivité et les mathématiques. *Bulletin de la Sabix* 19: 30–68.

Greenwood, Tim. [=Timothy William]. 2011. A reassessment of the life and mathematical problems of Anania Širakac'i. *Revue des Études Arméniennes* 33: 131–186.

Hewsen, Robert H. 1968. Science in seventh-century Armenia: Ananias of Širak. *Isis* 59 (1): 32–45.

Khudson, Kevin. 2017. Use your fingers: The abacus just might improve your arithmetic abilities. *Forbes* (online edition) article published on April 28, 2016. Retrieved from https://www.forbes.com/sites/kevinknudson/2016/04/28/use-your-fingers-the-abacus-just-might-improve-your-arithmetic-abilities/#5de9e66b48cc on August 11, 2017.

Kidwell, Peggy Aldrich, Amy Ackerberg-Hastings, and David Lindsay Roberts. 2008. *Tools of American mathematics teaching, 1800–2000*. Baltimore: Johns Hopkins University Press.

Knuth, Donald E. 1972. Ancient Babylonian algorithms. *Communications of the ACM* 15 (7): 671–677.

Martinovic, Dragana, Viktor Freiman, and Zekeriya Karadag (eds.). 2013. *Visual mathematics and cyber learning*. (Series *Mathematics education in the digital era*, vol. 1, series eds. Dragana Martinovic and Viktor Freiman). Dordrecht: Springer + Business Media.

Menninger, Karl. 1969. *Number words and number symbols: A cultural history of numbers*. Cambridge (MA) & London: MIT Press.

Ministère de l'Éducation et du Développement de la petite enfance (MEDPENB). 2016. Programme d'études: Mathématiques au primaire (1re année). Direction des services pédagogiques. Retrieved from http://www2.gnb.ca/content/dam/gnb/Departments/ed/pdf/K12/servped/Mathematiques/Mathematiques-1reAnnee.pdf on August 11, 2017.

Monaghan, John, and Luc Trouche. 2016. Mathematics teachers and digital tools. In Monaghan, John, Luc Trouche, and Jonathan M. Borwein. *Tools and Mathematics: Instruments for learning*. 357–384, Cham: Springer.

Moyer-Packenham, Patricia S. (ed.). 2016. *International perspectives on teaching and learning mathematics with virtual manipulatives.* (*Mathematics education in the digital era*, vol. 7, series ed. Dragana Martinovic, and Viktor Freiman). Dordrecht: Springer + Business Media.

Oppenheim, Adolf Leo. 1959. On an operational device in Mesopotamian bureaucracy. *Journal of Near Eastern Studies* 18 (2): 121–128.

Orbeli, Iosif Abgarovich [Орбели, Иосиф Абгарович]. 1918. Вопросы и решения вардапета Анании Ширакца, армянского математика VII века [Voprosy i resheniya vardapeta Ananii Shirkaca, armyanskogo matematika VII veka] (Questions and solutions of vardapet Ananiya of Shirak, an Armenian mathematician of the 7th century; in Russian). Petrograd. [Reprinted in Орбели И. А. *Избранные труды.* Ереван, 1963, 512–531.].

Pape-Carpantier, Marie. 1878. *Notice sur l'éducation des sens et quelques instruments pédagogiques.* Paris: Delagrave.

Pape-Carpantier, Marie. 1901. *Enseignement pratique dans les écoles maternelles* (9e édition). Paris: Hachette.

Proust, Christine. 2000. La multiplication babylonienne: la part non écrite du calcul. *Revue d'histoire des mathématiques* 6: 293–303.

Regnier, Jean-Claude. 2003. Le Boulier-Numérateur de Marie Pape-Carpantier. *Bulletin de l'APMEP* 447: 457–471.

Ritter, Jim. 2010. Translating rational-practice texts. In *Writings of early scholars in the Ancient Near East, Egypt, Rome, and Greece: Translating ancient scientific texts*, eds. Annette Imhausen and Tanja Pommerening, 349–383. Berlin/New York: de Gruyter.

Roberts, David Lindsay. 2014. History of tools and technologies in mathematics education. In *Handbook on the history of mathematics education*, eds. Alexander Karp and Gert Schubring, 565–578. New York, NY: Springer.

Roberts, David Lindsay, Allen Yuk Lun Leung, and Abigail Fregni Lins. 2012. From the slate to the web: Technology in the mathematics curriculum. In *Third international handbook of mathematics education*, eds. M. A. (Ken) Clements, Alan J. Bishop, Christine Keitel, Jeremy Kilpatrick, Frederick K. S. Leung (*Springer international handbooks of education*, vol. 27), 525–547. New York: Springer.

Sachs, Abraham S. 1947. Babylonian mathematical texts I. Reciprocals of regular sexagesimal numbers. *Journal of Cuneiform Studies* 1 (3): 219–240.

Simonov, Rem A. [Симонов, Рэм Александрович]. 2001a. *Естественнонаучная мысль древней Руси. Избранные труды* [*Estestvennonauchnaya Mysl' Drevneĭ Rusi: Izbrannye Trudy*] (The Scientific Mentality of Old Russia: Selected Works; in Russian). Moscow: Moscow State University of Printing.

Simonov, Rem A. [Симонов, Рэм Александрович]. 2001b. Древнерусский абак (по данным моделирования основы денежных систем) [Drevnerusskiĭ abak (po dannym modelirovaniya osnovy denezhnykh sistem)] (Ancient Russian abacus (according to the data obtained via emulation of the basis of monetary systems); in Russian), in Simonov 2001a, 29–37.

Simonov, Rem A. [Симонов, Рэм Александрович]. 2001c. Археологическое подтверждение использования на Руси в XI в. архаического абака ("счета костьми") [Arkheologicheskoe podtverzhdenie ispol'zovaniya na Rusi v XI v. arkhaicheskogo abaka ("scheta kost'mi")] (Archaeological proof of the use of the archaic abacus ("counting with seeds") in Russia in the 11th century; in Russian), in Simonov 2001a, 12–29.

Spasskiĭ, Ivan G. [Спасский, Иван Георгиевич]. 1952. Происхождение и история Русских счетов [Proiskhozhdenie i istoriya Russkikh schetov] (Origin and history of the Russian schyoty; in Russian). *Istoriko-matematicheskie issledovaniya* 5: 269–420.

Szendrei, Julianna. 1996. Concrete materials in the classroom. In *International handbook of mathematics education*, Part 1, eds. Alan J. Bishop, Ken Clements, Christine Keitel, Jeremy Kilpatrick, Colette Laborde, 411–434. Dordrecht: Springer.

Vasuki, K. [Mathivanan]. 2013. The impact of abacus learning of mental arithmetic on cognitive abilities of children. ALOHA Mental Arithmetic Scientific Report. Retrieved from http://www.alohaspain.com/public/file/ALOHA_Benefits_ScientificReport.pdf on August 11, 2017.

Varol, Filiz, and Dale Farran. 2007. Elementary school students mental computation proficiencies. *Early Childhood Education Journal* 35 (1): 89–94.

Wilderspin, Samuel. 1840. *The infant system, for developing the intellectual and moral powers of all children, from one to seven years of age*, 7th revised ed. London: Hodson.

Part II
Middle East, India, Greece, and Byzantine Empire in Antiquity and Middle Ages

Computation in Early Mesopotamia

Duncan J. Melville

Abstract The history of Mesopotamian mathematics begins around 3300 BCE with the development of written systems for recording the control and flow of goods and other economic resources such as land. Numeration was bound up with measurement and was a collection of concrete systems. One of the key developments over the subsequent thousand years or so was the gradual rationalization of these complex concrete systems and the consequent emergence of an abstract conception of number and techniques of computation that applied regardless of metrological category. Throughout their history Mesopotamian scribes organized knowledge in the form of lists. In mathematics there were also lists, but along with lists came metrological and mathematical tables, two-dimensional arrays of data that organized information both vertically and horizontally. A key example is tables giving lists of lengths of sides of square or rectangular fields, along with their areas; the problem of computation of areas remained a constant concern throughout the period covered here. In this chapter, we cover the development of Mesopotamian computation from the archaic period up to the edge of the emergence of the fully abstract sexagesimal computational system for which they are renowned, tracing, as far as can be seen with currently available sources, the long developmental process.

Keywords Mesopotamian mathematics · Sargonic mathematics
Early Dynastic · Archaic · Computation · Mathematical tables

Introduction

Origins are murky. The closer we approach beginnings the more fragmentary and partial the sources. Our own prejudices and preconceptions can lead us to misinterpret what evidence there is. We have to be particularly careful to guard against this tendency when studying the very early history of mathematics. It is easy to

D. J. Melville (✉)
St. Lawrence University, Canton, NY, USA
e-mail: dmelville@stlawu.edu

© Springer Nature Switzerland AG 2018
A. Volkov and V. Freiman (eds.), *Computations and Computing Devices
in Mathematics Education Before the Advent of Electronic Calculators*,
Mathematics Education in the Digital Era 11, https://doi.org/10.1007/978-3-319-73396-8_2

assume that mathematics, and especially elementary mathematics, is universal and therefore ancient categories of thought and concepts match ours. They do not.

The quest for a deeper understanding of how Mesopotamians conceived of and implemented their mathematics has been difficult, involved many scholars over decades of work, and is far from complete, particularly with respect to the third millennium, the subject of this chapter. The history of the development of the field from the pioneers of the 1930s to the 1990s is wonderfully told in Jens Høyrup's excellent "Changing Trends" paper (Høyrup 1996); for an updating of the story see (Melville 2016).

The best-known period of Mesopotamian mathematics is the Old Babylonian (ca. 2000–1600 BCE). From this period we have an abundance of mathematical texts using the sexagesimal place-value system and relatively clear hints of computational techniques and physical aids. However, the introduction of the abstract sexagesimal place value system late in the 3rd millennium (typically located in the Ur III period of 2100–2000 BCE) led to a radical disjunction in computational practice and we cannot suppose that instruments and aids in use in the Old Babylonian period and later were also used before then. A discussion of Old Babylonian practice appears elsewhere in this volume. The one tool we are certain third millennium scribes used to aid their computation is the mathematical list or table, a characteristic Mesopotamian technique of organizing data. A short table summarizing the standard modern periodization of early Mesopotamian history is included at the end of this chapter.

Much of the literature on early (that is, before 2100 BCE) Mesopotamian mathematics is technical and intended (only) for specialists. As an introduction to those seeking general orientation to the overall field, I recommend Eleanor Robson's chapter on Mesopotamian mathematics in the source book edited by Victor Katz (Robson 2007), and her book (Robson 2008). Robson's bibliographies will provide excellent pathways into the literature. For a deeper look at the third millennium in particular, the best starting point remains (Nissen et al. 1993).

One of the most important developments of third millennium Mesopotamian mathematics is the gradual emergence of an abstract conception of number from quantity notation tied to specific metrological units. Exactly how this occurred is still not clear, and historians argue greatly over how to interpret evidence at different stages of this development.

Another cause for debate is the interpretation of arithmetic, algorithmic procedures, and the possible role of geometric analysis. These topics are touched on below although a detailed description of the nuances of the various positions scholars have taken is beyond the scope of this paper.

Some Background

The period considered in this chapter takes place before the development of the abstract sexagesimal place-value system, the base-60 cuneiform system that allowed context-free computations, especially multiplication. In the third millennium BCE before the Ur III period, computation was context-dependent, and the development of an abstract concept of number, and of an abstract numerical script is a process that spans the thousand years of our concern.

Mathematical exercises in Mesopotamia were computational. Typical problems were structured so that the solution required computing some specific quantity. The happy accident of writing on clay means that the archaeological record of the extent and development of mathematics in early Mesopotamia is more complete than that for any other culture. However, it also means that what we have is only what they chose to write down. Anything not in the written record, such as oral instruction or possible auxiliary counting aids, is largely lost to us. Archaeology has not been particularly helpful in understanding third-millennium computational practices. We have tablets, some with archaeological context, many not; we know what students learned, but have much less idea of how they learned it.

Writing in Mesopotamia arose in response to bureaucratic demands: the need to record and control the flow of goods. Thus, from the beginning, writing and mathematics were deeply intertwined. The archaic period (ca. 3350–3000 BCE) presented scribes with a complex series of metrological systems, each with its own notation, and trainees had to learn to manipulate the different quantities and symbols, and explore the linkages between systems, for instance between the length and area systems when computing quantities to do with fields or houses.

The main tool for education of beginners was the auxiliary table. Metrological data were gathered into systematically organized tables that allowed students to learn and explore connections between systems. An example of such a table is given in Fig. 1. Its contents are discussed later in the chapter.

Another well-known characteristic of Mesopotamian mathematics was its algorithmic nature. Learning how to solve problems involved learning step-by-step procedures. Precisely how these computational algorithms were developed can not usually be determined, but various classes of procedures for dealing with different types of problems can be observed.

As the writing system, metrological notation, and society changed over the course of the third millennium, mathematical problems and procedures naturally changed as well. However, the centrality of writing (after all, the records we have are from scribes), the use of tables, and the algorithmic habit provide a constant theme.

In this chapter, we present a series of snapshots of mathematics and computational practices, analyzing tablets from several different locations and times. The Sargonic period (ca. 2340–2200 BCE) produced a number of interesting metrological geometric problems; the recent publication of table texts from the Early Dynastic IIIa and b periods (ca. 2600–2340 BCE) has provided some new insights

loose
fragments

Fig. 1 VAT 12593. Table of lengths and areas. *Source* Cuneiform Digital Library Initiative, P106078, http://cdli.ucla.edu/dl/photo/P010678.jpg. http://cdli.ucla.edu/search/archival_view.php? ObjectID=P010678

into length-area computations; there is an important collection of mathematical tablets from Šuruppak from around 2500 BCE, and the archaic period (ca. 3000 BCE) is represented by a large collection of tablets mostly from Uruk. This trajectory into the ever more distant past will help uncover constant themes as well as emphasize the differences in different periods.

Third-millennium Mesopotamia was an agrarian society with large urban settlements (Adams 1981) (see Fig. 2). For the administrators of a redistributive economy, two constant key computations were determining quantities of goods or rations for multiples of people, given the basic amount for a single person (in various categories), and finding the areas of fields, given the lengths of the sides. The latter computation was important for estimating harvest, seed, animal and labor requirements, taxation, and allotting land to individuals. These two kinds of

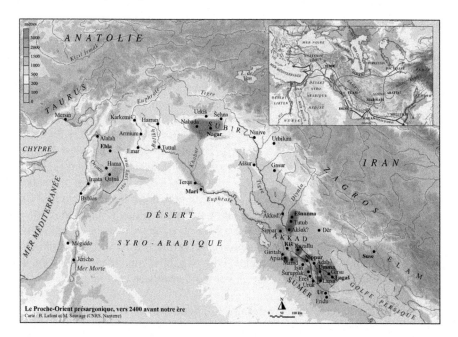

Fig. 2 Map of Near East. *Source* CDLI wiki: http://cdli.ox.ac.uk/wiki/lib/exe/fetch.php?media=
carte-ediii.jpg. http://cdli.ox.ac.uk/wiki/doku.php?id=cartes_atlas

problems gave rise to the two major types of (what we would term) multiplication
—repeated addition, and length-area conversions. Repeated addition operated in a
single metrological domain such as repeatedly adding quantities of grain, while
length and area computations required understanding the linkages between two
different systems, in this case the length and area systems. The abstraction of the
sexagesimal place-value system allowed the unification of calculation procedure for
these two kinds of 'multiplication'. For a detailed treatment of the terminology and
conceptualization of Mesopotamian arithmetic, see Høyrup (2002a, b).

Sargonic Mathematics (2340–2200 BCE)

The extant corpus of Sargonic mathematical tablets is very small, fewer than twenty
texts are known (a list of tablets with their publication histories is given in Foster-
Robson 2004). Almost all examples concern length-area computations of rectangular or square fields. Additionally, there is one example of a problem of division of
trapezoidal area that we will not discuss further here. [For more on that tablet (IM
58045), see Friberg (1990), Robson (2007), and Friberg (2014).]

The length-area problems come in two guises. In one, the task is to compute an
area given the lengths of the sides. In the other type the goal is to find the length of

one side of a rectangular field given the area and the length of the other side. The second problem is, of course, extremely artificial and unlikely to arise in actual surveying practice, but even the direct problem of computing the area tends to have a quite artificial construction. The sizes and shapes of the areas (fields) are also not necessarily realistic (Liverani 1990; see also Foster-Robson 2004) and it is clear that the problems are exercises in conversions between length and area units, a problem for which Sargonic metrology was almost uniquely unsuited.

In this period, the key length-unit was the *nindan* (ca. 6 m) and the key area unit was the *sar* of 1 *nindan* square (ca. 36 m^2). The *sar* provided a very clear linkage between the two systems. However, it was the only convenient unit linkage. The *nindan* was the largest length unit in use, sitting atop a complex system of smaller units. Multiples of the *nindan* were recorded with cuneiform discrete notation. On the other hand, the area *sar* was the smallest area unit, sitting at the bottom of a complex system of larger units. Since there were no subunits available, fractions of a *sar* were recorded with made-up sixtieths (*gin*) borrowed from weight metrology.

The relationships between the length units (slightly simplified) are given in (Table 1), and the area units in (Table 2). For more details on metrology see Powell (1990).

Here we give two representative examples of Sargonic length-area problems, one direct and one inverse (originally published as texts 27 and 29 in Limet 1973) (Figs. 3 and 4).

A translation of each problem is:

1. 11 *nindan*, 1 *kuš-numun*, 1 *giš-bad*, 1 *šu-bad*. Its area 1 *iku* ¼ *iku* 2 ½ *sar* 6 *gin* 15 *gin-tur*. It was found.
2. The average long sides are 2 40 *nindan*. What is the short side? The area is 1 *iku*. Its short side is 3 *kuš-numun*, 1 *giš-bad*, 1 *šu-bad*.

Note that the 'problems' contain a bare minimum of information, especially the first, which mostly consists of a length and an area. No task is stated, no procedure

Table 1 Sargonic length units	1 *nindan* = 6 *kuš-numun*
	1 *kuš-numun* = 2 *giš-bad*
	1 *giš-bad* = 2 *šu-bad*
	1 *giš-bad* = 3 *šu-du-a*
	1 *šu-bad* = 15 *šu-si*
	1 *šu-du-a* = 10 *šu-si*
	1 *šu-si* (ca. 17 mm)

Table 2 Sargonic area units	1 *sar* = 1 *nindan* × 1 *nindan*
	1 *iku* = 100 *sar*
	1 *eše* = 6 *iku*
	1 *bur* = 3 *eše* (ca. 6.5 ha)

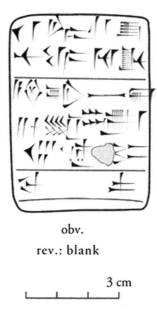

obv.

rev.: blank

3 cm

Fig. 3 Sargonic text 1. *Source* Cuneiform Digital Library Initiative, P213161, http://cdli.ucla.edu/
search/archival_view.php?ObjectID=P213161

3 cm

Fig. 4 Sargonic text 2. *Source* Cuneiform Digital Library Initiative, P213163, http://cdli.ucla.edu/
search/archival_view.php?ObjectID=P213163

is given, no shape is mentioned, and it doesn't even say that it is about a field. The
second problem at least states a question, but again does not mention a field or state
explicitly the shape. The term 'average' in the second example refers to a standard
procedure for estimating the area of a not-quite rectangular shape by multiplying the
average of the lengths with average of the widths.

The laconic and partial written evidence makes the task of the historian attempting to reconstruct ancient mathematical practices all the harder, although in the cases such as this, much of the context can be deduced from other examples.

The artificiality of the problems is clear. In the first problem, the length is given in what Friberg terms 'wide-span' numbers, that is, a quantity spanning a range of units from large to small. The area then comes out as a complicated awkward quantity. Additionally, the length, 11 *nindan*, 1 *kuš-numun*, 1 *giš-bad*, 1 *šu-bad*, with a series of 1s, is hardly chosen at random. For the second problem, the area is a simple quantity and the lengths again span the available units. It is also worth noting that the second field is around 960 m long and about 3 m wide.

In 2004 Foster and Robson published a new tablet from a private collection and took the opportunity to revisit the entire Sargonic corpus, summarizing the state of knowledge (Foster-Robson 2004); the next year, Friberg published another Sargonic tablet and also surveyed the overall corpus (Friberg 2005). Their interpretations are starkly different and there is no consensus among historians as to the correct view of Sargonic conceptions of mathematics.

Correct calculations offer little hint of how the answers were found; analysis of errors in mathematical computations can be helpful for historians. Sometimes the failure mode can reveal the underlying procedure. There is an error in the computation of the Foster-Robson tablet, and the authors use the mistake as the basis of their interpretation, arguing that the given answer "makes sense only if we assume that the scribe has treated the area measurement in *iku* as if it were in *sar*...and this type of error could only come about if scribes were expected to convert standard mixed metrological notation into sexagesimal multiples and fractions of a base unit" (2004, 6). Thus, Foster and Robson see "convincing evidence for sexagesimalisation" (2004, 1) in the Sargonic period. Hence, Foster and Robson provide a "summary of calculation" for the second problem listed above as: $1\,40 \div 2\,40 = 0; 37\,30 = 0; 30 + 0; 05 + 0; 02\,30$, that is, as a sexagesimal calculation with conversions back into metrological units as the last step, just as the procedure would have been done in the Old Babylonian period.

Friberg rejected the Foster-Robson interpretation of Sargonic mathematics, and in particular the computational errors as providing evidence for an early sexagesimalization, stating his goal at the beginning of his paper, "It has been claimed repeatedly by several authors... most recently Foster and Robson that sexagesimal numbers in place value notation must have been used in the complicated computations needed to solve the problems stated in the [Sargonic] metric division exercises and square-side-and-area exercises, always without explicit solution procedures. The aim of the present paper is to show that it is easy to explain those computations in less anachronistic ways" (2005, 1).

The less anachronistic way Friberg had in mind was to view of the problems as "metric geometry". Where Foster and Robson saw computation with numbers, Friberg saw manipulation of figures with given lengths and areas, hence his term "metric geometry". In this case, the Sargonic length-area problems are part of his sweeping re-assessment of Mesopotamian mathematics extending Høyrup's "cut-and-paste" geometrical interpretation of Old Babylonian quadratic mathematics (Høyrup 2002a, b) back to the early third millennium. Indeed, Friberg considers the metric-geometric area manipulations as the source of the later interest in quadratic problems.

Friberg referred to problems of the inverse type, as in the second example above, as "metric division problems", and stated, "the object of the exercise is not to divide a number by another number, but to divide a given area by a given length" (2005, 2). That is, a geometrical rather than arithmetic procedure.

For the second problem above, Friberg observed that 1 *iku* is a square of 10 × 10 *nindan* and then applied the following "factorization algorithm":

"1 *iku* = 10 *nindan* · 10 *nindan* (*a square with the side 10 nindan has the area 1 iku*)
= 40 *nindan* · 2 *nindan* 3 *kuš-numun* (*one side multiplied by 4, the other by ¼*)
= 2 40 *nindan* · 3 1/2 1/4 *kuš-numun* (*the length multiplied by 4, the side by ¼*)

Hence, the answer is that the short side is 3 1/2 1/4 *kuš-numun* = 3 *kuš-numun* 1 *giš-bad* 1 *šu-bad*" (2005, 6).

This presentation is still numerical. A clearer view of Friberg's geometric interpretation is given in Fig. 5 (where we have abbreviated the units *nindan* (n), *kuš-numun* (kn), *giš-bad* (gb), *šu-bad* (šb); the transliteration of the units here is slightly different from that of Friberg).

Friberg's metric geometry interpretation is detailed, complicated, and subtle and here we have given only a simple example without his detailed justifications. While his proposal is not without difficulties, it does act as a corrective to an excessive focus on metric quantities as numbers.

Sargonic scribes left no record of precisely how they conceived of their mathematical exercises nor did they explain how they achieved their results. If they visualized models of their fields to manipulate, they did not say so; if they had some kind of counting board for computation, they also did not say. A proper understanding of Sargonic mathematical processes awaits further research.

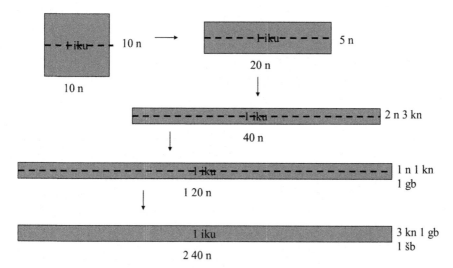

Fig. 5 Friberg's metric geometry

Mathematical Tables from the Early Dynastic III Period (2600–2340 BCE)

While the meagre Sargonic corpus of mathematics texts does not include any of the table texts that were such an important part of mathematical practice, the preceding Early Dynastic III period (ED III) includes five known examples, of which three have only recently been published and subjected to detailed analysis. All concern length-area metrology, relating lengths of sides and resulting areas of square or rectangular surfaces.

The most well-known is VAT 12593 (Deimel 1923, 82; Nissen et al. 1993; see Fig. 1 for a photograph). This is a large tablet from EDIIIa Šuruppak (ca. 2500 BCE) with a table of lengths and area of square fields of descending size from 1 *gešu* (that is, 600) *nindan* to 5 *nindan*, with corresponding areas decreasing from 3 *šar* 2 *buru* to ¼ *iku*. The reverse of the tablet is badly broken and presumably the table originally continued down to smaller sizes. The table is organized in three columns, first the lengths of the front of the field (*sag*), secondly the width, marked *sa₂* indicating it is equal to the front, and thirdly the areas, labelled *gan₂* in the first row to indicate that the areas represent fields (Tables 3 and 4).

For the lengths, the *aš* sign represents 1 *nindan*, the *u* sign 10 *nindan*, the *geš* 60 *nindan*, and *gešu* 600 *nindan*. For the areas, we have 1 *eše* = 6 *iku* and 1 *bur* = 3 *eše* as in the Sargonic texts discussed above. For the larger units, we have 1 *buru* = 10 *bur* and 1 *sar* = 6 *buru*.

Table 3 Transliteration of VAT 12593 (obverse)

Length	Width	Area
1 *gešu nindan sag*	1 *gešu sa₂*	3 *šar* 2 *buru gan₂*
9 *geš*	9 *geš sa₂*	2 *šar* 4 *buru* 2 *bur*
8 *geš*	8 *geš sa₂*	2 *šar* 8 *bur*
7 *geš*	7 *geš sa₂*	1 *šar* 3 *buru* 8 *bur*
6 *geš*	6 *geš sa₂*	1 *šar* 1 *buru* 2 *bur*
5 *geš*	5 *geš sa₂*	5 *buru*
4 *geš*	4 *geš sa₂*	3 *buru* 2 *bur*
3 *geš*	3 *geš sa₂*	1 *buru* 8 *bur*
2 *geš*	2 *geš sa₂*	8 *bur*
1 *geš*	1 *geš sa₂*	2 *bur*

Table 4 Transliteration of VAT 12593 (reverse)

Length	Width	Area
5 *u*	5 *u sa₂*	1 *bur* 1 *eše* 1 *iku*
4 *u*	4 *u sa₂*	2 *eše* 4 *iku*
3 *u*	3 *u sa₂*	1 *eše* 3 *iku*
2 *u*	2 *u sa₂*	4 *iku*
1 *u*	1 *u sa₂*	1 *iku*
5 *aš*	5 *aš sa₂*	¼ *iku*

The precise purpose of such a text is not exactly clear, and we do not have sufficient evidence to be certain of how these tables were used. Nor is it clear how the underlying computations of areas were made. It is unlikely that the original computations were performed in the order in which the information is presented in the table, from large quantities to small. That argues that the table is some kind of summary. It could have been written to show that certain simple length-area relationships had been memorized, or it could have functioned as a reference table as an aid to calculating more complicated situations, presumably by subdividing a surface into smaller, regular pieces. Computation of field areas presents us with a nice balance of arithmetic and geometry.

In their discussion of this tablet, Nissen, Damerow and Englund suggested the table was an exercise, rather than a practical aid, writing, "The exact purpose of this table of areas of square fields is not known. We may exclude the possibility that it served as some sort of table of calculations used to consult particular values. The list was more likely to have been written as an exercise containing easily determinable field surfaces every land-surveyor was required to know which could be added together in calculating complicated surfaces" (1993, 139). Certainly the well-known Old Babylonian multiplication tables were largely exercises in showing mastery of a topic, rather than aids for those with poor memories and perhaps this example falls into that genre.

The only other known EDIIIa table text, MS 3047, was published by Friberg in 2007. Similar in shape and format to VAT 12593, the obverse of the tablet contains a table of lengths and areas for rectangular shapes where the length of the rectangle is always 60 times its width. In contrast to VAT 12593, the values increase in size in each line and, rather surprisingly, the table concludes with a total area. Accounting texts from Šuruppak frequently include long lists of individual entries summarized by a total, but it is unusual to see this appearing in a mathematical text. The reverse of MS 3047 contains a mysterious table that Friberg suggested represented a geometric progression of areas, but it cannot yet be completely understood.

These two EDIIIa (2600–2500 BCE) tables are complemented by three ED IIIb (2500–2340 BCE) texts. The first published, A 681 (Luckenbill 1930), has a rather different layout from the two Šuruppak texts. Instead of a table divided into columns to be read across the tablet from left to right, this one gives lengths of squares and corresponding areas to be read down each column individually.

The next ED IIIb text to appear, CUNES 50-08-001, was also published in Friberg (2007). This is a large, multi-column tablet with five different tables of square areas, ranging from the very large 1 šaru (that is, 36000 nindan) to very small 1 šu-bad (1/24 nindan).

Most recently, we have the tablet published by Feliu (2012). This tablet contains two tables. The first is an almost exact duplicate of VAT 12593, the other computes areas of rectangular shape where one side is held fixed throughout the table while the other varies in each line.

These few tables, all ostensibly on the same topic, that of computation of rectangular and square areas, present us with a great diversity of practice. While some of them may have been informed or derived from actual surveying practice, that is, fields of realistic sizes, others, and especially CUNES 50-08-001 seem concerned with a theoretical extension of practical surveying beyond the bounds of the metrological system. Given that the extension to small area units comes from invoking sexagesimal fractions from weight metrology, we are again confronted with the question of how much a proto-sexagesimal metrological idea was in the air centuries before the development of the abstract sexagesimal place-value system. We are also left to question the extent to which tables of areas should be viewed as arithmetical calculations and how much as metrological geometric constructs. Friberg said of CUNES 50-08-001 that it was "a clear forerunner of the invention of sexagesimal numbers in place value notation" (2007, 426). On the other hand, Proust (forthcoming) has a detailed review of the corpus of five tablets, in which it is argued that a metric geometric approach is possible behind the older (EDIIIa) tablets while the systematic exploration found in CUNES 50-08-001 represented a shift in conceptualizing multiplication, and in particular the construction of areas from given linear measures, from a geometric orientation to an arithmetic one. That is, Proust sees these area tables as providing a marker of an increasing arithmetization of mathematical thinking.

Friberg's notion of metric geometry as detailed in (2005, 2007, 2014) requires sophisticated manipulation of shapes and a careful knowledge of metrological relationships. One begins with a simple shape and deforms it into the desired final result. Such procedures can yield complicated word problems, but are not needed for the simple tables from Šuruppak discussed above. Proust's proposed geometric approach to the area tables starts from the observation that the key linkage between lengths and areas at this time is not the *sar* of 1 square *nindan*, but the *iku*, the area of a square 10 *nindan* on a side. From this it is a simple matter to construct reference shapes for all larger area units and combine them to produce the desired squares for the tables. Not much more than counting is required, and certainly no abstract multiplication Proust (forthcoming). Thus a geometric approach obviates complicated calculations.

Mathematical Problem Texts from Šuruppak (ca. 2500 BCE)

The texts from Šuruppak date from around 2500 BCE and come from a narrow period, a few months to a few years. A large number of administrative texts have allowed a reconstruction of parts of the organization of the city (Pomponio and Visicato 1994; Visicato 1995, 2000). There is a term for 'scribe', and a hundred are known from the economic documents. Šuruppak was part of a group of half a dozen cities that were preparing for war at the time the documents were written. We can assume they lost as the city was then largely abandoned.

Among the administrative texts are a small number of mathematical ones, including a group that can be considered as the world's oldest known mathematical word problems. One of these problems is to compute the area of a very large square, 5 *gešu nindan* (ca. 18 km) on a side (TSS 188, Jestin 1937). As is the case with the later Sargonic problem texts discussed above, the actual tablet contains not much more than quantities. No task is specified, and no procedure explained (Fig. 6).

The text was analyzed by Friberg (2007, 148–149). The given answer is incorrect. Recall that the largest entry on the table VAT 12593 gave the area of a square of side 1 *gešu nindan* as 3 *šar* 2 *buru*. The problem on this tablet is to determine an area 25 times as great.

The relevant portion of the area metrology is given in the factor diagram below.

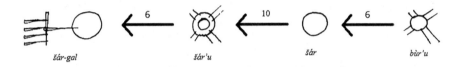

| *šár-gal* | *šár'u* | *šár* | *bùr'u* |

Friberg observed that the given answer is precisely what would have been obtained if the area of a 1 *gešu nindan* square had been incorrectly looked up in a table or recalled as 3 *šar* 3 *buru* instead of 3 *šar* 2 *buru*. Friberg briefly summarized the subsequent calculation in an equation without indicating the details of how the intermediate steps were conceived or carried out. His description implies an essentially arithmetical procedure.

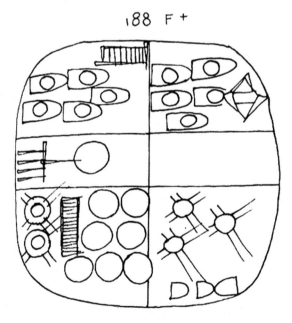

Fig. 6 TSS 188, computation of a large area. *Source* Cuneiform Digital Library Initiative, P010773. http://cdli.ucla.edu/dl/lineart/P010773_l.jpg. http://cdli.ucla.edu/search/archival_view.php?ObjectID=P010773

Starting from an incorrect base area also upsets all the nice linkages that occur for the larger units and it really should have been apparent that something had gone wrong if following Proust's subdivision of a diagram approach. However, the initial error could propagate effortlessly through the resulting computations if some kind of repeated addition or 'correspondence' procedure had been used. The issue arises because of the limitations of the specific notation in use during this period.

If the large surface was conceived as a 5 by 5 grid of smaller areas then computing the larger area becomes a simple problem in addition. If one square had area 3 *šar* 3 *buru* then two squares would have area 7 *šar*, 3 squares 1 *šaru* 3 *buru*, four squares 1 *šaru* 4 *šar* and five squares 1 *šaru* 7 *šar* 3 *buru*. Since five squares form a row or column, the calculation could then operate on rows to get the final result. In this model only simple addition and an understanding of the unit conversions inherent in the factor diagram is needed. All intermediate values are maintained in the standard metrological notation. Also, note that multiples of a given unit are recorded with repetitions of a unit, so tallying and bundling are the only requirements. These computations could have easily been carried out either in writing on some kind of temporary scratch pad surface, or using a simple additive counting board with symbolic markers for the different units.

We have emphasized the importance of core linkages between metrological systems. The use of correspondences based on simple linkages was first suggested by Friberg in his analysis of an Ebla mathematical text as a division problem. Friberg formulated the problem as: "Given that you have to count with 1 *gubar* [an Ebla capacity unit] for 33 persons, how much do you count with for 260,000 persons?" (Friberg 1986, 19). Friberg made the crucial observation that division of 260,000 by 33 was not possible using Eblaite notation and thus a different procedure must have been used.

A similar proposal for dealing with division problems from Šuruppak was given in Melville (2002) and Friberg (2007, 410–415) summarizes an improved version of the original suggestions for both the Ebla and Šuruppak problems. Below we give two examples, one simple, one more complicated, to illustrate the technique.

The tablet TSS 81 (Jestin 1937) reads (in the translation from the *Digital Corpus of Cuneiform Mathematical Texts*, see http://oracc.museum.upenn.edu/dccmt/): "40 sons of builders (each) received 2 ban as a flour gift. (Total) 3 *lidga*, 1 *barig*, 2 *ban* of flour" (Fig. 7).

The modern approach would be to take an abstract multiple ($40 \times 2 = 80$) of the base unit *ban* and then convert the 80 *ban* into the correct units. However, the capacity notation in Šuruppak precluded writing more than 5 *ban* (see Melville 2002 for details). Quantity notation was tied to the measurement system and not abstracted out as a number. Instead, the scribe must have worked up from the base correspondence of 1 person to 2 *ban* up to the correspondence of 40 people to the total quantity of 3 *lidga*, 1 *barig*, 2 *ban* using some kind of tabular arrangement and aggregating and bundling the quantity units.

A similar approach must also have been used for the ration 'division' problem from Šuruppak (TSS 50, Jestin 1937): "A granary of barley. Each man received 7 *sila* of grain. Its men: 4 *šaru* 5 *šar* 4 *gešu* 2 *geš* 5 *u* 1; 3 *sila* of barley remains" (Fig 8).

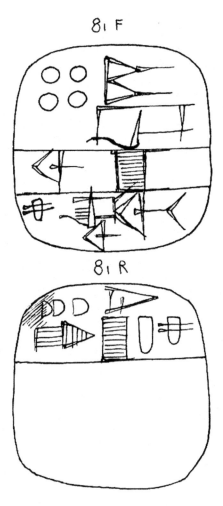

Fig. 7 TSS 81, ration computations. *Source* Cuneiform Digital Library Initiative, P010737. http://cdli.ucla.edu/dl/lineart/P010737_l.jpg. http://cdli.ucla.edu/search/archival_view.php?ObjectID=P010737

The scribe must have worked up from the correspondence of 1 man to 7 *sila* up through the various capacity units to the granary at the top. In fact, there is another copy of this problem, TSS 671 (Jestin 1937), or at least an attempted answer. In this case, the answer given is incorrect, but if some system of correspondences was used, the given solution required only one simple error in the middle of the computation. The challenge for the scribe was managing correspondences between two different metrological domains, each with different notation and different relationships between larger and smaller units.

While it is dangerous to generalize from such a modest sample, the fact that these division problems dominate the small corpus of mathematical problems suggests that the construction and solution of complex metro-mathematical division

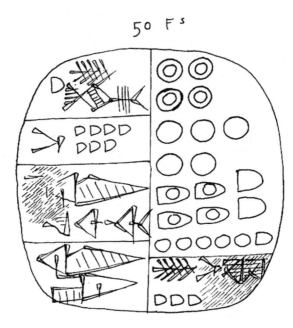

Fig. 8 TSS 50, ration computations. *Source* Cuneiform Digital Library Initiative, P010721. http://
cdli.ucla.edu/dl/lineart/P010721_l.jpg.   http://cdli.ucla.edu/search/archival_view.php?ObjectID=
P010721

problems was popular 4500 years ago. As Friberg observed, "the fact that three of the
four oldest known mathematical problem texts were concerned with exactly the same
kind of 'non-trivial' division problems must be significant: the obvious implication is
that the 'current fashion' among mathematicians about four and a half millennia ago
was to study non-trivial division problems involving large…numbers and 'non-regular'
divisors such as 7 and 33" (Friberg 1986, 22). The theory of correspondences suggests
that these problems may have been solved using simple repetition and bundling with a
collection of intermediate steps organized in a list or tabular fashion.

Archaic Mathematics (ca. 3000 BCE)

Although more is now known about the mathematics of the Archaic period around
3000 BC than was the case a few decades ago, much is still mysterious. The best
introduction remains (Nissen et al. 1993).

Writing and mathematics were introduced in mesopotamia for administrative
purposes, in order to monitor and record the flow of goods in a bureaucratic context.
As Foster observed, "Accountability means the obligation to keep records for
property that does not belong to the record-keeper. Although some have suggested
that Sumerian accountability grew out of the difficulty of managing large quantities

or transactions too complicated to remember, this misses the fundamental purpose of institutional record-keeping of all periods" (Foster 2005, 78). Thus, most of the sources we have derive from an institutional context and reflect institutional needs and priorities; what other kinds of mathematics there may have been in society have not left similar traces.

Another complication with the archaic texts is that it is often difficult to tell if a tablet represents an actual administrative document, a practice model document, or an exercise. In this context, Friberg used the term 'metro-mathematical text', commenting that "the term is appropriate because it is often difficult or impossible to distinguish a complicated and mathematically interesting administrative-economic text from a carefully designed mathematical exercise constructed in order to demonstrate the use of, and manipulation with, various kinds of measures and their numerical notations" (Friberg 1997, 2).

One indicator of artificiality that Friberg used was the presence of round or "almost-round" quantities, especially in areas of fields. One might expect fields in administrative archives to come in all shapes and sizes. The presence of lengths or areas that are conspicuously round (especially large and conspicuously round) or that differ from a round number either by a simple fraction or a small quantity suggests an artificial exercise. The frequency of occurrence of such measures convinced Friberg that many of these texts were educational and led to his geometric division interpretation extending back into the earliest script phases (Friberg 1997, 2014).

Friberg's argument is delicate and one must take care in its interpretation. One example that he uses in both papers (1997 and 2014) is an archaic text where the areas of five rectangular fields are computed and then the total is recorded (MSVO 1, 2, Englund and Grégoire 1991). The total area is written as 3 *šar* 5 *buru* 2 *bur* 2 *eše* 5 *iku*, not an obviously round quantity. However, converting this into a multiple of *iku*, Friberg noted that the total is 4193 *iku* or, as he wrote, "$7 \cdot 10 \cdot 60 \, iku - 7 \, iku$", commenting that, "This nearly round area number is hardly accidental" (2014, 4). Friberg is not suggesting that the quantity 4200 *iku* was written out, as that would not be possible in the contemporary notation, but that an underlying geometrical sense was used to construct problems where the solution comes very close to a regular figure. The point being that in problems involving fields, especially areas of fields, some kind of geometric representation may have been used, presumably marked on some temporary surface such as a dust board, the ground, or clay, but that this aid is not recorded on the final tablet that has come down to us. The only trace of the geometry is in the quantities chosen.

Early writing was largely curviform, drawn by a reed stylus onto wet clay. Quantity notation was somewhat different. The other end of the stylus was pressed into the clay either at an angle, producing a horizontal or vertical wedge shape, or straight down producing a circle. Two different sizes of stylus were used and the notation could combine them so, for example, producing a small circle inside a large wedge. These basic signs were decorated in various ways to extend to a repertoire of about 50 different quantity signs. How far this practical disjunction in way of writing quantities reflected a conceptual distinction is unclear.

Each sign represented a certain sized unit when referring to a particular metrological domain such as lengths, areas, capacities, time, or discrete goods. As there were more units than signs, many of the signs, especially the simple basic ones, appeared in several different contexts, where the same pair of signs could stand for different multiples depending on what was being measured. For example, the circular sign obtained by pressing the end of the stylus vertically into the clay represented 6 wedges in the capacity system, 10 wedges in the system used for discrete goods and 18 wedges in the area system. Unravelling this complex collection of metrological units and their relationships was a correspondingly difficult task. The main description of the results is in Nissen et al. (1993). Multiples of each basic unit were recorded with repetitions of the unit sign—three times a circle was shown as three circles. It is not known for certain how computations were carried out. All that was written down was the result of a computation, and there are no texts explaining arithmetic procedures. However, as we have seen above, there are some hints, but they are slight and open to contradictory interpretations.

Calculation

The question of precisely what kinds of computational aids and tools the Mesopotamian scribe could call upon, and exactly how certain kinds of calculations were carried out is problematic. The evidence is thin, indirect, and contradictory. We are considering a period of over a thousand years with evidence of slow but steady conceptual development. There is not necessarily a single answer. More than 20 years ago, with reference to the archaic period, Peter Damerow and Robert Englund wrote,

> How such calculations were performed is not understood in detail. Instruments that could have served as calculation aids are as yet not attested in archaeological finds, or have not been identified as such. Lexical lists from later periods, however, suggest that the Sumerians used tallying boards made of wood, which being perishable would not have been unearthed in excavation. There is also some evidence that the sign SANGA, designating the chancellor of an economic unit, derives from a pictogram depicting such a tallying board. This, in fact, is supported by the cuneiform sign ŠID which, also having developed from this sign, was employed as an ideogram with the meaning "account". (Nissen et al. 1993, 134)

Sadly, at least with respect to archaeology, not much has changed in the intervening years. There is still no incontrovertible evidence for any kind of counting board. However, we can say a little more.

First, the lexical evidence. There is a Sumerian term (*giš-šudum-ma*) translated as "tally-stick". However, the word only appears (twice) in much later Old Babylonian literary sources. The term is not attested in the third-millennium sources at all and, while it is indeed possible that Sumerians made use of tally-sticks, such would be more a record device than a calculation aid. The second term (*giš-nig-šid*) appears in lexical lists (long lists of related words) but does not appear outside of them. The Sumerian *giš* is a determinative signifying an object made out of wood, so if the term referred to a real object it would be some kind of wooden device aiding in accounting. One of the problems with interpreting the lexical texts is that in them Sumerian scribes strove for an idealized completeness that did not always reflect reality. The fact that such objects are not ever mentioned in administrative documents is problematic, perhaps more so than their absence from the archeological record.

The hint provided by the shape of the archaic sign *sanga* is more intriguing. Before the term "scribe" appeared in the middle of the third millennium, there was a category of officials associated with receipt and disbursement of goods called *sanga* who often needed to determine totals of numerous entries or ratios of ingredients to go towards making goods (beer, for example). The sign is very well attested and came in a number of variants, of which the most common type was:

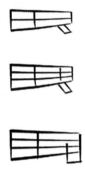

(CDLI signlist, http://cdli.ucla.edu/tools/SignLists/protocuneiform/archsigns.html)

The suggestive shape has led some to argue that the sign represents some kind of counting board with a receptacle for counters on the side. The farthest anyone has committed themselves in print is Jøran Friberg, who in a parenthetical aside remarked, "the sign *sanga* itself may be the picture of a box for number tokens". (Friberg 2007, 151)

What makes the weakness of the evidence more frustrating is that the third millennium seems bracketed by the use of counters. Before writing developed, there was a widespread system of 'tokens' in use in Mesopotamia, and it appears that at least some of these tokens acted as 'counters' recording quantities of types of goods. The exact details of the relationships of early counters to the goods is unclear and whether they merely acted as a record of quantities or were used for calculations is completely unknown (Lieberman 1980; Schmandt-Besserat 1992; Englund 1993; Friberg 1994). At the other end of the period, both Proust and Høyrup have argued for the presence of a counting board in Old Babylonian mathematics (Proust 2000; Høyrup 2002a, b). But this is after the development of the abstract sexagesimal system, and their arguments depend in part on analysis of errors in many-place computations which would not have occurred the same way in earlier periods. Perhaps the new kinds of computation called forth new technology.

Another approach is to consider the kinds of calculations scribes were called upon to perform. We have seen that the archaic system had a wide range of quantity notation with often narrowly specified subject matter. Over the course of the following thousand years, number was gradually abstracted out of quantity. However, within this range of systems, there were distinct similarities. Each metrological system contained multiple different units. In all cases, multiples of a unit were given by repetition of the unit, either repeating the whole sign or by modifying a base sign with a series of strokes. There were thus three kinds of operations of increasing difficulty that were needed: simple accumulation, replacement or bundling of units within a sequence, and relations between units in different systems.

In the first case, simple accumulation, 'addition' is merely copying. That is, what we might think of as '$2 + 2 = 4$' becomes in an accumulation or tallying system '$II + II = IIII$', with the particular sign for the unit varying with context. The next level is bundling the appropriate collection of a small unit into one of the next larger unit. This operation could be learned either solely through writing or with some sort of physical aids. What complicated the issue for Mesopotamian scribes as opposed to modern children confronting, say, number rods, is that each individual system contained multiple different sized bundles to scale from one unit to the other. Context was vital, and this the written system was able to express clearly.

The third, and by far the most difficult, problem was establishing linkages between metrological domains that used different kinds of multiples in the span of their units. We have seen examples of this in the problem of giving rations to people, where the capacity units do not nicely correspond to the discrete system used to count people, and, above all, in the core problem of computing the area of a field from measurements of the sides. Length and area units did not align well. The evidence we have is that scribes responded to this challenge by developing tables reflecting simple square and rectangular cases, and possibly diagrams, but not the use of counting boards.

The main basis of Sumerian scribal education lay in copying and memorization. Scribes developed their writing and linguistic skills through copying exemplars and gradually memorizing signs and forms. It is probable that mathematics was learned the same way. Proust's reconstruction of elementary mathematics education at Nippur during the Old Babylonian period shows that students learned lists of metrological signs, then mathematical tables (in this later period multiplication tables), and only entered on doing even simple calculations once they were thoroughly grounded in the intricacies of the metrological systems (Proust 2007). This pedagogical approach had deep roots.

If mathematical calculation, and in the third millennium this meant accumulation and correspondences, only came late in scribal education, it may well be that such calculations were carried out in written form. The professional identity of a scribe depended on the exclusive ability to write and, certainly in the archaic period, the complexity of quantity notation argues for a written form of computation, although this must remain speculative in the absence of better evidence.

Appendix: Periodization

The absolute chronology of Mesopotamia is an area of current vigorous scholarly debate, especially before the first millennium. The further back in time one goes, the larger the uncertainties. For convenience here we follow the middle chronology periodization adopted by the Cuneiform Digital Library Initiative (CDLI), as follows:

Period	Date
Uruk IV	~ 3350 to 3200
Uruk III	~ 3200 to 3000
Early Dynastic I–II	~ 2900 to 2700
Early Dynastic IIIa	~ 2600 to 2500
Early Dynastic IIIb	~ 2500 to 2340
Sargonic	~ 2340 to 2200
Ur III	~ 2100 to 2000
Old Babylonian	~ 2000 to 1600

References

Adams, R. Mc. C. 1981. *Heartland of cities. Surveys of ancient settlement and land use of the central floodplain of the Euphrates*. Chicago: University of Chicago Press.

Deimel, A. 1923. *Die Inschriften von Fara II: Schultexte aus Fara*. Leipzig: J. C. Hinrichs'sche Buchhandlung.

Englund, R.K. 1993. Review of the origins of script. *Science* 260 (5114): 1670–1671.

Englund, R.K., and J.-P. Grégoire. 1991. *The Proto-Cuneiform texts from Jemdet Nasr* (*Materialien zu den Frühen Schriftzeugnissen des Vorderen Orients*, vol. 1). Berlin: Gebrüder Mann Verlag.

Feliu, L. 2012. A new Early Dynastic IIIb metro-mathematical table tablet of area measures from Zabalam. *Altorientalische Forschungen* 39: 218–225.

Foster, B.R. 2005. Shuruppak and the Sumerian city state. In *Memoriae Igor M. Diakonoff*, ed. L. Kogan, et al. Winona Lake: Eisenbrauns.

Foster, B.R., and E. Robson. 2004. A new look at the Sargonic mathematical corpus. *Zeitschrift für Assyriologie* 94: 1–15.

Friberg, J. 1986. The early roots of Babylonian Mathematics, III. Three remarkable texts from ancient Ebla. *Vicino Oriente* 6: 3–25.

Friberg, J. 1990. Mathematik. In *Reallexikon der Assyriologie VII*, ed. D.O. Edzard, 531–585. Berlin and New York: De Gruyter.

Friberg, J. 1994. Preliterate counting and accounting in the Middle East. *Orientalistische Literaturzeitung* 89: 477–502.

Friberg, J. 1997. Round and almost round numbers in protoliterate metro-mathematical field texts. *Archiv für Orientforschung* 44 (45): 1–58.

Friberg, J. 2005. On the alleged counting with sexagesimal place value numbers in mathematical cuneiform texts from the Third Millennium BC. *Cuneiform Digital Library Journal* 2005:2 (http://cdli.ucla.edu/pubs/cdlj/2005/cdlj2005_002.html).

Friberg, J. 2007. *A remarkable collection of Babylonian mathematical texts. Manuscripts in the Schøyen collection: Cuneiform texts I*. New York: Springer.

Friberg, J. 2014. Geometric division problems, quadratic equations, and recursive geometric algorithms in Mesopotamian mathematics. *Archive for History of Exact Sciences* 68 (1): 1–34.

Høyrup, J. 1996. Changing trends in the historiography of Mesopotamian mathematics: An insider's view. *History of Science* 34: 1–32.

Høyrup, J. 2002a. A note on Old Babylonian computational techniques. *Historia Mathematica* 29: 193–198.

Høyrup, J. 2002b. *Lengths, widths, surfaces: A portrait of Old Babylonian algebra and its kin*. New York: Springer.

Jestin, R. 1937. *Tablettes sumériennes de Shuruppak au Musée de Stamboul*. Paris: E. de Boccard.

Lieberman, S.J. 1980. Of clay pebbles, hollow clay balls, and writing: A Sumerian view. *American Journal of Archaeology* 84: 339–358.

Limet, H. 1973. *Étude de documents de la période d'Agadé appartenant à l'Université de Liège*. Paris: Les Belles Lettres.

Liverani, M. 1990. The shape of Neo-Sumerian fields. *Bulletin on Sumerian Agriculture* 5: 147–186.

Luckenbill, D. 1930. *Inscriptions from Adab*. Chicago: The University of Chicago Press.

Melville, D. 2002. Ration computations at Fara: Multiplication or repeated addition? In *Under one sky: Astronomy and mathematics in the Ancient Near East* (London, 2001) (AOAT 297), ed. J.M. Steele, and A. Imhausen, 237–252. Münster: Ugarit-Verlag.

Melville, D. 2016. After Neugebauer: Recent developments in Mesopotamian mathematics. In *A mathematician's journeys: Otto Neugebauer and modern transformations of ancient science*, ed. A. Jones, C. Proust, and J. Steele, 237–263. Cham: Springer.

Nissen, H.J., P. Damerow, and R. Englund. 1993. *Archaic bookkeeping: Early writing and techniques of economic administration in the ancient Near East*. Chicago: University of Chicago Press.

Pomponio, F., and G. Visicato. 1994. *Early Dynastic administrative texts of Šuruppak*. Napoli, Istituto universitario orientale di Napoli, Dipartimento di studi asiatici.

Powell, M.A. 1990. Masse und Gewichte. In *Reallexikon der Assyriologie VII*, ed. D.O. Edzard, 457–530. Berlin and New York: De Gruyter.

Proust, C. 2000. La multiplication babylonienne: la part non écrite du calcul. *Revue d'histoire des mathématiques* 6: 293–303.

Proust, C. 2007. *Tablettes mathématiques de Nippur*. Paris: Institut Français d'Études Anatoliennes-Georges Dumézil.

Proust, C. (forthcoming). Early-Dynastic tables from Southern Mesopotamia, or the multiple facets of the quantification of surfaces. In *Mathematics and administration in the ancient world*, ed. K. Chemla, and C. Michel.

Robson, E. 2007. Mesopotamian mathematics. In *The mathematics of Egypt, Mesopotamia, China, India, and Islam*, ed. V. Katz, 57–186. Princeton, NJ: Princeton University Press.

Robson, E. 2008. *Mathematics in Ancient Iraq: A social history*. Princeton, NJ: Princeton University Press.

Schmandt-Besserat, D. 1992. *Before writing I–II*. Austin: University of Texas Press.

Visicato, G. 1995. *The Bureaucracy of Šuruppak*. Münster: Ugarit-Verlag.

Visicato, G. 2000. *The power and the writing*. Bethesda, MD: CDL Press.

Duncan J. Melville is a Professor of Mathematics at St. Lawrence University, NY, USA. His research centers on Mesopotamian mathematics and, in particular, the development of abstraction during the third millennium BCE. Publications include "After Neugebauer: Recent developments in Mesopotamian mathematics," in Alexander Jones, Christine Proust, and John Steele (eds.), *A Mathematician's Journeys: Otto Neugebauer and Modern Transformations of Ancient Science*, Springer (2016), 237–263, and "The Mathland mirror: On using mathematical texts as reflections of everyday life," in H. Neumann, et al. (eds.), *Krieg und Frieden im Alten Vorderasien 52e Rencontre Assyriologique Internationale International Congress of Assyriology and Near Eastern Archaeology*, Münster, 17–21. Juli 2006. AOAT 401. Münster: Ugarit-Verlag (2014), 517–526.

Computational Techniques and Computational Aids in Ancient Mesopotamia

Jens Høyrup

Abstract Any history of mathematics that deals with Mesopotamian mathematics will mention the use of tables of reciprocals and multiplication in sexagesimal place-value notation—perhaps also of tables of squares and other higher arithmetical tables. Less likely there is a description of metrological lists and tables and of tables of technical constants. All of these belong to a complex of aids for accounting that was created during the "Ur III" period (twenty-first c. BCE). Students' exercises from the Old Babylonian period (2000–1600 BCE) teach us something about their use. First metrological lists then metrological tables were learned by heart. These allowed the translation of real measures into place-value measures in a tacitly assumed basic unit. At an advanced level, we see multiplications, where first two factors and then the product are written in sequence on a clay tablet for rough work. Problem texts show us more about the use of the metrological tables and the tables of technical constants. Neither genre allows us to see directly how additions and subtractions were made, nor how multiplications of multi-digit numbers were performed. A few errors in Old Babylonian problem texts confirm, however, that multiplications were performed on a support where partial products would disappear once they had been inserted—in a general sense, some kind of abacus. Other errors, some from Old Babylonian period and some others from the Seleucid period (third and second c. BCE), show that the "abacus" in question had four or five sexagesimal levels, and textual evidence reveals that it was called "the hand". This name was in use at least from the twenty-sixth c. BCE until c. 500 BCE. This regards addition and subtraction from early times onward, and multiplication and division in Ur III and later. A couple of problem texts from the third millennium deals with complicated divisions, namely divisions of large round numbers by 7 and by 33. They use different but related procedures, suggesting that no standard routine was at hand.

J. Høyrup (✉)
Roskilde University, Roskilde, Denmark; Institute for the History of Natural Sciences, Chinese Academy of Sciences, Beijing, China; Max-Planck-Institut Für Wissenschaftsgeschichte, Berlin, Germany
e-mail: jensh@ruc.dk

© Springer Nature Switzerland AG 2018
A. Volkov and V. Freiman (eds.), *Computations and Computing Devices in Mathematics Education Before the Advent of Electronic Calculators*, Mathematics Education in the Digital Era 11, https://doi.org/10.1007/978-3-319-73396-8_3

Keywords Sexagesimal place-value system · Mathematical tables (Mesopotamia)
Abacus (Mesopotamia) · Scribe school curriculum (Mesopotamia)
Centennial system (Mari)

For Ying

Introduction: The Familiar

Any history of mathematics that deals with Mesopotamian or more narrowly with
Babylonian mathematics will speak of tables of reciprocals and multiplication in
sexagesimal place-value notation—perhaps also of tables of squares and other higher
arithmetical tables such as n^3 and $n^2 \times (n + 1)$.[1] It is possible though less plausible
that they also mention metrological tables and tables of technical constants.

Let us start by describing this system, postponing the discussion of its use and
general historical setting.

The underlying number system, as stated, was a sexagesimal place-value notation, whereas ours is a decimal place-value notation. In our notation, the digit "7"
may refer to the number seven, but just as well to 7×10, 7×10^2, ..., or to
7×10^{-1}, 7×10^{-2}, 7×10^{-3}, ...; what is actually meant is determined by its
location within the sequence of digits—its "distance from the decimal point".
Similarly, a Mesopotamian digit "7" may stand for 7, 7×60, 7×60^2, ..., as well
as 7×60^{-1}, 7×60^{-2}, ...

In the Mesopotamian notation, however, there was no analogue of the decimal
point and thus no way to determine absolute magnitude from the distance to it.
There was also no sign for zero, and in principle, "16 40" might thus mean not only
$(16 \times 60 + 40) \times 60^n$ but also $(16 \times 60^2 + 0 \times 60^1 + 40) \times 60^n$, etc.[2] This may

---

[1]See, for instance, the popularizations (Neugebauer 1934) and (Neugebauer 1957), on which many
general histories build. Since they are of no particular importance in what follows, I shall not return
to the higher arithmetical tables.

[2]Such intermediate zeroes only came in current use (most often not for a missing sexagesimal
place but for missing units or tens) in the Seleucid epoch (third to second century BCE), even
though two texts from around 1600 BCE [TMS XII and XIV, see Høyrup (2002a:15 n. 16)] indicate
them occasionally, and two ambiguous fragments from the intervening period seem to suggest
continuity rather than Seleucid reinvention. This is one of several indications that Mesopotamian
calculators did not think of their system solely as sexagesimal but also (perhaps predominantly) as
a "seximal-decimal" notation (just as Roman numerals may be thought of as "dual-quintal").

One or two lines in the extensive corpus of Seleucid astronomical texts may even contain a
final zero; the interpretation is quite dubious, however [Neugebauer 1955: I. 121, 166, 208]. In any
case, final zeros never came into in widespread, not to say general use.

With or without final zero, the Babylonian placeholder, a mere punctuation mark, was
something quite different from our zero. Our zero, beyond serving as placeholder, is also a number,
the outcome of a subtraction $a - a$. When encountering such subtractions, the Old Babylonian
texts might say "one is as much as the other" or "it is missing"—or they would, literally, treat the
outcome as not worth speaking about and not state any result (Høyrup 2002a: 293). The situation
never occurs in later texts.

seem odd to us, but we shall see that the inherent ambiguity in this floating-point notation probably created no problems in the context where it served.

In some of the text types that we shall discuss the order of magnitude plays no role—just as it plays no role when we look at a slide ruler whether "2.5" stands for 2.5, 25 or 0.25 (indeed, the same position of the slide rule gives us 2.5×4, 25×400 and 0.25×0.4) (Fig. 1). In such cases, we may render what appears as "16 40" as 16.40 (or 16..40 if we suspect an empty intermediate order of magnitude is intended). In other texts, a specific order of magnitude is certainly meant—just as 3.1416 and certainly not 314.16 is meant where the modern slide rule writes π. If "16 40" is to be interpreted as $16 \times 60^2 + 40 \times 60$, we shall translate it $16\grave{}\,40\grave{}$; if it is to be understood as $16 \times 60 + 40$, we shall write $16\grave{}40$, and if it stands for $16 + 40 \times 60^{-1}$, we shall write $16°40'$ (when it is not needed as a separator, "$°$" will be omitted). $30'$ thus means $\frac{1}{2}$, while $10''$ means $\frac{1}{360}$.[3]

This generalization of our modern degree–minute–second notation for time and angles (which descends via ancient Greek astronomy from the Mesopotamian system) has the advantage that no zeroes are written which are not in the original text (except those indicating missing units, without which the tens could not be identified as such); one may omit the pronunciation of the 'and' and keep them as tacit knowledge, just as the Mesopotamian calculators did with their knowledge about the intended order of magnitude—*they* wrote nothing corresponding to $\grave{}$, $°$ and $'$.

The place-value notation was not needed for and also hardly used for additions and subtractions; we shall return to that issue. Its purpose was to serve multiplication and division.

In our algorithm for multiplication, we make use of a multiplication table with 10×10 entries, which we learn by heart. The Mesopotamian calculators, however, did not need 60×60 entries. They were trained on tables where important "principal numbers" were multiplied by 1, 2, 3, ..., 19, 20, 30, 40 and 50. So, 18×37 would have to be found as $18 \times 30 + 18 \times 7$.

The term "division" may refer either to a type of question or to a procedure. The Mesopotamian calculators were fully familiar with the question "what shall I multiply by b in order to get a"—our equation $bq = a$, whose answer is $q = \frac{a}{b}$, but they had no *standard procedure* by which to produce directly the number q from the numbers a and b. Instead, if possible, they made use of a multiplication, finding

---

[3]This notation was introduced by Assyriologists in the early twentieth century. Later, various alternatives have been used, the most widespread of which will write $7\grave{}13°41'40''$ as 7,13;41,40. It is particularly advantageous in the analysis of mathematical-astronomical texts.

Fig. 1 A circular slide rule from ca. 1960
Author's photo

q as $a \times \frac{1}{b}$. For this purpose, they employed a table of reciprocals, called IGI,[4] copied so often in school that it was learned by heart—Fig. 2 shows the standard version.[5]

In most practical computation, the coarse grid provided by the standard table was sufficient. We have a few tables listing approximate reciprocals of "irregular numbers", that is, numbers that do not have a reciprocal that can be expressed as a finite sexagesimal fraction. They may have been computed as school exercises or as schoolmasters' experiments—we do not know; but in any case, they show that approximate reciprocals of irregular numbers *could* be determined. We also know a technique that was used to find the reciprocals of regular numbers that did not appear in the standard table. As a simple illustration, we may pretend that $A = 44.26.40$ does not appear and try to find $\frac{1}{A}$. We observe that the final part of the

---

[4]Sumerian is conventionally transliterated as small caps (sometimes as spaced writing if we believe to know the pronunciation and as small caps or capital letters if we use sign names).

[5]Since 1.12, 1.15 and 1.20 already appear to the right as IGI 50, IGI 48 and IGI 45, respectively, they are sometimes omitted to the left; moreover, some early tables have as their first line "Of sixty, its $\frac{1}{3}$ …". Originally, the table thus seems to have been thought of as fractions of 60 and not of 1, that is, as reciprocals; since the table served in floating-point calculations, this was of no consequence.

Of 1, its 2/3 [is]	40	27, its IGI	2 13 20
Its half [is]	30	30, its IGI	2
3, its IGI	20	32, its IGI	1 52 30
4, its IGI	15	36, its IGI	1 40
5, its IGI	12	40, its IGI	1 30
6, its IGI	10	45, its IGI	1 20
8, its IGI	7 30	48, its IGI	1 15
9, its IGI	6 40	50, its IGI	1 12
10, its IGI	6	54, its IGI	1 6 40
12, its IGI	5	1, its IGI	1
15, its IGI	4	1 4, its IGI	56 15
16, its IGI	3 45	1 12, its IGI	50
18, its IGI	3 20	1 15, its IGI	48
20, its IGI	3	1 20, its IGI	45
24, its IGI	2 30	1 21, its IGI	44 26 40
25, its IGI	2 24		

Fig. 2 Translation of the table of reciprocals

number is 6.40, which is the reciprocal of 9.[6] We therefore write A as a sum, $A = 44.20.0 + 0.6.40$, and find that $9 \times A = 6.39.0 + 0.1.0 = 6.40$. Now, 6.40 is still the reciprocal of 9, whence $9 \times 9 \times A = 1$. Therefore, $\frac{1}{A} = 9 \times 9 = 81$.

The selection of principal numbers for multiplication tables is closely related to the table of reciprocals and its use: the only irregular number to appear as a principal number is 7, while all two-place numbers of the right column of the table of reciprocals appear, as do a few others (2.15 and 4.30) that may be derived from entries in the standard table by doubling one side and halving the other.

Why?

In order to understand to the full how this system was used, we need to look at the purpose for which it was created. The place-value idea may have been in the air as a mere notation for centuries, but the *system* connecting notation and tables was a creation of the twenty-first century BCE,[7] a period known as "Third Dynasty of Ur" or, for simplicity, Ur III. Early in this period, an extremely centralized system of economic management was created, with overseer scribes directing labour troops and responsible for costs as well as produce. As an example, we may consider how to calculate the labour and barley values of a ditch with length l and rectangular cross section $w \times d$. In practical life, horizontal extensions were measured in units NINDAN (1 NINDAN ≈ 6 m), subdivided into 12 cubits, each of which consisted of 30 fingers; vertical extensions were measured in cubits. So, firstly, l and w were

---

[6]Jöran Friberg has introduced the very adequate name "trailing part algorithm" for the technique.
[7]Here and elsewhere, I follow the "middle chronology", as do most Assyriologists.

expressed in the "basic unit" NINDAN[8] as place-value numbers, and d was similarly expressed as a place-value number of cubits. To find the total volume as the product of these three numbers would now be a straightforward operation, since the unit of volume was NINDAN × NINDAN × cubit. In contrast, it would be quite laborious to find it directly, for instance, from $l = 8$ NINDAN 3 cubit, $w = 2$ cubit 15 fingers, $d = 2$ cubit 10 fingers. Once the volume had been found, the number of man-days required would follow from division by the amount of dirt a worker was supposed to dig out per day (that is, multiplication by its reciprocal), and the barley value from multiplication of the man-days by the daily barley wage of a worker—both again expressed in sexagesimal place-value multiples of basic units for volume, respectively, capacity measure. Once the place-value expression of a metrological value had been found, it would finally have to be reconverted into normal metrology, which would presuppose knowledge of the absolute order of magnitude to which the place-value numbers corresponded.

The conversion of the metrological units into place-value units and vice versa was made by means of "metrological tables". These would tell not only the conversion of the single units but also their multiples. For instance, the table for horizontal extension would start as shown in Fig. 3,[9] stating not only that a finger is 10 (namely 10″ NINDAN) but also that two fingers are 20, etc. These tables were copied so oft in school that future calculators knew them by heart; in this way, conversion of a composite expression like 8 NINDAN 3 cubit was reduced to an addition—there was no need to multiply 5 (the converted value of the cubit) by 3.

Such metrological tables existed for weight, capacity, horizontal and vertical extension and area (volumes were measured in area units, the standard area 1 SAR = 1 NINDAN^2 being presupposed to be provided with a default thickness of 1 cubit).

A final group of tables contains *technical constants*.[10] Some of these are norms for work—how much dirt is a worker supposed to dig out in a day or to carry a fixed distance in a day, etc. Others might serve in geometrical computation. For the circle area, we find the constant 5—to be understood as 5′: Under the assumption that the perimeter p is 3 times the diameter d, the area is indeed $\frac{1}{12}p^2 = 5' \times p^2$. For the diameter, we find the constant 20 (to be understood as 20′): $d = \frac{1}{3}p = 20' \times p$.

Technical constants that might turn up as divisors were chosen as regular numbers, preferably as numbers appearing in the table of reciprocals; that explains

---

[8]Obviously, any unit 60″ NINDAN would do in principle, but since the NINDAN was an existing unit abundantly used in practical life, we may take for granted that the calculators would think in terms of NINDAN and not, for instance, $\frac{1}{60}$ NINDAN.

[9]Translated from the edition in Proust (2008: 42). The actual specimen goes no further, but it is only the beginning of the ideal complete table, known in total from the combination of such fragments.

[10]These are less well-treated in the general literature than the arithmetical tables. A recent thorough analysis is Robson (1999).

1 finger	10	$^{1}/_{3}$ cubit	1.40
2 fingers	20	$^{1}/_{2}$ cubit	2.30
3 fingers	30	$^{2}/_{3}$ cubit	3.20
4 fingers	40	1 cubit	5
5 fingers	50	$1^{1}/_{3}$ cubit	6.40
6 fingers	1	$1^{1}/_{2}$ cubit	7.30
7 fingers	1.10	$1^{2}/_{3}$ cubit	8.20
8 fingers	1.20	2 cubits	10
9 fingers	1.30	...	

Fig. 3 Beginning of the metrological table for horizontal extension

why the reciprocals of such numbers would also turn up as principal numbers for multiplication.

Exactly as the floating-point calculations on the slide rule of an engineer fifty years ago, calculations in the place-value system could only serve for intermediate calculations, and they would normally leave just as few traces in the written record. We have the various tables and even evidence for the way they constituted an ordered curriculum during the Old Babylonian period (2000–1600 BCE)—remains from Ur III are very rare, just sufficient to show that the system had been created. We also have Old Babylonian student exercises of multiplication showing two factors and their product (but no intermediate calculations). However, the above description of the full combined use of the various tables is based on reconstruction and on Old Babylonian mathematical school texts,[11] not on real administrative records.

Addition and Abacus

As mentioned, we have no indication that the place-value notation was of any use for additions and subtractions. In particular, we have no exercise tablets with additions as we have for multiplications—if the multiplication goes beyond what follows directly from the multiplication tables and asks for the addition of partial products, these leave no traces on the tablet and must therefore have been manipulated in a different medium. For instance, the outcome of $1.03.45 \times 1.03.45$ is stated directly to be 1.07.44.03.45 (UET 6/2 222, in Robson (1999: 252))—certainly a calculation few if anybody would be able to perform by mere mental calculation combining multiplication table entries.

"... leave no traces"—or rather, leave only rare indirect traces (Høyrup 2002b). One of these is problem #12 of the text BM 13901 (ed. Neugebauer 1935: III, 3), where the outcome of the multiplication $10'50'' \times 10'50''$ is stated to be $1'57''$ $\underline{46}'''40''''$ instead of $1'57''\underline{21}'''40''''$ (since the problem is inhomogeneous of the second degree, we can see which absolute order of magnitude is intended).

---

[11]See, for instance, VAT 8389 #1, as discussed in Høyrup (2002a: 77–82).

25 (25‴) has thus been added erroneously in the calculation to 21 (21‴), and the only reasonable explanation for that is that 25 has arisen as a partial product and has been inserted twice instead of once—something that could never happen in our paper algorithm, where we see which steps have already been performed.

How could 25 arise as a partial product in the right order of magnitude? There seems to be only one straightforward way, namely by a calculation of 50″ × 50″ as (5 × 5) (10″ × 10″) = 25 × 1‴40″″ = 25‴ + 25 × 40″″ = 25‴ + 16‴40″″.

That may sound strange. We know multiplication tables with 50 as principal number and thus containing 50 × 50. However, if we compare the number of extant tables of this kind with the number of surviving copies of the table of reciprocals we see that it was hardly learned by heart—learning tables by heart was done by repeated copying. [12] So, the conclusion appears to be that at least this computation was made on an instrument where you had to remember where you were in the process because a step, once performed, became invisible—similarly to modern pocket calculators. The instrument could be some kind of reckoning board making use of counters, but it is difficult to exclude other possibilities (however, the numbers occasionally inscribed in empty spaces of mathematical tablets offer good evidence that writing on clay with subsequent deletion, in the style of a medieval dust abacus, was not the medium—it is also difficult to see why the number of "places" available on a support of this kind should be restricted). In any case, subtraction was spoken of (in different Sumerian words, and thus without *linguistic* continuity) during Ur III and in Seleucid times as "lifting up", which can hardly refer to anything but the removal of counters.

The text TMS XIX #2 (ed. Bruins and Rutten 1961: 103, pl. 29) provides us with supplementary evidence. Here, two errors are made. [13] In line 4, 14′48″53‴20″″ × 14′48″53‴20″″ is stated to be 3′39″[28‴]44″″26$^{(5)}$40$^{(6)}$, not 3′39″28‴43″″27$^{(5)}$24$^{(6)}$26$^{(7)}$40$^{(8)}$). In lines 6–7, 11″6‴40″″ is added to the number 3′39″[28‴]44″″26$^{(5)}$40$^{(6)}$, and the result is stated to be 3′50″36‴43″″40$^{(5)}$ instead of 3′50″35‴24″″26$^{(5)}$40$^{(6)}$. In the former case, a string "43 27" has been changed into "44 26", after which the repeated "4 26" causes the calculator to change "44 26 24 26 40" into "44 26 40" (the number is used further on and therefore cannot be a copyist's mistake). The second error is more complex, but even here it looks as if a unit has been misplaced in the order of fourths instead of that of thirds (see Høyrup 2002b: 196).

All in all, it thus seems that numbers were represented by counters placed on a counting board (in cases or whatever was used to keep together counters belonging to the same group) in such a way that a unit in one order of magnitude could easily

---

[12]Neugebauer and Sachs (1945: 12, 20) lists 14 standard tables of reciprocals but only one "single multiplication table" (the type that reflects training) with principal number 50. In Neugebauer (1935: I, 10–13, 36), the numbers are, respectively, 25 and 0.

[13]In both cases, Evert Bruins's transliteration differs from Marguerite Rutten's hand copy of the cuneiform, but since the tablet is one of those which the Louvre had mislaid, the transliteration is based on the hand copy and not on fresh collation with the tablet; the deviations must hence be due to erroneous readings or to misguided attempts to repair. I therefore build on the hand copy.

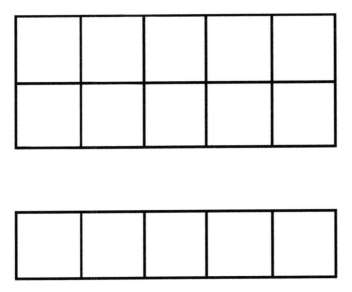

Fig. 4 Two possible configurations of the Old Babylonian abacus

be misplaced or pushed accidentally into a neighbouring order but not as easily into the tens of the same or a neighbouring order of magnitude. One possibility—probably the most obvious *for us*—is shown in the upper part of Fig. 4, but the configuration shown in the lower part is also possible, provided distinct counters were used for ones and tens (which would correspond to the written numerals).

This concerns how additive (and, we may assume, subtractive) operations were performed in the context of place value computation during the Old Babylonian period. As Proust, Christine (2000) has discovered, however, other error types offer further insight. In a large table of many-digit reciprocals from the Seleucid epoch (AO 6456), repeatedly two sexagesimal places are added together, 45.7 becoming 52, 40.14 appearing as 54, etc. This happens only in the interior of numbers of more than 4 digits, as if computations had been performed on an instrument of limited capacity. An Old Babylonian table listing continued doublings of 2.5 (N3958; 2.5 until 2.5×2^{39}) gives support to that interpretation: when the numbers grow beyond 5 places, they are written as two numbers separated by a separation character, apparently corresponding to calculations on two separate devices (in the end, when space in the column becomes scarce, the separation character is omitted). The first such number "10 + 6.48.53.20" can be reduced without difficulty to 10.6.48.53.20; soon, however, the right-hand part itself grows beyond five places, and the correct interpretation of "5.20 + 3.38.4.26.40" would be 5.23.38.04.26.40, but performing this operation correctly asks for meticulous book-keeping about places, and errors as those that abound in the Seleucid table are easily explained—as formulated by Proust (2000: 302), they are the "scars of recombination of two separate pieces" (or even, in one entry, three pieces).

Proust also suggests that the device may have carried the name "the hand", pointing to the term ŠU NU TAGA, "which the hand cannot grasp" used about 60^4, a five-place number, and referring to the present author for supplementary evidence. ŠU NU TAGA is known from Old Babylonian times, but the supplementary evidence spans most of Mesopotamian history: already in the mid-third millennium, ŠU.NIGIN, "the hand holds", designates the total of an account (below, we shall encounter more evidence from the same epoch); in the Old Babylonian text Db$_2$-146, an intermediate result is put "on your hand" and referred to afterwards as "your hand" (Høyrup 2002a: 258*f*), and the astronomical procedure text BM 42282 + 42294 (probably between the sixth and the fourth century BCE (ed. Brack-Bernsen and Hunger 2008)) prescribes that "you hold in your hand your year", which makes no sense unless the current year is inserted in a calculational procedure or device.

In texts from Ur III as well as the Seleucid epoch, we also see that subtraction is spoken of as "taking up" or "lifting up" (ZI in Ur III, NIM in Seleucid times). The introduction of a new word in late times probably means that the terms describe an extra-linguistic operation—presumably to "take up" counters from the counting board.

The early appearance of the name "hand" shows that this board is much older than the place-value system. However, the system for counting had been sexagesimal since the appearance of writing in the later fourth millennium BCE, only with distinct signs for 1, 10, 60, 600, 3600, 36,000 and 216,000; it was thus an absolute-value system. A counting board that would serve calculation in this system would therefore not only be of equal use for addition and subtraction of place-value numbers, it may even have inspired the very invention of the place-value notation—which however was to remain "in the air" until the full *system* with appurtenant tables was devised.

The Centesimal System and the Decadic Notation

An interesting further or parallel development was discovered a few years ago in Old Babylonian Mari and other cities in the Middle-Euphrates region, towards the Mesopotamian north-west: a place-value system with base 100 (Chambon 2012). It was only used for integers and served the counting of people and quantities of the capacity unit SILÀ. It uses the same basic signs for 1 and 10 as the sexagesimal place-value system, and it is therefore likely to be an adaption of the already known sexagesimal precursor;[14] however, a partially similar system is found in Ebla somewhat to the west around 2400 BCE, yet indicating the order of magnitude of places by mean of number words for 100, 1000 and 10,000 and using the signs of the Sumerian absolute-value counting within each place. The Ebla notation may

---

[14]Some Mari scribes were trained in sexagesimal place-value arithmetic in the early eighteenth century, so it was not unfamiliar. Moreover, the sign for units was differently oriented in the place value and in the traditional absolute-value systems (vertical respectively horizontal), and the centesimal notation agrees on this account with the sexagesimal place-value system.

also have been inspired by an abacus (of a type corresponding to the local spoken numeral system, which was decadic, that is, with base 10), but it can just as well have been a direct emulation of the way a number like 36,892 was spoken. Similarly, the later Mari system may have received inspiration from the Ebla notation and not only from the sexagesimal place-value system, or the similarity may be accidental, caused by the shared decadic spoken numerals.

Whereas the Sumerian spoken numeral system had been sexagesimal, the numerals of Semitic languages are indeed decadic, as are those of Indoeuropean languages. That is the underlying reason that Ebla, Mari and other north-western cities made use of the centennial system—the native language in this region was Amorite or Akkadian[15] or some other Semitic dialect. In the former Sumerian south, it is likely that the daily language was already Akkadian during Ur III, even though the official language of the state (and thus the language of scribehood) was still Sumerian; it certainly was when the Old Babylonian mathematical texts were written, but outside the area just mentioned the impact of general language on number writing was more modest: 5782 would be written 5 *līm* 7 *mē* 82 (5 thousand 7 hundred 82, 82 being written in the traditional Sumerian absolute-value system, as $60 + 20 + 2$), as it had been in Ebla.

We have no—and in all probability there *were* no—conversion tables between this almost-decadic number notation and the sexagesimal notation. In cases where the numbers did not enter calculations that would have no importance; however, if they were to be added (which might happen if, for example, they counted numbers of workers engaged in various parts of a larger project), we may speculate whether the counting board was used for this purpose with a different understanding of its structure; we may also guess that such non-standard ways to use the abacus might have served for operation on for instance capacity measures,[16] in particular before the implementation of the place-value system—but both suggestions remain mere conjectures.

Third Millennium Difficult Division

As mentioned already, we have evidence that Old Babylonian calculators were able to find approximate reciprocals of irregular numbers. However, we have no hints as to the methods that were used.

---

[15]Akkadian is the language whose main dialects in the second and first millennium are Babylonian and Assyrian.

[16]The fundamental capacity unit was a SILÀ (ca. 1 l). In Ur III and the Old Babylonian period, it was subdivided sexagesimally into 60 GÍN and the GÍN again in 180 ŠE. 10 SILÀ were 1 BÁN, and 6 BÁN constituted 1 BARIGA. 5 BARIGA, finally, made up a GUR, and GUR were counted in absolute-value sexagesimal numbers. So, for calculating grain quantities (where SILÀ would normally be the smallest unit taken into account), columns or cases with values 1, 10, 60 and 300 for successive cases or columns would be adequate.

From the mid-third millennium BCE, on the other hand, we have three texts that show something about how large round numbers could be divided by irregular divisors.

Two of the texts are from the city Šuruppak and can be dated to c. 2550 BCE (Høyrup 1982). They both deal with the distribution of a "storehouse" of barley to workers, each of whom receives 7 SILÀ. The "storehouse" of Šuruppak of the time was expected to contain 40ˋ (=2400) GUR.MA, each GUR.MA ("great GUR") consisting of 8ˋ(=480) SILÀ (that is, 1 "storehouse" = 1,152,000 SILÀ). The problem is thus to divide 2400 × 480 by 7. One of the texts (TMS 50) gives the correct answer 45ˋˋ42ˋ51 (=164,571) men, 3 SILÀ being "left on the hand", that is, left as remainder on the counting board. The other (TMS 671), however, finds 45ˋˋ36ˋ (=164,160) men. As it turns out, this is an intermediate result if the correct solution is found in the following way: first, we divide the number of GUR.MA in a storehouse by 7; that is, we find how many times 7 GUR.MA is contained in 40ˋ GUR.MA); the answer is 342 times, with a remainder of 6 GUR.MA. Then, we multiply by the number of times 7 SILÀ is contained in 7 GUR.MA, which is obviously 8ˋ = 480 times, getting 164,160— the very result obtained in the second text. This is thus as many men as will get 7 SILÀ each from the storehouse *if we forget about the remainder*. However, if we divide the remainder of 6 GUR.MA by 7 SILÀ, we find that 411 more men will receive their ration (in total thus 164,571 men, the result stated in the first text), with a remainder of 3 SILÀ.

It is impossible to find reasonable alternative procedures that also have the result stated in the mistaken text as an intermediate result. We may therefore be confident that this was how the result was reached; the analysis leaves open the question, however, how 40ˋ (=2400) and 8ˋ(=480) were divided by 7.

The third text (TM.75.G.1392) is from Ebla and from c. 2400 BCE; I follow Jöran Friberg's interpretation (1986: 16–21). The text appears to show a method for finding out how much grain has to be distributed to 260,000 persons, if 33 persons receive 1 *gú-bar*.[17]

It is stated (for simplicity, the sub-units are translated as fractions in the left column, while the middle column reduces these fractions; both follow Friberg) that

$3\frac{4}{120}$ *gú-bar*	$(=3\frac{1}{30}$ *gú-bar*$)$	for 100 persons
$30\frac{6}{20}$ *gú-bar*	$(=30\frac{3}{10}$ *gú-bar*$)$	for 1000 persons
$303\frac{4}{120}$ *gú-bar*	$(=303\frac{1}{30}$ *gú-bar*$)$	for 10,000 persons
$3030\frac{6}{20}$ *gú-bar*	$(=3030\frac{3}{10}$ *gú-bar*$)$	for 100,000 persons
$6060\frac{1}{2}\frac{2}{20}\frac{2}{120}$ *gú-bar*	$(=6060\frac{6}{10}\frac{1}{60}$ *gú-bar*$)$	for 200,000 persons
$1818\frac{24}{120}$ *gú-bar*	$(=1818\frac{2}{10}$ *gú-bar*$)$	for 60,000 persons

In all: 7879 *gú-bar* of barley for 260,000 persons.

---

[17] The *gú-bar* is a local Ebla unit; the transliteration is written in italics because it renders a syllabic writing of a Semitic word.

Since 33 persons receive 1 *gú-bar*, $3 \times 33 = 99$ persons receive 3 *gú-bar*. $100 = 99 + 1$ persons therefore should receive 3 *gú-bar* $+ \frac{1}{33}$ *gú-bar*, etc. Firstly we notice, however, that all values are slightly rounded: $\frac{1}{33}$ is replaced by $\frac{1}{30}$ and $\frac{10}{33} = \frac{30}{99}$ by $\frac{3}{10}$; in the final summation, $\frac{1}{2}\frac{2}{20}\frac{2}{120} + \frac{4}{20} = \frac{49}{60}$ is approximated as 1. Secondly, we observe that the successive values are not obtained by simple multiplication (by 10 respectively 2). Precisely how the values in the successive lines are found we cannot decide, but in any case we see that the division of 260,000 by 33 (or, in classical formulation, the measurement of 260,000 persons by 33 persons) is found through filling-out: first, by decupling and doubling, we go as far as possible, that is, until 200,000 persons; 60,000 persons remain, whose allocation is probably found by multiplying the allocation of 10,000 person by 6 (no rounding needed). Quite plausibly, the simpler divisions in the single lines were carried out in a similar way.

We further observe that the trick used in the Šuruppak texts is different from the method of the Ebla text (while, of course, its simpler divisions may or may not have been performed as fillings). The two texts do not present us with a standard way (and certainly not with an "algorithm") for performing divisions by irregular numbers; instead, they represent systematic exploration—in Friberg's words (1986: 22),

> the "current fashion" among mathematicians about four and a half millennia years ago was to study non-trivial division problems involving large (decimal or sexagesimal) numbers and "non-regular" divisors such as 7 and 33.

Nothing prevents, however, that such exploration could eventually lead to the creation of standard methods and that these would come to be used by the Old Babylonian calculators.

Long-Time Developments—Summary and Conclusion

Through accounting and metrologies, Mesopotamian mathematics can be followed back to the "proto-literate" period (c. 3300–3000 BCE) where writing was created (created, indeed, in order to serve in accounting, by providing the context that gave meaning to the numbers of the accounts). But we know nothing about the computational techniques in use by then.

Only Šuruppak, around 2550 BCE, provides us with some insights. Šuruppak presents us with evidence of several kinds that the "hand" reckoning board was in use, and it gives us the first example of the division by an irregular number. From Šuruppak, we also have the earliest table of squares, where the side is given in length metrology and the area measured in area units (Neugebauer 1935: I, 91).

Three more square tables come from the following century (Edzard 1969; Feliu 2012; Friberg 2007: 419–427); one of them also lists rectangular areas, one of the sides being constantly 1` NINDAN.

During the centuries preceding Ur III, we find several instances of notations that suggest ongoing groping for the place-value idea, but almost all contain mistakes showing that the *system* was not yet in existence (Powell 1976).[18] The system was only to be created during Ur III—and its complex combination of a number notation and the variety of table types without which it would be of no use shows that it was certainly a deliberate creation, not the outcome of accumulated accidental developments.

The Ur III state broke down around 2000 BCE, but the scribes of the less centralized Old Babylonian successor states were still trained in place-value calculation. After the collapse around 1600 of the final Old Babylonian state, the Babylon of the Hammurabi dynasty, we know less. Scholar-scribes were still taught some rudiments— Ashurbanipal, the last important Assyrian king (r. 668–631 BCE), who had originally been meant to become a high priest, boasts that he is able to perform multiplications and find reciprocals. That seems to be the high point of the mathematics he knows about: in same text, he claims to be able to read tablets "from before the flood", that is, from the mid-third millennium, which appears not to be true—but real scholar-scribes at his court could do it. Those who took care of mathematical administration after the collapse of the Old Babylonian state were hardly scholar-scribes—there is evidence that only the most basic vocabulary surrounding the place-value system was conserved in Sumerian. However, at the creation of mathematical astronomy from the seventh century BCE onward, the place-value system again came in use albeit within a very restricted environment. As we have seen, this environment still used the "hand" reckoning board, and it also knew the trailing part algorithm.

Mathematical astronomy survived at least until the late first century CE (Hunger and de Jong 2014); by then, mathematical administration had given up the cuneiform heritage since long. The disappearance of mathematical astronomy therefore entailed the final demise of the Mesopotamian calculation techniques, after their having been practised for more than 2000, some of them for at least 2500, perhaps 3400 years.

References

Brack-Bernsen, Lis, and Hermann Hunger. 2008. BM 42484+42294 and the Goal-Year method. *SCIAMUS*, 9: 3–23.

Bruins, Evert M., and Marguerite Rutten. 1961. *Textes mathématiques de Suse*. Paris: Paul Geuthner.

Chambon, Grégory. 2012. Notations de nombres et pratiques de calcul en Mésopotamie; Réflexions sur le système centésimal de position. *Revue d'Histoire des Mathématiques*, 18: 5–36.

Edzard, Dietz Otto. 1969. Eine altsumerische Rechentafel (OIP 14, 70). In *Lišān mithurti. Festschrift Wolfram Freiherr von Soden zum 19.VI.1968 gewidmet*, ed. W. Röllig, 101–104. Kevelaer: Butzon & Bercker/Neukirchen-Vluyn: Neukirchener Verlag des Erziehungsvereins.

---

[18]Whiting (1984) goes further than Powell in his claims, but his argument suffers from a lack of distinction between *sexagesimalization* and place value.

Feliu, Lluís. 2012. A new early dynastic IIIb metro-mathematical table tablet of area measures from Zabalam. *Altorientalische Forschungen*, 39: 218–225.

Friberg, Jöran. 1986. The early roots of Babylonian mathematics. III: Three remarkable texts from ancient Ebla. *Vicino Oriente*, 6: 3–25.

Friberg, Jöran. (2007). *A Remarkable Collection of Babylonian Mathematical Texts*. New York: Springer.

Høyrup, Jens. (1982). Investigations of an early Sumerian division problem, c. 2500 B.C. *Historia Mathematica*, 9, 19–36.

Høyrup, Jens. 2002a. *Lengths, widths, surfaces: A portrait of Old Babylonian algebra and its kin.* New York: Springer.

Høyrup, Jens. 2002b. A note on Old Babylonian computational techniques. *Historia Mathematica*, 29: 193–198.

Hunger, Hermann, and Teije de Jong. 2014. Almanac W22340a from Uruk: The latest datable cuneiform tablet. *Zeitschrift für Assyriologie und Vorderasiatische Archäologie*, 104: 182–194.

Neugebauer, Otto. 1934. *Vorlesungen über Geschichte der antiken mathematischen Wissenschaften. I: Vorgriechische Mathematik.* Berlin: Julius Springer.

Neugebauer, Otto. 1935. *Mathematische Keilschrift-Texte.* I–III. Berlin: Julius Springer (1935–1937).

Neugebauer, Otto. 1955. *Astronomical cuneiform texts: Babylonian ephemerides of the Seleucid period for the motion of the sun, the moon, and the planets.* London: Lund Humphries.

Neugebauer, Otto. 1957. *The exact sciences in antiquity*, 2nd ed. Providence, Rh.I.: Brown University Press.

Neugebauer, Otto, and Abraham Joseph Sachs. 1945. *Mathematical cuneiform texts.* New Haven, Connecticut: American Oriental Society.

Powell, Marvin A. 1976. The antecedents of Old Babylonian place notation and the early history of Babylonian mathematics. *Historia Mathematica*, 3: 417–439.

Proust, Christine. 2000. La multiplication babylonienne: la part non écrite du calcul. *Revue d'Histoire des Mathématiques*, 6: 293–303.

Proust, Christine. 2008. Avec la collaboration de Manfred Krebernik et Joachim Oelsner. *Tablettes mathématiques de la collection Hilprecht.* Wiesbaden: Harrassowitz.

Robson, Eleanor. 1999. *Mesopotamian mathematics 2100–1600 BC. Technical constants in bureaucracy and education.* Oxford: Clarendon Press.

Whiting, Robert M. 1984. More evidence for sexagesimal calculations in the third millennium B. C. *Zeitschrift für Assyriologie und Vorderasiatische Archäologie*, 74: 59–66.

Jens Høyrup born 1943, was educated as a particle physicist (1969). In 1971–1973 he taught physics at an engineering school. Since 1973 he has been connected to Roskilde University (since 2005 emeritus) as a historian of science, first in the Department of Social Sciences, since 1978 in Human Sciences. His main field of research has been history of pre-modern mathematics. He is permanent Honorary Research Fellow at the Institute for the History of Natural Science, Chinese Academy of Sciences, and for the time being visiting scholar at the Max-Planck-Institut für Wissenschaftsgeschichte, Berlin. His publications include: *In Measure, Number, and Weight. Studies in Mathematics and Culture.* New York: State University of New York Press, 1994; *Human Sciences: Reappraising the Humanities through History and Philosophy.* Albany, New York: State University of New York Press, 2000. *Jacopo da Firenze's* Tractatus Algorismi *and Early Italian Abbacus Culture.* Basel etc.: Birkhäuser, 2007. More in the list of references.

Interpreting Tables of the *Arithmetical Introduction* of Nicomachus Through Pachymeres' Treatment of Arithmetic: Preliminary Observations

Athanasia Megremi and Jean Christianidis

Abstract Nicomachus of Gerasa's *Arithmetical introduction*, a late antique arithmeticwork that was highly appreciated for its educational value throughout Late Antiquity and the Middle Ages, is traditionally characterized by historians of mathematics as theoretical, without any practical orientation at all. In this chapter, it is argued that the textual analysis of the work suggests otherwise. More specifically, the inclusion of tables that are accompanied by meticulously detailed instructions towards reading, constructing and using them, suggests that these particular sections of the work could be viewed in the context of a certain arithmetical *practice* and its *learning*. Furthermore, we will argue that the *Quadrivium* of the Byzantine scholar Georgios Pachymeres provides us with a case study of how the Nicomachean tables can be interpreted as devices that could be used in arithmetical problem-solving.

Keywords Nicomachus · Diophantus · Pachymeres · Problem-solving Arithmetic · Tables

Introduction: Studying Greek Mathematics in a Late Antique Context

Preliminary Remarks

There are many things to be said for the diverse, multifaceted intellectual tradition of Late Antiquity that have been brought to attention since this period has gained its

A. Megremi (✉)
Department of History and Philosophy of Science, National and Kapodistrian University of Athens, Athens, Greece
e-mail: amegremi@phs.uoa.gr

J. Christianidis
Department of History and Philosophy of Science, National and Kapodistrian University of Athens, Athens, Greece; Centre Alexandre Koyré, Paris, France
e-mail: ichrist@phs.uoa.gr

© Springer Nature Switzerland AG 2018
A. Volkov and V. Freiman (eds.), *Computations and Computing Devices in Mathematics Education Before the Advent of Electronic Calculators*,
Mathematics Education in the Digital Era 11, https://doi.org/10.1007/978-3-319-73396-8_4

own spotlight in historical research. Today we have gained further insight by tracing this tradition into the Middle Ages and examining its connection to Classical Antiquity, in a sense creating boundaries to better understand continuity and discontinuity. Other generic characteristics such as the relations between intellectual traditions that transcend culture, the rise of late antique cities and empire culture, syncretism, the evolution of relations between oral and written culture with a progressive emphasis on the latter, the commentary tradition of late antique authors and the creation of their own tradition, and the reorganization of knowledge and educational notions are all being studied in late antique context, giving them the chance to be understood in their own terms and not in comparison to Classical Antiquity, as was the common practice until the 1970s.[1]

The diversity of Greek mathematical culture is a research field that is currently receiving attention and gaining new appreciation (Lloyd 2012; Taub 2013).[2] Greek mathematics is traditionally associated with the emergence of a way of doing mathematics which is theoretical, abstract and demonstrative, i.e. postulatory-deductive. The constitutive role that Greek mathematics has played, because of this character, on the formation of Western mathematics is generally acknowledged. Yet, other types of mathematics, computational and procedural (or algorithmic), which are often associated in scholarship with other ancient mathematical traditions, are also present in Greek mathematical literature. This is especially the case of late antique mathematical elaborations such as we witness in the Heronian and the Ptolemaic corpuses in which both aspects, the demonstrative and the procedural/computational, are present.[3] This is also the case of texts containing series of problems with their solutions, such as Diophantus' *Arithmetica* (1893–1895, i) or Heron's *Metrica* (1903; 2014). Furthermore, schemata and tabular diagrams are characteristics of mathematical written culture that evolves during the period of the Later Roman Empire well into the Middle Ages, Western and Eastern. Evidence to this, in the case of the Eastern Roman Empire, one can trace up to the scholars and scholiasts of the Palaeologan Renaissance (second half of thirteenth and throughout the fourteenth century), such as Maximos Planudes (ca. 1255–ca. 1305)[4] and   Georgios Pachymeres

---

[1]This is how the editors of the collective volume (Bowersock, Brown, Grabar 2001, ix) describe, very briefly, what they mean by Late Antiquity: "The essays in this volume ... share the frank assumption that the time has come for scholars, students, and the educated public in general to treat the period between around 250 and 800 as a distinctive and quite decisive period of history that stands on its own. It is not, as it once was for Edward Gibbon, a subject of obsessive fascination only as the story of the unraveling of a once glorious and 'higher' state of civilization. It was not a period of irrevocable Decline and Fall; nor was it merely a violent and hurried prelude to better things". See also (Marrou 1977).

[2]For a more general survey of the recent studies on Greek mathematics, see Sidoli (2014b) and the bibliography provided there. Concerning especially the mathematics and science of Late Antiquity, the reader is referred to the survey of Alain Bernard (2018).

[3]See especially Heron's *Metrica* (1903; 2014) and Ptolemy's *Almagest* (1898–1903; 1984).

[4]See his commentary, compiled in the last decade of the thirteenth century, on the first two books of Diophantus' *Arithmetica* (Diophantus 1893–1895, ii, 125–255).

(1242–ca. 1310),[5] in whose compositions we see tabular forms and/or series or lists of mathematical content often used to make sense or emphasize or study the mathematical texts. The *Arithmetical introduction* of Nicomachus of Gerasa (first half of the second century CE) and the texts affiliated with it[6] can also be viewed through this tradition and be ascribed to it. Late Antiquity provides a pertinent context for understanding all these different traditions in Greek mathematics and the way in which mathematical practice and mathematical education subtly yet critically interacted.

In this chapter, we shall study one of the aspects of the *Arithmetical introduction* of Nicomachus, a text of late antique arithmetic with renowned educational value throughout Late Antiquity and the Middle Ages, which has been traditionally characterized as theoretical, without any practical orientation at all. Historians of philosophy have studied the text as an introduction to the arithmetical ideas of Neophythagorean thinkers (O'Meara 1989). The *Arithmetical introduction* is indeed an introduction to the study of the properties of numbers, that is, an introduction concerning the theory of numbers. Is theoretical arithmetic all that we can learn from this work though? We will argue that a textual analysis of the work suggests otherwise. For this purpose, the aspect of *representation* and *organization* of the content of the text will be elaborated. It concerns more specifically the *meticulously detailed instructions* provided by Nicomachus towards *reading*, *constructing* and *using* some of the tables included in the printed edition of his text bearing on pairs of numbers in relation.[7] The presence of such textual units in Nicomachus' treatise gives rise to several questions. Could it be possible to produce a case study where such textual units could be studied through the prism of a certain *functionality*? Do the surrounding text and accompanying tables have any heuristic value, and, if so, of what sort? Finally, could tables and text together be viewed as portraying underlying teaching purposes, and, if so, of what sort? We will argue that the activity of problem-solving provides us with a suitable and historically sensitive context within which the above-stated questions can be answered

---

[5]See (Pachymère 1940; Diophantus 1893–1895, ii, 78–122). Pachymeres' treatise was also compiled in the last decade of the thirteenth century.

[6]We refer by this expression to the works of Iamblichus (ca. 245–ca. 325), Domninus (fifth century CE), Boethius (ca. 475/7–ca. 526), Asclepius (b. ca. 465), John Philoponus (ca. 490–ca. 570), and Soterichus (probably eleventh century), which are mentioned in the bibliography.

[7]The edition of Nicomachus' treatise that is still used is Hoche's edition of 1866. All references in Nicomachus' text are given throughout this study according to this edition; the dot (.), when necessary, separates the page number from the line numbers. Hoche's edition contains six tables. Four of them are included in the part of the text where Nicomachus discusses the kinds of "numbers in relation", and their properties (1866, 51; 57; 77; 78). The *editio princeps* of the *Arithmetical introduction*, however, contains only one table (Nicomachus 1538, 28; see Fig. 1), coinciding with the table on p. 51 of Hoche's edition. For this reason, and because the table in question is the most generic—for example, the second of the aforesaid tables of Hoche's edition, (1866, 57), is in fact a part of it—our investigation will be focused on the study of this table.

affirmatively.[8] The rest of this chapter is devoted to achieving this goal. Furthermore, the historical value of our interpretation will be examined.

Basic Cultural Notions from Late Antiquity in Relation to the Study of Arithmetic

During the period of Late Antiquity,[9] there is a revival of Neopythagoreanism within the framework of Neoplatonic tradition. Schools of higher education (in philosophy, rhetoric and law) are established in every important city of the later Roman Empire. One would also point out the blooming of written culture, following the gradually larger emphasis on written texts, monographs and commentaries alike. Such characteristics reflect on an emanating cultural shift that will lead both to the medieval text tradition and also to medieval educational activity. In these contexts, one can also view the reorganization of knowledge through new prisms and the invention of more elaborate means of study of ancient knowledge. Comparative narratives are embedded in the cultural inheritance of the late antique scholars, showing their striving to adjust and create tools and methods that will better compliment their own notion of knowledge. As part of this tradition, one can perceive the gradual institutionalization of the four mathematical disciplines (arithmetic, music, astronomy and geometry) as the prerequisite knowledge for learning philosophy and, proceeding to the Middle Ages, theology. This stems from the vision of Neoplatonic writers such as Iamblichus and Proclus (410/412–485), deeply affecting the educational values of their intellectual descendants in whatever field this may be. Thus, the appropriation of methods of study and the defining of the material of study become a great part of the activity of scholars that involved themselves in educational practice.[10] So we are today in a position to know that arithmetic was indeed taught as part of what was later known in the medieval West as the *Quadrivium*, and specifically as its first part. We are also in a position to know that Nicomachus' *Arithmetical introduction* was the basis for the texts introducing arithmetic throughout Late Antiquity and the Middle Ages. Its value was twofold since it was considered to be important for learning and teaching

---

[8]Questions of this kind have not been discussed sufficiently from the viewpoint of an arithmetical problem-solving activity. To the best of our knowledge, only interpretations of tables as relational and representational functions, serving as counting/computing devices in astronomy, have been developed, as for example in the case of Ptolemy's table of chords (Sidoli 2004, 2014a). Regarding especially the tables in Nicomachus, there are no similar attempts apart from (Kappraff 2000) in which two Nicomachean tables (1866, 77; 78) are discussed from the viewpoint of their relevance with the general theory of proportions and its applications to architecture.

[9]Late Antiquity, roughly speaking, is the period from the third to the late eighth century.

[10]For a comprehensive discussion of the education in Late Antiquity see the studies of Hadot in (2005). See also (Marrou 1977; Watts 2006).

arithmetical theory and also for the mathematical concepts that had value for the study of philosophy, such as *unity, diversity, equality, inequality* and more.

Another characteristic of late antique culture is the establishing of networks of scholarship between the high schools of the Late Antique Empire. The connection between the schools of Athens and Alexandria is an excellent example of the interaction of knowledge between scholars and allows us to access to the establishment of the teaching material and its content (Watts 2006). This interaction and the relationships between teacher and student are a means of understanding the nature of late antique intellectual tradition. A *common language* between scholars is not only an important element for transmitting knowledge, but also for practicing philosophy. It also allows us to understand the nature and spirit of their studies and to gain insight into their purposes. The commentary tradition of the era strengthens this point as a reason and an outcome alike of this continuous exchange of learning and teaching experiences between scholars, whose mobility often includes at least one more stop, usually Rome, Antioch or Constantinople.

One last characteristic that one should mention in relation to the teaching and learning tradition of this period is the gradual progression from *the philosopher* as the central intellectual authority of society to *the scholar*, who will from thereon take his place. This is an important heritage to the upcoming medieval tradition and one that will define the role of the teacher in later times. A *scholar* is no longer the idealized icon of a "holy man" who practises philosophy as a way of life, as was the case of the *philosopher*. He is the polymath, well versed in science and letters, contributing to education as well as to knowledge, the protector of the past and the representative of society's finest. He will be the one entrusted with preserving and teaching knowledge, with diplomatic missions and with consulting high-ranked officials.

What We Know about the Teaching and Learning of Arithmetic in Late Antiquity and Some Added Remarks

Arithmetic is the first of the four mathematical disciplines to be studied. It is the subject of works, or portions of works, of such authors as Theon of Smyrna (early second century CE),[11] Nicomachus, Iamblichus,[12] and Domninus.[13] We know that Neoplatonic philosophers like Plotinus (204/5–270) and Proclus have concerned themselves with arithmetical concepts.[14] Although we have no reason to believe

---

[11] See Théon de Smyrne (1892, 2010).

[12] See Iamblichus (1894), Jamblique (2014).

[13] See Domninus of Larissa (2013).

[14] See especially Plotinus' sixth *Ennead*, which is devoted to numbers (Plotinus 1988; Slaveva-Griffin 2009), and Proclus' discussion of the concepts like *plurality, unity, limit, unlimited, one, many, quantity* (ποσόν), *equality*, etc. in the first prologue of his *Commentary on the first book of Euclid's Elements* (Proclus 1873, 32–37; Morrow 1970, 27–30).

that philosophers were interested in the art of arithmetical calculation (λογιστική), we do have indications that elements of "practical arithmetic" were in fact considered when addressing theoretical arithmetic. This is why Nicomachus in the beginning of the *Arithmetical introduction* explains that there are uses of numbers in everyday life and there are problems that demand the knowledge of arithmetical properties that have their own value, but ultimately these are not the subject of arithmetic (8.8–15). This observation shows that a discussion concerning the content of arithmetic was ongoing even at his time. It may be also indicative of the fact that computational and theoretical arithmetic could be affecting effectively each other, as one can see in the cases of Heron (fl. after 62 CE),[15] during this period, Diophantus (second half of the third century CE), or Theon of Alexandria (fl. in 360s and 370s CE), during the subsequent centuries.[16] The teaching of the two though was separated as the texts of Late Antiquity attest.

Moreover, we know that during Late Antiquity and the Middle Ages (a) the *Arithmetical introduction* and its contents were part of the precursory study for higher education; (b) that since the teaching of arithmetic was based on this specific work, it was also based on the mathematical tradition this work belongs to, that is, (Neo) pythagoreanism, which was very much alive during this period, and which left room for including elements taken from computational mathematics, thus reflecting in a sense the more general tendency of arithmetization that permeates mathematics during this period (Fowler 1999, 8–10); (c) that the character of this tradition was so common that ultimately it became part of the generic cultural educational tradition of late antique and medieval society; (d) thus, this tradition influenced the cognitive "collective consciousness" which carried with it the tools developed in its framework and adapted them to the teaching and learning processes.

Tables as Conveyors of Mathematical Knowledge in Nicomachus' *Arithmetical Introduction*

As a rule, Nicomachus' *Arithmetical introduction* has been approached by scholars in the past from the perspective of either philosophy or mathematical content. It has not been studied from the angle of the *teaching and learning process*, which is our broader interpretational context in this study. Within such a context, textual units as conveyors of knowledge can be regarded as possibly serving different purposes. In this sense, the tables of the *Arithmetical introduction* and the texts surrounding them could be approached through different perspectives: not only do they produce a visual stimulus for the mathematical information, categorizing and relating, but

---

[15]For Heron's dating, see Sidoli (2011), Masià (2015) and the recent edition of the *Metrica* (2014, 15–22) by F. Acerbi and B. Vitrac.

[16]See especially the works of Heron of Alexandria and Theon of Alexandria which are mentioned in the bibliography, and Diophantus' *Arithmetica* (Diophantus 1893–1895, i).

they can also be regarded as heuristic tools of investigation, deriving and demonstrating. The way theory of ratios is presented by Nicomachus in chapters 27–33 of the first book illustrates the features we are talking about. We will see how in this context the specific element of tabular representation is used.

Numbers in Relation and Their Classification into Kinds

After the first section of his book, where Nicomachus studies the properties of numbers *in themselves*, there follows a second section which studies the properties of numbers *related to one another*. This pertains to the presentation and explanation of the kinds of inequality, a basic property of things (in this case numbers) that are regarded in relation to one another. There are two kinds of inequality, the *greater* and the *lesser*. Each kind is further divided into five subkinds; the *greater* in *multiples, superparticulars,*[17] *superpartients,*[18] *multiple superparticulars,*[19] and *multiple superpartients;*[20] the *lesser* in *submultiples, subsuperparticulars, subsuperpartients, submultiple superparticulars, submultiple superpartients.* As the names themselves make clear, the kinds of the *lesser* are the inverse of the corresponding kinds of the *greater*.

The Text Introducing the Generic Table Showing the Numbers in Relation

In between presenting and explaining the multiples and the superparticurals, Nicomachus introduces a tabular diagram (διάγραμμα) in which, as he claims,

---

[17]A number A is called *superparticular* of another number B if A contains B in its whole and further a part of it (49.1–4). In modern terms, A is called "superparticular" of B if $A = B + \frac{1}{n} \cdot B$. For $n = 2$, we have the first kind of superparticulars, the "sesquialters"; for example, if $B = 6$, its sesquialter is 9, since $9 = 6 + \frac{1}{2} \cdot 6$. Similarly, for $n = 3$ we have the second kind of superparticulars, the "sesquitertians". And so on.

[18]A number A is called *superpartient* of another number B if A contains B in its whole and further parts of it more than one (55.12–14). In modern terms, A is called "superpartient" of B if $A = B + \frac{m}{m+n} \cdot B$ ($m > 1$). For example, if $B = 12$, $m = 2$ and $n = 1$, we find the number 20 ($20 = 12 + \frac{2}{3} \cdot 12$), called "superbitertian". Other kinds of "superpartients" are the "supertriquartans", the "superquadriquintans", etc.

[19]A number A is called *multiple superparticular* of another number B if A contains B more than once and further a part of it (59.7–10). In modern terms, A is called "multiple superparticular" of B if $A = kB + \frac{1}{n} \cdot B$. For example, for $k = 2$ and $n = 2$ we have the numbers which are called "double sesquialters". Thus, the "double sesquialter" of 8 is 20, since $20 = 2 \cdot 8 + \frac{1}{2} \cdot 8$. Similarly, the number 28 is "triple sesquialter" of the number 8.

[20]A number A is called *multiple superpartient* of another number B if A contains B more than once and further parts of it more than one (64.2–5). In modern terms, A is said "multiple superpartient" of B if $A = kB + \frac{m}{m+n} \cdot B$ ($m > 1$).

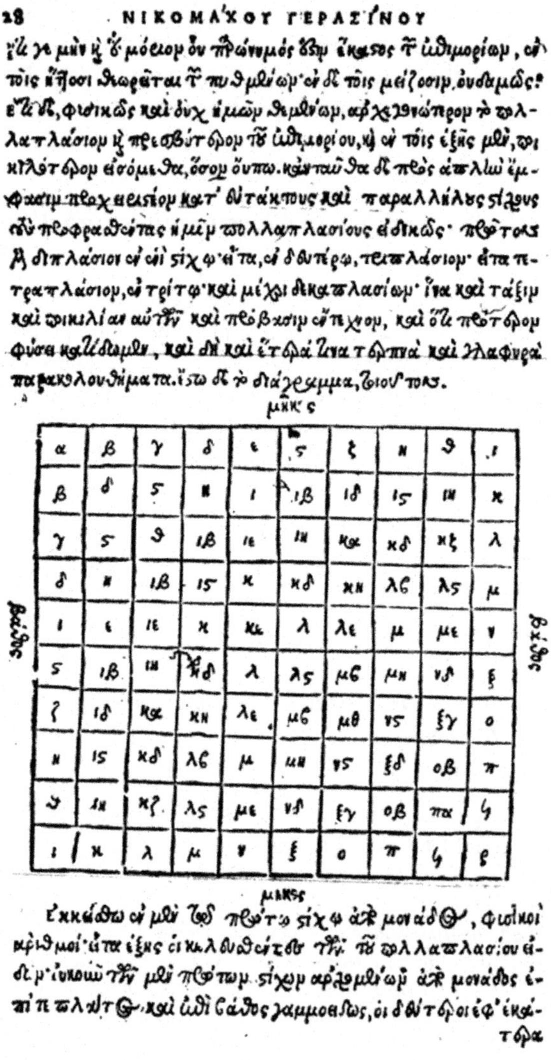

α	β	γ	δ	ε	ϛ	ζ	η	θ	ι
β	δ	ϛ	η	ι	ιβ	ιδ	ιϛ	ιη	κ
γ	ϛ	θ	ιβ	ιε	ιη	κα	κδ	κζ	λ
δ	η	ιβ	ιϛ	κ	κδ	κη	λβ	λϛ	μ
ε	ι	ιε	κ	κε	λ	λε	μ	με	ν
ϛ	ιβ	ιη	κδ	λ	λϛ	μβ	μη	νδ	ξ
ζ	ιδ	κα	κη	λε	μβ	μθ	νϛ	ξγ	ο
η	ιϛ	κδ	λβ	μ	μη	νϛ	ξδ	οβ	π
θ	ιη	κζ	λϛ	με	νδ	ξγ	οβ	πα	ϙ
ι	κ	λ	μ	ν	ξ	ο	π	ϙ	ρ

Fig. 1 The table of Nicomachus as it appears in the *editio princeps* of the *Arithmetical introduction* (Paris, 1538)

not only will we learn the primordial nature of the multiples in comparison to the superparticulars but also “other delightful and elegant accompaniments (παρακολουθήματα)”. The relevant passage is the following (50.21–51.23; see Fig. 1):

> That by nature and by no disposition of ours the multiple is more elementary and precedent in order than the superparticular we shall shortly learn, through a somewhat intricate process. And here, for a simple presentation (ἔμφασιν), we must prepare in well-ordered and parallel rows the multiples specified above, according to their varieties, first the double in one line, then in a second the triple, then the quadruple in a third, and so on as far as the

tenfold multiples, so that we may perceive their order and variety, their regulated progress, and which of them is naturally prior, and indeed other delightful and elegant accompaniments. Let the diagram be as follows.

1	2	3	4	5	6	7	8	9	10
2	4	6	8	10	12	14	16	18	20
3	6	9	12	15	18	21	24	27	30
4	8	12	16	20	24	28	32	36	40
5	10	15	20	25	30	35	40	45	50
6	12	18	24	30	36	42	48	54	60
7	14	21	28	35	42	49	56	63	70
8	16	24	32	40	48	56	64	72	80
9	18	27	36	45	54	63	72	81	90
10	20	30	40	50	60	70	80	90	100

Right after the setup of the table, Nicomachus explains some of these accompaniments and moreover how the kinds of superparticulars are created (γένεσις). In the passage below, the creation of the first kind of superpaticulars, that is, the sesquialters, is described (53.9–14):

> [...] In comparison with the second row reading either way, which begins with the common origin 4 and runs over in cross-lines to the term 2 in each row,[21] the rows which are next in order beneath display the first kind of the superparticular, that is, the sesquialter, between terms occupying corresponding places.

In the above passages, one can distinguish three different textual forms in Nicomachus' presentation of the matter: *prose*, the *table* and *sequence* of information. The *table*[22] is an easily discernible form. The *prose* along the table has a twofold character: it is complimentary to the table but can also stand-alone, and it can be read apart from the table if necessary. The *sequence* of information described in the text gives a multifaceted, highly layered text. Combining all three characteristics allows one to *learn* and at the same time *work through* all relative information. This is being accomplished by the immersion of the reader[23] into the text. The extremely detailed text is organized to the possibility of the reader to be able to connect with the text, creating a context, and at the same time practise all acquired information,

---

[21]Our translations from Nicomachus are adapted from (Nicomachus of Gerasa 1926). Nicomachus refers in this passage to both the second row and the second column of the table that intersect on number 4, a term considered as "common origin". Producing over the "common origin" the two lines we find the number 2, next in order of which is number 3. The pair (2, 3) displays, says the text, the first kind of superparticular ratio, which is the sesquialter. The above is magnificently displayed in the corresponding table from the 1521 edition of Boethius' *Arithmetica* (see Fig. 2).

[22]In the rest of this chapter, we refer to this table as the "generic table", meaning by this that it displays all kinds of relations (ratios) referred to in Sect. Numbers in Relation and Their Classification into Kinds.

[23]We use the term "reader" in a broad sense, meaning any attendant or receiver of the text.

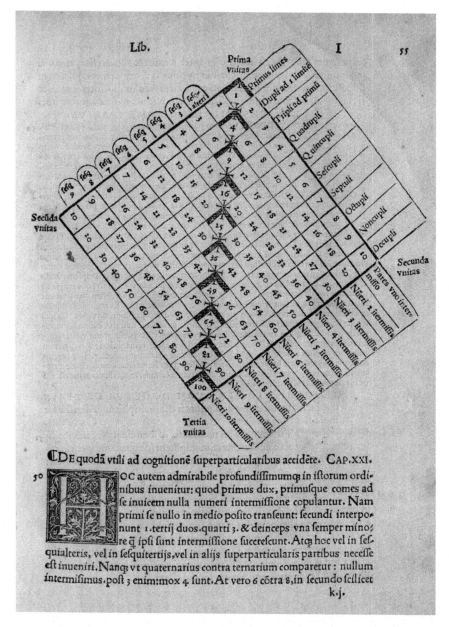

Fig. 2 'Generic table' as it appears in the magnificent 1521 edition of Boethius' *Arithmetica* (Boethius 1521, f. 55r)

so it is organized to reach these ends. We describe this as *functionality* of the text. This functionality can be of use to the reader for a plurality of purposes.[24]

Discussion of the Text

Our first remark regarding the above text is its instructive character. This character is quite obvious in the way Nicomachus explains how to construct the table and then how one can produce superparticulars from the comparison of the rows of multiples. More specifically, concerning the latter, the text explains how the sesquialters, which are the first kind of superparticulars,[25] are created from the multiples: one has to select "the second row", i.e. the doubles, and compare it to "the row which is next in order", i.e. the triples, thus one will find "the first kind of the superparticular, that is, the sesquialter".

This leads us to our second remark, that is, how pairs of superparticulars (of any kind) can be located searching through the table. The key here is the phrase "terms occupying corresponding places" (ὁμοταγεῖς πρὸς ὁμοταγεῖς). This simple wording is what allows the reader to make sense and use of the rows of the table: each row and its concomitant form pairs of numbers in a specific relation, provided that those numbers are ὁμοταγεῖς (that is, they occupy corresponding places). Similar canonical schemata are applied to all kinds of numbers in relation.

There are diverse properties and/or determinations that one can *observe* in the table or maybe *use* the table to find.[26] If one is driven enough to go through a thorough investigation of the table, then one would gain great advantage; but Nicomachus asserts that since one may drift off from ratios so he will not pursue this investigation:

> An ambitious person might find many other pleasing things displayed in this diagram, upon which it is not now the time to dwell, for we have not yet gained recognition of them from <our> introduction ... (55.1–3)

This makes further clear the instructive character of Nicomachus' text. The prompting of the reader to further explore, the insistence on multiple examples and verbs such as "might find" are also indicative of the instructive nature of the text.

There is more linguistic evidence from the Nicomachean text that invigorates our interpretation of the text. The limited nature of this chapter does not allow us to thoroughly examine every linguistic element. We should mention that words and expressions like "regulated progress" (51.9), "we may perceive" (51.10), "the

---

[24]The functionality of a text may be explicitly stated by the author but this is not the case here, in which Nicomachus' intentions, whatever they may be, are implied.

[25]Cf. footnote 17.

[26]For example, the terms on the diagonal of the table are all squares; the terms surrounding each square, on both sides, are "heteromecic"; the sum of two successive squares and twice the heteromecic number between them is always a square.

cross-lines like the letter chi (χ), ending in the term 3" (52.15), "as is evident in the diagram" (52.20–21), "these numbers are detected in the structure of the diagram in places just above the quadruples" (53.5–7), "display" (53.13), "the following feature of the diagram, moreover, is of no less exactness; the terms at the corners are units" (54.7–9), strongly suggest an interrelation between the prose and the table, creating a textual unit, a coherent argument, that nevertheless allows a *creative reading* of the Nicomachean text. Our view of this creative reading is that every reader or conveyor of the text can reproduce it emphasizing on a different function without making an injustice to the spirit of Nicomachus' text, and this is what we call *functionality*. This is not lost to the ancient readers of the *Arithmetical introduction* as they choose to reproduce the table or not, and readapted it or not.[27]

The interrelation between the three textual units (table, prose and sequence) will not be further examined here since it is not directly connected to our purpose. Concluding from the description of the Nicomachean text, we could argue that the table is being presented by Nicomachus as *supporting* and *verifying* prior theoretical statements and accounts but there is also a sense of *practicing procedures* or *sequences*. He supports the table with adequate vocabulary in his prose and literally prompts the reader to practise and see for himself the usefulness of the table. A *teaching and learning* context is essential for such an interpretation, since it not only includes the view of the teacher (in this case, the author) but also the view of the learner (be it reader, attendant, student, or user). The view of the learner is important when trying to understand the functionality of a text because the learner is in the receiving end of the didactic interaction. That is why we use the term "teaching and learning", so as to show the dimension of the learning part and its importance in the transmission of, in this case, mathematical knowledge.

The Function of Tables in Nichomachus

Our study of the textual forms suggests that the *functionality* of the textual units, related to the generic Nicomachean table, is presented in the following ideas: (a) the discourse in hand (wording, expression, structure) contains elements that mediate between the reader and the author facilitating comprehension and contextualizing information; (b) the formation of an object such as a table as well as the listing of information of sequences creates a mental picture of information, helping the reader to visualize and memorize efficiently; (c) the simultaneous study of different elements of the text allows the reader to create a "map of information" (mapping) that can be used to his inclination or intention.

One could argue that a table connected to mathematical material is a presentation or a carrier of specific mathematical information. Tables traditionally are used in mathematics for computational and representational reasons. They may also be

---

[27]The ways ancient and Byzantine commentators and readers of Nicomachus refer to the tables and the surrounding text, are briefly discussed in the Appendix.

thought part of an argument; they can be used either to derive conclusions or to evaluate information, or even represent an argument visually. So, in addition to the use as a computational instrument, the function of a table may be (a) relational, (b) visual, (c) investigational, and/or (d) heuristic.[28]

The more apparent of these functions is the *visualization* of the affirmations stated. Visualization makes any affirmation more evident, tangible and easily recalled, creating a mental picture of the text. *Investigation* of the material becomes technical in a sense both because of the emphasis on details as well as on acquisition of technique, of skill (τεχνολόγησις). A *relational function* can be explained as an explicit set of rules that form arguments. Lastly, the *heuristic* properties of a table can be understood in the sense of derivation and/or verification of predicates, suppositions and conclusions. The above functions characterize all[29] Nicomachean tables.

Nicomachus chooses to use tables even when a table is not necessary to make a pure theoretical presentation of the theory of ratios. It is normal to assume that a table, as part of an argument, can be used as heuristic and representational, as it is claimed in the Nicomachean text, but it is also normal to assume that it can be at the same time used as having computational value, even if it is not stated explicitly in the text; this can be furthermore supported if the wording, that is part of the argument, leaves room for a multilayered reading, as in the case of the Nicomachean text. In the last part of this chapter, we shall try to shed more light to the heuristic aspect of the Nicomachean tables, presenting their possible use in problem-solving by the Byzantine scholar Georgios Pachymeres, through a concrete example.

Arithmetical Problem-Solving: Contextualizing the Heuristic Tool Function of the Nicomachean Generic Table

In the last part of the previous section, we discussed several functions conveyed by the tables that Nicomachus constructs and discusses in the chapters of the *Arithmetical introduction* dealing with numerical ratios. In this section, we are going to further investigate one of these functions, namely the function of the table as a heuristic tool. As already said, the fact that Nicomachus does assign such a role to the table is manifested by the wording of his text. For example, in chapter 19.20 of the first book he concludes the first discussion of the generic table constructed in 19.9 by giving his reader the following advice to look at the table in order *to find by himself* other relationships that the numbers may have to one another: "An ambitious person *might find* (εὕροι) many other pleasing things displayed in this diagram" (55.1–2, our emphasis). The last paragraph of the passage from chapter 19.14

---

[28]For a recent analysis of the functionality of tables, especially in Greek mathematical astronomy, see Sidoli (2004, 2014a).

[29]Nevertheless, take into consideration footnote 7.

quoted in Sect. The Text Introducing the Generic Table Showing the Numbers in Relation above, in which Nicomachus explains how pairs of numbers in sesquialter ratio can be discovered in the generic table, provides another clear-cut example. The text reads: "In comparison with the second line ..., the lines which are next in order beneath *display* (παρεμφαίνουσι) the first species of the superparticular, that is, the sesquialter, *between terms occupying corresponding places*" (53.9–14, our emphasis). The personal pronoun ἡμῖν is implied here, accompanying the active verb παρεμφαίνουσι (show, display, exhibit) as indirect object,[30] which means that the saying is addressed *to us*, the readers of the table, whoever the readers might be. Therefore, the meaning of the passage is that, if one wants to find numbers in sesquialter ratio one has to locate in the table the appropriate rows and pick the numbers that occupy corresponding places in them.

Similar examples could be added, showing that Nicomachus assigns to the tables a heuristic function. Notwithstanding what the goal, or rather the goals, the tables as heuristic tools might have for Nicomachus himself, it is plausible to presume that the range of such a functionality did not remain unaffected by the process of appropriation of the *Arithmetical introduction* in the centuries that followed its composition. In the rest of this study, we shall advance the idea that Pachymeres, a Byzantine scholar of the late thirteenth century, endowed this heuristic function with a dimension never attested before. More precisely, by examining the chapter on Arithmetic of his *Quadrivium*, we shall propose that the generic table of Nicomachus could serve as a means for teaching how to solve arithmetical problems.

Georgios Pachymeres is one of the leading scholars of the early Palaeologan Renaissance, that is, the period of literary flourishing which followed the restoration of the Byzantine Empire, after the recapture of Constantinople by the Byzantines in 1261. The best known of his works is the history of the first fifty years of the Palaeologan dynasty, from 1258 to 1308, but apart from this he also wrote works on philosophy, rhetoric, theology, philology and mathematics, mostly related to his teaching activities (Constantinides 1982, 61; Golitsis 2007). His work on mathematics is the *Treatise of the four mathematical sciences: arithmetic, music, geometry, and astronomy*, the most important quadrivium that was produced by the Byzantines.[31] Pachymeres' *Treatise* (hereafter *Quadrivium*) is a highly erudite composition, based on a variety of ancient works. The sources from which he drew are the following (Constantinides 1982, 157): for the book on arithmetic, the main sources are Nicomachus, Diophantus and Euclid; for music, Ptolemy and Porphyry; for geometry, Euclid (not only the *Elements* but the *Optics* as well); and for astronomy, Aratos, Archimedes, Aristotle, Cleomedes, Euclid, Ptolemy and Theon.

---

[30]Cf. the sentence "Nature shows us (ἡ φύσις ... ἡμῖν παρεμφαίνει) that the doubles are more original than the triples ...", in chapter 4.3 of the second book, where the pronoun ἡμῖν does appear (79.10–14).

[31]The text of the *Quadrivium* is preserved in at least fourteen manuscripts, one of which, *Angelicus* 38 (C. 3. 7), is autograph of Pachymeres himself (Harlfinger 1971, 357 n. 3). The complete edition of the *Quadrivium* was prepared by P. Tannery but published posthumously by E. Stéphanou (Pachymère 1940).

We see, therefore, that Pachymeres' *Quadrivium* comprises a significant portion of the core of ancient Greek mathematics. It must be noticed, however, that, besides erudition, what characterizes Pachymeres' composition is that he does not imitate slavishly his models, as Herbert Hunger has remarked (Hunger 1978, ii, 246). On the contrary, we do find in the work marks of novelty, indicating the author's freshness and undogmatic attitude in dealing with his sources. This spirit of freshness is wonderfully exemplified by Pachymeres' idea to include in the first book of the *Quadrivium*, dealing with arithmetic, a series of problems stemming from the first book of Diophantus' *Arithmetica* and, furthermore, to combine ingeniously this material with material provided by Nicomachus' *Arithmetical introduction*, producing this way an unconventional intermixture of two sources, which were believed already from the late antique times to be so different from each other.[32]

The book on arithmetic of the *Quadrivium* consists of 74 chapters and it is structured as follows: the first five chapters contain a preface of philosophical content. Chapters 6–24 constitute the main Nicomachean part of the book. Next comes the Diophantine part, which covers the chapters 25–44. After two chapters dealing with the square and oblong (heteromecic) numbers, the book concludes with chapters 47–74 which form its Euclidean part.

The Diophantine part of the book, as said before, covers twenty chapters.[33] In the first chapter, Pachymeres presents the technical terms of the "arithmetical theory" that Diophantus exposes in his introduction to the *Arithmetica*, and the operations with these terms. Regarding this chapter it is interesting to point out that, although Diophantus presents in his text all this material as making part *of a coherent machinery*, several parts of which are interconnected in a suitable manner *so as to produce a general Way* (ὁδός, method) to be followed for solving arithmetical problems (Christianidis 2007), in Pachymeres' text not all parts of the machinery are exposed. In particular, Pachymeres does not discuss, nor does he even mention the *equation* that arises from the problem, according to the method of solution taught by Diophantus. Clearly the aims of the two authors in composing their works were not identical.

In chapters 26–44, Pachymeres solves a number of arithmetical problems corresponding to the eleven first problems of the first book of Diophantus. However, the solutions are not the same as Diophantus'. Although Diophantus' basic method of solution is algebra, Pachymeres, as we shall see, solves the problems numerically.[34] Before embarking on this study, though, it is interesting to point out that the

---

[32]Cf. the anonymous late antique commentary on Nicomachus, published by Tannery in the second volume of his edition of Diophantus, where we read: "For Diophantus, in the thirteen books of the *Arithmetica*, teaches the measured number (μετρούμενον ἀριθμόν), while the goal of Nicomachus is to teach the measuring number (μετροῦντα ἀριθμόν)" (Diophantus 1893–1895, ii, 73.25–28).

[33]This part of the *Quadrivium* was first published by Tannery in the second volume of his edition of Diophantus (Diophantus 1893–1895, ii, 78–122).

[34]Diophantus' and Pachymeres' methods are two of several methods that were used in Antiquity and the Middle Ages for solving problems. Incidentally, it is not uncommon to encounter worked-out problems to be solved, in the same book, by more than one method.

problems that Pachymeres creates out of the eleven problems of Diophantus are dozens. Thus Pachymeres establishes a lengthy series of problems, which is structured according to two criteria: on the one hand, the arrangement in which Diophantus presents his problems; on the other hand, for those problems of which the enunciation involves ratio, the order in which Nicomachus discusses, in his *Arithmetical introduction,* the kinds of ratios. The seriality of Pachymeres' problems is investigated in depth in Megremi and Christianidis (2014, 2015), so we refer the interested reader there. We cannot avoid mentioning, however, that the complex scheme by which Pachymeres structured his series of problems is another mark of his unconventional and innovative spirit.

Let us see how the two methods of solution are displayed in the example of the second problem of Diophantus, which asks a given number to be divided into two numbers having a given ratio (Diophantus 1893–1895, i, 16.24–25). This generic enunciation is exemplified by a single instantiated version, in which the given number is 60 and the given ratio is 3:1 (Diophantus 1893–1895, i, 16.26–27). Pachymeres discusses problems modelled on the aforesaid problem of Diophantus in six chapters of his book (27–32), each chapter dealing with a different kind of ratio (multiple, superparticular, superpartient, multiple superparticular and multiple superpartient) and comprising more than one problem. In this way, he creates and solves twenty-five problems, which can all be seen as instantiated versions of the generic enunciation of Diophantus' second problem. Below we present the problem of Diophantus facing one of the twenty-five problems of Pachymeres.

Diophantus' problem I.2[a]	Pachymeres' first problem of chapter 27[b]
It is required to divide a proposed number into two numbers having a given ratio.	The arithmetical problems related to the multiplications are as follows; if we are prescribed to divide the given number into <two numbers having> double, or triple, or whatever multiple ratio, such that the part has to the part this ratio.
Now, let it be proposed to divide 60 into two numbers having a triple ratio.	
Let it be assigned the lesser <sought-after number> to be 1 *arithmos*; then the greater will be 3 *arithmoi*, and the greater is thrice the lesser. It is further required the two together to be equal to 60 units. But the two added together are 4 *arithmoi*. Therefore, 4 *arithmoi* are equal to 60 units; therefore the *arithmos* is 15 units.	So, if we are prescribed to divide in double ratio, it is necessary to take the subtriple of the whole and make it the lesser term, of which we take the double, that is to say the remainder, <which is> the greater term. And the problem is solved.
Thus, the lesser will be 15 units and the greater 45 units.	For example, if we are prescribed to divide the number 24 into <two numbers having> a double ratio, we are looking for its subtriple, and it is 8. Its double, that is the remainder, 16, becomes the greater term, and the problem is satisfied.
	Indeed, 16 is the double of 8, and 16 and 8 <together yield> 24.

[a]See (Diophantus 1893–1895, i, 16.24–18.6)
[b]See (Pachymère 1940, 50.12–21; Diophantus 1893–1895, ii, 85.28–86.12)

The solution of Diophantus, to begin with, is carried out through (1) naming the unknowns (the lesser is assigned to be "1 *arithmos*", corresponding to our 1*x*, the greater "3 *arithmoi*", corresponding to 3*x*); (2) making operations with the named unknowns (the two together, that is, the "1 *arithmos*" and the "3 *arithmoi*", make "4 *arithmoi*"); (3) establishing and solving an equation, framed in the language of the names (4 *arithmoi* are equal to 60 units); and (4) calculating the numerical values of the two sought-after numbers by using the solution of the equation. Clearly, the method of solution used here is algebra.[35]

None of the above features appears in Pachymeres' solution: no names are assigned to the unknowns, no operations are performed on the names, and no equation is created. Any operation is performed on concrete numbers, be they given, or resulting from a previous step of the resolutory procedure. Undoubtedly, Pachymeres' solution is not algebraic; it is a numerical solution, in the strict sense of the term.

Indeed, the solution is developed according to the following steps:

1. First, one must determine the submultiple ratio corresponding to the next of the given multiple ratio. In the present case, the given ratio is double; therefore, the ratio in question is the subtriple.
2. Second, one has to find the subtriple of 24, which is 8.
3. The number thus found is set the lesser of the two numbers we are seeking.
4. For the next step, Pachymeres proposes two variants:

    (a) In the first variant, the double of the number found in step 2 must be taken, which gives 16.
    (b) In the second variant, the number found in step 2 is subtracted from 24, the remainder being 16.

5. The number thus found is set as the greater of the two numbers we are seeking.

The solution can be represented in a concise form by the following diagram (the italicized numbers are those the finding of which the enunciation of the problem calls for):

Given: 24 (the number to be divided), double ratio
First variant

$$24 \xrightarrow{\text{take the subtriple}} 8 \xrightarrow{\text{take the double}} 16,$$

Second variant

$$24 \xrightarrow{\text{take the subtriple}} 8 \xrightarrow{\text{take the remainder } [24-8]} 16.$$

---

[35]Diophantus' method of solution is discussed in depth in Christianidis 2007; Christianidis and Oaks 2013.

The variation in the fourth step put aside, the solution relies on the following conditions:

(1) Since the given ratio is double, one has to know that in order to carry through the first step of the procedure *he must work with the subtriple ratio.*

(2) Knowing the number 24, *he must find what number forms with it a pair in subtriple ratio.* That is, he must find the number 8, since the pair (8, 24) is in subtriple ratio.

Of the above two prerequisites, the former depends on the type of the problem. The latter, however, is not related to the problem. It is a knowledge that the practitioner should have mastered in advance or must have been suitably prepared to achieve. Thus, to solve problems of the type discussed here by the method taught by Pachymeres, the practitioner must be able to answer questions regarding ratios of the following type:

(Q) Knowing one of the two terms of a pair (A, B) in a specific ratio, find the other term of the pair.

This question is in fact implied in all problems involving ratio that Pachymeres solves. And indeed its answer is not always as trivial as in the problem discussed above.

For example, in chapter 30 Pachymeres solves a problem of the same type as the above, but now the given number has to be divided into two, in multiple super-particular ratio. More precisely, the problem asks to divide the number 56 into two numbers, in double sesquialter ratio (Pachymère 1940, 53.25–26; Diophantus 1893–1895, ii, 90.19–21). In modern terms, and resorting to the more convenient to the modern reader fractional setting proposed in footnote 19, two numbers A and B $(A > B)$ are sought such that $A + B = 56$, and $A = 2 \cdot B + \frac{1}{2}B$. Pachymeres presents the solution in a very concise form, but we can easily recognize the path he follows:

1. Since the given ratio is double sesquialter, one must know that to carry out the problem he has to work with subtriple sesquialter ratio.
2. He must find the subtriple sesquialter of 56, which is 16. $(56 = 3 \cdot 16 + \frac{1}{2}16)$
3. The number thus found is set as the lesser of the two sought-after numbers.
4. Next, the number found in step 2 must be subtracted from 56, the remainder being 40.
5. The number thus found is set as the greater of the two sought-after numbers.

In a synoptic way, the solution of the problem can be represented as follows:

Given: 56 (the number to be divided), double sesquialter ratio

$$56 \xrightarrow{\text{take the subtriple sesquialter}} 16 \xrightarrow{\text{take the remainder } [56-16]} 40.$$

By virtue of the above, the question that naturally arises is the following: how is it possible to cope methodically with the task involved in the second step (enabling the passage from number 56 to number 16), not having at our disposal the concept of fraction and its arithmetic? In other words, how is it possible to determine the number x of the pair $(x, 56)$ in subtriple sesquialter ratio? In the second part of this study, we claimed that the text of Nicomachus provides us with the necessary material to deal with such issues. Let us be more precise.

Nicomachus treats the multiple superparticular ratios in chapter 22 of the first book of the *Arithmetical introduction* (1866, 59.7–63.21). The pairs in triple sesquialter ratio, in particular, are produced, in regular order, from the "basic pair", called πυθμήν, which is the corresponding ratio in its lowest terms. In this case, the basic pair is (7, 2). From this pair, the entire series of triple sesquialter pairs can be produced by taking the successive multiples of the terms: (14, 4), (21, 6), (28, 8), (35, 10), (42, 12), (49, 14), (56, 16), (63, 18) and so on. As for the basic pair, it is determined as follows: since the triple sesquialter is the first kind of the triple superparticulars, and since the basic pairs of the triple superparticulars are found by comparing with the successive even and odd numbers starting from 2 (i.e. 2, 3, 4, 5, ...), the odd and even numbers which start from 7 and advance by three (i.e. 7, 10, 13, 16, 19, ...), first with first, second with second, and so on,[36] the basic pair of the triple sesquialter ratios is (7, 2). Now, knowing the basic pair, it is easy to locate the corresponding columns in the generic table, and thus find the series of triple sesquialter pairs.

1	**2**	3	4	5	6	**7**	8	9	10
2	**4**	6	8	10	12	**14**	16	18	20
3	**6**	9	12	15	18	**21**	24	27	30
4	**8**	12	16	20	24	**28**	32	36	40
5	**10**	15	20	25	30	**35**	40	45	50
6	**12**	18	24	30	36	**42**	48	54	60
7	**14**	21	28	35	42	**49**	56	63	70
8	*16*	24	32	40	48	*56*	64	72	80
9	**18**	27	36	45	54	**63**	72	81	90
10	**20**	30	40	50	60	**70**	80	90	100

The pair (56, 16) we are seeking is seen in the eighth row in the table

---

[36]The table below displays the basic pairs (πυθμένες) of the subspecies of the triple superparticular ratios:

7	10	13	16	19	22	...
2	3	4	5	6	7	...

That is, (7, 2) is the πυθμὴν of the series of the triple sesquialters, (10, 3) is the πυθμὴν of the series of the triple sesquitertians, (13, 4) is the πυθμὴν of the series of the triple sesquiquartans, and so on. The table does not appear in Nicomachus' text, it has been transmitted though as marginal addendum to it (see the critical apparatus on page 62 of Hoche's edition).

Things become even more arduous as we advance to more composite ratios. For example, in chapter 32 Pachymeres solves a problem of the same type as the above, but now the ratio involved belongs to the kind of multiple superpartient.[37]

Namely, the problem asks to divide the number 23 into two numbers in double superquintisextan ratio (Pachymère 1940, 55.31–56.4; Diophantus 1893–1895, ii, 93.16–23). In modern terms, two numbers A, B ($A > B$) are sought so that $A + B = 23$ and $A = 2 \cdot B + \frac{5}{6}B$. Pachymeres presents the solution in a very concise form, but we can easily recognize the path he follows:

1. Since the given ratio is double superquintisextan, one must know that the ratio he has to work with is the subtriple superquintisextan.
2. Next, the subtriple superquintisextan of 23 must be found, which is 6. ($23 = 3 \cdot 6 + \frac{5}{6}6$)
3. The number thus found is set as the lesser of the two sought-after numbers.
4. Next, the number found in step 2 must be subtracted from 23, the remainder being 17.
5. The number thus found is set as the greater of the two sought-after numbers.

Synoptically, the solution of the problem can be represented as follows:

Given: 23 (the number to be divided), double superquintisextan ratio

$$23 \xrightarrow{\text{take the subtriple superquintisextan}} 6 \xrightarrow{\text{take the remainder } [23-6]} 17.$$

Again, we are facing the same type of question as before: how can we determine the number x of the pair $(x, 23)$ in subtriple superquintisextan ratio? Although not detailed in the treatment of the multiple superpartient ratios, the text of Nicomachus gives us again the key to answer the question. The basic pairs of the subspecies of the triple superartient ratios are produced by comparing the odd numbers, from 11, with the successive even and odd numbers, from 3, one by one. Thus the pair (11, 3) is the πυθμὴν of the series of the triple superbipartient; the pair (15, 4) is the πυθμὴν of the series of the triple supertriquartan; the pair (19, 5) is the πυθμὴν of the series of the triple superquadriquintam, and so on. The πυθμένες are shown in the table below (not appearing in the sources):

---

[37]As said in Sect. The Text Introducing the Generic Table Showing the Numbers in Relation (footnote 20), a pair (A, B) forms a multiple superpartient ratio if A contains B more than once and further parts of it more than one. As the first or the second component of the name varies, the subkinds of the multiple superpartients are produced. Thus by making the multiple component double, triple, quadruple etc., one gets double superpartiens, triple superpartients, quadruple superpartients, etc. Similarly, by varying the superpartient component, one produces multiple superbipartients, multiple supertripartients, multiple superquadripartients, etc. Now, by combining the two aforesaid processes, one gets the various subkinds of the multiple superpartient ratios.

11	15	19	23	27	31	...
3	4	5	6	7	8	...
triple superbipartient	triple supertriquartan	triple superquadriquintam	triple superquintisextan	...	...	...

The pair (23, 6) we are seeking is the πυθμὴν of the series of the triple superquintisextan ratios, the other ratios of the series being produced by taking the successive multiples of the two terms of the πυθμὴν.

All problems we discussed so far from Pachymeres' book on arithmetic correspond to one problem of Diophantus, namely the problem 2 of the first book. As a consequence, notwithstanding what the process by which one answers the question Q is in each case, the solution is always conducted by following the same basic algorithm. We shall conclude this study by examining another problem, in which a different algorithm is applied. Below we reproduce the first problem of chapter 35 of Pachymeres' book on arithmetic, and the corresponding problem 3 of Diophantus' first book:

Diophantus' problem I.3[a]	Pachymeres' first problem of chapter 35[b]
To divide a proposed number into two numbers in a given ratio and difference.	Again, if we are prescribed to divide <a number> in superpartient ratio and with an excess of so many units, we shall solve the problems in this manner.
Now, let it be proposed to divide 80 into two numbers such that the greater is thrice the lesser and further exceeds <it by> 4 units.	
Let it be assigned the lesser <sought-after number> to be 1 *arithmos*; then the greater will be 3 *arithmoi* and 4 units. And the greater is thrice the lesser and furthermore exceeds <it by> 4 units. Now, I want the two to be equal to 80 units. But the two added together are 4 *arithmoi* and 4 units. Therefore, 4 *arithmoi* and 4 units are equal to 80 units.	Well, let <us take> first the superbitertient, and let the 18 be proposed, which we are prescribed to divide in superbitertient ratio with an excess of two units.
	I subtract the excess and I look for the subdouble superbitertient of the remainder 16; it is 6. I make the latter the lesser <term>, and the remaining plus the excess, 12, the greater <term>, and the problem is solved.
I subtract likes from likes; the remaining 76 units, therefore, are equal to 4 *arithmoi*; and the *arithmos* becomes 19 units.	For 12 is made up of the excess of two units and a remaining <amount> which is superbitertient with respect to 6.
To the numerical values:[c] the lesser will be 19 units and the greater 61 units.	

[a]See (Diophantus 1893–1895, i, 18.8–20)
[b]See (Pachymère 1940, 58.25–31; Diophantus 1893–1895, ii, 97.8–17)
[c]We translate in this manner the Diophantine phrase "ἐπὶ τὰς ὑποστάσεις". For a recent discussion on the meaning of this phrase see (Christianidis 2015)

The problem asks to divide a given number into two numbers such that the greater exceeds a given multiple of the lesser by a given amount. In modern terms, we would write it as follows:[38]

---

[38]The reader should keep in mind that the modern writing distorts the true character of the problem. The problems that both Diophantus and Pachymeres solve are *problems in arithmetic*,

$$a = X + Y, X = \lambda \cdot Y + b,$$

where a is the given number (80 and 18 in Diophantus and Pachymeres, respectively), b the excess (4 and 2, respectively) and λ the given ratio (triple and superbitertient,[39] respectively).

Diophantus' solution is a solution by algebra: the lesser sought-after number is named "1 *arithmos*", which makes the greater one to get the name "3 *arithmoi* and 4 units". The two named terms are added to get "4 *arithmoi* and 4 units", next this expression is equated with the 80 units to get the equation "4 *arithmoi* and 4 units are equal to 80 units", the equation is then simplified, solved, and from the solution the numerical values of the unknown numbers are calculated.

Again, nothing of the above occurs in Pachymeres' solution. As in the examples already discussed, Pachymeres' solution is purely numerical, in which no names are assigned to the unknowns, no operations are performed on the names, and no equation is created.

The solution is developed on the basis of the following steps:

1. First, we subtract the excess 2 from 18: $18 - 2 \rightarrow 16$.
2. Second, we must determine the submultiple superbitertient ratio corresponding to the next of the given, superbitertient, ratio. It is the subdouble superbitertient.[40]
3. We have to find the subdouble superbitertient of 16, which is 6.
4. The number thus found is set the lesser of the two numbers we are seeking.
5. We subtract the number found in step 3 from 16: $16 - 6 \rightarrow 10$.
6. We add the excess 2 to the number found in step 5: $10 + 2 \rightarrow 12$.
7. We set the number found in step 6 the greater of the two numbers we are seeking.

In a concise way, the algorithm applied can be represented as follows:

Given: 18 (the number to be divided), 2 (the excess), superbitertient ratio

$$18 \xrightarrow{\text{subtract the excess}[18-2]} 16 \xrightarrow{\text{take the subdouble superbitertient}} 6$$

$$16 \xrightarrow{\text{take the remainder }[16-6]} 10 \xrightarrow{\text{add the excess }[10+2]} 12.$$

In pursuing the solution, we are again facing the question: Given a number A how one can find another number B such that the two numbers are in a particular ratio? In the present case, we have to find the subdouble superbitertient of 16. Again, the text of Nicomachus provides us with the means to answer the question.

---

not algebraic problems, as the modern symbolic writing entails, since the enunciations do not contain equations neither any algebraic terms.

[39]Two numbers A and B are in superbitertient ratio if $A = B + \frac{2}{3}B$.

[40]Two numbers A and B are in double superbitertient ratio if $A = 2 \cdot B + \frac{2}{3}B$.

Indeed, in chapter 23 of the first book, Nicomachus explicitly mentions the pair (12, 6) which is in double superbitertient ratio. He writes (1866, 64.8–16):

> From what has already been said it is not hard to conceive of the varieties of this relation, for they are differentiated in the same way as, and consistently with, those that precede, double superbitertient, double supertripartient, double superquadripartient, and so on. For example, 8 is the double superbitertient of 3, 16 of 6, and in general the numbers beginning with 8 and differing by 8 are double superbipartients of those beginning with 3 and differing by 3, when those in corresponding places in the series are compared.

What Nicomachus says in this passage is that the basic pair of the pairs in double superbitertient ratio is (8, 3) and the other pairs in this ratio can be found by taking the multiples of the two terms of the basic pair. Thus, knowing the basic pair one could locate the corresponding columns in the generic table, and find the series of double superbitertient pairs.

1	2	**3**	4	5	6	7	**8**	9	10
2	4	*6*	8	10	12	14	*16*	18	20
3	6	**9**	12	15	18	21	**24**	27	30
4	8	**12**	16	20	24	28	**32**	36	40
5	10	**15**	20	25	30	35	**40**	45	50
6	12	**18**	24	30	36	42	**48**	54	60
7	14	**21**	28	35	42	49	**56**	63	70
8	16	**24**	32	40	48	56	**64**	72	80
9	18	**27**	36	45	54	63	**72**	81	90
10	20	**30**	40	50	60	70	**80**	90	100

The pair (6, 16) we are seeking is found in the second row of the table

We do not know if Pachymeres actually found the number 6 by the procedure presented above, that is, by using the table. There is no conclusive textual evidence to support this claim. However, for solving the problem by employing the method he teaches, the question how to find the subdouble superbitertient of 16 needs to be answered. In other words, finding the number 6 out of the given number 16 and the given subdouble superbitertient ratio is a mandatory step for the solution via this method. Indeed, this is explicitly recognized by Pachymeres himself who writes "*I look for* the subdouble superbitertient of the remainder 16; it is 6" (our emphasis). But how can the number 6 be found? We explained that one could give an account of Pachymeres' answer to this question by resorting to the generic table of Nicomachus. Although such a reconstruction is a hypothesis, it is historically plausible. As we have seen, Nicomachus' text allows us to interpret the tables it contains in the context of a certain practicality. Besides, the function of the generic table as a means for finding numbers in ratio is attested in the writings of a number of late antique readers and commentators of Nicomachus.[41]

---

[41]See more in the Appendix.

Conclusion

In this study, we have intended to argue that through a *teaching and learning context* and using *textual evidence* from the *Arithmetical introduction* we can gain insight to the ways a table can convey, in this case, arithmetical information, as well as how it was embedded in the educational practice of its time, through the problem-solving activity. Thus, we can answer the questions raised at the beginning of our chapter and draw the conclusion that it is possible to produce a case study where reading, constructing and using tables can be studied through the prism of functionality, as Georgios Pachymeres' example shows us. A table can be viewed in a teaching and learning context along with the text that surrounds it, as having a special practical purpose, that of facilitating a problem-solving procedure. The functional role of the table in this case is perceived through the lens of the reader (any possible receiver) and attested to by the precise case study we have presented. Through this case study, we can understand the reception of the Nicomachean work during the thirteenth century in the Eastern Roman Empire and realize the uses of the Nicomachean arithmetical theory in revealing an extended use of the tables included in the *Arithmetical introduction* as heuristic tools.

Acknowledgements The authors thank Jeffrey Oaks, who kindly read this paper and suggested improvements. They also thank the anonymous referees for their remarks and corrections.

Appendix: Tables and the ancient readers of Nicomachus

We claimed in Sect. Discussion of the text that Nicomachus' treatise allows the reader to proceed to a creative reading of the textual units related to the tables. This claim is further supported by the way ancient and Byzantine scholars have approached these particular textual units of the Nicomachean text.

Boethius in his Latin translation and adaptation of Nicomachus' material creates in fact his own version of an introduction to Arithmetic. The purpose of his text can be detected through his emphasis on tabular forms. For each kind of multiple superparticular ratio, for example, Boethius provides each time a new, two-row tabular form of the kind in hand (Boèce 1995, 51; 59; 63; 64; see Fig. 3).

Iamblichus, about two centuries earlier, creating his own version of an introduction to Arithmetic, does refer to the process of finding pairs of numbers in relation, and even alludes to the existence of a table on which one can see this process in the making by also using terms such as "horizontally and vertically" (Iamblichus 1894, 39.23; Jamblique 2014, 108.10–11), "in the form of the letter gamma" (Iamblichus 1894, 40.11, 40.13; Jamblique 2014, 108.21, 108.22), etc. Yet, Iamblichus does not include a table in his text, and he does not prompt the reader to follow the processes described in one. Iamblichus, to whom "pure" Neopythagorean ideas have been attributed, is more concerned with the theoretical aspect of the statements concerning numbers in relation.

Lib.   I   63

latur.quā vero partē cōparati numeri clauserit: fecundū fuperparticularē cō=
paratione habitudinéqʒ vocabitur. Horū autē exēpla huiufmodi funt. Du=
plex fefquialter eft: vt quincʒ ad duo.habent enim 5, binariū numerū bis &
eius mediā id eft 1.Duplex vero fefquitertius eft:feptenarius ad ternariū cō=
paratus.At vero nouenari⁹ ad quaternariū:duplex fefquiquartus.Si vero 11
56 ad 5:duplex fefquiquintus.⊂Et hi feper nafcētur:difpofitis in ordinē a bina 1
rio numero omnibus naturaliter paribus imparibufcʒ terminis : fi cōtra eos
oēs a quinario numero impares cōparētur.vt
primū primo:fecūdū fecūdo,tertiū tertio,cau
te & diligēter apponas.vt fit difpofitio talis.

2	3	4	5	6	7	8	9	10	11
5	7	9	11	13	15	17	19	21	23

⊂Si vero a duobus paribus omnibus difpofitis terminis:illi qui a quinario 2
numero inchoātes,quinario numero rurfus fe=
fe tranfiliunt comparentur:omnes duplices fef
quialteros creant, vt eft fubiecta defcriptio.

2	4	6	8	10	12
5	10	15	20	25	30

⊂Si vero a tribus inchoent difpofitiones:& tribus fefe tranfiliant,& ad eos 3
aptentur qui a feptenario inchoantes, feptenario fefe numero tranfgrediun=
tur:omnes duplices fefquitertij,habita di=
ligenter comparatione, nafcuntur. vt fub=
iecta defcriptio monet.

3	6	9	12	15	18	21
7	14	21	28	35	42	49

⊂Si vero omnes in ordinē quadrupli difponantur:hi qui naturalis numeri 4
quadrupli funt,vt vnitatis quadruplus, & duorum, triumcʒ & quatuor,atcʒ
quinarij,& cæterorum fefe fequentium, vt ad eos aptentur a nouenario nu=
mero inchoantes, femper fefe nouenario
præcedentes:tunc duplicis fefquiquartæ
proportionis forma texetur.

4	8	12	16	20	24
9	18	27	36	45	54

⊂Ea vero fpecies huius numeri,quæ eft triplex fefquialtera,hoc modo pro= 5
creatur:fi difponātur a binario numero omnes in ordinem pares, & ad eos
feptenario numero inchoantes,feptenario
fefe fupergredientes , folito ad alterutrum
modo comparationis aptentur.

2	4	6	8
7	14	21	28

⊂Si aūt a ternario numero ingreffi: cūctos naturalis numeri triplices difpo 6
namus,& eis a denario numero denario fefe fuper
gredientes ordine cōparemus : omnes triplices fef=
quitertij in ea terminorū cōtinuatione prouenièt.

3	6	9	12
10	20	30	40

⊂CAP.VICESIMIQ VARTI COMMENTARIVS.

55 A M ad eas,quæ ex prioribus conftant:progreditur inæqualitares.primi=
tufcʒ quidnā fit multiplex fuperparticularis exponit.nam, cʒ is eft maior
numerus:qui minorē plufcʒ femel amplectitur,adhuc autē & minoris ali=
quam partem,cʒ autem plufcʒ femel:multiplex dicitur,cʒ vero præter inte
grum ambitū,partem aliquā:fuperparticularis. Eam ob rem fi bis conti=
net & fecundā partem:duplus fefquialter.duplus quidem:cʒ bis continet.
fefquialter:cʒ eius fecundā partem.vt (exempli caufa)5 ad 2.nam 5 conti=
net binariū bis:& infuper vnitatem,quæ 2 eft medietas. cʒ fi ter cōtineat,
                                                                        l.iij.

Fig. 3 f. 63r of the 1521 edition of Boethius' *De institutione arithmetica*, displaying the multiple superparticular ratios in small two-row tables

Commentators and scholiasts also have a different approach. From the critical apparatus of Hoche's editions, we can see that there are indeed quite a few tables concerning the kinds of ratios included in marginal scholia found in the manuscripts used in his edition (1866, 58; 62). Clearly for the scholiasts, the emphasis was on understanding how the pairs of each kind of the "greater" or the "lesser" can be found or created. On the other hand, Ioannes (John) Philoponus' exegesis of the *Arithmetical introduction* (1864, 1867) does include a series of instructions and an exposition of the pairs of the kinds of ratios, but not a table, and reasonably so. Philoponus' exegesis works as a complement to the Nicomachean text and as such it explicitly refers to the tables included in the text of the *Arithmetical introduction* but does not reproduce them, nor does it create others for that matter (Philoponus 1864, 38–43). One last example of the various ways of reading can be found in the last decade of the thirteenth century in the quadrivium of Georgios Pachymeres. Pachymeres' version of arithmetical theory includes, in the case of theory of ratios, both a series of mathematical methods and a series of problems that are connected to ratios, a unique view considering time and locality. This quite interesting blending of theoretical arithmetic and problem-solving was examined in the last part of the present chapter.

References

Sources

Asclepius of Tralles. 1969. *Commentary to Nicomachus' Introduction to Arithmetic*. (Edited with an Introduction and Notes by Leonardo Tarán). Philadelphia: The American Philosophical Society.

Boèce. 1995. *Institution Arithmétique*. Texte établi et traduit par J.-Y. Guillaumin. Paris: Les Belles Lettres.

Boethius. 1521. *Divi Severini Boetii Arithmetica duobus discreta libris. Adjecto commentario Gerardi Ruffi mysticam numerorum applicationem perstingente declarata*. Paris: apud S. Colinaeum.

Diophantus. 1893–1895. *Diophantus Alexandrinus opera omnia*, 2 vols. Edidit et Latine interpretatus est P. Tannery. Leipzig: Teubner.

Domninus of Larissa. 2013. *Encheiridion and spurious works*. Introduction, critical text, English translation and commentary by P. Riedlberger. Piza, Roma: Fabrizio Serra Editore.

Héron d'Alexandrie. 2014. *Metrica*. Introduction, texte critique, traduction française et notes de commentaire par Fabio Acerbi et Bernard Vitrac. Pisa, Roma: Fabrizio Serra Editore.

Heron of Alexandria. 1903. *Rationes dimetiendi et Commentatio dioptrica*. Recensuerunt L. Nix et W. Schmidt. Leipzig: Teubner.

Iamblichus. 1894. *Nicomachi Arithmeticam introductionem liber*. Edidit H. Pistelli. Leipzig: Teubner.

Jamblique. 2014. *In Nicomachi Arithmeticam*. Introduction, texte critique, traduction française et notes de commentaire par N. Vinel. Pisa, Roma: Fabrizio Serra Editore.

Nicomachus. 1538. *Nicomachi Geraseni Arithmetica libri duo*. Paris: In officica Christiani Wecheli.

Nicomachus. 1866. *Nicomachi Geraseni Pythagorei Introductionis Arithmeticae libri II.* Recensuit R. Hoche. Leipzig: Teubner.

Nicomachus of Gerasa. 1926. *Introduction to Arithmetic.* Translated into English by Martin Luther d'Ooge; with studies in Greek arithmetic by Frank Egleston Robbins and Louis Charles Karpinski. New York: Macmillan.

Pachymère. 1940. *Quadrivium de Georges Pachymère, ou Σύνταγμα τῶν τεσσάρων μαθημάτων: ἀριθμητικῆς, μουσικῆς, γεωμετρίας καὶ ἀστρονομίας.* Ed. P. Tannery; texte revisé et établi par E. Stéphanou. Città del Vaticano: Biblioteca Apostolica Vaticana.

Philoponus, John. 1864. *ΙΩΑΝΝΟΥ ΓΡΑΜΜΑΤΙΚΟΥ ΑΛΕΞΑΝΔΡΕΩΣ (ΤΟΥ ΦΙΛΟΠΟΝΟΥ) ΕΙΣ ΤΟ ΠΡΩΤΟΝ ΤΗΣ ΑΡΙΘΜΗΤΙΚΗΣ ΕΙΣΑΓΩΓΗΣ.* Primum edidit R. Hoche. Leipzig: Teubner.

Philoponus, John. 1867. *ΙΩΑΝΝΟΥ ΓΡΑΜΜΑΤΙΚΟΥ ΑΛΕΞΑΝΔΡΕΩΣ ΤΟΥ ΦΙΛΟΠΟΝΟΥ ΕΙΣ ΤΟ ΔΕΥΤΕΡΟΝ ΤΗΣ ΑΡΙΘΜΗΤΙΚΗΣ ΕΙΣΑΓΩΓΗΣ.* Primum edidit R. Hoche. Berlin: Apud S. Calvary.

Plotinus. 1988. *Ennead VI.6–9.* Translated by A. H. Armstrong. Cambridge, Mass.: Harvard University Press.

Proclus. 1873. *Procli Diadochi, in primum Euclidis Elementorum librum commentarii.* Ex recognitione G. Friedlein. Leipzig: Teubner.

Proclus. 1970. *A Commentary on the First Book of Euclid's* Elements. Translated with introduction and notes by Glenn R. Morrow. Princeton, NJ: Princeton University Press.

Ptolemy. 1898–1903. *Syntaxis Mathematica.* Edidit J. L. Heiberg, 1 vol. in 2 parts. Leipzig: Teubner.

Ptolemy. 1984. *Almagest.* Translated and Annotated by G. J. Toomer. London: Duckworth.

Soterichus. 1871. *Soterichi ad Nicomachi Geraseni introductionem arithmeticam de Platonis psychogonia scholia,* in *Gymnasium zu Elberfeld. Jahresbericht über das Schuljahr 1870–1871,* Elberfeld, pp. i–iv and 1–6.

Theon of Alexandria. 1936–43. *Commentaires de Pappus et de Théon d'Alexandrie sur l'Almageste,* ed. A. Rome. Città del Vaticano: Biblioteca Apostolica Vaticana. [*Tome II. Théon d'Alexandrie, Commentaire sur les livres 1 et 2 de l'Almageste* (1936). *Tome III. Théon d'Alexandrie, Commentaire sur les livres 3 et 4 de l'Almageste* (1943).].

Théon de Smyrne. 1892. *Exposition des connaissances mathématiques utiles pour la lecture de Platon.* Traduite pour la première fois du grec en français par J. Dupuis. Paris: Librairie Hachette.

Théon de Smyrne. 2010. *Lire Platon. Le recours au savoir scientifique: Arithmétique, musique, astronomie.* Texte présenté, annoté et traduit du grec par J. Delattre Biencourt. Toulouse: Anacharsis.

Modern Studies

Bernard, A. 2018. Greek mathematics and astronomy in late antiquity. In *The Oxford Handbook of Science and Medicine in the Classical World,* ed. P. Keyser, and J. Scarborough, 869–894. Oxford: Oxford University Press.

Bowersock, G. W., Brown, P., and O. Grabar (eds.). 2001. *Interpreting late antiquity: Essays on the postclassical world.* Cambridge, Mass., London: The Belknap Press of Harvard University Press.

Christianidis, J. 2007. The way of Diophantus: Some clarifications on Diophantus' method of solution. *Historia Mathematica, 34,* 289–305.

Christianidis, J. 2015. The meaning of hypostasis in Diophantus' *Arithmetica.* In *Relocating the History of Science: Essays in Honor of Kostas Gavroglu,* ed. T. Arabatzis, J. Renn, and A. Simões, 315–327. Heidelberg, New York, Dordrecht, London: Springer.

Christianidis, J., and J. A. Oaks. 2013. Practicing algebra in late antiquity: The problem-solving of Diophantus of Alexandria. *Historia Mathematica, 40,* 127–163.

Constantinides, C. N. 1982. *Higher education in Byzantium in the thirteenth and early fourteenth centuries (1204–ca. 1310)*. Nicosia: Cyprus Research Centre.

Fowler, D. 1999. *The mathematics of Plato's academy: A new reconstruction* (2nd ed.). Oxford: Clarendon Press.

Golitsis, P. 2007. George Pachymère comme didascale. Essai pour une reconstitution de sa carrière et de son enseignement philosophique. *Jahrbuch der Österreichischen Byzantinistik, 58,* 53–68.

Hadot, I. 2005. *Arts libéraux et philosophie dans la pensée antique. Contribution à l'histoire de l'éducation et de la culture dans l'antiquité*, seconde édition. Paris: Librairie Philosophique J. Vrin.

Harlfinger, D. 1971. *Die Textgeschichte der pseudo-aristotelischen Schrift Περὶ ἀτόμων γραμμῶν*. Amsterdam: Hakkert.

Hunger, H. 1978. *Die hochsprachliche profane Literatur der Byzantiner*, 2 vols. München: C. H. Beck.

Kappraff, J. 2000. The arithmetic of Nicomachus of Gerasa and its applications to systems of proportion. *Nexus Network Journal, 2,* 41–55.

Lloyd, G. E. R. 2012. The pluralism of Greek 'mathematics'. In *The history of mathematical proof in ancient traditions*, ed. K. Chemla, 294–310. Cambridge: Cambridge University Press.

Marrou, H.-I. 1977. *Décadence romaine ou antiquité tardive? IIIe-VIe siècle*. Paris: Seuil.

Masià, R. 2015. On dating Hero of Alexandria. *Archive for History of Exact Sciences, 69,* 231–255.

Megremi, A., and J. Christianidis. 2014. Georgios Pachymeres, reader of Nicomachus: The arithmetical theory of ratios as a tool for solving problems. *Neusis, 22,* 53–85. [In Greek].

Megremi, A., and J. Christianidis. 2015. Theory of ratios in Nicomachus' *Arithmetica* and series of arithmetical problems in Pachymeres' *Quadrivium*: reflections about a possible relationship. *SHS Web of Conferences* 22: # 00006. Available online at http://dx.doi.org/10.1051/shsconf/20152200006.

O'Meara, D. J. 1989. *Pythagoras revived: Mathematics and philosophy in late antiquity*. Oxford: Clarendon Press.

Sidoli, N. 2004. *Ptolemy's mathematical approach: Applied mathematics in the second century*. Doctoral dissertation. University of Toronto.

Sidoli, N. 2011. Heron of Alexandria's Date. *Centaurus, 53,* 55–61.

Sidoli, N. 2014a. Tables in Ptolemy. *Historia Mathematica, 41,* 13–37.

Sidoli, N. 2014b. Research on ancient Greek mathematical sciences, 1998–2012. In *From Alexandria, Through Baghdad. Surveys and Studies in the Ancient Greek and the Medieval islamic Mathematical Sciences in Honor of J. L. Berggren*, ed. N. Sidoli, G. Van Brummelen, 25–50. Berlin, Heidelberg: Springer-Verlag.

Slaveva-Griffin, S. 2009. *Plotinus on number*. Oxford, New York: Oxford University Press.

Taub, L. 2013. On the variety of "genres" of Greek mathematical writings: Thinking about mathematical texts and modes of mathematical discourse. In *Writing science: medical and mathematical authorship in ancient Greece*, ed. M. Asper & A.-M. Kanthak, 333–365. Berlin, Boston: De Gruyter.

Watts, E. J. 2006. *City and school in late antique Athens and Alexandria*. Berkeley, Los Angeles, London: University of California Press.

Athanasia Megremi received her Ph.D. on "Late Antique Arithmetic and Commentary Scholarship" in July 2016. She is a temporary lecturer of history of mathematics at the National and Kapodistrian University of Athens, at the Department of History and Philosophy of Science. She studies the history of late antique and medieval mathematics. Her representative publications include: "Mathematical commentary in Late Antiquity: two examples of commentary practice in the *Arithmetic introduction* of Nichomachus of Gerasa," in *Science and Technology, Historical and Historiographical Studies*, in I. Mergoupi-Savaidou et al. (eds), 2013 (in Greek), "Georgios Pachymeres, reader of Nicomachus: The arithmetical theory of ratios as a tool for solving problems," *Neusis* 22, 2014 (with J. Christianidis, in Greek), "Theory of ratios in Nicomachus' *Arithmetica* and series of arithmetical problems in Pachymeres' *Quadrivium*: reflections about a possible relationship," *SHS Web of Conferences* 22, 2016 (with J. Christianidis).

Jean Christianidis, a professor of history of mathematics at the National and Kapodistrian University of Athens, and head of the Department of History and Philosophy of Science, partner member at the Centre Alexandre Koyré (Paris), and editor-in-chief of the Greek journal of the history and philosophy of science and technology *Neusis*, studies the history of Greek mathematics. Among his recent publications are "Practicing algebra in late antiquity: The problem-solving of Diophantus of Alexandria," *Historia Mathematica* 40, 2013 (with J. Oaks), "A new analytical framework for the understanding of Diophantus's Arithmetica I-III," *Archive for History of Exact Sciences* 66, 2012 (with A. Bernard), "Solving problems by algebra in late antiquity: New evidence from an unpublished fragment of Theon's commentary on the Almagest," *Sciamvs* 14, 2013 (with I. Skoura), and "Situating the Debate on 'Geometrical Algebra' within the Framework of Premodern Algebra," *Science in Context* 29, 2016 (with M. Sialaros). He is also the editor of the collection *Classics in the History of Greek Mathematics* (Kluwer, 2004).

The So-Called "Dust Computations" in the *Līlāvatī*

Takanori Kusuba

Abstract The rules and examples in mathematical texts written in Sanskrit in medieval India are composed in verse form, so they are in a terse style. The texts are silent in terms of actually carrying out computations. In order to understand the rules, students solved problems given either by the author or the commentator. When students read the rules, they learned the procedure of calculation. When they read examples, they learned how to set down given numbers. The present chapter analyzes how those who read the Sanskrit mathematical works, in particular the *Līlāvatī*, performed computations.

Keywords Indian mathematics · Bhāskara II (Bhāskarācārya) · *Līlāvatī* Arithmetical operations · Dust board

Introduction

The sources used to study Indian mathematics in this chapter are Sanskrit texts written from AD 500 onwards. The earliest extant text is the *Āryabhaṭīya* by Aryabhata, who was born in AD 476.[1] These texts are included in the texts on mathematical astronomy, or *siddhāntas*. Some of them contain a few chapters devoted solely to mathematics. Among the four chapters of the *Āryabhaṭīya*, Chapter 2 is devoted to mathematics. It gives only rules composed in verse using metrics called *āryā*.[2] A commentary on the text written by Bhāskara I in 629 is extant. The commentator gives many examples in form of stanzas. His contemporary, Brahmagupta, wrote the

---

[1]This does not mean that mathematics was not studied before the *Āryabhaṭīya*. Staal states clearly: "It is unlikely that they (mathematicians who wrote in Sanskrit) studied books. In India, no *paṇḍita* or traditional scholar does" (2010, p. 44).

[2]The first and third quatrain each contain 12 syllabic instants, the second 18 and fourth 15. In addition, 1 is allotted to a short vowel, and 2 to a long one.

T. Kusuba (✉)
History and Philosophy of Science, Osaka University of Economics, Osaka, Japan
e-mail: kusuba@osaka-ue.ac.jp

© Springer Nature Switzerland AG 2018
A. Volkov and V. Freiman (eds.), *Computations and Computing Devices in Mathematics Education Before the Advent of Electronic Calculators*,
Mathematics Education in the Digital Era 11, https://doi.org/10.1007/978-3-319-73396-8_5

Brāhmasphuṭa-siddhānta (hereafter *BSS*) in 628. This text includes a chapter on arithmetic, in which only rules are provided with no details being given for computations, and one chapter on algebra. After the *BSS*, Sanskrit mathematics came to consist of two major fields: *pāṭīgaṇita* and *bījagaṇita*. The former usually deals with arithmetical calculations, and the latter with equations. The *Pāṭīgaṇita* by Śrīdhara (ca. 800, hereafter *PG*) was the first *pāṭīgaṇita* text. It contains rules with examples. However, only a portion is edited based on the only known copy of the manuscript. It is not known how influential it was. After Śrīdhara, Āryabhaṭa II (fl. ca. 10th century AD) wrote the *Mahā-siddhānta* (hereafter *MS*), and Śrīpati (fl. 1039) wrote the *Siddhānta-śekhara* (hereafter *SS*). The most influential Sanskrit mathematical text in India is the *Līlāvatī* by Bhāskara (born 1114) commonly known as Bhāskara II or Bhāskarācārya. It is included in the *Siddhānta-śiromaṇi*, but considered a separate text. The number of commentaries and the distribution of surviving manuscripts suggest it was very widely read. The book contains what the author's predecessors discussed. The mathematical texts written by the author's successors were influenced by the book. We could say that the *Līlāvatī* is representative of mathematics texts in medieval India. Below is a list of medieval Indian texts.

Author	Text	*Pāṭīgaṇita (arithmetical calculations)*	*Bījagaṇita (algebra)*
Brahmagupta	*Brāhmasphuṭa-siddhānta*	Chapter 12	Chapter 18
Śrīdhara	*Pāṭīgaṇita*		
Śrīdhara	*Bījagaṇita* (lost)		
Āryabhaṭa II	*Mahā-siddhānta*	Chapter 15	Chapter 18
Śrīpati	*Siddhānta-śekhara*	Chapter 13	Chapter 14
Bhaskara II	*Siddhānta-śiromaṇi*	*Līlāvatī*	*Bījagaṇita*

Note: Śrīdhara's rule for solution of quadratic equations, in the form of a stanza, is quoted by later authors. However, his treatise on the *bījagaṇita* is lost

My aim in this section is limited. I want to provide references to Indian mathematics concerning its (1) general history and (2) translations. As far as its general history is concerned, Datta and Singh wrote their *History of Hindu Mathematics: A Source Book* in 2 volumes in 1935–1936. For a brief historiography, *see also* Gupta (2002). Plofker (2009a) provides a valuable description of the history of Indian mathematical traditions. As for translations, in 1817 Colebrooke translated the 12th and 18th chapters of the *BSS*, *Līlāvatī* and *Bījaganita*, into English. This translation still remains useful today. The second chapter on mathematics in the *Āryabhatīya*, with Bhāskara I's commentary on it, was translated into English and studied by Keller in 2006. Some primary sources from excerpts from the *Vedas* to the writings of Kerala School are partly translated by Plofker (2007). Hayashi published an edition of a text attributed to Srīdhara with an English translation and notes in 2013. These publications are useful to understand the framework and features of *pāṭīganita* texts.

Līlāvatī: Structure of the Work

The *Līlāvatī* is a text on *pāṭīgaṇita* composed in Sanskrit metrics. Sanskrit is the traditional language of scholarship in India. The Sanskrit term *pāṭīgaṇita* is a compound word. The term *pāṭī* is believed to be derived from *paṭṭa*, which means "board," and *gaṇita*, past passive particle of the verb *gaṇ* (to count or to compute), means "mathematics" or "computation." Therefore, the *pāṭīgaṇita* texts illustrate mathematical computations made using a board. Datta and Singh claimed that the calculations were performed on sand spread on a board. They pointed out that Bhāskara refers to the Sanskrit term *dhūlīkarman* (dust-computation).[3]

The *Līlāvatī* is a textbook for students of astral sciences. It consists of basic operations and practical mathematics. The basic operations consist of eight fundamental operations to be performed with positive integers and fractions, namely, addition, subtraction, multiplication, division, raising to second and third powers, and extraction of square and cube roots.[4] The rules of addition of zero, square of zero, multiplication and division by zero are also given.[5] The format of exposition in the *Līlāvatī* is as follows:

- Rule (in verse).
- Example (in verse).
- Setting down of numbers.
- Author's own commentary (in prose).

Bhāskara does not provide the reader with intermediary calculations and gives only answers. This study investigates how calculations, as taught in the *Līlāvatī*, are carried out. For the English translation of the rules, I quote the latest version by Plofker (2007).

Setting Down of Numbers for Calculation

At the outset, numbers given in the example are set down. In what order are the numbers set down? The rule and example for addition and subtraction may provide a clue:

> The sum or difference of the numerals according to their places is to be made, in order or in reverse order.[6]

---

[3]Datta and Singh (1935 [2001], p. 129). They refer to a stanza number in one of the works of Bhāskara. The term appeared in the auto-commentary on the stanza.

[4]In his *Līlāvatī*, Bhāskara II does not give rules for negative numbers, but he does provide them in his *Bījagaṇita*.

[5]The rules for calculations with zero are given for the first time in the *BSS* (Chapter 18, stanzas 31–36). *See* Colebrooke (1817 [2005], pp. 339–340). Details are given in the subsequent text.

[6]Plofker (2007, p. 449). This rule stanza is numbered 12. At the beginning of the auto-commentary on an example of multiplication, Bhāskara quotes the first quarter of the rule stanza. Generally,

In an example for addition in the *Līlāvatī*, numbers 2, 5, 32, 193, 18, 10, and 100 are given. It is well known that Indians used a decimal place value numbering system. The numbers are stated in Sanskrit words as *dvi* (2), *pañca* (5), *dvātrimśat* (2–30), *trinavatiśata* (3–90–100), *aṣṭādaśa* (8–10), *daśa* (10), and *śata* (100), respectively. Bhāskara uses cardinal numbers. Here 193 is represented as *tri* (3) *navati* (90) *śata* (100). This manner reflects the order of writing down numerals. The digits written in the units column are mentioned first. In an example of multiplication below, Bhāskara writes a multiplicand 135 as *pañca* (five)-*tri* (three)-*eka* (one) without decimal place value names. The numbers are set down from right to left, that is, from units to tens, then to hundreds, and so on. This is "in order" as mentioned above. Therefore, "in reverse order" means that the numeral in the greatest decimal position is mentioned first.

An example of addition and subtraction

Calculate 2 + 5 + 32 + 193 + 18 + 10 + 100. Subtract the sum from 10,000. Answer: 360; 9640.

The numbers to be added should be placed vertically with their least significant digits aligned. In order to follow the procedure, the numbers which are newly written in each step are in italics.

2	2	2	2
5	5	5	5
32	32	32	32
193	193	193	193
18	18	18	18
10	10	10	10
100	100	100	100
	20	*160*	*360*

Step 1. Addition of units: 2 + 5 + 2 + 3 + 8 + 0 + 0 = 20.
Step 2. Addition of tens: 3 + 9 + 1 + 1 + 0 + 2 = 16.
Step 3. Addition of hundreds: 1 + 1 + 1 = 3.[7]

---

when a rule is referred to in a commentary the first portion is quoted. The stanza number is not provided by the author, but probably added by an editor for convenience of reference.

[7] According to Datta and Singh (1935 [2001], p. 132, footnote 1), the *Manorañjana* explains the process as follows: the sum of units is 2 + 5 + 2 + 3 + 8 = 20, the sum of tens is 3 + 9 + 1 + 1 = 14, and the sum of hundreds is 1 + 1 = 2. After that, the addition of 20, 140, and 200 is performed. Datta and Singh do not mention who the author of the book is. Presumably this is a commentary on the *Līlāvatī* by Rāmakṛṣṇa from the 15th century. This procedure does not require rubbing out. Datta and Singh say that this method was never used.

Multiplication[8]

> One should multiply the last [i.e., most significant] digit in the multiplicand by the multiplier, [and then the other digits], beginning with the next to last, by [the same multiplier] moved [to the next place].[9]

In the English translation several remarks are added in square brackets, as above, to make the translation clearer, so that the reader may understand the rule. A *verbatim* translation of the Sanskrit verse reads as follows:[10]

> The last in the multiplicand (*guṇyāntyam*) digit (*aṅkam*) by the multiplier (*guṇakena*) one should multiply (*hanyād*) by moved (*utsāritena*) similarly (*evam*) the penultimate and so on (*upāntimādīn*).

The verse gives an algorithm. It teaches that computations should be carried out in three steps:

Step 1: multiply the last digit in the multiplicand by the multiplier.
Step 2: rub out the last digit.
Step 3: move the multiplier one place to the right.

Example of Multiplication

The multiplicand is 135, the multiplier is 12.[11]

In the setting down of the numbers, the digit 5 in the multiplicand is written first, then 3 and 1 are written. Next, the rightmost digit 2 in the multiplier is written above the leftmost digit of the multiplicand, and then 1 is written to its left.

[Original configuration]	[Multiplier]	1	2		
	[Multiplicand]		1	3	5
Multiply the leftmost digit of the multiplicand (1) by the leftmost digit of the multiplier (1). Write the product (1) [under the leftmost digit of the multiplier]	[Multiplier]	1	2		
	[Multiplicand]	*1*	1	3	5

(continued)

---

[8]According to Gaṇeśa Daivajña, who wrote a commentary on the *Līlāvatī* in 1545, Indians seem to recite multiplication tables. He refers to a portion of a multiplication table from 1×1 to 1×10, 2×1 to 2×10, and 3×1 to 3×10, and calls it the "well-known reciting".

[9]Plofker (2007, p. 449, stanza 14).

[10]The original text reads *gunyāntyam ankamgunakena hanyād utsāritenaivam upāntimādīn.*

[11]The multiplier (12) is referred to with a word which literally means "the Sun." When numbers are given in a versified example, they are frequently indicated by words conventionally accepted as denoting those numbers. This system is very convenient for composing verses. Many synonymous words are used in order to satisfy the meter. A Sanskrit term *aṅka*, which means "number," indicates 9. For a list of word–numerals *see* Datta and Singh (1935[2001], vol. 1, pp. 54–57).

(continued)

Multiply the leftmost digit of the multiplicand (1) by the second (and, eventually, the rightmost) digit of the multiplier (2). Rub out the leftmost digit of the multiplicand (1).[a] Write the product (2) in this position	[Multiplier]	1	2		
	[Multiplicand]	1	2	3	5
Move the multiplier to the right for one position. The multiplicand becomes 35	[Multiplier]		*1*	*2*	
	[Multiplicand]	1	2	3	5
Multiply the leftmost digit of the multiplicand (3) by the leftmost digit of the multiplier (1). Add the product (3) to 2	[Multiplier]		1	2	
	[Multiplicand]	1	*5*	3	5
Multiply the digit 3 by the second (and, eventually, the rightmost) digit 2 of the multiplier. Rub out the leftmost digit of the multiplicand (3). Write the product (6) in this position	[Multiplier]		1	2	
	[Multiplicand]	1	5	*6*	5
Move the multiplier to the right for one position. The multiplicand becomes 5	[Multiplier]			*1*	*2*
	[Multiplicand]	1	5	6	5
Multiply the last digit (5) in the multiplicand by the leftmost digit of the multiplier (1). Add the product (5) to 56	[Multiplier]			1	2
	[Multiplicand]	1	*6*	*1*	5
Multiply the last digit (5) by the second (and, eventually, the rightmost) digit 2 of the multiplier. Rub out the last digit 5 in the multiplicand. Add the product (10) to the two rightmost positions	[Multiplier]			1	2
	[Multiplicand]	1	6	*2*	*0*

[a]The Sanskrit terms used for multiplication, originally meaning "killing" or "destroying," may imply rubbing out

Thus, we obtain 1620 as the product. Srīdhara calls this method "door-junction," i.e., junction of two doors.[12] Bhāskara gives different methods, presumably in order to provide easier or quicker computation:

> Or, the multiplicand is [set] down repeatedly, corresponding to the [separate] parts of the multiplier, [and] multiplied by those parts; [and the results are] added up. Or, the multiplier is divided by some [number], and the multiplicand is multiplied by that [number] and by the quotient; [that is] the result. Those are the two ways of dividing up the number in this manner. Or [when the multiplicand is] multiplied by [the multiplier's] separate digits, [the product] is added up [from those]. Or, [if the multiplicand is] multiplied by the multiplier decreased or increased by a given [number], [the product should be] increased or decreased by the multiplicand times the given [number].[13]

Bhāskara explains how to execute multiplication in his own commentary as follows. According to the first method, the multiplier is broken up into two or more parts whose sum is equal to it. Then the multiplicand is multiplied separately by those. The results are added together. For example, if one needs to calculate 135 × 12, he

---

[12]Derivation of this term is obscure. Fillioza refers to a type of wooden plank used in South India. Each plank is placed in a groove where it slides (2010, p. 47).

[13]Plofker (2007, p. 449, stanzas 14–16).

represents this product as $135 \times (4 + 8)$. Then $135 \times 4 = 540$ and $135 \times 8 = 1080$ are calculated separately, and added. According to the second method, the multiplier is broken up into two or more aliquot parts. For example, $135 \times 12 = 135 \times (12 \div 3) \times 3 = 540 \times 3 = 1620$. The mathematical text is not written in terms of ordering, but according to topics. One more method is "[when the multiplicand is] multiplied by [the multiplier's] separate digits, [the product] is added up [from those];" it is illustrated by the following example: $135 \times 12 = 135 \times 10 + 135 \times 2 = 1620$. Literally, the multiplier is "divided into places" (*sthānavibhāga*), respectively multiplied, and the partial products are added according to their "places" (*yathāsthānayute*). The partial products are placed as follows:

1	3	5	
	2	7	0

The last method mentioned in the excerpt corresponds to the following operations:

$$135 \times 12 = 135 \times (12-2) + 135 \times 2 = 1620; \text{ and}$$
$$135 \times 12 = 135 \times (12+8) - 135 \times 8 = 1620.$$

In his commentary mentioned above Ganeśa refers to the "door-junction method."[14] This is the lattice multiplication called the *gelosia* multiplication in medieval Europe. In the commentary, no figures are given. However, another Ganeśa, the author of *Ganitamañjalī*, provides figures.[15] It must be noted that the *gelosia* method does not involve rubbing out figures. An example of the multiplication of 135 by 12, using the *gelosia* method is given below.

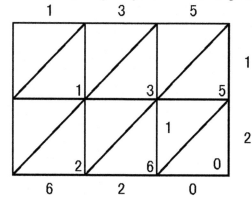

---

[14]Colebrooke (1817 [2005], p. 7) and Datta and Singh (1935 [2001], p. 145).

[15]*See* p. 29 in the edition of Hayashi (2013b) and the example of "multiplicand 125, multiplier 15." No information about this other Ganeśa is known. His father Dhundhirāja is believed to be a nephew of Jñānarāja (fl. AD 1503).

In his rule, Srīdhara states "multiply successively in the inverse or direct order." In his example, the multiplicand is 1296 and the multiplier is 21. This is the direct process explained in a commentary on the *Pāṭīgaṇita* of unknown date. For the convenience of the reader, I present the procedure in the following table:

Instructions	Configuration on the counting surface				
[The operands should be placed one above the other]				2	1
	1	2	9	6	
Multiply the first digit in the multiplicand, 6, by the first digit in the multiplier, 1. The product, 6, is to be placed under 1				2	1
	1	2	9	6	6
Then multiply 6 by 2. The product is 12. Rub out 6 under 21 and place 2 there; 1 is placed under 9				2	1
	1	2	9	2	6
			1		
Move 21 for one position to the left [The multiplicand thus becomes 129]			2	1	
	1	2	9	2	6
			1		
Then multiply 9 by 21. The product of 9 by 1 is added to 2 ...			2	1	
	1	2	9	1	6
			2		
... and the product of 9 by 2 is added to 2. Rub out 9			2	1	
	1	2	0	1	6
			2		
Move 21 one place to the left [The multiplicand becomes 12]		2	1		
	1	2	0	1	6
		2			
Multiply 2 by 21. 2 is multiplied by 1, ...		2	1		
	1	2	2	1	6
		2			
... 2 is multiplied by 2 and the product 4 is added to 2. Rub out 9		2	1		
	1	6	2	1	6
Move 21 for one place to the left [The multiplicand becomes 1]	2	1			
	1	6	2	1	6
1 is multiplied by 21 and the product is added to 6.	2	1			
	2	7	2	1	6

It must be noted that the third supplementary row is necessary in this direct process.[16] An anonymous commentator of *PG* says that the "door-junction method in the inverse way is easy to be done" (*vilomagatyā kavāṭasandhihsukara*). Bhāskara II selected the inverse way.

---

[16]Hayashi (2013a, p. 294).

Division

> In division, the divisor [multiplied by some number] is subtracted from the last [digit(s) of the] dividend; that multiplier is the result. But when possible, one should divide after having reduced the divisor and dividend by a common [factor].[17]

Example of Division

Divide 1620 by 12.

Instructions	Configuration						
Quotient[1]							
The divisor 12 is placed below the dividend 1620. Divide 16 by 12	Dividend	1	6	2	0		
	Divisor	1	2				
The quotient is 1 and the remainder is 4. Rub out 16 in the dividend and write remainder 4	Dividend		4	2	0	1	
	Divisor		1	2			
Move 12 for one place to the right[2]	Dividend		4	2	0	1	
	Divisor			1	2		
Divide 42 by 12. The quotient is 3 and the remainder is 6. Rub out 42 and write remainder 6	Dividend			6	0	1	*3*
	Divisor			1	2		
Move 12 for one place to the right	Dividend			6	0	1	3
	Divisor				*1*	*2*	
Divide 60 by 12. The quotient is 5. Rub out 60	Dividend					1	3	5
	Divisor				1	2	

[1]Bhāskara does not mention where to write the quotient, so the quotient 1 is placed in a separate column
[2]This operation is not mentioned in the rule

This operation must begin with the largest decimal digits.

Square

As other mathematicians before him, Bhāskara begins this rule with the definition of a square.

> The product of two equal [quantities] is called the 'square.' Now the square of the last [digit] is set down, and so are the subsequent digits multiplied by the last [digit] times two, each [produced] above [the place] of its own [digit]. [Then] when one has moved [to a fresh location] the quantity [to be squared], disregarding the last [digit], [the procedure is performed] again.[18]

---

[17]Plofker (2007, p. 450, stanza 18).
[18]Plofker (2007, p. 450, stanza 19).

Example of Squaring

Square 297. The intermediate results of computations are presented in the following table.

Instructions	Configurations on the counting surface				
Square the last digit 2: $2^2 = 4$. [Bhāskara does not mention where to set down the product, but the commentator Gaṇeśa says "above itself," therefore, the obtained 4 should be placed above the digit 2]	4				
	2	9	7		
Double the last digit, 2, [the result] is 4, which is [to be] multiplied by the subsequent digit, 9. The result is 36. Place 6 above 9 and add 3 to 4	7	6			
	2	9	7		
The 4 is multiplied by the digit subsequent to 9, that is, by 7, the number of units; [the result] is 28. Then write 8 above 7 and add 2 to 6	7	8	8		
	2	9	7		
One round of operations is completed. Rub out the last digit, 2, and move [the lower operand] for one position to the right	7	8	8		
		9	7		
Square the last digit, 9: $9^2 = 81$. Add 81 to the number above the last digit 9, that is, to 88	8	6	9		
		9	7		
Double the last digit, 9, that will be 18; it is to be multiplied by the subsequent digit, 7. The product is 126. Place 6 above 7 and add 12 to 69	8	8	1	6	
		9	7		
Another round of operations is completed. Rub out the leftmost digit, 9, and move [the remaining digit] for one position to the right	8	8	1	6	
					7
Square the last digit, 7: $7^2 = 49$. Place 9 above the last digit, 7. Add 4 to 16. Erase 6 and write 20	8	8	2	0	9
					7
Erase 7 [The result is set on the instrument]	8	8	2	0	9

The mathematical basis of this method is as follows. If x is written as a number abc where a, b, and c occupy the positions of hundreds, tens, and units, that is, $x = 100a + 10b + c$, then $x^2 = a^2 \cdot 10^4 + 2ab \cdot 10^3 + 2ac \cdot 10^2 + b^2 \cdot 10^2 + 2bc \cdot 10 + c^2$.

Bhāskara gives different methodologies as he did in the case of multiplication:

> Or the square of two [separate] parts [of a given number] is their product multiplied by two, added to the sum of the squares of those parts. Or, the square is the product of [two equal] numbers [separately] increased and decreased by a given [quantity], added to the square of that given [number].[19]

He gives two methods. The first is if $x = a + b$, then $x^2 = 2ab + (a^2 + b^2)$. The second is $x^2 = (x - a)(x + a) + a^2$. Bhāskara gives examples and explains these procedures in his auto-commentary:

---

[19]Plofker (2007, p. 450, stanza 20).

$9^2 = (4+5)^2 = 2 \times 4 \times 5 + 4^2 + 5^2;$
$14^2 = (6+8)^2 = 2 \times 6 \times 8 + 6^2 + 8^2 = (10+4)^2 = 2 \times 10 \times 4 + 10^2 + 4^2 = 196;$ and
$297^2 = (297-3)(297+3) + 3^2 = 294 \cdot 300 + 9 = 88,200 + 9 = 88,209.$

He also asks for the square of 10,005, and gives the answer only. It seems reasonable to suppose that Bhāskara may have used the two following methods:

1. $10,005^2 = (10,000 + 5)^2 = 2 \times 10,000 \times 5 + (10,000^2 + 5^2).$
2. $10,005^2 = (10,005 + 5)(10,005 - 5) + 5^2.$

Square Root

Having subtracted the [largest possible] square from the last 'odd [decimal] place' [i.e., from the one- or two-digit multiple of the highest even power of 10 contained in the given number], multiply the square-root [of the subtracted quantity] by two. When the [next] 'even place' is divided by that,[20] after subtracting the square of the [integer] quotient from the next 'odd place' after that, set down two times the quotient in the 'row' [of successive digits of the answer]. When the [next] 'even place' is divided by the 'row,' after subtracting the square of the quotient from the next 'odd place,' set down that result multiplied by two in the 'row;' [do] this repeatedly. Half of the 'row' is the [desired] square-root.[21]

An "odd place" is one which represents an odd power of ten (e.g., 10^1, 10^3, 10^5, etc.), and an "even place" is one which represents even powers of ten (e.g., 10^0, 10^2, 10^4, etc.).

The calculation involves four steps:

Step 1: one has to subtract the largest square from the number in the leftmost "odd place" and to double the root;
Step 2: one has to divide the number in the "even place" by it;
Step 3: one has to subtract the square of the quotient from the number in the previous "odd place;"
Step 4: one has to set down the doubled quotient in the row containing the root and divide the number in the "even place" by it.

An example of the procedure is shown in Table 1.

The mathematical rationale of this method is as follows: $(10a + b)^2 = a^2 \cdot 10^2 + 2ab \cdot 10 + b^2.$

---

[20]Here, to operate upon the "next place" means to use the remainder from the previous operation with an additional least significant digit brought down from the subsequent decimal place.

[21]Plofker (2007, p. 451, stanza 22).

Table 1 Extraction of the square root of 88,209

Operation						(Partial) result	Comment
Mark odd and even places[1]	o	e	o	e	o		
Set the number	8	8	2	0	9		Given number
From the digit in leftmost odd position (8) subtract the largest square smaller than 8 (2^2)	*4*	8	2	0	9		Remainder
Place the partial result (2) in a separate position						2	
Double the (partial) result						4	Root
Divide 48 in even place by the doubled root (4); since the quotient is larger than 10, take the largest digit (9) as the substitute of quotient; subtract the product of the quotient and the (partial) root (36) from 48209	*1*	2	2	0	9		Remainder
Subtract the square of the substitute of quotient ($9^2 = 81$) from 122 (starting in odd place)		*4*	*1*	0	9		Remainder
Double the substitute of quotient and add it to 4 (interpreted as 40)						40 + 18 = 58	Root
Divide 410 starting in even place by 58; the result is 7 and the remainder is 4				*4*	9	7	Quotient
Subtract 7^2 from 49					0		Remainder
Double the quotient (7) and add it to 58 (interpreted as 580)						580 + 14 = 594	Root
Divide the result of addition (594) by 2						297	Result
Square root: 297							

[1] I use the letters "o" and "e" for odd and even places, respectively

Cube

> And the product of three equal [quantities] is defined as [cube]. The cube of the last [digit] is set down, and then the square of the last multiplied by the first and by three, and then the square of the first multiplied by three and by the last, and also the cube of the first. All [those] are added up according to their different place-values; [that] is the cube [of a two-digit number]. [Or if one] considers that [quantity] as [split into] two parts, the last [part may be divided] in the same way repeatedly. Or, in finding squares and cubes, the procedure may be performed [starting with] the first digit.[22]

The mathematical basis of this method is as follows: let a denote the last digit, and b the first, $(10a + b)^3 = a^3 \cdot 10^3 + 3a^2b \cdot 10^2 + 3ab^2 \cdot 10 + b^3$. The positional decimal calculation without zero is done in this way.

---

[22]Plofker (2007, p. 451, stanzas 24–25).

Example of Cube

Calculate 125^3. First of all, one should find the cube of 12. Now, 1 is the last digit and 2 is the first digit. Therefore, according to the procedure, the following operations should be performed.

1^3	1			
$3 \times 1^2 \times 2$		6		
$3 \times 2^2 \times 1$		1	2	8
2^3	1	7	2	8

So, we calculate that $12^3 = 1728$, yet the number 125 is a combination of 12 dozens and 5 units; the same algorithm can be applied to it.

12^3	1	7	2	8			
$3 \times 12^2 \times 5$		2	1	6	0		
$3 \times 5^2 \times 12$				9	0	0	
5^3					1	2	5
	1	9	5	3	1	2	5

So, 125^3 is equal to 1,953,125. Bhāskara mentions a method starting with the first digit. Let us try it here.

5^3			1	2	5	
$3 \times 5^2 \times 2$		1	5	0		
$3 \times 2^2 \times 5$		6	0			
2^3		8				
	1	5	6	2	5	

That is, $25^3 = 15,625$. Next, the following computations should be performed.

25^3			1	5	6	2	5
$3 \times 25^2 \times 1$		1	8	7	5		
$3 \times 1^2 \times 25$		7	5				
1^3		1					
	1	9	5	3	1	2	5

Bhāskara provides two alternative methods:

> Or the quantity is multiplied by [each of its] two parts, multiplied by three, and added to the sum of the cubes of the parts. [Or] the cube of the square root, multiplied by itself, is the cube of the square number.[23]

---

[23]Plofker (2007, p. 451, stanza 26).

The mathematical basis for the first method is: $(a + b)^3 = 3ab(a + b) + a^3 + b^3$. For example, let us take $a = 4$ and $b = 5$. Then $9^3 = (4 + 5)^3$, $ab(a + b) = 4 \times 5 \times 9 = 180$, $3ab(a + b) = 180 \times 3 = 540$, and $3ab(a + b) + a^3 + b^3 = 540 + (4^3 + 5^3) = 540 + 189 = 729$. In another example we have $27^3 = (20 + 7)^3$, $3 \times 20 \times 7 \times 27 = 11,340$; $20^3 + 7^3 = 8,343$; $11,340 + 8,343 = 19,683$.

The second method is based on the identity $a^3 = ((\sqrt{a})^3)^2$. For example, $4^3 = ((\sqrt{4})^3)^2 = (2^3)^2$; $64^3 = (8^3)^2$; $9^3 = ((\sqrt{9})^3)^2 = (3^3)^2 = 27^2 = 729$, etc.

Cube Root

> The first [decimal place] is a cube place; then there are two non-cube [places], and so forth. When one has subtracted from the highest cube [place] the [greatest possible] cube, the [cube] root is put down separately. Divide the next [place] by the square of that [cube root] multiplied by three; set the result in the 'row.' Subtract the square of that [quotient] times three, multiplied by the last [digit of the root], from the next [place]; subtract the cube of the quotient from the next [place] after that. [Proceeding] in the same way repeatedly, the 'row' becomes the cube root.[24]

In the square root extraction procedure, the "odd places," that is, positions of units, hundreds, dozens of thousands, and so on were considered positions related to the squares of the units, dozens, hundreds, etc., of the root. Similarly, in the procedure of cube root extraction the position of units was considered the first "cube position" corresponding to the positions of units of the root; the position of tens was the "first non-cube position," and the position of hundreds was the "second non-cube position." The second "cube position" was that of thousands; it corresponded to the position of dozens of the root. In Table 2 symbol c stands for "cubic positions" (i.e., 10^0, 10^3, 10^6, etc.), $n1$ for the "first non-cube positions" (i.e., 10^1, 10^4, 10^7, etc.), and $n2$ for the "second non-cube positions" (i.e., 10^2, 10^5, 10^8, etc.).

The computation involves three steps. In the c place, subtract a cube. In the $n2$ place, divide by three times the square. In the $n1$ place, subtract the product of three times the square and the quotient. The mathematical basis of this method is as follows: $(10a + b)^3 = a^3 \cdot 10^3 + 3a^2b \cdot 10^2 + 3ab^2 \cdot 10 + b^3$.

Eight Operations of Fractions

Bhāskara described eight operations with fractions. At the outset he presents the rules for the reduction of fractions having different denominators to fractions having a common denominator.[25] To make b/a and d/c have a common denominator, we can multiply the numerator and the denominator of one fraction by the denominator

---

[24]Plofker (2007, p. 452, stanzas 28–29).

[25]Usually the denominators were called "divisors".

Table 2 Extraction of the cube root of 1,953,125

Operation	c	n2	n1	c	n2	n1	c	(Partial) result	Comment
Mark cube and two non-cubes	1	9	5	3	1	2	5		Given number
From the digit in leftmost cube position (1) subtract the largest cube (1^3)		9	5	3	1	2	5		Remainder
Place the partial result (1) in a separate position							1		
Divide the digit in the leftmost $n2$ position (9) by tripled square of the partial result (3×1^2); subtract 1 from the quotient (3) to obtain the next digit of the root (2).[1] Subtract 2· (3×1^2) from 9		3	5	3	1	2	5		
Produce the partial result								$1 \times 10 + 2 = 12$	Root
Multiply the tripled square of the digit obtained at the previous step by the first partial result (i.e., calculate $3 \times 2^2 \times 1 = 12$), subtract it from 35 in positions $n2 - n1$		2	3	3	1	2	5		Remainder
Subtract 2^3 from 233 in positions $n2 - n1 - c$.		2	2	5	1	2	5		Remainder
Divide 2251 in positions $n2 - n1 - c - n2$ by tripled square of the partial result $(3 \times 12^2 = 432)$. Quotient is 5 and remainder, 91				9	1	2	5		
Produce the (final) result								$12 \times 10 + 5 = 125$	Root
Multiply the tripled square of the digit obtained at the previous step by the second partial result (i.e., calculate $3 \times 5^2 \times 12 = 900$), subtract it from 912 in positions $c - n2 - n1$					1	2	5		
Subtract 5^3 from 125 in positions $n2 - n1 - c$							0		
Cube root: 125									

[1] The quotient is 3. However, tripled square of the quotient multiplied by the digit computed at this step should be subtracted from 9 and therefore should be less than 9. Take the largest number that (a) satisfies these conditions and (b) is equal or less the obtained quotient. This number is 2

of the other, that is, transform the fractions into bc/ac and ad/ac. The notation used to write down numbers with fractional parts was as follows: the integer part was written above, the numerator was written under it, and the denominator was written under the numerator. There was a special symbol (dot) placed above the numerator meaning that the fraction had to be subtracted from, and not added to, the integer part. For example, $2 + \frac{1}{4}$ and $2-\frac{1}{4}$ would be presented as follows:

$$
\begin{array}{cc}
2 & \overset{\bullet}{2} \\
1 & 1 \\
4 & 4
\end{array}
$$

The symbol • was used for negative numbers in the *bījaganita* texts.[26]

Eight Operations with Zero[27]

The Indians used the decimal place-value system both in notation and in calculation. A symbol indicating a void was essential:

In addition, zero [produces a result] equal to the added [quantity], in squaring and so forth [it produces] zero. A quantity divided by zero has zero as a denominator; [a quantity] multiplied by zero, is zero and [that] latter [result] is [considered] "[that] times zero" in

---

[26]*Bījaganita* 5. *See also* Colebrooke (1817 [2005], p. 131).

[27]The operations first appeared in chapter 18 of the *BSS*. It can be explained as follows. In the setting down process, the first syllable of each technical term is used to indicate the number given in the example. For instance, in an example of series where the first term (*ādi*), say, 1, and the common dfference (*uttara*), 2, are given, the setting down is: ā 1 u 2. This sort of setting down is also used in the mathematical field of *bījaganita*. In chapter 18, stanza 48, a formula for the quadratic equation, $x = (\sqrt{(4ac + b^2)}-b)/2a$, when $ax^2 + bx = c$; $a, b, c > 0$, is given (*see* Colebrooke 1817 [2005], p. 346). In the setting down process the first syllables of *rūpa* (quantity), of *yāvattāvat* (literally meaning "as much as," but used for an unknown), and of *varga* (square) were used. The equation $ax^2 + bx = c$ was set down in the form:

yāva	a	yā	b	rū	0
yāva	0	yā	0	rū	c

When an equation $a_1x^2 + b_1x + c_1 = a_2x^2 + b_2x + c_2$ was set down as:

yāva	a_1	yā	b_1	rū	c_1
yāva	a_2	yā	b_2	rū	c_2

In order to use the formula, the following operation $(a_1 - a_2)x^2 + (b_1 - b_2)x = c_2 - c_1$ $(a_1 \neq a_2)$ had to be completed. This would be set down as follows:

yāva	a_1-a_2	yā	b_1-b_2	rū	0
yāva	0	yā	0	rū	c_2-c_1

When any of a_i, b_i, or c_i were negative, the sign referring to the negative values was used; when any of them was equal to zero, a zero sign was used, or an empty space was left. I do not discuss here what symbol Bhāskara II used for zero. One of the Sanskrit terms denoting zero in the word–numeral system was *bindu*, which means "dot" or "drop".

subsequent operations. A [finite] quantity is understood to be unchanged when zero is [its] multiplier if zero is subsequently [its] divisor, and similarly [if it is] diminished or increased by zero.[28]

Bhāskara gives the following rules. Let a be any number, then $a + 0 = 0 + a = a$, $a^2 = 0$, $\sqrt{0} = 0$, $0^3 = 0$, and $\sqrt[3]{0} = 0$. When a is divided by 0, it is called *khahara*, that is "zero-divisor" or "zero-denominator." $a \times 0 = 0$. When $a \times 0$ is used in calculation, it is a "zero-multiplier." Bhāskara may have thought that $a \times 0 \div 0 = a \div 0 \times 0 = a$. He goes on to give an example and solves the equation $(x \times 0 + (x \times 0)/2) \times 3 \div 0 = 63$, giving an answer of 14.

Concluding Remarks

In this chapter I have presented documentary evidence that illuminates how the nine numerals and zero were used in medieval India. The main feature of such computations was the use of numerals. When students calculated they rubbed out numerals at each step and wrote the obtained results in the same place. In multiplication, for instance, they wrote the digits of the multiplicand first and later replaced them with the digits of the product. This study could support the theory which states that it must have been easy to rub out the operands and replace them with newly obtained values. Datta and Singh believed that the computations that involved rubbing out were performed using dust boards. However, one can raise questions concerning the use of a dust board. What did the dust board look like? A modern statue of Bhāskara II portrays him holding a board on his knees. However, it would have been difficult to perform computations with a dust board in this way. Hayashi (2013a, p. 293) presumes that the calculations were made on a sort of laptop blackboard with a piece of chalk or on the ground with a stick. According to the study by Sarma, not only schoolboys but also court scribes used wooden boards. The wooden tablet is a well attested writing medium in ancient India.[29] It is well known that the Hindu numeral system was transferred to Arab, and later, Latin scholars.[30] Equipment like a dust board is referred to as *takht* in Arabic or *tabula* in Latin.[31] Uqlīdisī, who wrote the earliest extant Arabic work on Indian arithmetic (AD 952/3), states: "In Book Four we present all the arithmetic of the Indians that has been done on the *takht*, but here with no *takht* and no erasing; we carry it out on a sheet of paper, thus dispensing with the dust and the board. Nevertheless the

---

[28]Plofker (2007, p. 453, stanzas 45–46). Cf. *MS* 15.10.11, *SS* 14.6.

[29]Cf. Sarma (1985, pp. 1–2).

[30]For the history of its transmission, *see* Kunitzsch (2003) and Burnett (2006).

[31]For example, in the *Liber Abbaci* of Leonardo Pisano, *see* Boncompagni's edition, p. 7: "in tabula dealbata in qua littere leuiter deleantur." Sigler (2002) translates this as "in the chalk table in which the letters are easily deleted."

working is simpler, quicker and of less cost because here we need only an inkpot and sheet of paper."[32] In India computations without rubbing out appeared in the 15th century.

The Sanskrit word that means "study" is *abhyāsa*, which literally means "repeated practice." Mnemonic verses assisted Indian students who studied mathematics, because they memorized the rules and the examples.[33] The *Līlāvatī* consists of various meters. The book is also a good text for stanzas.

Acknowledgements Annotations to Japanese translation of the *Līlāvatī* by Hayashi Takao were useful to me.

References

Āpaṭē, Vināyaka Gaṇeśa. (ed.). 1937. *Līlāvatī* edited by Ganeśa's *Buddhivilāsinī* and Mahīdhara's *Līlāvatīvivarana*. Anandāśrama Sanskrit Series, No. 107, Poona.

Burnett, Charles. 2006. The semantics of Indian numerals in Arabic, Greek and Latin. *Journal of Indian Philosophy* 34: 15–30.

Colebrooke, Henry Thomas. 1817 [2005]. *Classics of Indian mathematics: Algebra, with arithmetic and mensuration, from the Sanskrit of Brahmagupta and Bhaskara*. Translated by H. T. Colebrooke, with a foreword by S. R. Sarma. Delhi: Sharada Publishing House (Originally published in 1817).

Datta, Bibhutibhushan, and Singh, Avadhesh Narayan. 1935 [2001]. *History of Hindu mathematics*, Vol. 1. Lahore: Motilal Banarsi Das; reprint: Delhi, Bharatiya Kala Prakashan.

Fillioza, Pierre-Sylvain. 2010. Modes of creation of a technical vocabulary: the case of Sanskrit mathematics. *Ganita Bhāratī* 32 (1–2): 37–53.

Gupta, Radha Charan. 2002. Chapter 18: "India". In *Writing the history of mathematics: Its historical development*, ed. Joseph W. Dauben, and Christoph J. Scriba. Basel: Birkhäuser.

Hayashi, Takao. 2013a. The *Ganitapañcaviṃśī* attributed to Srīdhara. *Revue d'histoire de mathématiques* 19: 245–332.

Hayashi, Takao (ed.). 2013b. *Ganitamañjalī* of Ganeśa. *Indian Journal of History of Science*, 1–122.

Keller, Agathe. 2006. *Expounding the mathematical seed: A translation of Bhāskara I on the mathematical chapter of the Aryabhatīya*. Basel: Birkhäuser.

Kunitzsch, Paul. 2003. The transmission of Hindu–Arabic numerals reconsidered. In *The enterprise of science in Islam: New perspectives*, ed. J.P. Hogendijk and A.I. Sabra, 3–21. Cambridge, London: The MIT Press.

Plofker, Kim. 2007. Mathematics in India. In *The mathematics of Egypt, Mesopotamia, China, India and Islam: A sourcebook*, ed. V.J. Katz, 385–514. Princeton University Press: Princeton.

Plofker, Kim. 2009a. *Mathematics in India*. Princeton: Princeton University Press.

Plofker, Kim. 2009b. Sanskrit mathematical verse. In *The Oxford handbook of the history of mathematics*, ed. E. Robson, and J. Stedall, 519–536. Oxford: Oxford University Press.

Saidan, Ahmad Salim. 1978. *The arithmetic of Al-Uqlīdisī: The story of Hindu-Arabic arithmetic as told in Kitāb al-Fuṣūl fī al-Ḥisāb al-Hindī*. Dordrecht: Springer.

---

[32]Translated by Saidan (1978, p. 36).

[33]For metric verse in Sanskrit texts, *see* Plofker (2009b).

Sarma, Sreeramula Rajeswara. 1985. Writing material in ancient India. *Aligarh Oriental Series* 5: 1–22.

Shukla, Kripa Shankar (ed.; trans. by). 1959. *Pāṭīganita*. Lucknow: Lucknow University Press.

Sigler, Laurence E. (Trans. by). 2002. *Fibonacci's Liber Abaci*. New York: Springer.

Staal, Frits. 2010. On the origins of zero. In *Studies in the history of Indian mathematics*, ed. C.S. Seshadri, 39–53. New Delhi: Hindustan Book Agency.

Kusuba Takanori 楠葉隆徳 is a Professor of History and Philosophy of Science at Osaka University of Economics. He works on the history of Indian and Arabic exact sciences. He published a critical edition with English translation and commentary of the last two chapters (on combinatorics and magic squares) of the *Gaṇitakaumudī* written by Nārāyaṇa Paṇḍita in 1356; *Arabic Astronomy* in Sanskrit with David Pingree, (Leiden 2002); a critical edition with the English translation and mathematical commentary of the *Gaṇitasārakaumudī*, a middle Indic mathematical text composed in the early fourteenth century, with S. R.Sarma, T. Hayashi and M. Yano (Delhi 2009); articles with N. Sidoli on the Arabic revisions of works by Menelaus and Theodosius.

Reading Algorithms in Sanskrit: How to Relate Rule of Three, Choice of Unknown, and Linear Equation?

Charlotte-V. Pollet

Abstract Texts in India were transmitted in the context of oral transmissions. A consequence is that their transmission implied memorization. In the written texts that are available to us, it is difficult to locate a precise section on a specific topic if the reader does not know in advance where this section is supposed to be written. A verbal knowledge of the text is required prior to a written one. Algebraic texts in Sanskrit are thus shaped like lists of operations where the progress of the algorithm seems, at first sight, more valued than its understanding. Nevertheless, the prescription of algorithmic operations delivers some clues as to what the author expected his readers to do or to understand. This chapter presents an excerpt from the *Bījagaṇitāvataṃsa* written by Nārāyaṇa in 14th-century India. The example prescribes an algorithm to set up a linear equation by means of a Rule of Three. Yet, the algorithms presented in the general rule and the one used in the commentary are slightly different. These differences reveal a deep understanding and a deductive interpretation of the rule by the commentator. Tabular setting, choosing unknowns, and Rule of Three played a key role in the elaboration of the reasoning.

Keywords Algorithm · Unknown · Rule of three · 14th-century India
Nārāyaṇa

Introduction

In the Indian context, we know that mathematical texts composed in Sanskrit emerged from an environment of orality (Filliozat 2004), where ancient and medieval mathematical texts were versified for the purpose of memorization and recitation. A voluminous mass of texts has been transmitted from generation to generation without first being written. Filliozat (2004, p. 149) assumed that texts

C.-V. Pollet (✉)
Center for General Education, National Chiao-Tung University,
1001 University Road, Hsinchu 30050, Taiwan
e-mail: charlotte.pollet7@gmail.com

© Springer Nature Switzerland AG 2018
A. Volkov and V. Freiman (eds.), *Computations and Computing Devices in Mathematics Education Before the Advent of Electronic Calculators*,
Mathematics Education in the Digital Era 11, https://doi.org/10.1007/978-3-319-73396-8_6

were composed in the context of instruction where the master explained orally the *sūtra* (that is, treatises usually written as poems) and the disciples memorized the *shlokas* (two-line strophes) of the *sūtra* and the content of the oral explanation. He analyzed two examples, one borrowed from the *Śulba-sūtra* (Sen and Bag 1983, verses II. 64–65) probably composed between 700 and 500 BC, and one from the *Āryabhaṭīya* (verse 2) by Āryabhaṭa from the 5th century AD with an excerpt of its commentary by Bhāskara I (7th century AD) composed in AD 629 (Shukla 1976). Indian mathematical texts are literary texts written in metrical form imitating poetic texts. The metrical form is a device for memorization.

It could be that the mathematical texts available now are also the result of a tradition of teaching. Datta and Singh (1935, vol. 2, p. 126) gave some hypothetical description of teaching for the field of mathematics called *pāṭīgaṇita* (literally "board-computation," sometimes translated as "arithmetic"). Datta and Singh do not provide information for the other field, *bījagaṇita* (literally "seed-computation," sometimes translated as "algebra"). They seem to assume that the difference between *pāṭīgaṇita* and *bījagaṇita* learning was a matter of degree of difficulty. It is quite difficult to be more specific regarding type and purpose of education or level of understanding. This raises the question of the role of written texts—which were produced in big quantities in India—in an oral tradition of transmission. The consequence is that we also do not know precisely how people were supposed to work with the algorithms prescribed in written texts.

Texts are shaped like lists of operations where fidelity to the content and to the execution of the algorithms seems, at first sight, more valued than innovation for new algorithms. Repetition and memorization were often involved to work with this type of text. Nowadays, these pedagogical processes are despised for not encouraging learner understanding. As it is a question of memorization and recitation, it is often believed that practitioners who read algorithms composed as lists of instructions just needed to follow the prescription without understanding them. The idea is that an algorithm is a practical recipe, which requires no reasoning. This chapter will challenge this idea. Algorithms are not simple lists of operations. They are texts whose specificity is that they are not narration. They are not mere descriptions of mathematical recipes and their results. They are first witnesses of nonverbal knowledge and they require a context to be understood. That is to say they imply specific action and specific understanding. The same operation can imply different understandings. This chapter will examine an example where observation of algorithms shows a predilection for tabular representation, the use of a certain type of equation as a model, and construction based on algorithms containing multiplication and division. The prescription of algorithmic operations delivers some clues as to what the author expected his readers to do or to understand.

The question is thus addressing the function of tabular setting. Can tabular settings be considered as computing devices? It is probable that Indian mathematicians were performing their computations on sand or on boards covered with dust. This was inferred from the word *pāṭīgaṇita*, which is also a synonym of *dhūlī-karma* (literally "dust-work"). This indicates that figures may have been written on dust spread on a board (Datta and Singh 1935, vol. 2, p. 123). Yet, we have no trace of ancient representation or description of the use of these materials. Indian dust writings may represent given parameters and variations of these parameters. They register transformed information and play a role in the way calculations were performed. Yet, it is difficult to directly assimilate tabular setting to a computing device. The problem is that we also have no trace of these devices, only tabular settings in written texts remind us of their existence. Tabular settings inserted in written texts cannot be manipulated, but they are tracks and records of manipulations. By studying an example, we can thus raise the question of the status of tabular setting as mere illustration.

This example is an excerpt from the *Bījagaṇitāvataṃsa* written by Nārāyaṇa in 14th-century India. Information on Nārāyaṇa is scarce. He composed two books, each dealing with the two major fields of Indian mathematics: *Gaṇitakaumudī* (*Moonlight of Mathematics*, abbreviated later as GK) and *Bījagaṇitāvataṃsa* (*Garland of Seed-Mathematics*, abbreviated later as BGA). The colophonic verse of the GK indicates that he was working in the middle of the 14th century and the distribution of the available manuscripts leads us to think that he lived presumably somewhere in Northern India (Kusuba 1993, pp. 1–3). I will present a simple example and a *sūtra* on how to set up an algebraic equation borrowed from the second part of the BGA.

Description of Nārāyaṇa's Algorithms

The extant piece of the BGA is composed of two parts. This text is now still incomplete. Both available parts are composed of versified *sūtra*s accompanied by prose examples with their solutions given in the commentary. Each *sūtra* states a mathematical rule in a very concise style. The commentary explains and illustrates the rule. The commentary is attributed to Nārāyaṇa himself (Hayashi 2004). The first part of the BGA enumerates objects and operations involved in the construction of polynomials. The second details a procedure for setting and solving linear equations and gives a list of 40 examples to illustrate it. The 40 problems of the second part of the BGA refer to the same vocabulary to enunciate operation. Despite its variations due to data problems, the procedure is systematic. This is an algorithm which works with symbolic values, more than with numerical ones. Constants and unknowns are symbolized by syllables of the Sanskrit alphabet.

This second part of the BGA attracted my attention due to its prescription for the setting up and solving a linear equation (verses 3–7 in the Appendix to this chapter). It is followed by several examples (the Appendix provides only example 1 referred as E1) with four solutions given in the form of commentary (the Appendix only provides two of the four solutions, namely 1a and 1b). Each of the examples contains one or several small horizontal tables ordering the different terms of polynomials or equations. Strangely, the unknown chosen for the example does not exactly follow the general rule presented in the *sūtra*. The commentator often modifies a parameter concerning the unknown in order to ensure the algorithm generates systematically a linear equation. Moreover, the commentator refers automatically to an operation which is not mentioned in the *sūtra* and which guides the order of terms in tables, namely a Rule of Three (*trairāśika*), that is, a series of operations where a multiplication is followed by a division applied to objects that have a relation of proportion. The non-required operation is used more than twice: to set up each polynomial in the equation and also to solve it. It seems that there is a rule on one side and a practice on the other.... This is what I will explain.

I have transformed the order of presentation of the Sanskrit text in order to highlight the discrepancy between the *sūtra* and its commentary. The translation presented here is based on a critical edition of two manuscripts available to me thanks to the courtesy of Professor Hayashi Takao. I have used copies of two manuscripts from the Sarasvati Bhavana Texts collection preserved in Benares Sanskrit College, referred to hereafter as B1 and B2. I will start first with example E1 and insert the *sutra*, broken down into parts, inside the translation of its commentary. Following the algorithm, I will describe three aspects: (1) choice of unknown, (2) tabular settings, and (3) use of the Rule of Three. I will explain later how they are related. Below is a translation of E1 followed by the *sūtra* and a comparison with its commentary.

> (E1)[1] A merchant has eight horses of the same price and six hundred *rūpas*, and another merchant has [horses] measured by the Sun (twelve)[2] and a debt of two hundred *rūpas*.[3] The two [merchants] have equal properties. What is the price of a horse?

The example involves two merchants and their properties. The first one has 8 horses and 600 rupees, the second 12 horses and a debt of 200 rupees. One is asked the price of a horse and the value of the properties of each merchant. In this example the price of a horse is what we need to know. Literally, it is unknown. The *rūpa* is a monetary value used to name the constant. I transcribe the unknown quantities, the properties, as x_1 and x_2, and the price of a horse as x. We also need to assume y_1 and y_2, the prices of 8 and 12 horses, respectively. If so, we can transcribe the situation as:

---

[1] For the verse, I use the numbering used by Hayashi (2004). I added the letter E to distinguish verses which are examples from verses which are rules. The latter are marked by a number in brackets only. The numbering of the commentary into several parts is mine. Here, for this first example, I distinguish two parts 1a and 1b. The translation of verses is borrowed from Hayashi (2004) with slight modifications of mine. For the purpose of this chapter, I exceptionally replace Sanskrit terms kept by Hayashi with their modern interpretations and keep the Sanskrit terms in round brackets. Square brackets are used to mark words that were added to the Sanskrit text for the purpose of readability. Therefore, it reads "unknown" for *avyakta*, "as much as" for *yāvattāvat*, and "constants" for *rūpa*. The translation of the commentary is mine.

[2] Sanskrit uses metaphorical substitutes to name quantities for the purpose of metric versification. In the present case, "the Sun" means 12.

[3] Here I do not translate the word *rūpa*, in order to keep its evocation of monetary value. Later, in the *sūtra* and algorithm, the term is directly translated as "constant" for mathematical purposes.

$$x_1 = x_2$$
$$x_1 = y_1 + 600$$
$$x_2 = y_2 - 200$$

Choice of Unknown

Interestingly, the first verse of the *sūtra* starts with the identification of unknowns.

Verse (3): "Having assumed the value of the unknown (*avyakta*) [quantity] to be marked by one or several 'as much as' (*yāvattāvat*), increased or decreased by some constants (*rūpa*)."

The *sūtra* requires one to first state what is unknown (or literally "invisible," *avyakta*) and to symbolize this quantity by the first Devanagari[4] letter "*yā*" of the expression "*yāvattāvat*" (literally "as much as"). If there is one unknown $1x$ then "*yā* 1" is recorded, and $2x$ is recorded as "*yā* 2." If there is a second unknown, another Devanagari letter is used. In the present case, everything is done in order to have equations with only one unknown.[5] Indeed, the commentator makes an assumption as to what the unknown should be: "*Here the price of a horse is the unknown*" and he sticks to this choice for each of the solutions (1a and 1b here) presented in example E1. In 1a, he even explicitly expresses the unknown with the initial letter *yā* of *yāvattāvat*: "*Its value is a yāvattāvat, yā 1.*" Then a constant, marked by the Devanagari letter *rū*, can be added (literally "increased" in translation) or subtracted ("decreased") from it to construct expressions of the form *yā* A *rū* B as stated in the *sūtra*.

The situation is clear in 1a. The two expressions that come to mind are then *yā* 8 + *rū* 600 and *yā* 12 − *rū* 200. Surprisingly, this is not exactly what happens in the commentary. First, the commentator constructs an expression by means of a Rule of Three instead of directly recording the expressions *yā* 8 + *rū* 600 and *yā* 12 − *rū* 200 in solution 1a. Second, in solution 1b, the commentator creates another situation where $x = s + 1$, which is described later. That is, the commentator makes "variations" on the unknown: "*Moreover, the price of a horse is assumed to be yā 1 rū 1.*" This solution was not asked for in the statement of problem E1. It seems that there is a distinction between "types" of unknowns. One is the unknown quantity one looks for from the statement of the problem—namely the one I noted x. The other is a "variation" on the unknown chosen by the commentator to operate on the algorithm—namely the one I noted s. Now, x and s can be the same ($x=s$, in solution 1a) or different ($x = s + 1$, in solution 1b) for the same problem (E1). Solution 1b was not required by the example and thus brings into question the

---

[4]Devanagari is an alphasyllabary alphabet of India and Nepal.

[5]In the available parts of the BGA, there is only one case of several unknowns (E30). The commentator introduces this exceptional example as "an example of a certain person."

justification of these two situations ("variations" on the unknown and use of the Rule of Three). I suggest that the two situations are related, as we will see later.

Tabular Settings

Once the quantity and "variations" are identified, the procedure is stated as follows:

> Verse (4): "one should perform the computation upon that value according to the statements of the questioner. In order to obtain the result, the two sides (*pakṣa*) should be made equal (to each other) carefully."

The commentator performs the operations required by the procedure according to values stated in the problem. In 1a, the two "sides" are immediately equal. The commentator puts the two "sides" one above the other in a tabular setting (*nyāsa*) in order to state their equality. The finality (*artha*) of this setting is the "equal subtraction" or "uniform subtraction" (*samaśodhana*). The recurrent expression is: *samaśodhanārtha-nyāsa*, that is, a tabular setting of the final stage of uniform subtraction.

In the example above quoted from E1, the setting of a *samaśodhana* is composed of two *pakśa*. The term *pakśa* means literally "wings," or "sides." In the present example, the "sides" are placed one above the other. In many examples from the BGA, the "sides" are not represented. The author refers to them as "*sama iti pakṣau*," "when the two sides are equal" (E5, E11, E19, E20, E24, E26, E27, and E32), an operation leading to the equation. Does the term refer to the position on the tabular setting or to the object placed in the setting? Is the term abstracted from its original spatial meaning? Is it a step toward conceptualization? The BGA does not provide enough elements for answers to these questions.

We only know that there are graphical representations of equality between polynomials. This setting down in two rows is a way of stating the equality between the two expressions. The terms in the position of unknowns and constants can then be subtracted. But the equation resulting from the subtraction does not appear in a tabular setting. The subtraction of the constant and the unknown are treated separately and rhetorically. Thus, in E1, the expected expression resulting from the subtraction of *yā* 4 from *rū* 800 never appears. There is nothing equivalent to $4x - 800 = 0$ or $4x = 800$. The way to solve the equation is given directly, that is, in the tabular setting we see two polynomials but not the final equation.

So, in the solution 1a given in manuscript B2, one sees the Sanskrit text as shown in Fig. 1.[6] This can be transcribed into modern terms as: $8s + 600 = 12s - 200$ (the short oblique line above a quantity in the Sanskrit text

---

[6]Manuscript B1 ends in the middle of the last sentence of problem E1. Therefore, we cannot see its solution.

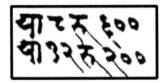

Fig. 1 Manuscript B2, folio 43

yā 8	rū 600
yā 12	rū -200

Fig. 2 Transcription of manuscript B2, folio 43

Fig. 3 Manuscript B1, folio 3b

Fig. 4 Manuscript B2, folio 19

means the quantity is negative. I could not reproduce this sign with words, so I transcribe it as "-") (Fig. 2). What we see in E1 cannot be taken as a model for every setting. There are various ways to represent tabular settings. There are "variations" between manuscripts B1 and B2 as well as between problems. Tabular settings of polynomials are framed. However, it often happens that where there is a table in one of the manuscripts, the other manuscript places the quantities in a line between two sentences, or when an operation is presented in a tabular setting in several examples it is still placed in a line in the next example of the same type of operation.

For example, in B1, in part I of E10, the example concerning the multiplication of unknown quantities by a constant is presented in the following way: a tabular setting with the result stated as shown in Fig. 3.

On the other hand, in B2, the operation is presented in one line with its result in a sentence on the next folio, as shown in Fig. 4.

| yā 8 | rū 608 |
| yā 12 | rū 188 |

Fig. 5 Transcription of B2, problem 1b

I translate E10 as:
"*Setting down for the first example*:

| yāva 2 | yā 0 | rū 1 |
| yāva 0 | yā 2 | rū 0 |

What is produced in the summation is yāva 2 yā 2 rū 1."[7]
The notation is not stable and there is a large diversity of shapes for the table.[8] The result of the subtraction is always rhetorical.

The Rule of Three

Interestingly, the equality mentioned in the previous section is given after a development based on the Rule of Three by the commentator. In 1b, the Rule of Three is briefly mentioned, then setting up is composed as shown in Fig. 5.

This results from the "variation" on the unknown where $x = s + 1$. So:

$$x_1 = 8(s+1) + 600 = 8s + 608$$
$$x_2 = 12(s+1) - 200 = 12s - 188$$
$$x_1 = x_2 \rightarrow 8s + 608 = 12s - 188$$

No further details are given regarding the computation in 1b. Yet, the commentary on the Rule of Three in 1a gives some clues about what is performed. The Rule of Three is a computation of an unknown quantity z from three known quantities, a, b, and c, when there is a proportional relationship among them. That is, $a:b = c:z$ and $z = bc/a$. Here a, b, c, and z are usually called respectively: "the standard quantity" or "argument" (*pramāṇa-rāśi*), "the fruit of standard quantity" (*pramāṇa-phala-rāśi*), "the requisite quantity" or "requisition" (*icchā-phala-rāśi*),

---

[7]One may consider the following explanation done with the modern transcription: $(2x^2 + 0x + 1) + (0x^2 + 2x + 0) = 2x^2 + 2x + 1$.

[8]Among the different examples in part II, tabular settings are varied. Sometimes a frame is used for the results of the equation (E9, E10, E17, E18, E34, and E40), sometimes for fractions (E6, E35, E37, and E38), sometimes for the Rule of Five (E8, E9, E10, and E11), and sometimes for series (E14 and E19). Conversely, sometimes there are no tabular settings (E7, E13, E15, E16, E22, E29, and E36). Or, like in E1, the *pakṣa* are represented (E1, E2, E3, E4, E5, E18, E20, E31, E32, E33, and E39) and sometimes explicitly named.

while z is *iccha-phala*, "the fruit of the prerequisite." The proportional relationship on which a Rule of Three is based is expressed as follows: "when b is obtained from a, what is obtained from c?" or, "if b is for a, what is for c?" The BGA reads:

atra trairaśikam‖ yadyekāśvasya maulyam yāvattāvanmānam tadāṣṭānām kimiti‖

"Here is a Rule of Three: if the price of a horse is the value of [one] *yāvattāvat*, then what is [the price] of eight [horses]?"

Then, in the BGA, the data is set up in one line with blank spaces in between, as follows: *a b c*. In the context of algebra, the term in the middle is of a different nature: a and c can be two constants and b, an unknown quantity, or vice versa. But only the setting down is given with its solution. The procedure is not detailed and it seems it is supposed to be known through the versified rule quoted above. In an attempt to provide more details for this example, I add into the following paragraphs the operations required for the Rule of Three.

First, the first "side" of the equation will be set up as $(8s + 600)$ according to the following quotation:

"if the price of one horse is the value of (one) 'as much as' (*yāvattāvat*), then what is (the price) of eight (horses)?"

pramāṇa (standard) = 1
phala (fruit) = $y\bar{a}$ 1 (or $1s$ in modern notation)
iccha (requisite) = 8
Those are set up in one line as follows: 1 $y\bar{a}$1 8.
Multiplying the second by the last and dividing by the first gives: $\frac{1s \times 8}{1} = 8s$.
Then, following the data of the statement of the problem, 600 rupees are added, which gives us the first side: $8s + 600$.
Concerning the second side $(12s - 200)$:

"if (the price) of one horse is $y\bar{a}$ 1, then what is (the price) of twelve (horses)?"

pramāṇa (standard) = 1
phala (fruit) = $y\bar{a}$ 1 (or $1s$ in modern notation)
iccha (requisite) = 12
Those are set up in one line as follows: 1 $y\bar{a}$1 12.
Multiplying the second by the last and dividing by the first gives: $\frac{1s \times 12}{1} = 12s$.
Then, following the data of the statement of the problem, 200 rupees are removed. The second side is: $12s - 200$.
The two sides are made equal: $8s + 600 = 12s - 200$. These two polynomials are set in a framed table.
The interesting point is that the Rule of Three works as a function would do. The question is "if $4s$ produces 800, then what is produced from $1s$?" It does not really contain information on the value of s. The verbal transcription means that there is something like a function F such that $F(4x) = 800$, and it is known that $F(t) = kt$, for some value of k. This question "how much is $F(x)$?" can be answered by "$F(x) = F(4x)/4 = 200$." Unknowns and their "variations" recall, from a cognitive

point of view, variables for functions. Presumably, the same thing was done for 1b. Then the equations are solved by the same process.

Verse (5): "One should subtract the unknown (*avyakta*) from one [side] and the constant (*rūpa*) from the other, and divide the remainder of the constant (*rūpa*) by the remaining unknown."

After quoting the verse of *sūtra* again in solution 1a in the commentary, the commentator strictly follows the instructions:

the unknown (*avyakta*) of the first side is subtracted from the unknown (*avyakta*) of the second side. The remainder is *yā* 4. The constant (*rūpas*) of the second side are subtracted from the constant (*rūpas*) of the first side. The remainder is 800.

This can be transcribed into modern notation as:

$$8s + 600 = 12s - 200$$
$$12s - 8s = 200 + 600$$
$$4s = 800$$

At the end of the procedure, after subtracting the two sides from one another, the expression corresponds to $4s - 800$. Its solution is also found with the Rule of Three: if $4s$ produces 800, what is produced for $1s$?

pramāṇa (standard) $= 4s$
phala (fruit) $= 800$
iccha (requisite) $= 1s$

Those are set up in one line as follows: *yā* 4 *rū* 800 *yā* 1.

Multiplying the second by the last and dividing by the first we get: $\frac{1s \times 800}{4s} = 200$. Thus, $s = 200$.

The *sūtra* quoted above calls for the division of the remainder of the known quantity by that of the unknown. One would have: $s = \frac{800}{4}$. Instead, the commentator uses again the Rule of Three: $4:800 = 1:s \rightarrow s = 200$. In 1b, same process is suggested to find the solution as $4s = 796$, then $4:796 = 1:s \rightarrow s = 199$.

The procedure then ends with the root for the equation:

Verse (6–7): "The value of the unknown quantity becomes known indeed [in this way]. Or else, when there are many of the unknown (*ājñāta*) (quantities), one should assume [them] to be 'as much as' (*yāvattāvat*) multiplied by two, etc., or divided [by them], or otherwise increased or decreased by constants (*rūpas*)."

Once he knows the value of the variant unknown, the commentator substitutes this in the different "sides" and finds the results. According to the "sides," the unknown is multiplied by, or divided by, or added to, etc., some constants. I transcribed it for 1a: as $s = 200$ and $x = s$, thus, $x_1 = 8 \times 200 + 600 = 2200 = x_2$. However, the commentator does not show any steps for this part of the procedure. The result is given directly:

Having raised the 'as much' (*yāvat*) by this, the property of the first [merchant] produced is 2200. The property of the second is 2200.

The same applies to 1b, once $s = 199$ is found, the result is immediately given. If $s = 199$ and $x = s + 1$ then $x = 200$ can be easily deduced.

Relation Between Rule of Three and Unknowns

In the solution to 1b of E1, the reference to the procedure is made through the phrase "*as before, with the rule of three.*" Why is the commentator adding this new solution, which is not required as a procedure, and neither is required in the *sūtra*? In solutions 1a and 1b, the commentator computes the same value twice, using two independent and identical algorithms but with different "variations" of the unknown. The same sequence of operations constructed as an imitation of the Rule of Three is used mechanically, leading to the same results. It could be that the commentator sought to show possible and basic alternative approaches to the problem. However, the alternatives are not concerned with the algorithm or the solution. Instead, they are concerned with the mathematical objects. Regardless of the nature of the objects, the procedure is the same.

The Rule of Three

Trairāśika is translated as "the three quantity operation" by Hayashi (2004, p. 454), "the rule of three terms" by Datta and Singh (1935, vol. 1, chap. 12, p. 203), or the "Rule of Three" (capitalized) by Sarma (2002) and Keller (2011). A specific emphasis is put on this rule because a verse enunciates it in the middle of the commentary:

> The standard and the requisite [quantities put down] (*pramāṇeccha*) in the first and the last (i.e., the third) [places] are of equal categories, but the fruit (*phala*) [quantity put down] in the middle [place] is of a different category. That [middle term] multiplied by the last and divided by the first is the fruit of the requisite.

The same statement of the Rule of Three appears twice in Nārāyaṇa's works (GK, vol. 60). Keller showed for other mathematical texts that this formulation is syntactically rigid (2011, p. 8). It is more or less expressed in the same way.

In the solution of mathematical problems, the Rule of Three and its derivatives (Rule of Five, etc.) were not new. Sarma comments (2002): "The writers in Sanskrit were well aware of the theory. Commenting the rule given by Āryabhaṭa, Bhāskara I notes that this rule encompasses Rules of Five, Seven and others because there are special cases of the rules of three itself. Bhāskara II even declares that the Rule of Three pervades the whole field of arithmetic with its many variations just as Viṣṇu pervades the entire universe through his countless manifestations." The metaphor states that the Rule of Three is spread out in the whole of arithmetic as one rule which unifies diversified procedures.

It is a curious way to establish equations. Sarma insisted on the "mechanical" aspect of the procedure. According to him, it offers quick solutions to nearly all problems concerning commercial transactions (2002, p. 134). In the BGA, the rule does not imply an economy of procedures as the procedure extends using it. The same procedure is used throughout the problems. Naming the procedure is sufficient in indicating what kind of algorithm is required: a multiplication followed by a division. This procedure, by a continuous stream of iterations of operations, leads directly to the establishment of the two sides of the equation and its solution. Indeed, the Rule of Three could be used for the sake of facility. However, another interpretation can be proposed.

The procedure given in the verse quoted by the commentator provides an order in which the operations should be carried out. First, the "fruit" multiplies the "requisite" and, second, the result is divided by the "standard." In the solution of this problem, the Rule of Three is systematically used when a multiplication followed by a division is required. According to Keller, in the commentary to the *Āryabhaṭīya*, the rule seems to provide a mathematical grounding for procedures involving these two operations. It is used in geometry to highlight the existence of proportional entities (2006, vol. I, p. xxxvi).

Keller (2006, vol. I, pp. xxxv, li) also suggested that the Rule of Three could be used to give a new reading of an algorithm. Using the reading of the *Āryabhaṭīya* by Bhāskara I (629 CE), she formulated the hypothesis that this rule could help in "re-reading" a given procedure as a set of known procedures. This might have also been a method intended to ground or prove the newly read procedure. She showed that an explanation, or proof of verification, can consist of providing a "re-interpretation" of a given procedure via the Rule of Three. Without the rule, an algorithm could appear at first sight to be a sequence of arbitrary operations. The Rule of Three provides an argument to justify the sequence; it gives a meaning to the sequence, a way to "re-interpret" it within the frame of a known rule. It is a mathematical tool which enables specific problems to be "re-interpreted" as abstract and general cases (Keller 2012). The Rule of Three stresses simultaneously a mathematical property and introduces a computation.

Choosing the Unknown for a Linear Equation

While Bhāskara explains and re-interprets the procedure, it seems that the commentary of the BGA only describes it. Elements here are not sufficient for the conclusion that Nārāyaṇa is proving or explaining a general procedure. There are, however, several elements that argue in favor of verification or justification. First, it seems that regardless of what the mathematical objects are, the author is attempting to create a link between objects of different natures: horses and prices in rupees, constants, unknown quantities, and "variations" on unknowns. The first two occurrences of the Rule of Three imply two constants and an unknown. The third occurrence implies two unknowns and one constant. The relation of proportion becomes the link between different objects, which are not linked to each other. In this respect, seeing an un-prescribed Rule of Three and "variations" on the unknown together makes sense. The Rule of Three also helps to distinguish constants from unknowns by putting them in opposition. It provides an explanation for their provenance when constituting the "sides" of the equation. It provides a justification and verification for the origin of the terms and for the roots of an equation. It provides a description of terms physically represented on a counting device and ready to be used to execute the algorithm. Therefore, this practice observed in the commentary is also related to a specific practice of verification.

It could be the verification of an algorithm by means of showing that the process does not change despite the change in objects (unknown, "variation," constant), or it could be the confirmation of obtaining an identical result by using two independent procedures. It is a way to convince the reader that the algorithm is correct and clearly shaped. The Rule of Three appears to be a systematization of all these "variations." Algorithms can be seen as continuous streams of operations, whose appearance is chaotic or meaningless. The commentator wants the procedure to fit into the model given in verses 3–7. The mechanization of the procedure through a well-known rule (Rule of Three) seems like a channel or an anchor in the gush of operations. It shows the consistency of the sequence of operation, and frames the sequence into a convenient pattern. It is also a way to express the orthodoxy of the procedure hidden under the tangible diversity of objects and operations. This is related to the fact that the commentator voluntarily sticks to the model of a linear equation.

The commentator finds a way to avoid simultaneous equations in several unknowns. If I follow the notation I proposed for the translation of data of the problem in Section "Choice of Unknown", the commentator sets out that $x = s$ and he has to compute y_1 and y_2 according to s with a Rule of Three:

$$1/s = 8/y_1 \rightarrow 8s/1 = y_1 \rightarrow y_1 = 8s$$

Thus, $x_1 = 8s + 600$.

$$1/s = 12/y_2 \rightarrow 12s/1 = y_2 \rightarrow y_2 = 12s.$$

Thus, $x_2 = 12s - 200$.
As $x_1 = x_2 \rightarrow 8s + 600 = 12s - 200$.

We see that using this system he reduced the problem to a linear equation in one unknown. In the 40 problems, a linear equation is a model of reference. The first step of the procedure that consists of reducing several unknowns to one "variation," or of selecting one unknown to be the "variation," is crucial to creating a linear equation. Etymologically, equations are conceived as equalities (*samaśodhana*). However, they are not always represented this way, either when they are described in the discourse of the commentator or in the settings. At no moment is the equality explicitly stated in the text, nor is there a stable or specific representation of it. It is not the "acme" of the procedure, like it would be in the case of Li Ye's algorithms, as we see in my other contribution to this book (Chap. 8). Instead of having two distinct steps, establishing and solving, what we see is a continuous stream of operations starting with choosing the unknown and ending with the roots. This objective of choosing the type of equation implies a specific reflection on the unknown. The tabular setting relates polynomial equations to the Rule of Three where the Rule of Three offers a systematic algorithm and a justification as well as a signification of the setting. This Rule makes sense of algorithms.

Conclusion

In the BGA, the division gives a meaning and an origin to the terms of the equations and their roots, and this practice is to be related to a specific practice of verification. There are traces of an investigation on mathematical objects (unknowns, constants, "variation" on unknowns, and real objects like horses and monetary values) and the relation among objects is set according to their correspondences. Here the linear equation as a model and an equation is conceived equally. We see here a simple algorithmic approach. Yet, it implies a deep understanding and a deductive inter-pretation of the algorithm itself, where tabular settings play a key role.

In none of the examples lie justifications of the role of the Rule of Three. Texts are made for working, but how and why they communicate prescriptions to readers is far from being universal. The practitioner must know what is in question in order to execute the algorithm, so there is a prerequisite of a specific type of reasoning to execute an algorithm. There is a meaning behind the written list of operations. The reader must know how to find their way in the text and how to interpret the text. This shows that practitioners understood what they were doing, and that practices are embedded in a specific cultural environment (Pollet 2012).

Appendix

This critical edition is based on manuscripts B1 and B2 from Benares Sanskrit College. The first part of the BGA was the object of a brief description and tran-scription into modern mathematical language published by B. Datta in 1933. This same part was the object of a Sanskrit edition by K. S. Shukla in 1970, based on a single incomplete manuscript from Lucknow, which is a copy of the Benares manuscript used by Datta. These two manuscripts ended at the beginning of part II, in the middle of the first example E1, and were never translated into English. Another manuscript was found by David Pingree (1970/1995) in the Sanskrit collection of the Benares Sanskrit College, which is also incomplete but extends up to the middle of the commentary on the 40th example of linear equations. The new part was published by Takao Hayashi in 2004. Hayashi presented a complete edition of the Sanskrit text of this second part, with his corrections and a translation of the sūtras. The transliteration presented below is borrowed from Hayashi. The reader can consult Hayashi (2004) for the details of the critical edition. The text in bold face represents his translations. The content of the commentary, however, is conveyed using modern mathematical transcription and is accompanied by Hayashi's commentary, but not translated. Translation of commentary from B2 is thus mine.

yāvattāvaccihnitamekaṃ vā bahumitaṃ tu parikalpya /

rūpāḍhyaṃ vā rūponitamathavāvyaktamānamiti // 3 //

māne tasminnevodde śālāpavatsamācaret karma /

phalasiddhyai tau pakṣau tulyau kāryau prayatnena // 4 //

ekasmād avyaktaṃ viśodhayedanyatastu rūpāṇi /

śeṣeṇāvyaktena ca samuddharedrūpaśeṣamiha // 5 //

avyaktasya ca rāśermānaṃ vyaktaṃ prajāyate nūnam /

ajñāteṣu bahuṣu vā yāvat tāvaddvikādi saṃgunitam // 6 //

bhaktaṃ rūpairyuktaṃ vivarjitaṃ vā prakalpayedevam /

nijabuddhyā vijñeyaṃ kvacidavyaktasya ca mānam // 7 //

[...]

udāharaṇāni /

samānamaulyā vaṇijo ṣṭaghoṭā

ekasya rūpāṇi śatāni ṣaṭca/

ṃaṃ śate 'nyasya ca vājino 'rka-

mitāḥ samasvau ca kimśva maulyam// (E1) //

ādyo 'ṣṭayukto dalitaḥ pareṇa

tulyo bhavedāpi tathāparasya /

trighnasya purvaścaturūnitasya

samo bhavenme vada vājimaulyaṃ // (E2) //

atrāśvamaulyamajñātaṃ / tasya mānaṃ yāvattāvat yā 1 /

atra trairāśikam / yadyekāśvasya maulyaṃ yāvattāvanmānaṃ tadā-

ṣṭānāṃ kimiti / nyāsaḥ 1 yā 1 8 /

ādyantayoḥ pramāṇecche samajātī phalaṃ tvitarajāti /

madhye tadantyatāḍitamā dyahṛdicchāphalaṃ bhavati //

iti trairāśikena jātamaṣṭānāṃ maulyaṃ yā 8 / etadrūpaśataṣatke

prakṣipya jātamādyasya dhanaṃ yā 8 rū 600 / punaryadyekasyā-

śvasya yā 1 tadā dvādaśānāṃ kimiti / 1 yā 1 12 / jātaṃ dvādaśā-

nāmaśvānāṃ mūlyaṃ yā 12 / etadṛṇe rūpaśatadvaye prakṣipya

jātamanyapuruṣasya dhanaṃ yā 12 rū 200/ pakṣāvetau samāviti

śodhanārthanyāsaḥ

yā 8 *rū* 600	
yā 12 *rū* 200	

ekasmādavyaktaṃ viśodhayedanyatastu rūpāṇi /

iti prathamapakṣāvyaktaṃ dvitīyapakṣāvyaktācchodhitaṃ śeṣaṃ yā 4 /

dvitīyapakṣarūpāṇi prathamapakṣarūpebhyo viśodhya śeṣaṃ 800 /

punastrairāśikam / yadi yāvaccatuṣṭayasyāṣṭau śatāni maulyaṃ tadā

yāvadekasya kimiti yā 4 rū 800 yā 1 / labdhaṃ yāvattāvanmānaṃ

200 / etadekasyāśvasya mūlyam 200 / anena yāvadutthāpya jātaṃ

prathamasya dhanam 2200 / dvitīyasya 2200 /

atha kalpitamaśvamūlyaṃ yā 1 rū 1 / prāgvattrairāśikena labdha-

mūlyayoḥ svasvadhanena yathoktavadyuktayoḥ samaśodhanārtha-

nyāsaḥ

yā 8 *rū* 608	
yā 12 *rū* 188	

prāgvallabdhaṃ yāvattāvanmānaṃ 199 / anena yāvadutthāpya jāta-

maśvamūlyaṃ tadeva 200 //

Translation

(3–7) **Having assumed the value of the unknown** (*avyakta*) **(quantity) to be marked with a singular or plural** *yāvattāvat*, **increased or decreased by** *rūpa*s **[if necessary], one should perform computation upon that value according to the statements of the questioner. In order to obtain the result, the two sides** (*pakṣa*) **should be made equal [to each other] carefully. One should subtract the unknown from one [side] and the** *rūpa*s **from the other, and divide the remainder of the** *rūpa*s **by the remaining unknown. The value of the unknown quantity becomes known indeed [in this way]. Or else, when there are many unknown [quantities], one should assume [them] to be** *yāvattāvat* **multiplied by two, etc., or divided [by them], or otherwise increased or decreased by** *rūpa*s. **So the value of the unknown should be known by one's own intellect according to the case.**

[…]

Some Examples

(E1) **A merchant has eight horses of the same price and six hundred** *rūpa*s, **and another merchant has [horses] measured by the sun (twelve) and a debt of two hundred** (*rūpa*s). **The two [merchants] have equal properties. What is the price of a horse?**

(1a) Here the price of the horse is unknown (*ājñāta*). Its value (*māna*) is a *yāvattāvat*, *yā*1. Here [there] is a Rule of Three (*trairāśika*): if the price of one horse is the value of [one] *yāvattāvat*, then what is [the price] of eight [horses]?

Setting down: 1 *yā*1 8.

The standard and the requisite [quantities put down] (*pramāṇeccha*) in the first and the last (i.e., the third) [places] are of equal categories, but the fruit (*phala*) [quantity put down] in the middle [place] is of a different category. That [middle term] multiplied by the last and divided by the first is the requisite fruit.[9]

With a Rule of Three, the price of eight [horses] produced is *yā* 8. Having added (*prakṣipya*) this to six hundred *rūpas*; the property (*dhana*) of the first [merchant] produced is *yā* 8 *rū* 600. Furthermore, if [the price] of one horse is *yā*1, then what is (the price) of twelve [horses]? 1 *yā*1 12. The price for twelve horses produced is *yā* 12. Having added this to the debt (*ṛṇa*) which is two hundred *rūpas*, the property (*dhana*) of the other man produced is: *yā* 12 *rū* – 200. Since these two sides (*pakṣa*) are the same, the setting down aiming at the [uniform] subtraction (*śodhanārthanyāsa*) is:

| *yā* 8 | *rū* 600 |
| *yā* 12 | *rū* -200 |

One should subtract (*viśodayed*) the *avyakta* from each other, then the *rūpas*.[10]

The unknown (*avyakta*) of the first side is subtracted (*śodhita*) from the unknown (*avyakta*) of the second side. The remainder is *yā* 4. The constant (*rūpas*) of the second side is subtracted from the constant (*rūpas*) of the first side. The remainder is 800.

Once again, there is a Rule of Three. If it consists of four *yāvat* and eight hundred [*rūpas*], what is the price for one *yāvat*: *yā* 4 *rū* 800 *yā* 1. The value of a *yāvattāvat* is obtained: 200. This is the price of one horse.

Having raised (*utthāpya*) the *yāvat* by this, the property of the first [merchant] produced is 2200. [The property] of the second is 2200.

(1b) Moreover, the price of a horse is assumed to be *yā* 1 *rū* 1. As shown before, with a Rule of Three, the two prices are obtained with each one's property according to what is said previously for the two sums. The setting down aiming at the uniform subtraction is:

| *yā* 8 | *rū* 608 |
| *yā* 12 | *rū* -188 |

As previously, the value of a *yāvattāvat* is obtained: 199. Having thus found a *yāvattāvat*, the price of a horse is produced, which is precisely 200.

---

[9]The verse is also found in *Gaṇitakaumudī* verse 60, translated by Singh Paramanand (1998, p. 47, Rule 60): "*Pramāna* (i.e., the argument) and *icchā* (i.e., the requisition) are of the same denomination (and should be set down) in the first and last places. *Phala* (i.e., the result) is of a different denomination (and it should be set down) in the middle, that (placed in the middle) multiplied by the last and divided by the first happens to be *icchāphala* (i.e., the desired result)."

[10]Citation from the BGA II, verse 5ab. Hayashi (2004, p. 440): "One should subtract the unknown from one (side) and the (known) number from the other."

References

Primary Sources

Nārāyaṇa, *Bījagaṇitāvatamsa* (dates unknown). Manuscripts B1 (call no. 35579) and B2 (call no. 98699). Xerox copies provided by Benares Sanskrit College, Sarasvati Bhavana Texts collection.

Singh, P. 1998/2002. The *Gaṇita Kaumudī* of Nārāyaṇa Paṇḍita. *Gaṇita Bhāratī* 20, 1998: 25–82 (Chaps I–III); 21, 1999: 10–73 (Chap IV); 22, 2000: 19–85 (Chaps V–XII); 23, 2001: 18–82 (Chap XIII); 24, 2002: 35–98 (Chap XIV).

The Śulba-sūtra of Baudhāyana, Āpastamba, Kātyāyana and Mānava, with Text, English Translation and Commentary. Ed. S.N. Sen and A.K. Bag, New Delhi, Indian National Science Academy, 1983.

Āryabhaṭīya of Āryabhaṭa with the commentary of Bhāskara I and Someśvara, critically edited with the Introduction and Appendices by K.S. Shukla, New Delhi, Indian National Science Academy, 1976.

Secondary Sources

Datta, Bibhutibhushan. 1933. The algebra of Nārāyaṇa. *Isis* 19: 472–485.

Datta, Bibhutibhushan, and Avadhesh Narayan Singh. 1935. *History of Hindu Mathematics, a source book. Part I and II*. Bombay, India: Asia Publishing House.

Filliozat, Pierre-Sylvain. 2004. Ancient Sanskrit Mathematics: An oral tradition and a written literature. In *History of science, history of text*, ed. K. Chemla, vol. 238, 137–160. Boston Studies in the Philosophy of Science. Springer.

Hayashi, Takao. 2004. Two Benares manuscripts of Nārāyaṇa Pandita's *Bījagaṇitāvatamsa*. In *Studies in the history of the exact sciences in honour of David Pingree*, ed. by C. Burnett, J. Hogendijk, K. Plofker, and M. Yano, 386–496. Leiden-Boston: Brill.

Keller, Agathe. 2006. *Expounding the Mathematical seed*, vol. I and II. Basel-Boston-Berlin: Birkhäuser Verlag.

Keller, Agathe. 2011. George Peacock's arithmetic in the changing landscape of the history of mathematics in India. *Indian Journal of History of Science* 46 (2): 205–233.

Keller, Agathe. 2012. Dispelling Mathematical doubts. Assessing Mathematical correctness of algorithms in Bhāskara's commentary on the Mathematical chapter of the Aryabhatiya. In *The history and historiography of mathematical proof in ancient tradition*, ed. by K. Chemla, 287–508. Cambridge University Press. Available online: http://halshs.archives-ouvertes.fr/.

Kusuba, Takanori. 1993. *Combinatorics and Magic Squares in India: A study of Nārāyaṇa Pandita's Gaṇitakaumudī*, Chapters 13–14. Brown University.

Pingree, David. 1970/95. *Census of the exact sciences in Sanskrit, Series A*, 5 vols. Philadelphia: American Philosophical Society.

Pollet, Charlotte. 2012. Comparison of algebraic practices in medieval China and India. Ph.D. dissertation. National Taiwan Normal University/Université Paris 7 (Not published).

Sarma, Sreeramula Rajeswara. 2002. Rule of Three and its 'variation' in India. In *From China to Paris: 2000 years transmission of Mathematical ideas*, ed. by Y. Dold-Samplonius, J. W. Dauben, M. Flokerts, B. Van Dalen, vol. 46, 133–156. Stuttgart: Franz Steiner Verlag (*Boethius*, Band 46).

Charlotte-V. Pollet is an assistant professor at National Chiao-Tung University (Hsinchu, Taiwan). She teaches philosophy and history of sciences. After studying philosophy, Sanskrit and Chinese languages in University Lille-3 (France), she obtained a dual Ph.D. in mathematics from National Taiwan Normal University and in history and philosophy of sciences from University Paris-7 in 2012. Her research interests are history of mathematics (algebra, combinatorics, and geometry), cross-cultural transmission of knowledge and the role of education in transmitting scientific knowledge, especially in Chinese and Indian worlds. She is the author of the forthcoming monograph *The Empty and the Full: Li Ye and the Way of Mathematics*. She also works as educator and has received several awards from France and Taiwan for her work in pedagogy of philosophy.

Part III
Far East: China, Korea, Japan

Chinese Counting Rods: Their History, Arithmetic Operations, and Didactic Repercussions

Alexei Volkov

Abstract The chapter provides a history of Chinese counting rods (*suan* 算/筭) and a description of operations performed with them. The author discusses the representation of positive and negative numbers with counting rods of two colors or of two different cross-sections used in the procedure of solution of linear simultaneous equations described in the mathematical treatise *Jiu zhang suan shu* 九章筭 術 (Computational Procedures of Nine Categories) completed no later than the early first century CE and explained in the commentary on it written by Liu Hui 劉 徽 in 263. He also describes arithmetical operations performed with the counting rods mentioned in the mathematical treatise of the first millennium CE *Sun zi suan jing* 孫子筭經 (Master Sun's treatise on computations) and in the treatise *Shen dao da bian li zong suan hui* 神道大編曆宗算會 (Grand Compendium of Divine Dao: [Chapters presenting methods] of lineages [of experts] in calendrical computations and of assembly of mathematicians [of the past]) completed by Zhou Shuxue 周述 學 in 1558. The author claims that the "algorithmic style" of the traditional Chinese mathematics was intimately related with the use of the counting rods and that the transition from this instrument to the beads abacus (*suanpan* 算盤) that took place in the first half of the second millennium CE coincided with the general decline of Chinese mathematics.

Keywords History of Chinese and Japanese mathematics · History of counting devices in China and Japan · Chinese counting rods (*suan* 算/筭) Chinese algorithms for arithmetic operations · Chinese beads abacus (*suanpan* 算盤)

The author is grateful to two anonymous referees for their useful suggestions. All the remaining problems are his responsibility.

A. Volkov (✉)
National Tsing-Hua University, Hsinchu, Taiwan
e-mail: alexei.volkov@gmail.com

© Springer Nature Switzerland AG 2018
A. Volkov and V. Freiman (eds.), *Computations and Computing Devices in Mathematics Education Before the Advent of Electronic Calculators*, Mathematics Education in the Digital Era 11, https://doi.org/10.1007/978-3-319-73396-8_7

Introduction

This chapter is devoted to the counting instrument most intimately related to the formation of Chinese mathematical tradition at its early stages as well as to its highest achievements. This instrument, the counting rods, however, fell into complete oblivion in the second half of the second millennium CE, and its disappearance followed the decline of the "high" Chinese mathematics, the tradition especially marked by an outstanding development of polynomial algebra. Until now, the reasons for both events as well as the possible connection between them remain obscure. The present chapter is devoted to this instrument and, as its title suggests, it will focus on three interrelated aspects: the history of the counting rods, the way they were used in traditional Chinese mathematics, and the didactical practices that they defined, shaped, or influenced.

At least a very brief introduction to the history of Chinese mathematics appears to be in order here. This tradition was sufficiently mature by the time when the earliest extant mathematical treatises were compiled, that is, approximately by the third century BCE; apparently, private mathematics schools existed at that time, and a system of state mathematics education was established later, in the first half of the first millennium CE; it included a state university where mathematics was taught and mathematics examinations were regularly conducted. The instruction and examinations were briefly described in contemporaneous documents that allow their reasonably reliable reconstructions. In the second half of the first millennium CE, the system of state mathematics education was borrowed, with some modifications, by the authorities of Korea and Japan; the majority of mathematics textbooks used in their educational institutions were Chinese. The mathematics instruction in China was several times interrupted due to political reasons; it stopped by the end of the Tang dynasty 唐 (618–907) only to be rebuilt during the reign of the Song dynasty 宋 (960–1279).[1] Attempts to revive the system made after the fall of the Song dynasty were not successful.[2]

The mathematical texts used for instruction in governmental and private institutions of the first millennium CE were compiled, with very small number of exceptions, as collections of problems accompanied by algorithms for their solution; in this respect, Chinese mathematics was similar to the mathematical traditions of ancient Egypt and Mesopotamia. The algorithms were designed for a particular counting instrument, the counting rods; this instrument was used to represent integers, (decimal and common) fractions, and their configurations (in particular, two-dimensional arrays) and to perform operations with them. The simplicity of the instrument combined with its utmost versatility resulted in a number of mathematical methods unheard of in other parts of the world; among them were operations with negative numbers, solution of simultaneous linear equations with

---

[1] For Chinese terms, I use the *pinyin* transliteration of modern Mandarin reading of Chinese characters written in their traditional (not simplified) form.

[2] For more details and references, see Des Rotours (1932); Siu and Volkov (1999); Volkov (2014).

unlimited number of unknowns created by the first century CE, and methods of (numerical) solution of simultaneous polynomial equations of higher degrees in less than five unknowns developed by the early second millennium CE. These methods were intimately related to the use of counting rods and could not be transferred to the Chinese abacus (*suanpan* 算盤) that started getting popular in China in the early second millennium CE (or, as some authors claim, even earlier, in the late first millennium) and eventually became the *only* instrument used for solution of mathematical problems.

Counting Rods: Design and History

The Material and the Shape

The material the rods were made of, as well as their shape and size, varied considerably. Historical sources contain mentions of counting rods made of bamboo, bone, ivory, iron, and jade.[3] The objects considered the most ancient specimens of counting rods were recently excavated from tombs dated of the second or first centuries BCE; some of them are made of bamboo and some of bone.[4]

The earliest extant description of the Chinese counting rods *suan* 算/筭[5] is found in the chapter entitled "Lü li zhi" 律歷誌 ("Record on pitch-pipes and calendar") of the *History of the [Early] Han Dynasty* ([*Qian*] *Han shu* [前]漢書) compiled in the first century CE;[6] it reads as follows:

> […] 其算法用竹。徑一分長六寸。二百七十一枚而成六觚。為一握 。
>
> […] These methods of computations with counting rods use bamboo [sticks]. The diameter [of one stick] is one *fen*, and the length is six *cun*. [There are] 271 sticks, [they] form a hexagonal [bundle], and make one handful.[7]

---

[3]Needham (1959, pp. 71–72).

[4]Li (1955a); Mei (1983, pp. 58–59); Li and Du (1987, p. 8); Volkov (1998); Lam and Ang (2004 [1992], p. 45).

[5]There exist two Chinese characters used to refer to the counting rods, 算 and 筭; the reading of both of them is "*suan*" in modern Mandarin dialect. However, their meanings were not fully identical, and the difference is discussed in a section below. Throughout this paper, when citing an original text, I use the version of the character found in the edition cited.

[6]Although the historical treatise is authored by Ban Gu 班固 (32–92 CE), the most part of the chapter in question is to be credited to the authorship of Liu Xin 劉歆 (46 BCE–23 CE), as the introductory remarks of Ban Gu suggest; see Vogel (1994, pp. 141, 145, 147).

[7]*HS, juan* 21a, p. 2b. One *cun* at that time was approximately equal to 2.765 cm, according to Wu (1937, p. 65). The extant specimens of the counting rods currently preserved in Beijing University have round cross-sections; their length varies from 11 to 13 cm; see Han (2012); Dauben et al. (2013, p. 931, n. 4).

According to this description, the counting rods were (presumably round)[8] bamboo sticks six *cun* 寸 (16.59 cm) long and 1 *fen* = 1/10 *cun* (0.2765 cm) in diameter. A standard set contained 271 rods; as a number of modern historians noticed, 271 is a "figurative" number corresponding to the arrangement of round-shaped rods in a bundle forming in the cross-section a figure with 9 hexagonal layers and one central rod (Fig. 1 shows the structure of such a bundle with one central element and three layers).[9] This geometrical configuration formed by the cross-sections of the 271 counting rods diagram was well known to medieval Chinese mathematicians; a similar diagram can be found, for example, in the *Suan xue bao jian* 算學寶鑑 (Precious Mirror of the Science of Computations) authored by Wang Wensu 王文素 (1465–?).[10] Since each of the six sectors of the bundle contains $1 + 2 + \cdots + 9 = 45$ counting rods, the total amount of rods in the bundle equals to six times 45 rods plus one central rod, that is, 271. The distance between extreme points of every two opposite corners of the outer layer of the rods would be equal to 19 diameters of one rod, that is, to $19 \cdot 0.2765 = 5.2535$ cm, so it was indeed possible to hold such a bundle in one's hand.

The numbers featured in this configuration, six (the number of sides) and nine (the number of layers), played prominent role in the numerological speculations referred to in the same chapter of the *History* and originating from the *Book of Changes* (*Yi jing* 易經). The latter classic associated the numbers 6 and 9 with the cosmic forces Yin and Yang, respectively; according to traditional Chinese philosophy, the whole universe was generated and kept in motion by these two forces. The calculator holding a bundle of 271 counting rods was, metaphorically speaking, holding the entire universe in his hand.

A statement concerning the use of the rods for the purposes of calendrical astronomy appears in their description found in the section "[Characters having the key element] 'Bamboo' (*zhu* 竹)" of the dictionary *Shuo wen jie zi* 說文解字 compiled in 100 CE:[11]

---

[8]At first sight, the round cross-section of the rods is suggested by the use of the term *jing* 徑 found in their description; this term appears in mathematical texts in relation to circles and is conventionally translated as "diameter." However, the term *jing* was also used by some ancient authors to refer to the segment cutting a square through its center and parallel to two of its sides; see, for example, the commentary of Zhao Junqing 趙君卿 (fl. ca. 3rd c. CE) on the *Zhou bi suan jing* 周髀算經 (Computational treatise on the Gnomon of the Zhou [dynasty]) (*ZBSJ* 2001, p. 33). The term *jing* thus can be tentatively interpreted as referring to the shortest segment crossing a flat figure and passing through its center of symmetry. The conclusion about the round cross-sections of the rods mentioned in the *Han shu* is made on the basis of the reconstruction of the structure of the bundle mentioned in this source (see below).

[9]Li (1937 [1983], p. 63); see also Needham (1959, p. 71). The pictures provided by Li Yan show a cross-section of the bundle in which the sections of rods are shown as black and white dots, the number of the rods represented with black dots equals to 271, while the number of those represented with white ones equals to 60. The function of the white dots is not explained. Li Di (1997, p. 54) and Lam and Ang (2004 [1992], p. 44, Fig. 2.3) provide versions of Li Yan's drawing in which the white dots are removed.

[10]Wang (1524, p. 348).

[11]On this dictionary, see Bottéro and Harbsmeier (2008).

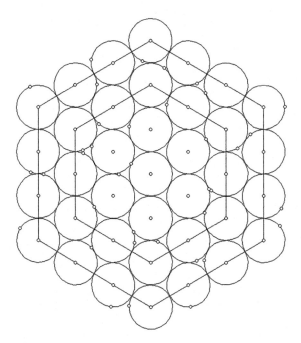

Fig. 1 Bundle formed by one central counting rod and only three layers of rods altogether containing 1 + 6·(1 + 2 + 3) = 37 rods

筭。長六寸。計歷數者。从竹从弄。言常弄乃不誤也。[12]

Counting rods *suan* 筭 are six *cun* long,[13] [they are used] to compute the numerical values of calendars. [The character *suan* 筭 is composed] of the characters *zhu* ["bamboo"] and *nong* ['to operate with, to handle']. This means: [if one] constantly operates [with the instrument], then [there will be] no mistake.

It is followed by an explanation of the meaning of the word *suan* 算:

算。數也。从竹从具。讀若筭。[14]

[The meaning of the word] "computations" (*suan* 算) is [the same as "operations with (?)] numbers" (or "numerical [divination]"?). [The character *suan* 算 is composed] of the characters *zhu* ["bamboo"] and *ju* ["instrument, tool"]. The reading [of this word is] the same as of the word *suan* 筭.[15]

Zhen Luan 甄鸞 (b. ca. 500–d. after 573), a prominent scholar active at the courts of the Liang 梁 (502–587) and Northern Zhou 北周 (557–581) dynasties,[16] in his

---

[12]*SWJZ, juan* 5.1, p. 9b.

[13]That is, 13.824 cm, since 1 *cun* at that time was equal to 2.304 cm (Wu 1937, p. 65).

[14]*Ibid.*

[15]It is sometimes difficult to decide whether the rods mentioned in earlier sources were used for mathematical computations, for simple "one-by-one" counting, for divination, or for other purposes.

[16]On Zhen Luan, dates of his life, and his activities, see Volkov (1994), Li (1997, pp. 257–258).

commentary on the treatise *Shu shu ji yi* 數術記遺 (Procedures of "numbering" recorded to be preserved [for posterity]) by Xu Yue 徐岳 (b. before 185–d. after 227) provides a slightly different description of the instrument[17]:

積算。今之常算者也。以竹為之。長四寸以放四時。方三分以象三才。 [...][18]

The [method of] "accumulation of counting rods" [refers to] the counting rods constantly [used] nowadays. [They are] made of bamboo, [their] length is four *cun*. [This number] metaphorically refers to the Four Seasons; one side of the square [in cross-section] is three *fen*, [it] symbolically represents "Three substances" [= Heaven, Earth, and Man].

The length of these bamboo rods was 4 *cun* long; to find their length expressed in modern units, one needs to know when exactly Zhen Luan completed his commentary, because 1 *cun* equaled to 2.951 cm before 565 CE and 2.668 cm in 565–580;[19] the length of the rods would be equal approximately to 11.8 cm in the first case, or to 10.67 cm in the second. One side of the square cross-section of these rods equaled to 3 *fen* (that is, about 0.885 or 0.8 cm, respectively). Both numbers "3" (of 3 *fen*) and "4" (of 4 *cun*) were once again related to numerological cosmology: the latter was related to the "Four Seasons" (*si shi* 四時), and the former, to the "Three Substances" (*san cai* 三才, the Heaven, the Earth, and the Man).[20] Similar symbolical correspondences appeared in the descriptions of other counting instruments found in the treatise.[21] The exact purpose of the symbolic representations embodied in the instrument described by Zhen Luan is unknown, but the mentioned above statement about the power of the instrument expressed through numerical symbolism of numbers 6 and 9 makes it plausible to read a similar message in the design of the instrument.

One more description of the instrument is found in the chapter "Lü li zhi" 律歷誌 ("Record on pitch-pipes and calendar") of the *History of the Sui* [dynasty] (*Sui shu* 隋書) completed in 636 CE;[22] it states that the bamboo counting rods are of 3 *cun* long,[23] that is, approximately 8.004 or 8.853 cm if the Sui dynasty measures were used, or 9.33 cm, if the length was given in the Tang dynasty measure units when the *History of the Sui* [dynasty] was compiled; the "width" of the rods was 2 *fen*,[24] that is, 0.5336, 0.5902, or 0.622 cm, depending on the value of the unit *fen*.[25]

Interestingly, the counting rods described by Zhen Luan and the rods for *negative* numbers described in the *History of the Sui* [dynasty] (see below) differ

---

[17]*SJSS* (2001, p. 447).

[18]*SJSS* (2001, p. 448).

[19]Wu (1937, p. 192).

[20]*SJSS* (1963, p. 542).

[21]See Volkov (2000).

[22]A translation and discussion of this excerpt is found in Section "The Reason for Alternation of Orientation of the Counting Rods" below.

[23]*SS* (1973, vol. 2, p. 387).

[24]Most probably, by the "width" the author means the diameter of the circle in the cross-section if the rods were designed as oblong cylinders, or the side of the square or of the equilateral triangle in the case if their cross-sections were square or triangular, respectively.

[25]*SS* (1973, *juan* 16, p. 3b). For the values of the units of length used here, see Wu (1937, p. 65).

considerably from those mentioned in the *History of the* [*Early*] *Han Dynasty*; they have square cross-section, are shorter and considerably thicker, and thus are similar to the Japanese counting rods *sangi* 算木 that appeared much later. One can suggest that the rods mentioned by Zhen Luan must have been of two different colors to represent positive and negative numbers (like the Japanese counting rods much later), since they had identical (square) cross-section.

The Time of Appearance of the Counting Rods

Starting from approximately the mid-first millennium BCE, Chinese texts mention the instruments called *ce* 策 ([counting (?)] bamboo slips), *suan* 算/筭 (counting rods), *chou* 籌 (bamboo tokens [for counting (?)]), and their combinations, such as *chousuan* 籌算, *chouce* 籌策, *suanchou* 算籌; modern scholars often consider all of them referring to the counting rods used for mathematical operations commonly called *suanzi* 算子 beginning from the Song 宋 dynasty (960–1279). This identification may be correct for the rods *suan* 算/筭 mentioned in the excerpt from the *History of the* [*Early*] *Han* [*Dynasty*] (*Han shu* 漢書) quoted above and less certain in other cases. One and the same text could contain mentions of different types of rods referred to with the same term; for instance, the mathematical treatise *Jiu zhang suan shu* 九章筭術 (Computational Procedures of Nine Categories)[26] compiled slightly earlier than the *Han shu*[27] contains two mentions of counting rods *suan* used in operations of square and cube root extraction,[28] while in two other cases the same term *suan* refers to a conventional (most likely, abstract) unit used in taxation system.[29] Mentions of the counting rods appear in the commentary written

---

[26]The title of this treatise has various renderings, the *suan shu* has been understood by some authors as "art of mathematics" (see, for ex., Dauben et al. 2013), while others suggested "computational prescriptions" (Martzloff 1997, p. 124) or "procédures mathématiques" [i.e., "mathematical procedures"] (Chemla and Guo 2004); the latter interpretation is certainly better, given that in the text of the treatise the term *shu* is systematically used to refer to computational procedures provided for solutions of problems. The term *jiu zhang* is almost unanimously understood as referring to the number of chapters of the treatise; this interpretation is based upon the modern meaning ("chapter") of the term *zhang*. However, as the author of these lines argued in a number of publications, the term *jiu zhang* historically referred to nine types of mathematical methods, i.e., to the curriculum of "traditional mathematics," and not to the number of chapters (this interpretation is confirmed by the existence of versions of the treatise bearing the same title but comprising different numbers of chapters).

[27]The exact time of compilation of the treatise is unknown. The preface of Liu Hui 劉徽 (263 CE) mentions Zhang Cang 張蒼 (b. before 252–d. 152 BCE) and Geng Shouchang 耿壽昌 (fl. 57–52 BCE) as its compilers, see *SJSS* (1963, p. 91), Chemla and Guo (2004, pp. 54–55, 127); see also Martzloff (1997, p. 129, nn. 29–30).

[28]*SJSS* (1963, pp. 150, 153).

[29]See problem 5 of *juan* 3 and in problem 4 of *juan* 6 (*SJSS* 1963, pp. 134, 124–125; *SJSS* 2001, pp. 111, 146–147); for translations, comments, and relevant references see Berezkina (1957, pp. 458–459, 481–482, 529, n. 7); Chemla and Guo (2004, pp. 289–291, 499–501, 989); Dauben

by Liu Hui 劉徽 (fl. 263 CE) on the same treatise some three hundred years later; all mentions of the counting rods in his commentary are related to solutions of simultaneous linear equations with counting rods.[30] Much more often, the treatises of the late first millennium BCE–early first millennium CE do not mention the counting rods at all, but use instead the term *zhi* 置, "to represent a given number with the counting instrument" (literally, "to put/set [a given number]"); the term *zhi* is systematically used in the *Jiu zhang suan shu*, as well as in the *Zhou bi suan jing* 周髀算經 (*Mathematical Treatise on the Gnomon of Zhou [dynasty]*) compiled in the early first century CE on the basis of the writings of the end of the first millennium BCE,[31] and in the philosophical treatise *Huainan zi* 淮南子 compiled in 139 BCE;[32] in the latter treatise, the counting rods *suan* are also mentioned at least once.[33] Interestingly enough, the term *zhi* is used in solutions of several problems in the mathematical treatise *Suan shu shu* 筭數書 (*Writing on the cal-culations [performed with] counting rods*) compiled no later than the early second century BCE, while the term *suan* occurs only in its title.[34] This distribution of the term in extant sources may suggest that at early stages the term *suan* was not the most common appellation of the bamboo rods used for computation. This con-jecture is indirectly supported by the fact that the list of mentions of the instrument *suan* 算 (or 筭) in ancient Chinese texts provided by Needham (1959, pp. 70–72) contains several references to objects whose design *differed* from that described in the *History of the Early Han [Dynasty]*; they apparently were used for the purposes different from computations described in mathematical treatises, in particular, as tokens for simple one-by-one counting. For instance, the *Yi li* 儀禮 (*Rituals and Ceremonies*), a Confucian classic compiled in the first century BCE–first century CE on the basis of earlier texts[35] in the chapter titled "Xiang she li 鄉射禮" (Ceremony of the County Archery [Competition]) mentions the use of counting tallies also referred to as *suan* 筭 used for simple one-by-one counting of the successful shots during the archery competitions and not for representation of

---

et al. (2013, pp. 277–279, 671–673). The term certainly could not refer to actual tokens used for operations, since the amounts of *suan* assigned to administrative units (counties *xiang* 鄉 and prefectures *xian* 縣) were rather large: for example, in problem 5 of *juan* 6, the numbers of *suan* assigned to six prefectures were equal to 42000, 34272, 19328, 17700, 23040, and 19136, respectively.

[30]*SJSS* (1963, pp. 225, 237–238, 240).

[31]According to Cullen (1996).

[32]*HNZ* (1989, vol. 1, p. 129); Volkov (1997, p. 144).

[33]The *Huainan zi* contains one mention of the counting rods *suan*, in Chapter 2 "Chu zhen xun" 俶真訓; the second occurrence of this term (in Chapter 20 "Tai zu xun" 泰族訓) can be understood as "[result of] computations." For translations of these excerpts, see Le Blanc and Mathieu (2003, pp. 60 and 949), respectively. Needham (1959, p. 71, note b) reports that there are mentions of the counting rods in Chapter 14 of the same treatise; he apparently means the instrument *chou* 籌 mentioned twice in this chapter.

[34]See *SSS* (2001); for translations of this treatise, see Cullen (2004), Dauben (2008).

[35]Boltz (1993, p. 237).

numbers on a counting surface.[36] Interestingly, the addendum to the chapter entitled *Ji* 記 (Notes) compiled, as some scholars suggest, after the text of the chapter had been written,[37] does not mention the counting tallies as *suan* but speaks about a set of 80 "arrow-tallies" (*jianchou* 箭籌) instead.[38] One can suggest that the reason for changing the term may have been that by the time when the "Notes" were compiled the word *suan* 筭 had become generally used for the standard counting rods described in the *History of the Early Han* [dynasty].

As an evidence of antiquity of the counting rods, J. Needham quotes the phrase *shan shu bu yong chou ce* 善數不用籌策 from Chapter 27 of the philosophical treatise *Dao de jing* 道德經 credited to the authorship of the legendary philosopher Lao zi 老子 traditionally believed to be active in the sixth century BCE;[39] in Needham's translation, the phrase reads "Good mathematicians do not use counting rods."[40] The conventional dates of lifetime of the philosopher combined together with the terms used by Needham to translate the excerpt (in particular, "mathematicians" for *shu* 數 and "counting rods" for *chou ce* 籌策) seemingly suggest that the use of counting rods for mathematical purposes had been established in antiquity, presumably prior to the mid-first millennium BCE.[41] If the expression

---

[36]Couvreur (1951, pp. 142–143).

[37]Boltz (1993, p. 236).

[38]箭籌八十。長尺有握。握素。(*YL, juan* 5b, p. 39a), that is, "the [number of] *jianchou* is 80, the length [of each *jianchou*] is 1 *chi* [and it] has a handle (or 'grip') [from which the bark] was removed (or 'without color')." Couvreur (1951, p. 175) suggests that the length of each *jianchou* was 20 cm, but the exact length of the object can be determined only if one knows the exact date of compilation of the text, since the length of 1 *chi* may have varied considerably throughout the first millennium BCE (Wu 1937). Couvreur translates *jianchou* 箭籌 with the French word "fiches," while Steele's translation of this term as "arrows" (Steele 1917, vol. 1, p. 120) looks somewhat misleading since it can make the reader think that he speaks about actual arrows for shooting; the latter hypothesis should be ruled out if one takes into account the length of these objects.

[39]Needham referred to the time when the treatise was compiled as "Warring States period (-fourth and -third centuries)." Since the conventional dates for the Warring States period (provided on p. 875 of his volume) are 480–221 BCE, Needham probably meant that he agreed with the opinion shared by a number of sinologists who believed that the compilation of the treatise was completed much later after the conventional dates of life of its legendary author.

[40]Needham (1959, p. 70).

[41]Needham (1959, p. 71, n. a) claimed that "a misprint in the *Thai-Phing Yu Lan* [i.e., *Tai ping yu lan* 太平御覽.-A.V.] encyclopaedia caused de Lacouperie to make nonsense of this reference." Needham most likely refers to the quote from the *Tai ping yu lan* found in Albert Terrien de Lacouperie's (1845–1894) paper (1883) on p. 330. As footnote 20 on p. 302 of the same paper suggests, Terrien de Lacouperie used an edition of 1807 preserved at that time in the Library of Royal Asiatic Society (London); I was unable to get access to this edition. The translation of the phrase under consideration suggested by Terrien de Lacouperie ("good mathematician ought not to use counting stalks") does not seem particularly wrong, so the misprint mentioned by Needham was most likely related to the source of this quotation. On Terrien de Lacouperie and his work, see de la Grassiere (1896).

shan shu 善數 (or *shan shu zhe* 善數者)[42] is understood as "skillful calculator(s)"
(or "[those who] are good at operations with numbers"), then the text claims that
those people do not use bamboo tallies (*chou* 籌) and bamboo slips/chips (*ce* 策),
presumably because they perform the necessary calculations mentally. Apparently,
J. Needham's rendering is based on the interpretation of the term *shan shu* [*zhe*] that
involves the modern meaning "number" of the word *shu* 數 and, most importantly,
on the assumption stating that the use of the counting instruments (in particular,
counting rods) began in China in unspecified ancient times. However, the word *shu*
in the context of the Daoist treatise may have had a wider meaning, referring to
"counting/calculation" in a very broad sense, in particular, to "fate-calculation," that
is, to divination using instruments referred to as *chou* and *ce*. The phrase thus may
have not been necessarily mentioning mathematical computations performed with
or without counting rods, as Needham and other modern translators assumed, but
referring to unspecified practices of divination, in particular the calculation of the
outcome of military operations. One should not forget the meaning "stratagem" of
the term *ce* 策 present in the phrase under consideration, which appears fitting
rather well within the context of planning military operations, and it was exactly the
term *chou ce* 籌策 that referred to the instrument for planning military campaigns
of a skilfull strategist of the late third century BCE mentioned in the early first
century BCE treatise *Shi ji* 史記 (Records of the Grand Historian) quoted by
Needham himself (1959, p. 71). One may conclude that none of the possible
interpretations of this short quote from the Daoist classic is better than the others.

The understanding of the words of Lao zi suggested by Needham, however, is
similar to the one adopted by some authors in China in the first millennium CE; for
instance, Liu Hui (fl. 263 CE), in his commentaries on the *Jiu zhang suan shu,*
quoted the same statement from the *Dao de jing* (changing the word *ce* to *suan*) in a
purely mathematical context:[43]

數而求窮之者。謂以情推。不用籌算。[44]

To use an infinitesimal procedure in order to look for the numerical value [of the volume of
pyramid] is what is called "conduct a deduction on the basis of the structure [of geometrical
figures]" "without using bamboo sticks and counting rods."[45]

---

[42]Two extant versions of the treatise (the so-called Heshang Gong's 河上公 (fl. ca. 180–157 BCE)
and Wang Bi's 王弼 (226–249 CE) editions, respectively) do not contain the character *zhe* 者,
unlike three other versions (Mawangdui manuscripts A and B, ca. 250 BCE, and Fu Yi's 傅奕
(ca. 555–639 CE) version).

[43]Liu Hui also used the term *ce* 策 in referring to counting rods elsewhere, see *SJSS* (1963, p. 237,
ln. 4).

[44]*SJSS* (1963, p. 168).

[45]Chemla and Guo translate it as "Lorsque l'on cherche à aller jusqu'au bout des quantités (*shu*),
cela signifie que l'on le déduit à l'aide de la situation (géométrique) (*qing*), cela n'implique pas de
calcul avec les baguettes" (2004, p. 433), while Dauben, Guo, and Xu paraphrase the excerpt as
"When mathematics is used to deal with something that is inexhaustible, reasoning should be used,
not counting rods" (2013, p. 557).

Apparently, the same understanding of the phrase from the Daoist classic is meant in the commentary of Zhen Luan on the aforementioned treatise *Shu shu ji yi* 數術記遺 who also modified its text and used the term *suanchou* 筭籌 instead of *chouce* 籌策.[46]

A passage dated of the year 542 BCE from the *Zuo zhuan* 左傳 (*Commentary of [Master] Zuo* [on the historical annals titled *Springs and Autumns*]), according to the modern scholarship, originally compiled in the fifth–fourth century BCE and edited during the Han dynasty (206 BCE–220 CE),[47] mentions an interpretation of a Chinese character perceived as a combination of the numeral "2" and three numerals "6" represented in "counting-rods" form.[48] However, the archaic form of the character in question also can be considered a combination of the *written* forms of the numerals 2 and 6. Moreover, if the number "666" was indeed written in the counting-rods form, as the authors suggest, the basic principle of alternation of orientations of the counting rods in odd and even positions would have been blatantly neglected.[49]

Another often reiterated evidence for the antiquity of the instrument is the "counting-rods numerals" found in the inscriptions made on animal bones and tortoise shells dated in the fourteenth–eleventh centuries BCE and on the coins of the of the fifth–third centuries BCE.[50] However, R. Djamouri argued that the most ancient graphic forms mentioned by Needham could not be considered independent numerals but rather parts of the particular characters representing months and necessarily comprising the element meaning "moon."[51] Martzloff, who shared the viewpoint of Needham, claimed that the "rod-numerals" representing numbers less than ten are found on the coins dated from the fifth–third centuries BCE onward; however, the examples he provides are of the early first century CE.[52] Chen Liangzuo emphasizes that many problems concerning the writing of the numerals on the pre-Qin coins remain open and provides a table with 52 numerals larger than 10 found on the coins that contains 31 numerals having at least one digit larger or equal to 5.[53] Yet only *one* numeral among them (namely, 657) represents the digits in the way the counting rods would do (e.g., the digit 6 is represented as a combination of one horizontal and one vertical stroke as in Table 1), while in 30 (!) cases standard written numerals (and not "rod-numerals") are used for digits from 5 to 9. Moreover, Chen suggests that the combinations of a stroke and *n* strokes

---

[46]*SJSS* (1963, p. 546, ln. 6); *SJSS* (2001, p. 450). For a translation and discussion, see Volkov (1997, pp. 155–158).

[47]Cheng (1993).

[48]Mikami (1934 [1926], p. 47); Needham (1959, p. 8); Chen (1978, p. 284).

[49]See also Chen (1978, pp. 284, 316, notes 110–111).

[50]Needham (1959, pp. 14, 70).

[51]Djamouri (1994, p. 20).

[52]Martzloff (1997, pp. 185–186, Fig. 12.4). Martzloff writes that the rod-symbols similar to those shown in this figure "are also attested well before" the first century CE, but does not provide references to support his claim.

[53]Chen (1978, pp. 278–279).

orthogonal to it, $n = 1, \ldots, 4$, conventionally interpreted as "rod-numerals" for digits 6, ..., 9, actually may have represented numbers 11, ..., 14 misinterpreted by later scholars.[54] The most crucial difference between the counting rods as described in the mathematical treatises and the so-called "rod-numerals" is that in the system of "rod-numerals" the units are *horizontal* and tens *vertical*, while units and tens are represented with the *vertical* and *horizontal* counting rods, respectively.

To summarize, one cannot decide whether the rods mentioned in the most ancient texts or found during archeological excavations were used for mathematical computations, simple one-to-one counting, divination, magical, or ritual practices. One can, however, safely assume that the methods of representation of numbers with counting rods and the procedures for arithmetical operations discussed below appeared no later than the early second or even third century BCE, yet there is still not enough evidence to claim that they existed much earlier. Moreover, the extant sources discussed in this paper suggest that neither the shape, material, dimensions of the instrument, nor the methods of representation of positive and negative numbers were firmly established even as late as the mid-first millennium CE.

The Counting Board: Myth or Reality?

The extant mathematical texts suggest that the operations with counting rods were conducted on a flat surface, e.g., on a table. Liu Hui's commentary on problem 18, chapter 8 of the *Jiu zhang suan shu* mentions a special piece of cloth *zhan* 氈 used for operations with rods and presumably supposed to be placed on any flat surface.[55] It is not known whether the decimal positions were marked on this cloth as rectangular or square cells, or they had imaginary boundaries. As the algorithms of multiplication, division, and extraction of roots described in the *Sun zi suan jing* and *Jiu zhang suan shu* suggest, the consecutive cells in one (horizontal) row were treated as decimal positions of a number. More specifically, to represent a number, one position in a row was selected to set the units of a given number, the first to the *left* of it for tens, the next one leftward for hundreds, and so forth, while the positions to the right of that of the units were used for 10^{-1}, 10^{-2}, ..., respectively. It remains unknown whether the calculators, on the basis of the consideration of the numerical data and operations involved, temporarily fixed or marked the position of units (thus imposing a provisional assignment of the powers of 10 to all the positions in the row) before representing the numbers, or the positions were marked on the counting surface beforehand, as some later Japanese depictions of the counting surface may suggest (see below). Since the range of represented numbers

---

[54]Chen (1978, p. 278).

[55]*SJSS* (1963, p. 236); Chemla and Guo (2004, p. 651) suggest "un tapis de feutre," while Dauben, Guo, and Xu translate it as "a felt cloth" (2013, p. 1001).

may have been varying, the latter option would have restricted considerably the capacity of the instrument.

Theoretically, *any* flat surface, even without a cloth with marked cells, may have been used to perform the calculations, yet some modern authors insisted that operations were performed on a special (wooden) counting *board* with drawn cells. In China, the theory of wooden counting board was supported by influential historian of mathematics Li Yan 李儼 (1892–1963);[56] J. Needham shared his opinion.[57] Later, U. Libbrecht even provided a reconstruction of the counting board which had "the form of a chessboard:"[58] he believed that the board was called *suanpan* 算盤, i.e., was referred to with the name later used in China for the abacus (on a "counting board" depicted in a *Japanese* treatise and referred to with the same Chinese characters 算盤 see below). In his paper (1982), he reiterated his claim: "All computations were made on the counting board. [...] When making difficult computations, several counting boards were used."[59] Both Needham and Libbrecht, however, failed to produce any convincing evidence proving that such an instrument ever existed;[60] in their works, they reproduced the same picture of the "counting board" copied from a mathematical treatise of the late sixteenth century.[61] My study of the picture shows that this "board" was a badly executed image of a beads abacus (see Appendix). Nowadays, the theory of the wooden "counting board" used for arithmetical operations is strongly doubted by a large number of scholars.[62]

To justify these doubts, at least partly, one can remark that, since the length of the counting rods mentioned in historical sources was approximately equal to 10 cm or even longer, the width of the hypothetical board should have been at least 1 m 30 cm if, for example, the number 1,644,866,437,500, found in problem 24 of

---

[56]Hua (1987, p. 58).

[57]Needham (1959, pp. 62, 69–72).

[58]Libbrecht (1973, p. 488).

[59]Libbrecht (1982, p. 207).

[60]Libbrecht (1982, p. 207) provides two quotations to support his claim. The first one says "Put (*chih* [= *zhi*] 置) on the counting board 36,783" (the source is not specified). Here, Libbrecht makes his conclusion of the use of the hypothetical counting board on the basis of the mere presence of the term *zhi* 置 in the text, but this term does not specify the instrument used. The second example of Librecht reads "*Chih yü shang fang* [= *zhi yu shang fang*] 置於上方, put it on the first board." Once again, the source of the quotation is not provided, and the phrase quoted undoubtedly means "place [the operand] in the upper row" or "place [the mentioned value] in [the position of] linear term [located] above"; the meaning of the term *fang* used here depends on the context of the cited instruction, but it certainly refers to a position on the counting surface ("upper side," i.e., the upper row, or the position of linear term of a polynomial equation), and not to a "board."

[61]Needham (1959, p. 70, Fig. 66); Libbrecht (1982, p. 217).

[62]Hua (1987, p. 60); Martzloff (1997, p. 209). See also Wang Ling's opinion expressed in his doctoral dissertation (1956): "The table, the ground or any flat surface can be made use of as the board" (as quoted in Martzloff 1997, p. 209).

Chapter 4 of the *Nine Categories*, had to be represented on it.[63] Such a remarkably large instrument certainly could not have remained unnoticed and unmentioned in the literary and historical works featuring mathematicians and astronomers (as it was the case for the counting rods mentioned on numerous occasions), whereas fragments of it would have been discovered by archaeologists. However, as Martzloff observed, no mention of such an instrument has been ever found in ancient and medieval Chinese sources, neither any remains of wooden counting boards have been excavated.[64] Conversely, in several literary works, there are mentions of the counting rods spread on a table, on a bed, or on the ground.[65] These data strongly suggest that "counting board" never existed in China. Indeed, a piece of cloth (e.g., the one mentioned y Liu Hui) or even a sheet of paper with marked square or rectangular cells should have been much more convenient for the purposes of the counting with rods due to the facility with which it could be used, transported, and stored. The latter conjecture is corroborated by the fact that in Japan sheets of paper were used (see Section "Two Comments on the Computations with the Counting Rods in Japan" below).

Time of the Extinction of the Counting Rods in China

The algebraic notation using the graphical images of counting rods was extensively employed by mathematicians of the Southern Song dynasty (1127–1279), Yang Hui 楊輝 (fl. 1261–1274), Qin Jiushao 秦九韶 (1202–1261?), Li Ye 李冶 (also known as Li Zhi 李治, 1192–1279?), and Zhu Shijie 朱世傑 (fl. 1299–1303). Later, An Zhizhai 安止齋 (dates unknown; active in the fourteenth century) in his *Xiang ming suan fa* 詳明算法 (*Detailed Clarifications of Computational Methods*, 1373),[66] Ding Ju 丁巨 in his *Ding Ju suan fa* 丁巨算法 (*Computational Methods of Ding Ju*, 1355),[67] and Jia Heng 賈亨 (dates unknown; active in the fourteenth century) in his *Suan fa quan neng ji* 算法全能集 (Omnipotent collection of computational methods)[68] also mentioned computations performed with the instrument. The picture showing the counting rods (*suanzi* 筭子) next to the abacus

---

[63]*SJSS* (1963, p. 155); for translations, see Berezkina (1957, p. 471); Chemla and Guo (2004, p. 379).

[64]Martzloff (1997, p. 209).

[65]Hua (1987, p. 58).

[66]This treatise contains detailed explanations of operations of multiplication and division performed with counting rods accompanied by numerous diagrams representing the configurations of rods on the counting surface (An 1373a, pp. 1354–1368; An 1373b, pp. 1578–1591). See also Li (1999, pp. 362–365).

[67]Ding (1355); esp. see pp. 3b–4b where the author provides a solution of simultaneous linear equations with the counting rods. For a general introduction of this treatise, see Li (1999, pp. 348–357).

[68]See Guo (1993, vol. 1, pp. 1313–1346); Li (1999, pp. 357–362).

(*suanpan* 筭盤) in the reading primer *Kui ben dui xiang si yan za zi* 魁本對相四言雜字 (*Miscellaneous Characters of the Basic Four-words-in-couples*) (1371) reproduced in its re-edition *Newly edited Four-words-in-couples* (*Xin bian dui xiang si yan* 新編對相四言) of 1436, at first glance, provides an evidence that even children were equally well familiar with both instruments in the fourteenth and fifteenth centuries. However, this conclusion is most probably premature, since the "counting rods" represented in this children's primer constitute a mantic figure rather than a configuration on the counting surface.[69] Conversely, the *Suan xue bao jian* 算學寶鑑 (Precious Mirror of the Science of Computations) of Wang Wensu 王文素 (b. ca. 1465–?) completed in 1524 explicitly mentions the counting rods and discusses operations with them; see his explanations of arithmetical operations below. Moreover, the treatise *Shen dao da bian li zong suan hui* 神道大編曆宗算會 (Grand Compendium of Divine Dao: [Chapters presenting methods] of lineages [of experts] in calendrical computations and of assembly of mathematicians [of the past]) by Zhou Shuxue 周述學 (dates unknown; active in the mid-sixteenth century) completed in 1558 not only offers detailed descriptions of multiplication and division performed with the counting rods, but also provides numerous depictions of configurations of the rods on the counting surface (see below). In turn, Mei Wending 梅文鼎 (1633–1721) in his *Li suan quan shu* 曆算全書 (Complete Writings on Calendrical Computations) makes it clear that by the late seventeenth century the counting rods were no longer used.[70] His book, therefore, sets the upper boundary for the time when the counting rods disappeared. The lower boundary can be set as the late sixteenth century, given the evidence of the above-mentioned book of Zhou Shuxue.

Two Comments on the Computations with the Counting Rods in Japan

By the late seventeenth century, the counting rods were used in Japan for solving algebraic equations of higher degrees with the (modified) Chinese "Procedures of Celestial Element" (*tian yuan shu* 天元術). The coefficients of the equations were represented with counting rods called in Japanese *sangi* 算木 (lit. "[pieces of] wood [for] counting"); these rods had square cross-sections and were wider and shorter than their Chinese counterparts.[71] According to A. Horiuchi, a piece of paper (presumably, with the drawn cells) was used to place the rods upon.[72] A picture appearing on the first page of the edition of 1783 of the *Seijutsu sangaku zue* 正術

---

[69]Similar mantic figures are shown and discussed in Kalinowski (1994, pp. 43–44, 57, 59).

[70]*LSQS, juan* 29; for translations, see Vissière (1892); Jami (1994). See also Jami (1998).

[71]According to Smith and Mikami (1914, p. 23), the *sangi* were wooden prisms 5 cm high with one side of the square base equal to 7 mm.

[72]Horiuchi (1994, p. 97).

Fig. 2 Picture of a master and two (?) disciples. From the *Seijutsu sangaku zue* 正術算學圖會 (Collection of pictures [related] to the science of computation [featuring] the standard/correct methods) by Miyake Katataka 三宅賢隆 (1663–1746). Edition of 1783, p. 1a

算學圖會 (Collection of pictures [related] to the science of computation [featuring] the standard/correct methods) originally published in 1716 by Miyake Katataka 三宅賢隆 (1663–1746) shows an individual performing operations with counting rods on a sheet of paper put on the ground with a grid 5×5 drawn on it (Fig. 2).[73] As one can expect, the size of the surface is rather large (it can be very approximately estimated as having dimensions 60×80 cm). Surprisingly, the caption of the picture claims that the name of this counting device is 算盤 (spelled *suanpan* in Chinese and *soroban* in Japanese); that is, it is the same as that of the beads abacus.

Interestingly enough, the composition of this picture is reminiscent of the picture of a master and two disciples found in the Chinese mathematical treatise *Suan fa tong zong* 算法統宗 (*Unified Origins of Computational Methods*) (Cheng 1592); see Appendix.

The Japanese treatise *Sanpō tengen shinan* 算法天元指南 (Compass [for understanding the technique of] "Celestial Element" [used in] computational methods) authored by Satō Shigeharu 佐藤茂春 (dates of life unknown; active in the late seventeenth century)[74] for the first time published in 1698 and then edited and published by Fujita Sadasuke 藤田貞資 (1734–1807) in 1795 under the title *Kaisei tengen shinan* 改正天元指南 (Corrected *Compass [for understanding the technique of]* "Celestial Element")[75] contains a picture titled "Soroban zu" 筭盤圖 (Diagram of the counting tray/abacus). The pictures in the former and the latter editions are not identical.

---

[73]Smith and Mikami (1914, p. 46) claimed that the first edition of the book appeared in 1716. The edition of 1795 used by them remained unavailable to me.

[74]Mikami (1974 [1913], p. 182), uses the transliteration "Moshun" of Satō's first name; Smith and Mikami (1914) use "Shigeharu" but mention "Moshun" as one of two possible readings (p. 287).

[75]Mikami (1974 [1913], p. 185) claims that this work was published in 1792; Smith and Mikami (1914, p. 184) reiterate this date and mention an alternative date of publication, 1793.

Fig. 3 Picture featuring a "counting tray/abacus" (*soroban* 算盤) in Satō Shigeharu's 佐藤茂春 *Sanpō tengen shinan* 算法天元指南 (*Compass* [*for understanding the technique of*] "*Celestial Element*" [*used in*] *computational methods*) (preface 1698) (*STS* (1698, chapter 1, p. 4a). Hua (1987, p. 58) redraws this picture with considerable modifications

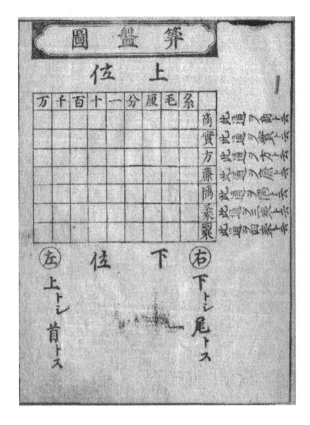

As Hua Yinchun pointed out, this "board" can be used only for one particular operation, namely for numerical solution of higher degree equations, and would not be suitable for performing the operation of multiplication.[76] More specifically, the board shown in Fig. 3 contains seven rows and nine columns that may have been used for representation of decimal places from 10^{-4} to 10^4 of the coefficients of polynomial equations of fifth degree; the board shown in Fig. 4 contains 8 rows and 11 columns that may have been used for representation of decimal places from 10^{-5} to 10^5 of the coefficients of polynomial equations of sixth degree (on the representation of polynomials with the counting rods see below). The positions of the powers of the unknown (speaking in modern terms) and the names of the decimal places are marked on both diagrams; however, it is unlikely that the positions of powers were *actually* marked on the surface of such an instrument, be it a special wooden board or a sheet of paper with drawn or printed cells, because in this case it would have been impossible to use it for operations with polynomials of degrees different from those specified on it. Moreover, if the actual instrument had only 9 or 11 columns, it would have been impossible to use it for operations with coefficients

---

[76]Hua (1987, pp. 58–59).

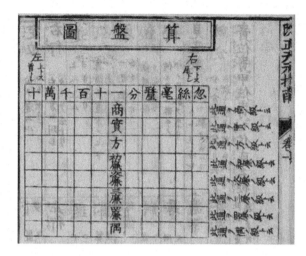

Fig. 4 Picture featuring a "counting board" in a new edition of Satō Shigeharu's book published in 1795 by Fujita Sadasuke 藤田貞資 (1734–1807) under the title *Kaisei tengen shinan* 改正天元指南 (Corrected "Compass [for understanding the technique of] 'Celestial Element'"). Mikami (1974 [1913], p. 185) claims that this work was published in 1792; Smith and Mikami (1914, p. 184) reiterate this date and mention a slightly different date of publication, 1793 (n. 1). This picture was reproduced in Smith and Mikami (1914, p. 24, Fig. 6), but its caption claimed that it was taken from the "*Tengen shinan* of 1698." Hua (1987, p. 59) redraws this picture; he replaces the traditional characters with their simplified versions used nowadays in Mainland China and even reproduces the erroneous "10" (十) in the upper left corner misprinted instead of "100,000" (十萬), exactly as it was in the original diagram of 1795. On Fujita Sadasuke, see Mikami (1974 [1913], pp. 168–170, 183, 187); Smith and Mikami (1914, pp. 183–184)

involving powers of 10 superior to 10^4 and/or inferior to 10^{-4} in the former case, and superior to 10^5 and/or inferior to 10^{-5} in the second case, respectively.

Operations with Counting Rods

Representation of Numbers with Counting Rods

The method used to represent numbers with counting rods is described in three Chinese texts of the first millennium CE: the *Sun zi suan jing* 孫子算經 (Master Sun's treatise on computations), *Xiahou Yang suan jing* 夏侯陽算經 (Xiahou Yang's treatise on computations), and in the manuscript P. 3349 with the bogus title *Suan jing* 笇[= 算]經 (Treatise on computations) found in Dunhuang monastery.

The preface of the anonymous (despite its title) treatise *Sun zi suan jing*[77] contains the following instruction concerning the representation of numbers with counting rods:

---

[77]The treatise was compiled most likely between the late third and the early fifth century CE; see Lam and Ang (2004 [1992], p. 28); Martzloff (1997, p. 124). For a detailed presentation of the

凡筭之法。先識其位。一從十橫。百立千僵。千十相望。萬百相當。[...]六不積。五不隻。

[As for] the universally [used] method of [computations with] the counting rods,[78] first of all [one should] be aware of their positions.[79] Units are vertical, dozens are horizontal; hundreds stand up, thousands lie down. Thousands and dozens look [similar] to each other; tens of thousands and hundreds match each other. [...] [If a digit equal or superior to] six [is obtained at some step], do not pile up [counting rods to represent it]; [if only] five [is obtained], do not [set] a sole [counting rod to represent it].[80]

Similar instructions are found in the preface to the *Xiahou Yang suan jing*:[81]

夫乘除之法 。先明九九 。一從十橫 。百立千僵 。千十相望 。萬百相當 。滿六已上 。五在上方 。六不積筹 。五不單張 。

For methods of multiplication and division, first of all [one should] be clear about the "nine nines" [= multiplication table]. Units are vertical, dozens are horizontal; hundreds stand up, thousands lie down. Thousands and dozens look [similar] to each other; tens of thousands and hundreds match each other. [When] completed six or more, five is placed on the upper side.[82] [If a digit equal or superior to] six [is obtained at some step], do not pile up counting rods [to represent it]; [if only] five [is obtained], do not extend [it] as solitary [piece]. [83]

---

textual history of the treatise, see Lam and Ang (2004 [1992], pp. 28–32); see also *SJSS* (1963, pp. 276–277), *SJSS* (2001, p. 287). The treatise was translated into Russian (Berezkina 1963), French (Schrimpf 1963, pp. 446–456; chapters 2 and 3 only), and English (Lam and Ang 2004 [1992]).

[78]Lam and Ang (2004 [1992], p. 193) interpret the universal quantifier *fan* 凡 as meaning "common [methods of computation]"; this understanding may suggest existence of some other ("uncommon") methods. Compare with Berezkina's rendering "Правило, которое [употребляется] всякий раз при умножении" (Rule that [is used] every time for multiplication) (1963, p. 23); note, however, that the original text does not mention multiplication.

[79]The term "position" (*wei* 位) is ambiguous; it can mean a position of one digit in each operand, but also can refer to a position of an operand on the counting surface (it is used in the latter sense in the methods described in this treatise immediately after the discussed excerpt.) Lam and Ang (ibid.) prefer the latter interpretation and translate *wei* as "the positions [of the rod numerals]" while Berezkina renders *wei* as "разряды" ("positions [of digits]", *ibid.*) thus preferring the former interpretation. The explanation provided right after the phrase in question seems matching better Berezkina's interpretation.

[80]*SJSS* (1963, p. 282), *SJSS* (2001, p. 262). Lam and Ang (2004 [1992], p. 194, note 2) translate this phrase "6 is not accumulation [of rods] and 5 is not a single [rod]" and add that this rule should be placed in the paragraph on the rod numerals, as Qian Baocong suggested (*SJSS* 1963, p. 558). Berezkina in her translation did not change the position of this phrase (most likely, because it makes sense in the paragraph devoted to multiplication where it is found in received editions of the treatise), see Berezkina (1963, pp. 23; 45, n. 22).

[81]There are reasons to believe that this treatise is a later compilation produced by Han Yan 韓延 in the late eighth century; see *SJSS* (1963, pp. 551–553). There exists an annotated Russian translation of the treatise (Berezkina 1985); see also Schrimpf's translation of all the problems of the treatise (1963, pp. 468–482) and Berezkina (1980, pp. 52–54).

[82]That is, for a digit $n > 5$ a counting rod representing five should be placed (horizontally or vertically) above the remaining $(n - 5)$ rods.

[83]*SJSS* (1963, p. 558); *SJSS* (2001, p. 463). Berezkina suggested "пять не является единичной протяженностью" ("five is not a unitary continuity") (1985, p. 298) which sounds like word-by-word translation of Chinese characters.

Digit	In positions for 10^{2n}	In positions for 10^{2n+1}
1		
2		
3		
4		
5		
6		
7		
8		
9		

Table 1 Handwritten representations of counting rods (from the treatise of the sixteenth century *Shen dao da bian li zong suan hui* 神道大編曆宗算會)

Finally, the *Suan jing* 筭經 (Treatise on computations) found in Dunhuang contains the following line:

六不積聚。五不單張。

[If a digit equal or superior to] six [is obtained at some step], do not pile up numerous [counting rods to represent it]; [if only] five [is obtained], do not extend [it] as solitary [piece].[84]

One can notice that this line is textually closer to that found in the *Xiahou Yang suan jing* than to the *Sun zi suan jing*.[85]

According to these three descriptions, if a digit k belonged to the interval $1 \leq k \leq 5$, it was represented with k counting rods placed vertically in the positions of units, hundreds, etc., that is, in the positions for 10^{2n}, $n = 0, 1, 2, \ldots$, and placed horizontally in the positions of dozens, thousands, etc., that is, in the positions for 10^{2n+1}. If the digit k belonged to the interval $6 \leq k \leq 9$, it was represented with $(k - 5)$ rod(s) oriented according to the same convention with an additional rod placed orthogonally above it/them. It goes without saying that for zero the position was supposed to be left empty. The configurations of counting rods used to represent digits are shown in Table 1.

---

[84]P. 3349; Li (1955b, p. 28).

[85]Libbrecht (1982, p. 217), compared this line of the Dunhuang treatise only with that found in the *Sun zi suan jing*.

The Reason for Alternation of Orientation of the Counting Rods

According to majority of modern authors, the alternation of the orientations of the counting rods in the position for 10^{2n} and 10^{2n+1} was used to avoid mixing up the rods representing digits in two consecutive decimal positions. For example, if the digits were all represented with vertically oriented rods, a number such as 12 (| ||) could have been easily mistaken for 3 (|||), while with alternating orientations the same number 12 is represented as —||. This "pragmatic" explanation is adopted by almost all modern historians of mathematics.[86] However, the alternation may appear superfluous, especially if the counting surface had marked cells, or if all the rods were placed horizontally. Instead, it can be tentatively suggested that the alternation of the orientations of the counting rods may have been related to the fact that the very notions of the "horizontal" and "vertical" were embedded into the general framework of symbolical cosmology and put into correspondence with the Earth and the Heaven, and, ultimately, with two complementary and mutually opposed primary forces Yin and Yang. An evidence of this particular semantics of the orientation of the rods used for divination is found in the sixth-century CE compendium *General Explanation of the Five Elements* (*Wu xing da yi* 五行大義): the rods representing the "heavenly trunks" (*tiangan* 天干) are placed vertically, while those used for the "earthly branches" (*di zhi* 地支) are set horizontally.[87] Even though the *Compendium* was compiled relatively late, the principles of disposition of the divination rods certainly had been established much earlier. The alternation of the orientations of the counting rods thus may have originally followed the general symbolical scheme rather than pursuing a purely pragmatic end.

A Remark on the Range of Numbers that Could Be Represented with the Counting Rods

Despite the seeming simplicity of the instrument, it could be used to represent numbers of a rather wide range. A simple computation shows that with the standard set of 271 counting rods mentioned in the *History of the Han* [*dynasty*] (see Section "The Material and the Shape" above) any natural number in the range 1, 2, ..., 255...54 (the latter number contains 53 digits "5") could be represented, even though the representation of the largest number of the range with the counting rods measuring 10 cm would require a counting surface of more than 5.5 m long. Historically, however, even larger numbers were treated by Chinese mathematicians

---

[86]See, for example, Needham (1959, p. 9); Yamazaki (1962, p. 126); Mei (1983, p. 59); Chemla (1994a, p. 3); Horiuchi (1994, p. 96).

[87]Kalinowski (1996, p. 77); see also Kalinowski (1994, pp. 43–44, 57, 59).

—see, for example, the calculations of the number of possible configurations on the board for the Chinese "encirclement chess" (*weiqi*, 圍棋 better known in the West under the Japanese name *Go*碁) by Shen Gua (or Shen Kuo) 沈括 (1031–1095); the latter calculated this number as equal to 3^{361} and then evaluated it (unfortunately, wrongly) as approximately equal to 10^{208}.[88] The number 3^{361} cannot be represented with the standard set of 271 counting rods, since it comprises 173 digits in its decimal notation and requires 486 counting rods for representation.

Negative Numbers

The earliest mention of the counting rods used for dealing with negative numbers in a mathematical text is found in the commentary of Liu Hui (completed in 263 CE) on the algorithm of solution of simultaneous linear equations found in the *Jiu zhang suan shu* (see above). The commentary mentions two ways to represent positive and negative coefficients of the equations with the counting rods: (1) to use colored (red and black, respectively) counting rods;[89] (2) to use counting rods with different (triangular and square, respectively) cross-sections.[90] Some sources also mention black and white rods,[91] most likely used for the same end. The explicit description of the rods with different cross-sections is found in the chapter entitled "Records on musical notes and calendar" ("Lü li zhi" 律曆志) of the *Sui shu* 隋書 (*History of the Sui* [*dynasty*]), but the idea goes back to more ancient times since it appears in earlier texts (see below). The relevant excerpt of the *History* reads:

正策三廉。積二百一十六枚。成六觚。乾之策也。負策四廉。積一百四十四枚。成方。坤之策也。觚、方皆徑十二。天地之大數也。[92]

> "The tallies [for] "actual [things]" (*zheng ce*) [= the rods for positive numbers] have three sides; their total number is 216, [they] form a [bundle with] hexagonal [cross-section]; [they] are the tallies of *Qian*.[93] The tallies [for] "due [things]" (*fu ce*) [= the rods for negative numbers] have four sides; their total number is 144, [they] form a [bundle with] square [cross-section]; [they] are the tallies of *Kun*.[94] The diameters of both the hexagon and of the square are 12 [units]; this is the great number of the Heaven and Earth.[95]"

---

[88]Brenier (1994, pp. 100 ff).

[89]Red color traditionally represented South and thus the (positive) cosmic force Yang 陽, while the black color represented North and the negative force Yin 陰.

[90]The numbers of sides (three and four) were numerical symbols of the cosmic forces Yang and Yin, respectively; see below.

[91]Needham (1959, p. 71). Apparently, black represented Yin, and white, Yang.

[92]*SS 1973, juan* 16, p. 3b.

[93]*Qian* is the symbol representing the cosmic principle Yang and the Heaven.

[94]*Kun* is the symbol representing the cosmic principle Yin and the Earth.

[95]This statement can be understood as referring to the fact that 12 = 3·4, where 3 is the "Number of the Heaven", and 4, the "number of the Earth". Another possibility to understand the link between the numbers 3, 4, and 12 is to interpret 12 as the perimeter of the right-angled triangle with the sides (3, 4, 5).

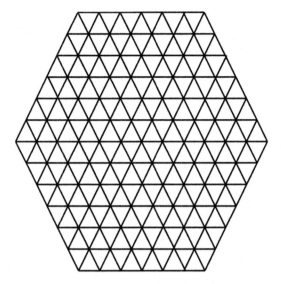

Fig. 5 A bundle of 216 counting rods with triangular cross-sections

The *History of the Sui* [*dynasty*] thus explains the numbers of sides of the polygons in the cross-sections of the rods within the framework of the "numerological cosmology": the rods with triangular cross-section represent number 3 associated with the "positive cosmic principle," Yang, while the rods with square cross-section represent number 4 associated with the "negative cosmic principle," Yin. The "positive principle" Yang was associated with the (round) celestial sphere and thus with a circle, while the "negative principle" Yin was related to the Earth which in traditional cosmology had a square shape. For the approximate value π = 3 used in the mathematical treatise *Jiu zhang suan shu,* the ratio of the perimeters and areas of a square and an inscribed circle, 4:π, coincided with the ratio 4:3. The cross-sections of the bundles of the rods with triangular and square cross-sections reconstructed by Li Yan (1931) are shown in Figs. 5 and 6.

The tallies representing the opposite cosmic forces Yin and Yang were mentioned as early as in the so-called "Commentary on the Attached Verbalizations" (*Xi ci zhuan* 繫辭傳) of the *Book of Changes* (*Yi jing* 易經):[96]

乾之策二百一十有六。坤之策百四十有四。凡三百有六十。當期之日。[97]

---

[96]The date of compilation of the *Xi ci zhuan* is uncertain; as W. Peterson suggests: "There is no direct evidence that it existed before the founding of the Han dynasty [i.e., 221 BCE.-A.V.], but it is apparently quoted or paraphrased in writings from the second century B.C. [...] [There are pieces of evidence that] strongly suggest that by the middle of the second century B.C. the 'Commentary on the Attached Verbalizations' existed and was circulating [...]." (Peterson 1982, pp. 75–76)

[97]*YJ, juan* 7, p. 29a.

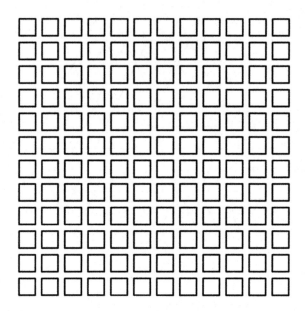

Fig. 6 A bundle of 144 counting rods with square cross-sections

> [The number of] tallies [corresponding to the hexagram] *Qian* [representing Heaven] is
> 216; [the number of] tallies [corresponding to the hexagram] *Kun* [representing Earth] is
> 144. Altogether there are 360 [tallies], corresponding to the [number of] days of [one] year.

The numbers 216 (= $2^3 \cdot 3^3$) and 144 (= $2^4 \cdot 3^2$) as well as their ratio (216:144 = 3:2)
apparently were related to numerological speculations featuring numbers 2 (representing Yin and Earth) and 3 (representing Yang and Heaven). Even though the
Xi ci zhuan does not mention that the tallies representing negative and positive
cosmic forces had square and triangular cross-sections, the perfect match of their
description with that found in the *Sui shu* strongly suggests so. There are two
reasons to believe that the same tallies were used to represent negative and positive
numbers. Firstly, the time of compilation of the *Xi ci zhuan* (that probably took
place in the third century BCE and at any rate no later than the early second century
BCE) was close enough to the time when the prototype(s) of the *Jiu zhang suan shu*
featuring negative numbers was/were presumably compiled. Secondly, the commentary of Liu Hui on problem 3 of Chapter 8 of the *Jiu zhang suan jing* contains
the following description of the counting rods used to represent positive and negative numbers:[98]

正算赤。負算黑。否則以邪正為異。[99]

---

[98] *SJSS* (1963, vol. 1, pp. 225–226).

[99] *SJSS* (1963, vol. 1, p. 225).

Counting rods [representing] "standard" [= actual, positive] [numbers] are red; counting rods [representing] "due" [= negative] [numbers] are black. Or else, the oblique and upright [cross-sections of the rods can be used] to differentiate [them].

The interpretation of the first method of representation of positive and negative numbers is straightforward, while the description of the second method was mis-interpreted by a number of modern authors, mainly due to the conciseness of the text. Some authors believed that a supplementary counting rod was put in a slanted position on the last digit of the number thus meaning that the represented number was negative or that all the rods representing a negative numbers were placed in slanted position.[100] Both hypotheses would be in conflict with the apparent parallelism between "positive" *zheng* and negative *fu* in the first part of the excerpt and the *xie* (oblique) and *zheng* (upright) in the second part: if indeed *xie* were meaning "be placed obliquely on the counting surface," it would have been referring to positive and not negative numbers. A comparison with the description in the *Sui shu* unambiguously demonstrates that Liu Hui commentary refers to the rods with triangular (referred to as "oblique") and square (referred to as "upright") cross-section mentioned in the *History of the Sui [dynasty]*.[101] However, it is true that in the mathematical texts of the Song (960–1279) and early Yuan (1279–1368) dynasty an extra stroke placed obliquely across the last digit of a number indicated that this number is negative, but this was only a method of *graphical* representation of the positive and negative numbers in manuscript or printed texts, and not the actual setting of the counting rods on the counting surface, as several authors duly pointed out.[102] The other way to distinguish between the positive and negative coefficients in written or printed texts would be to use red and black ink, respectively. Red and black colors of the counting rods and, later, of the written or printed numbers apparently followed the traditional association of the colors with the so-called Five Elements of Chinese natural philosophy: the red color was traditionally associated with the element "Fire," while the black color, with the element "Water"; the former is the ultimate manifestation of the cosmic force Yang, and the latter, of the force Yin. One more option was to write the aforementioned characters "positive/actual", *zheng*, and "negative/due", *fu*, next to the numbers.[103]

---

[100]Needham (1959, p. 91); Martzloff (1997, p. 202); Yabuuti (2000, p. 24); Chemla and Guo (2004, pp. 866–867, n. 44) mention this interpretation as possible.

[101]This reconstruction was suggested by Li Yan (1931, p. 36) (Needham 1959 mentions it on p. 90) and reiterated in Li (1955b, p. 142); Berezkina (1957, p. 570); see also Volkov (2001) (Chemla and Guo 2004, p. 867, n. 44 refer only to the latter author).

[102]Smith and Mikami (1914, p. 25); Horiuchi (1994, p. 97); Dauben et al. (2013, p. 931, n. 4).

[103]Libbrecht (1973, p. 70).

Arithmetical Operations with the Counting Rods

The extant mathematical texts of the beginning of the first millennium CE contain the algorithms for the following operations to be performed with the counting rods:

- two of four basic arithmetical operations (multiplication and division) for integer numbers (these operations can be performed with numbers including integer and fractional part expressed as decimal fraction);
- the arithmetical operations with common fractions;
- algorithms for extraction of square and cube roots that can be extended to the algorithms of numerical solution of quadratic and cubic equations;[104]
- algorithms for solution of simultaneous linear equations (a generally valid algorithm instantiated with problems containing up to five unknowns).[105]

Multiplication

Multiplication in the *Sun zi suan jing*

Multiplication procedure is described in the *Sun zi suan jing* as follows:[106]

凡乘之法。重置其位。上下相觀。上位有十步至十。有百步至百。有千步之千。以上命下。所得之數列於中位。言十即過。不滿自如。上位乘訖者先去之。下位乘訖者則俱退之。六不積。五不隻。上下相乘至盡則已。[107]

The method generally [used] for multiplication [is as follows]: set [on the counting surface the digits of the multiplicands placed in decimal] positions in [two] layers, observe [the

---

[104]There is no description of the algorithm of solution of quadratic equations in the *Jiu zhang suan shu*; however, the latter treatise contains a technical term that allows to reconstruct the procedure apparently used in Problem 19 of Chapter 9 of the treatise, see *SJSS* (1963, pp. 255–256); *SJSS* (2001, p. 191); for translations and analysis, see Berezkina (1957, p. 582, n. 23); Berezkina (1980, pp. 215–216); Chemla and Guo (2004, p. 892); Dauben et al. (2013, pp. 1131–1137). The *Zhang Qiujian suan jing* 張丘建算經 (Computational treatise of Zhang Qiujian) compiled in the fifth century CE contains two problems to be solved with quadratic equations, see *SJSS* (2001, pp. 321, 326); for translations see Berezkina (1969, pp. 47, 49), and for an analysis see Berezkina (1969, pp. 75–76; 1980, pp. 216–219). (Full) cubic equations are (numerically) solved in the *Qi gu suan jing* 緝古筭經 (Computational treatise on the continuation [of traditions] of ancient [mathematical methods]) compiled by Wang Xiaotong 王孝通 in 626 CE, see *SJSS* (2001, p. 447); for translations, see Berezkina (1975) and Lim and Wagner (2017); see also Berezkina (1980, pp. 227–232).

[105]See *SJSS* (2001, pp. 173–183); for translations, see Berezkina (1957, pp. 498–506); Chemla and Guo (2004, pp. 617–659); Dauben et al. (2013, pp. 905–1033).

[106]The procedure was translated and discussed by Berezkina (1963) and Lam and Ang (2004 [1992]); below, I provide my own translation and interpretation.

[107]*SJSS* (2001, p. 262).

configuration formed by the number] above and [the number] below.[108] [If among] the positions [of the number placed] above there are dozens, then move [the lower operand] to the [position of] dozens; [if among] the positions [of the number placed] above there are hundreds, then move [the lower operand] to the [position of] hundreds; [if among] the positions [of the number placed] above there are thousands, then move [the lower operand] to the [position of] thousands, [and so on].[109]

[Use the digits] above to "give orders" (*ming* 命) to [the digits] below,[110] and set the obtained values in the [decimal] positions of the middle [row]. Explanation: [if] 10 [or more counting rods are accumulated in some position at some step], then move [one unit to the next position]; [if the current position] does not contain enough [rods] to make it up to 10, then it remains as it is.[111]

[When you] completed the multiplication [by a digit found] in a position in the upper [row], [take] the first [digit and] remove it. [When you] completed the multiplication [of the digits] in [all the] positions in the lower [row], move backwards [= to the right] all its [digits]. {[When you need to set] the digit "six", do not pile up [six counting rods]; [when you need to set] the digit "five", do not [set] a solitary rod.}[112] [Numbers in] upper and lower [rows] are [thus] being multiplied; [when the number above] is eliminated, the [operation] is completed.[113]

---

[108]The literal meaning of this instruction is not very clear. The words *shang xia xiang guan* 上下相觀 can mean two different things: (1) this is a requirement stating that the two numbers should "look at" each other; in this case, "look at" is a technical term explained in the subsequent phrases; (2) the word *xiang* is used to say that the two numbers should "act" or should "be acted upon" similarly, in this case they are "observed together" by the operator. Berezkina renders this instruction with a rather obscure phrase "... [so that the numbers] in upper and lower [lines] would be placed accordingly" ([так, чтобы числа] в верхней и нижней [строках] были соответственно расположены) (1963, p. 23) and provides a word-by-word translation "[in such a way that] their upper and lower [decimal positions] would be mutually seen through" ([чтобы] верхние и нижние [разряды.-A.V.] взаимно просматривались) in note 18 on p. 45 without further explanations. Lam and Ang suggest "the upper and lower position facing each other" (2004 [1992], p. 193).

[109]This instruction was apparently misinterpreted by Berezkina (1963, p. 23; 1980, p. 90). Lam and Ang suggested the following interpretation of the verb *bu* (literally meaning "to walk", "to move [step by step]", here "to shift [counting rods] position by position"): "if there are tens in the upper position then the correspondence is with the tens, [i.e., the units of the lower numeral are below the tens of the upper numeral];" they repeated this interpretation for the cases of hundreds and thousands (2004 [1992], pp. 193–194).

[110]That is, to multiply them. This is a tentative translation; the term *ming* had different technical meanings in other texts, in particular, in the *Jiu zhang suan shu*. Berezkina (1963, p. 23) simply renders it as "multiply"; Lam and Ang suggest "the upper commands the lower" (2004 [1992], p. 194). An example provided in the treatise uses another term (*hu*) in a similar context; see below.

[111]Berezkina (1963, p. 45, n. 20) and Lam and Ang (2004 [1992], p. 194) understand this instruction in a different way; they suggest that the text mentions the case when the product obtained at some step results in a number superior to 9 and thus containing two digits.

[112]Lam and Ang (2004 [1992]) suggest that this instruction should be removed from this algorithm and placed in the section discussing the representation of numbers with counting rods (see above) (p. 194, n. 3).

[113]This algorithm was discussed by a number of authors (e.g., Yamazaki 1962, pp. 126–127; Chemla 1996, p. 120; Martzloff 1997, p. 217); it was misinterpreted by Needham (1959, pp. 62–63).

In the example provided below, the number 128 is multiplied by 81;[114] the performed operations are as follows:

At the first step, both operands are set in the upper and the lower rows of a three-row table formed by three rows of a counting surface; to help the reader, the decimal digits are represented here in their modern (Arabic) forms on the left side, see Figs. 7–16.

Since the upper operand contains a digit in the position of hundreds, the lower operand is to be shifted leftward for two decimal positions. In general case, the lower operand will be shifted leftward for n positions ($n > 0$) if the upper operand A belongs to the range $10^n \leq A < 10^{n+1}$; for a negative n, the lower operand should be moved rightward for $(-n)$ positions. The situation on the counting surface after this step is as in Fig. 8.

The digit in the position of hundreds of the upper row, 1, multiplies the digit of highest order of the lower operand, 8, and the result (8) is set in the middle row above the digit 8 just processed (Fig. 9).

The same digit, 1, of the upper operand multiplies the next (and, in this case, the last) digit of the lower operand, 1, and the result is placed in the middle row above it (in our case, the product, one-digit number 1, will be put in the third position from right of the middle row, as in Fig. 10).

The digit 1 of the upper row will not be used for the operation any longer; therefore, it is removed (Fig. 11).

The lower row operand is to be shifted rightward in such a way that its units will be placed under the first non-empty decimal position of the number in the upper row (in our case, it will be shifted rightward for one position, as in Fig.12).

The digit in the leftmost non-empty position of the upper row, 2, multiplies the digit in the leftmost position of the lower row, 8; the product, 16, should be placed in the middle row above the lower multiplicand, 8. To do so, one should place 6 units in the position occupied by digit 1 and 1 unit in the position occupied by digit 8. The resulting configuration will look as in Fig. 13.

At the next step, the digit 2 from the upper row multiplies the digit 1 of the lower row, and the result is placed above the latter digit 1; since all the digits of the number in the lower row are processed, the digit 2 has to be removed and the lower operand should be moved rightward for one position (Fig. 14).

Now, the digit in the first (and, in this case, the last) non-empty position of the upper row, 8, multiplies the digit in the leftmost position of the lower row, 8; the product, 64, should be placed in the middle row. To do so, one should add the digit 4 (the units of 64) to "2" in the middle row right above "8" in the lower row and

---

[114]Berezkina provides diagrams showing the configuration of rods on the counting surface in the process of solution of the first problem of the treatise (multiplication of 81 by itself) (Berezkina 1980, p. 91); Lam and Ang (2004 [1992]) provide an example of multiplication of 7239 by 23 (p. 58) and solution of the first problem (p. 59); they offer pictures of configurations of counting rods as well as a transcription in modern numerical notation. Martzloff in his monograph (1997) reproduced the multiplication of 81 by 81 with a mistake (or misprint): according to his diagram on p. 217, the partial result obtained at the first step, 80·80, equals to 640 and not to 6400.

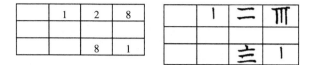

	1	2	8
		8	1

Fig. 7 The first step: both operands are set in the upper and the lower row

	1	2	8
8	1		

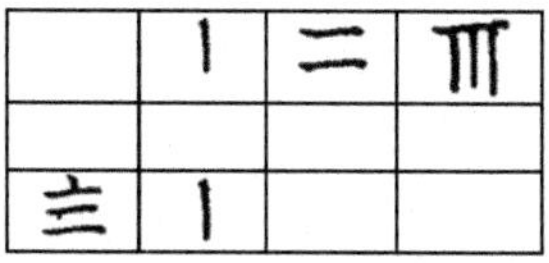

Fig. 8 The lower operand is shifted leftwards for two decimal positions

	1	2	8
8			
8	1		

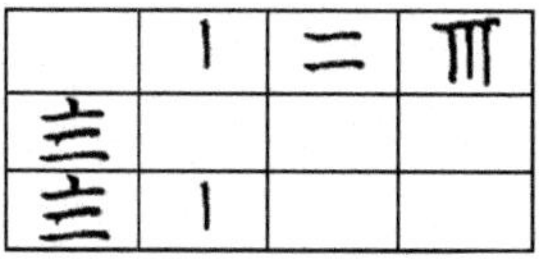

Fig. 9 The digit "1" in the upper row multiplies the digit "8" in the lower row, and the result is set in the middle row

	1	2	8
8	1		
8	1		

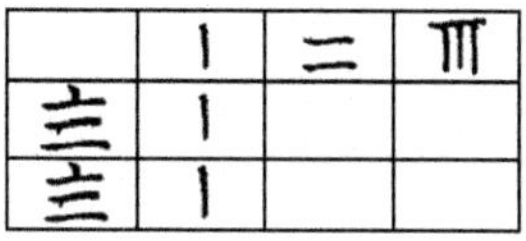

Fig. 10 The digit "1" in the upper row multiplies the digit "1" in the lower row, and the result is set in the middle row

		2	8
8	1		
8	1		

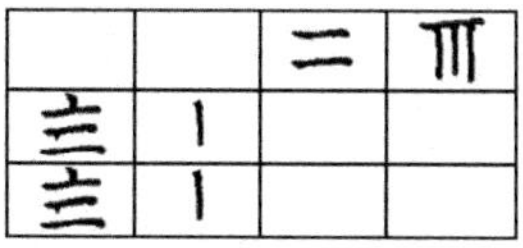

Fig. 11 The digit "1" in the upper row is removed from the counting surface

		2	8
8	1		
	8	1	

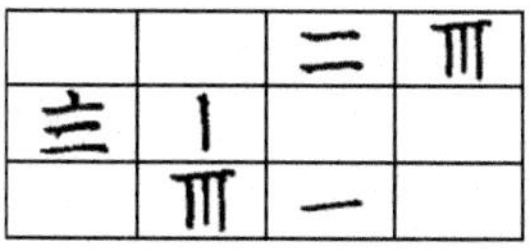

Fig. 12 The lower row operand is to be shifted rightward for one position

then add 6 (the dozens of 64) to "7" in the middle row. Since 6 + 7 = 13, one unit should be carried to the position of tens of thousands in the middle row, and its fourth position will become empty (Fig. 15).

		2	8
9	7		
	8	1	

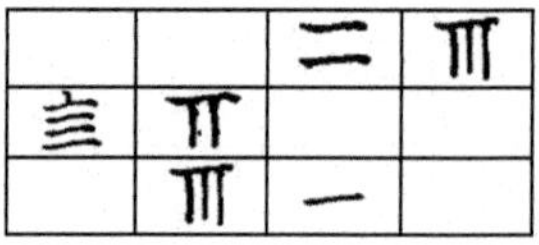

Fig. 13 The digit in the leftmost non-empty position of the upper row, 2, multiplies the digit in the leftmost position of the lower row, 8; the product, 16, is placed (i.e., added to 81) in the middle row

		2	8
9	7	2	
	8	1	

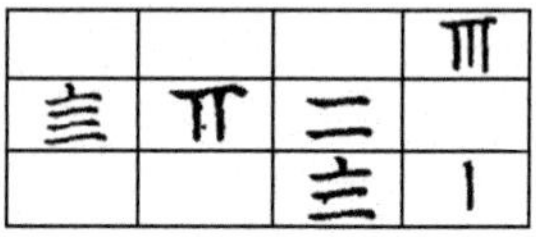

Fig. 14 The digit 2 from the upper row multiplies the digit 1 of the lower row, and the result is placed in the middle row

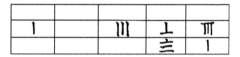

				8
1		3	6	
			8	1

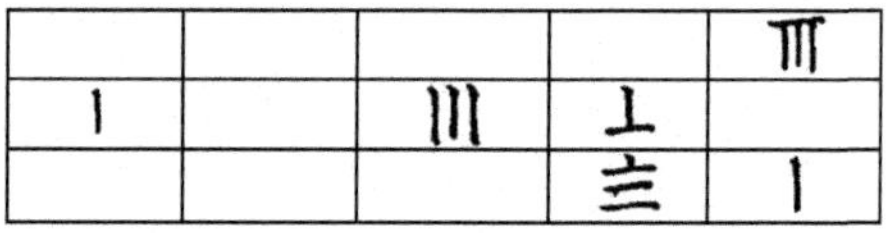

Fig. 15 The digit "8" of the upper row multiplies the digit "8" of the lower row; the product is placed in the middle row

1		3	6	8
			8	1

Fig. 16 Final configuration on the counting surface: the product is found in the middle row

At the next (and, in this case, the last) step, 8 from the upper row 8 multiplies 1 from the lower row, and the product, 8, is placed in the middle row above the digit "1"; since the operation with the digit 8 in upper row is completed, it is to be removed (Fig. 16).

Since the upper row does not contain digits any longer, the operation is completed, and the product is found in the middle row.

In the description of the operation of multiplication found in the extant sources, it was not specified whether the orientations of the counting rods should change every time when the operand placed in the lower row was moved to right or to left; technically, the orientations may have been kept intact to make the operations less time-consuming. In the example shown above, the orientations change, while in their reconstruction Lam and Ang (2004 [1992], pp. 57–59) apparently assumed that the orientations of the rods were supposed to remain unchanged. The diagrams showing the positions of the counting rods found in Yang Hui 楊輝's *Cheng chu tong bian ben mo* 乘除通變本末 (1274, p. 1053) and in the aforementioned treatise *Shen dao da bian li zong suan hui* 神道大編曆宗算會 (1558) of Zhou Shuxue 周述學, however, demonstrate that the counting rods were supposed to be oriented according to the parity of the positions in which they were placed (for more details see below).

An Example of Multiplication Provided in the *Sun Zi Suan Jing*

The first chapter of the treatise contains a list of operations comprising two parts, one is a multiplication of n by m, for $n = 9, 8, \ldots, 1$ and $m = n, n - 1, \ldots, 1$ and a multiplication of the obtained result by itself, while the second part is a division of the obtained product by m. For each n, the sum $S_n = \sum_{i=1}^{i=n} ni$, its square $(S_n)^2$, and the quotient $(S_n)^2/n$ are calculated. At the end of this list, all the sums S_i are added up to obtain the number $S = 1,155$; this number is multiplied by itself and divided by 9. The chapter ends with another list of 11 pairs of operations containing a multiplication of $4 \cdot 3^i$ by 3^{i+1} and division of the result by $2 \cdot 3^i$ for $i = 1, 2, \ldots, 11$; the result of each pair of operation thus amounts to a multiplication of 3^{i+1} by 2.

The purpose for which this list of problems was compiled and placed in the beginning of the treatise has never been identified. Berezkina stated only that "[t]he purpose of this table is unknown";[115] no discussion of this list of operations is offered in the section devoted to the first chapter of the treatise in Lam and Ang (2004 [1992], pp. 76–77). It can be conjectured that these computations may have been used by mathematics instructors to design problems to practice multiplication and division; the computation of the sums S_i and S most likely provided the instructor and/or learners with an easy way to check the correctness of the performed operations.

The first pair of operation and division is styled as two problems on multiplication and division, respectively. The problem on division will be discussed later; in this section, I shall provide a translation and discussion of the problem on multiplication. It reads as follows:

九九八十一。自乘德幾何。

荅曰。六千五百六十一。

術曰。重置其位。以上八呼下八。八八六十四。即下六千四百於中位。以上八呼下一。一八如八。即於中位下八十。退下位一等。收上位八十。以上位一呼下八。一八如八。即於中位下八十。以上一呼下一。一一如一。即於中位下一。上下位俱收。中位即得六千五百六十一。

Nine [times] nine is eighty-one. [If one] multiplies [it] by itself, how much it is?

Answer: 6561.

Procedure: "set [on the counting surface the digits of the multiplicands placed in decimal] positions in [two] layers."[116] Use the [digit] 8 in the upper [row] to "call" (*hu* 呼) the [digit] 8 below. Eight [times] eight is 64, so put down (*xia* 下) 6400 in the middle position [= row] (*wei* 位). Use the [digit] 8 in the upper [row] to "call" the [digit] 1 in the lower [row]. One [time] 8 is 8, so put down 80 in the middle position [= row] (*wei*). Move backwards [= to

---

[115]Berezkina (1963, p. 48, n. 37).

[116]This is the opening statement of the procedure described above quoted *verbatim*.

the right] [the number] in the lower position [= row] (*wei*) for one rank. Remove [*shou* 收] [the counting rods representing] 80 from the upper lower position [= row] (*wei*). Use the [digit] 1 in upper position [= row] (*wei*) to call 8 in the lower [row]. One [time] 8 is 8, so put down 80 in the middle position [= row] (*wei*). Use the [digit] 1 in the upper [row] to "call" the [digit] 1 in the lower [row]. One [time] one is one, so put down 1 in the middle position [= row] (*wei*). Remove [counting rods] from both the upper and lower positions [= rows] (*wei*), then obtain [the result] 6561 in the middle position [= row] (*wei*).

This description of the procedure does not mention that the lower operand has to be shifted leftward; this shift is however supposed to be performed, as the instruction "[m]ove backward [= to the right] [the number] in the lower position [= row] for one rank" suggests. This can be explained at least in two ways: (1) The original text was altered some time later, and a part of it was lost; (2) the algorithm of multiplication used here was borrowed from *another* source. This second option does not look improbable, especially given that the text of the procedure contains several occurrences of the term *hu* 呼 "to call" not used in the text of the algorithm provided in the beginning of the chapter (see above) where the term *ming* 命 "to give orders" was used instead.

It appears that the restrictions naturally imposed on the number of counting rods at the disposal of the calculator and the size of the counting surface, as well as the large number of the intermediate operations involved even in such simple operations as multiplication and division, stimulated the mathematicians to search for simplified and faster algorithms. Since in some of these methods only two rows were used, it was conjectured that these methods, with certain modifications, were later adopted for the abacus computations (in which the operands and the result are placed in one and the same row). Examples of such methods of multiplication are provided in the next sections.

Operations of Multiplication in the Mathematical Compendium of Zhou Shuxue

The aforementioned mathematical compendium *Shen dao da bian li zong suan hui* 神道大編曆宗算會 compiled by Zhou Shuxue 周述學 in 1558 remains practically unknown to the Western reader, and only a small number of publications in Chinese have been devoted to the treatise and its author (see, for instance, Li 1999, pp. 483–485; Yang 2003). At the time when Zhou was writing his book, the counting rods were falling into oblivion, and the abacus (*suanpan* 算盤) was gaining popularity among Chinese mathematicians; suffice it to mention the treatise *Suan fa tong zong*

		2	3	6
3	4	2		

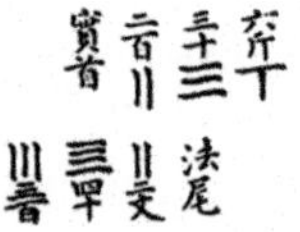

Fig. 17 Initial positions of the operands (236 and 342) on the counting surface (right) and their transcription with modern (Arabic) numerals (left)

6		2	3	6
3	4	2		

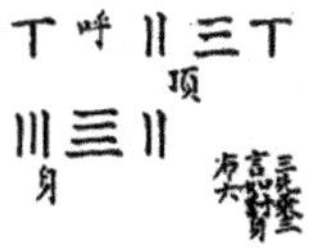

Fig. 18 First step of the procedure: multiplication of 3 of the lower operand by 2 of the upper operand. (Note a misprint in the original caption: "3 multiplies 3" instead of "2 multiplies 3.")

6	8	4	3	6
3	4	2		

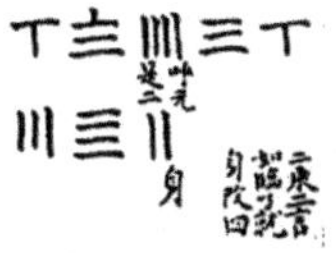

Fig. 19 Upper line contains a combination of digits 684 (the product of 342 and 2) and 36 (the two remaining digits of the upper operand)

算法統宗 (Unified Origins of the Computational Methods) entirely devoted to the operations with the *suanpan* published by Cheng Dawei 程大位 (1533–1606) in 1592. Zhou's treatise contains detailed explanation of the operations of multiplication and division; I will briefly present two methods of multiplications described in his treatise and will discuss his method of division in the Section "Division." A detailed discussion of Zhou's methods is found in (Volkov 2018).

To explain the first algorithm of multiplication, Zhou multiplies 236 by 342; the first diagram he provided shows the position of the two operands on the counting surface (Fig. 17).

The author provides captions explaining the meaning of the digits represented as configurations of counting rods. Each step of the procedure is accompanied by a diagram of the same kind; for instance, the second step is shown in Fig. 18: the first digit of the upper operand, 2, is marked with the character *ding* 頂 (lit. "top of the head"); this digit multiplies (lit. "calls" *hu* 呼) the first digit, 3, of the lower operand marked in the diagram with the character *shen* 身 "body." The result of this multiplication, 6, is placed above the operand 3.

The described procedure differs from the one featured in the *Sun zi suan jing* and discussed above, since the products of the digits are placed in the line occupied by the upper operand (according to the method described in the *Sun zi suan jing*, the products should be placed in the line located between the operands); however, despite this difference, the two algorithms are practically identical. When all the three digits of the lower operand are multiplied by the digit "2" of the upper operand, this digit has to be removed, as shown in Fig. 19.

After that, the lower operand should be shifted to the right for one position; interestingly enough, at this moment the counting rods representing its digits should change their orientation (Fig. 20).

6	8	4	3	6
	3	4	2	

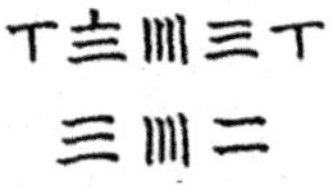

T 圭 Ⅲ 三 T
三 Ⅲ 二

Fig. 20 Lower operand is shifted to the right for one position, and the orientations of its counting rods are alternated in all three positions

8	0	7	1	2
		3	4	2

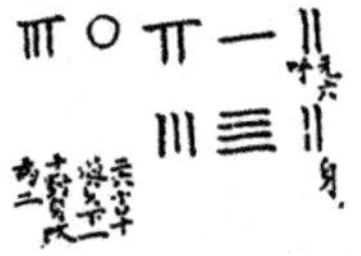

Fig. 21 Final configuration. The product (80712) occupies the upper row. No further instructions are provided concerning the lower operand (it is not specified that it should be removed from the surface). (Note the use of the round symbol for an empty position on the counting surface (i.e., zero); this symbol is used in the transcription of the procedure but is not supposed to be drawn on the counting surface)

2	3	6					
		3	4	2			

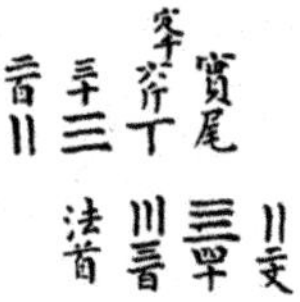

Fig. 22 Second method of multiplication; the original positions of the operands

2	3	6	*2*	*4*			
		3	4	2			

Fig. 23 Digit 6 above multiplies digit 2 below

The treatise provides pictures of configurations of the counting rods for every step of the procedure;[117] at the final step, the configuration looks as in Fig. 21.

The second procedure of multiplication is also supposed to be performed with the operands placed on the counting surface one above the other. However, unlike the first method, the position of units of the result is not determined in advance by the positions of the operands. At the first step, the upper operand is shifted to the left (Fig. 22).

The first stage of multiplication comprises a sequence of three steps; only the final disposition of the counting rods supposed to be obtained at the end of this stage is shown in the diagram. The detailed instructions are provided under the diagram. The operations to be performed are as follows:

---

[117]The interested reader can find the complete description of the procedure in Volkov (2018).

2	3	6	2	*5*	*2*		
		3	4	2			

Fig. 24 Digit 6 above multiplies digit 3 below

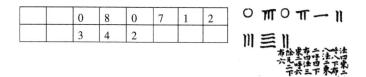

2	3	*2*	*0*	5	2		
		3	4	2			

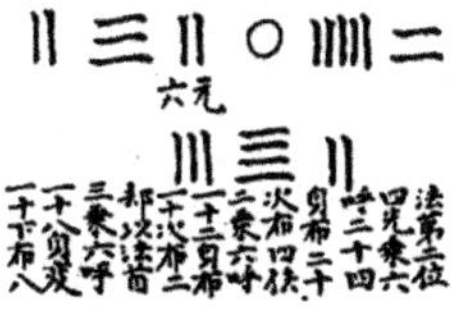

Fig. 25 Configuration on the counting surface after the first three steps

		0	8	0	7	1	2
		3	4	2			

Fig. 26 Final configuration on the counting surface. Note two round symbols for zero in the resulting product of the multiplication, one of them in the leftmost position

Step 1: Digit 6 of the upper operand multiplies digit 4 of the lower one; the result, 24, is placed in the upper row above the digits 4 and 2 (Fig. 23).

Step 2: Digit 6 of the upper row multiplies digit 2 of the lower one; the result, 12, is placed above the digit 2 (here 5 is the sum of the digit 4 that was placed on the surface earlier and the newly obtained digit 1 of the product 12) (Fig. 24).

Step 3: Digit 6 of the upper row multiplies digit 3 of the lower one; the result is 18. The digit 6 is removed from the surface, 2 is added to 18, and the sum, 20, is placed above 34 (Fig. 25).

One of two operands is shifted, and the process continues as before. The final configuration of the counting rods on the surface is shown in Fig. 26.

The result of the operation, 80712, is found in the upper row.[118]

Division

The procedure of division is described in the *Sun zi suan jing*; it was translated and discussed by Berezkina (1963, pp. 23–24) and by Lam and Ang (2004 [1992], pp. 194–195). Three horizontal rows of the counting surface are used (the divisor is

---

[118]The complete translation of the description of this method is found in Volkov (2018).

placed in the lower row, the dividend is placed in the middle one, and the result is placed in the upper row.)

Division in the *Sun zi suan jing*: the General Algorithm

The procedure of division is described as follows:

凡除之法。與乘正異。乘得在中央。除得在上方。假令六為法。百為實。以六除百。當進之二等。令在正百下。以六除一。則法多而實少。不可除。故當退就十位。以法除實。言一六而折百。為四十。故可除。若實多法少。自當復退。故或步法。十者置於十位。百者置於百位。{上位有空絕者法退二位。}[119] 餘法皆如乘時。實有餘者。以法命之。以法為母。實餘為子。[120]

[As for] the universally used method of division, it is a rectification of what has been changed by multiplication.[121] [When] multiplying, [one] obtains [the result] in the central [row of the counting surface]; [when] dividing, [one] obtains [the result] in the upper [row of the counting surface]. Suppose that six is taken as divisor, [and] 100 as dividend. [If one] divides 100 by six, [one] should advance [the divisor] for two ranks to make it [stay] exactly under the hundred. [Now one has to] divide one by six, the divisor is [relatively] large while the dividend is [relatively] small, and [one] cannot divide. Therefore [the divisor] should be retreated to occupy the position of dozens. [Now one] divides the dividend with the divisor; [he] says "one [time] six" and breaks the hundred. [The remainder] will be 40. Therefore it is possible to divide. If the dividend is [relatively] large and the divisor is [relatively] small, [it] can stay as it is in the position corresponding to hundreds, and is not supposed to be additionally retreated. This is why [according to] some [versions of the method] the divisor moves step by step [as follows]: [if the dividend has] tens, set [it] in the position of tens; if the dividend has hundreds, set it in the position of hundreds.[122] {[If] the [digits in the upper row (lit. "position")] are interrupted with an empty

---

[119]This is a commentary printed in smaller characters. Lam and Ang (2004 [1992], p. 195, n. 4) claim that this commentary was added by Li Chunfeng 李淳風 (602–670); however, this is rather unlikely, given that all other commentaries authored by Li Chunfeng were opening with a mention of his name. Instead, it is rather likely that this commentary was added by the author of the text himself.

[120]*SJSS* (1963, pp. 282–283); *SJSS* (2001, p. 262).

[121]Berezkina (1963, p. 23), interprets *yi* 異 as "opposite" and suggests "Правило, которое [употребляется] всякий раз при делении, прямо противоположно умножению:…" (The rule that [is used] every time for division, is directly opposite to multiplication). Lam and Ang (2004 [1992], p. 194) translate "In the common method of division (*fan chu zhi fa* 凡除之法), this is the reverse of multiplication." On the use of the term "common" see above; what the word "this" refers to in their rendering remains unclear. As Berezkina before them, they consider *yi* 異 as referring to an "opposite" operation, while the term *zheng* 正 vanishes from their translation. The authors of both translations apparently consider *yi* as the only predicate in this statement, and interpreted *zheng* as an additional term that qualified the predication ("directly opposite", as Berezkina suggested) and, in principle, was not very important (this is probably why Lam and Ang omitted it). However, it can be suggested that the terms *zheng yi* form an expression in which both parts are equally important.

[122]For a different interpretation of this phrase see Lam and Ang (2004 [1992], p. 195).

1		
		6

Fig. 27 The operands 1 and 6 are placed on the counting surface

1		
6		

Fig. 28 The divisor 6 is shifted leftward twice

1		
	6	

Fig. 29 The divisor 6 is shifted for one position to the right

space, the divisor advances[123] for two positions.} The remainder and the divisor both are [found in the same positions as] during the multiplication. Use the divisor to establish it [= the fractional part]: take divisor as denominator, the remainder will be the numerator [of the fractional remainder].

The algorithm of division contains a reference to a particular case, a division of 100 by 6. Due to its conciseness (and, as I argued above, a mistake reproduced in the extant versions of the treatise), it was slightly differently interpreted in Berezkina (1963) and Lam and Ang (2004 [1992]). The configurations generated on the counting surface would be as follows:

At the first step, the operands are placed on the counting surface as shown in Fig. 27. The upper row is left empty; here, the result will be eventually produced.

At the first step, the divisor 6 is supposed to be shifted leftward twice to occupy the position exactly under the leftmost digit of the dividend.

Interestingly enough, the anonymous author insists that this step has to be performed, even though it is absolutely obvious that the digit 6 in the lower row is superior to the digit 1 in the middle row, and a retreat of 6 for one position to the right will eventually take place. This element of the description can only mean that the operation was described in a particular "easy" case, a division of 100 by 6, but the author did not want to remove the first step, the displacement of the divisor to the position under the leftmost digit of the dividend which is necessary in the general case (Fig. 28).

---

[123]The received versions of the treatise read "the divisor retreats (*tui* 退) for two positions"; however, the previous phrase suggests that the text contains an error, and the word *jin* 進 (to advance) was erroneously replaced by *tui* 退 (to retreat). The word "retreat" in this position would have lead to a mathematically wrong procedure.

	1	
	4	
	6	

Fig. 30 The first digit of the quotient, 1, is obtained

	1	
	4	
		6

Fig. 31 The divisor 6 is "retreated" to the position of units

	1	6
		4
		6

Fig. 32 The final step: the quotient (16) and the remainder (4) are found

Since six is superior to one, the divisor has to be shifted for one position to the right. The situation on the counting surface will look as in Fig. 29.

At this step, one has to evaluate the quotient of division of the number in the middle row, 100, by the number in the lower row, 60 (in this particular case, this division is apparently perceived as division of 10 by 6 and hence the mention of the element "one time six [equals to six]" of the multiplication table mentioned at this stage). The first digit of the quotient, 1, is thus obtained; it is supposed to be placed above the current position of the leading (actually, the only) digit of the divisor, as in Fig. 30. The product 6 is subtracted from 10, and the remainder is mentioned as 40 (and not just four!)

Again, the divisor 6 should be "retreated" to the position of units, and the configuration looks as in Fig. 31.

At this step, the quotient of division of 40 by 6 should be evaluated with the help of the multiplication table (which the learners were supposed to know by heart). This quotient, 6, was supposed to be placed in the upper row above the divisor, as well as the remainder, 4 (Fig. 32).

The description of the procedure suggests that the computations stop at this stage with the result equal to 16 (integer part) and 4/6 (fractional part); the simplification of the fractional part is not mentioned. One cannot but notice a very particular choice of the numerical parameters introduced in the text of the (presumably, general) algorithm: the division of 100 by 6 yields a periodical fraction 16.(6) (or, in other words, makes the procedure continue indefinitely if the computation of the fractional part of the result is required). It could have been certainly very easy for the compilers to change the parameters to obtain an integer answer, or an answer comprising a relatively "simple" fractional part, as it was the case in the problem placed by the compilers almost immediately after the procedure (see the next section).

The Division in the *Sun Zi Suan Jing*: An Example

The same treatise contains a problem with a detailed solution which, apparently, was supposed to provide additional explanations as to how the algorithm was supposed to be performed. This problem reads as follows:

> [There are] 6561 [objects], nine persons share them. The question is: how much [one] person will obtain?
>
> Answer: 729.
>
> The procedure [is as follows]:
>
> Set 6561 in the middle row, this is the dividend. Set 9 persons in the lower [row], [this will be] the divisor. In the upper [row] set 700, use 7 in the upper [row] to 'call' [i.e., multiply] 9 in the lower [row]. 7 [times] 9 is 63, therefore remove 6300 from the middle row. Retreat [the number] in the lower [row] for one position. Then set 20 in the upper position. Use 2 in the upper [row] to 'call' [i.e., multiply] 9 in the lower row. Two [times] nine is 18, therefore remove 180 from the middle row. Again, retreat [the number] in the lower position for one rank. Then set 9 in the upper position. Then use 9 in the upper [row] to 'call' [i.e., multiply] 9 in the lower row. Nine [times] nine is 81, therefore remove 81 from the central position. The central position is completely empty. Remove [the operand] in the lower position. The [number] obtained in the upper position is the number [of objects] to be obtained [by each person].[124]

One can notice that this description of operations does not follow the description of the general algorithm precisely, even though the general idea is still the same. We do not read here, for instance, how the divisor is supposed to be shifted leftward; instead, we are told right away that 700 should be placed in the upper row. But the words "retreat [the number] in the lower [row] for one position" prove that the divisor 9 was shifted leftward and placed under the digit 5 before the first operation was performed. The quotient is integer, and the learner apparently did not have to face problems of the kind related to the division of 100 by 6. This problem on division provides a paradigmatic example for a long list of problems featuring operations of division and listed immediately after it in the first chapter of the treatise; none of them contains a detailed solution of this kind.

An Example from the Treatise of Zhou Shuxue

An interesting example of division performed with the counting rods is found in the aforementioned treatise of Zhou Shuxue (1558). In his example, the dividend 80712 is divided by 236; the operations apparently follow the description found in the *Sun zi suan jing*. The interesting element of this example is that it is more "realistic" than the two examples provided in the *Sun zi suan jing*: here, the divisor is not just one-digit number, and accordingly, the evaluation of the partial quotients is more complicated. A detailed discussion of this example is found in Volkov (2018).

---

[124]*SJSS* (2001, p. 263); for translations, see Berezkina (1963, p. 25), Lam and Ang (2004 [1992], p. 196).

Conclusions

The extant data suggest that the advent of the counting rods can be safely dated in the third century BCE at the latest. The counting rods provided the Chinese mathematicians with a powerful tool which allowed them dealing with the (potentially infinite) hierarchy of (positive and negative) powers of 10 and, later, with arbitrary powers (positive as well as negative) of variables (in the case of polynomials) or unknowns (in the case of equations). The crucial role played by the counting rods in the shaping of Chinese mathematics was stressed by many authors, see, for example, Lam (1986, 1987a, b). In particular, the counting rods and their graphical images allowed the representation of the mathematical objects having two-dimensional structure, such as the operations with common fractions represented as pairs of natural numbers, the solution of simultaneous linear equations with an arbitrary number of unknowns, the extraction of square and cube roots, and the solution of higher degree equations. Prior to the mid-thirteenth century, combinations of these features produced various algebraic methods, in particular the treatment of simultaneous nonlinear equations in up to four unknowns of arbitrary degree. The underlying idea of this kind of symbolical algebra was the old principle of the (exponential) relationship between the linearly ordered positions of the counting surface and the successive powers of a number. The counting surface and the counting rods thus became a powerful tool for a quasi-symbolic algebra, and not only for numerical computations.[125]

The "quick" versions of arithmetical operations originally designed for the counting rods and purported to economize the place and diminish the number of the rods to be used were later modified and used to operate with the (beads) abacus. The convenience and speed of simple arithmetic operations performed with the abacus is often considered one of the main reasons why the latter instrument became popular among various categories of users (merchants, artisans, functionaries, etc.) and finally even among (some) professional mathematicians. However, the fact that it finally replaced the counting rods cannot be explained solely on the basis of the convenience of performing the simplest arithmetical operations: the coexistence of the counting rods and the abacus for several centuries in China, in Korea, and, probably, in Vietnam precludes from making conclusions about the efficiency (as it can be argued, rather limited) of the abacus as the main reason for the disappearance of the counting rods. One can suggest that the extinction of the counting rods and the resulting domination of the abacus were related to the considerable changes in the social and cognitive fabric of the traditional Chinese mathematics in the fourteenth–sixteenth centuries and, in particular, to the deterioration of the channels of transmission of mathematical expertise.

---

[125]See, for instance, Hoe (1977); Chemla (1982); Lam (1982); Horiuchi (1994, pp. 97–109); Martzloff (1997, pp. 258–271).

Fig. 33 Picture titled 師生問難 ("Master and disciple(s): asking difficult [questions]") repro-duced from an edition of the *Suan fa tong zong* of 1883 and identical with Fig. 66 in Needham (1959, p. 70)

Appendix: On the picture of a counting instrument
from the *Suan fa tong zong* 算法統宗

To support his thesis about the existence of the "counting board" in China, J. Needham reproduced a picture from the *Suan fa tong zong* 算法統宗 (*Unified Origins of the Computational Methods*, 1592) by Cheng Dawei 程大位 (1533–1606).[126] The picture has a caption reading "Shi sheng wen nan" 師生問難 ("Master and disciple(s): asking difficult [questions]"); it features one relatively old individual (apparently, the "Master") and two younger persons gathered around a table on which a rectangular object is placed (Fig. 33).

The "Master" and the person standing on his right seem to be engaged in a conversation, while the person on the "Master's" left is standing rather passively; he looks smaller and his picture is not drawn very well, if compared with two others; one may suggest that this person is a servant, while the person on the Master's right is a disciple. The reason for reproducing this picture in Needham's volume was related to his statement about the existence in China of a counting instrument he referred to as "counting board": the object on the table was, according to Needham, a specimen of such a "board." In this Appendix, I will argue that the object drawn in this picture actually was a beads abacus (*suanpan*).

Needham's caption accompanying this picture is somewhat ambiguous; it reads "[…] a view of a counting-board (frontispiece of the *Suan Fa Thung Tsung* [i.e., *Suan fa tong zong*], + 1593)." It can be understood in two ways: (1) The picture comes from the original edition of 1593; (2) the picture comes from an unspecified (revised) edition of the version of the treatise originally published in 1593.[127] It turns out that the correct understanding is the second one. Indeed, the picture reproduced by Needham (and later copied by Libbrecht)[128] is identical with the one published in the edition of the treatise dated of 1883 and shown in Fig. 33, while the original edition of 1653 contains a somewhat different picture shown in Fig. 34. In the picture of 1883, the object on the table looks like a thick board with painted (or carved) square (or rectangular) cells. The number of columns equals to 10, one of them (on the reader's left) is not well drawn; the number of horizontal rows clearly equals to five. The horizontal dimension of the object can be roughly estimated as 50–60 cm, and the vertical one, as 30–40 cm. Most interestingly, the object is clearly rather thick; its height can be roughly estimated as 4–5 cm. All cells on the board are *empty*, and no counting rods are shown either on the table or in the hands of the individuals, even though depicting the rods would not be a difficult task for the block-carver. Chen Liangzuo's analysis of the picture provided by Needham (that is, the one shown in Fig. 33) led him to the conclusion that it is

---

[126]Needham (1959, p. 70, Fig. 66).

[127]Cheng Dawei published two version of his treatise; the one published in 1592 comprised 17 *juans* ("scrolls"), while the version published in 1593 contained 12 *juans*.

[128]Libbrecht (1982, p. 217), Fig. 6 refers to Needham (1959, p. 70).

Fig. 34 Picture titled 師生問難 ("Master and disciple(s): asking difficult [questions]") from the edition of the *Suan fa tong zong* of 1593 (later also reproduced in the edition of 1784)

Fig. 35 Picture titled 師生問難 ("Master and disciple(s): asking difficult [questions]") from the edition of the *Suan fa tong zong* of 1758

impossible to decide whether the depicted object is an abacus or a "counting board."[129]

The picture published in the edition of 1593 and of 1784 (Fig. 34) is slightly different. It was originally executed much better than the one shown in Fig. 33 (the faces of the "Master" and the disciple on the left look more realistic), but the quality of the picture from 1593 edition and of its copy printed in 1784 available to me did not allow to identify the object on the table. However, given that the pictures shown in Figs. 33 and 34 came from the opening pages of a treatise featuring the operations with the abacus and lacking any explicit mention of the counting rods, and taking into account the fact that by the time of its compilation the counting rods were almost completely forgotten in China, it would be reasonable to conjecture that the object intended to be depicted in the picture was an abacus, and not a conjectural "counting board," as J. Needham suggested.

The decisive piece of evidence is found in the edition of the treatise of 1758 preserved in Waseda University Library (Japan). This blockprinted picture shown in Fig. 35 is apparently identical with those found in the editions of 1593 and of 1784. The picture leaves no doubts that the instrument placed on the table is a Chinese beads abacus (*suanpan*); the most important detail that allows identify it is a clearly visible horizontal bar separating two sections of the abacus. The vertical bars for sliding beads are also shown rather clearly. The beads, however, are drawn as short horizontal lines between the vertical bars; this may have been the reason why the carvers who produced the printing block for the picture shown in Fig. 33 simplified the original picture and replaced the bars and beads with rectangular grid.

Another observation appears relevant here. The extant editions of the treatise in 17 *juans* contain a picture with the same caption, ("Master and disciple(s): asking difficult [questions]") shown in Fig. 36, but they do not contain the picture shown in Fig. 35 or its modifications; conversely, the editions in 12 *juans* do not contain pictures similar to that shown in Fig. 36 and instead often (but not always) provide generic versions of Fig. 34 (such as Figs. 33 and 35).[130]

The composition of this picture is somewhat similar to that of pictures shown in Figs. 33, 34, and 35: the Master is sitting and two disciples are standing; the Master turns to the disciple on his right; the Master is sitting under a large tree; the positions of the two disciples are similar to those shown in the 12-*juan* version, etc. However, there are some details that make the two pictures different: in particular, the one from the 17-*juan* edition contains such elements lacking in the shorter one as a deer behind the disciple on the Master's right and the (immortality?) mushroom growing in front of the Master's seat; in turn, the desk featured in the 12-*juan* edition does not appear in the 17-*juan* one. If the meeting of the Master with the disciples happens, according to the picture from the 12-*juan* versions, most likely in

---

[129]Chen (1977, p. 243).

[130]For instance, the edition of 1593 produced by the Wensheng tang 文盛堂 publishing house (currently preserved in Waseda University Library, Japan) does not contain the picture of the teacher and disciples, while the edition printed in the Dunhua tang 敦化堂 publishing house in the same year does contain the picture shown in Fig. 34.

Fig. 36 Master and two disciples (note the same caption as in Figs. 33, 34 and 35). From *juan* 13 of the 17-*juan* edition of 1675 (currently preserved in the library of Waseda University, Japan)

the inner court of the Master's house, the picture from the 17-*juan* version suggests that the meeting is taking place in a secluded place, "among brooks and pine trees," as the inserted text claims. What is the most important for the present discussion is that the disciple on the Master's right holds a clearly discernible abacus; interestingly enough, it is executed with some strange irregularities: the numbers of beads on some bars are clearly superior to the conventional two and five. It appears plausible to conjecture that the pictures shown in Figs. 35 (printed in 12-*juan* versions) and 36 (printed in 17-*juan* versions) were supposed to illustrate one and the same situation; this conjecture can be supported by the identical captions and similar positions of the portrayed individuals. If this conjecture is correct, it could provide an additional argument for the identification of the instrument depicted in 12-*juan* versions as an abacus and not as a "counting board."

Bibliography

Part A: Primary Sources

An Zhizhai 安止齋. 1373a. *Xiang ming suan fa* 詳明算法 (*Detailed Clarifications of Computational Methods*). In Guo 1993, vol. 1, pp. 1345–1397.

An Zhizhai 安止齋. 1373b. *Xiang ming suan fa* 詳明算法 (*Detailed Clarifications of Computational Methods*). In Jing 1994, vol. 1, pp. 1573–1619.

Cheng Dawei 程大位. 1592. *Suan fa tong zong jiao shi* 算法統宗校釋 (Unified Origin of Computational Methods, with Emendations and Explanations). Mei Rongzhao 梅荣照 and Li Zhaohua 李兆华 (Eds.). Hefei: Anhui jiaoyu.

Ding Ju 丁巨. 1355. *Ding Ju suan fa* 丁巨算法 (Computational methods of Ding Ju). In Guo 1993, vol. 1, pp. 1299–1314.

Guo Shuchun 郭書春 (ed.). 1993. *Zhongguo kexue zhishu dianji tonghui* 中國科學技術典籍通彙 (Comprehensive compendium of classical texts related to science and technology in China). *Shuxue juan* 數學卷 (Mathematical section). Zhengzhou: Henan jiaoyu chubanshe 河南教育出版社.

HNZ 1989 – Liu Wendian 劉文典 (ed.). *Huainan zi honglie jijie* 淮南子鴻烈集解 (Assembled Explanations of the Vast and Luminous Book [Compiled by] Masters of Huainan). Beijing: Zhonghua shuju, 1989.

HS – Ban Gu 班固. *Han shu* 漢書 (*History of the [Early] Han [Dynasty]*). *Qin ding Si ku quan shu hui yao* 欽定四庫全書薈要 edition, 1779.

Jing Yushu 靖玉樹 (ed.). 1994. *Zhongguo lidai suanxue jicheng* 中國歷代算學集成 (Collection of historical [documents on] mathematics in China). Jinan: Shandong renmin chubanshe.

LSQS – Mei Wending 梅文鼎, *Li suan quan shu* 歷算全書 (Complete Writings on Calendrical Computations). *Qin ding Si ku quan shu* 欽定四庫全書 edition, 1782.

P. 3349 – *Suan jing* 筭[=算]經 (Treatise on computations). Bibliotheque Nationale de France, Manuscript Pelliot 3349.

SJSS 1963 – Qian Baocong 錢寶琮 (ed.), *Suan jing shi shu* 算經十書 (Ten Treatises on Computations). Beijing: Zhonghua shuju.

SJSS 2001 – Guo Shuchun 郭書春 and Liu Dun 劉鈍 (eds.), *Suan jing shi shu* 算經十書 (Ten Treatises on Computations). Taibei: Jiu zhang.

SS 1973 – *Sui shu* 隋書 (History of the Sui [dynasty]). Beijing: Zhonghua shuju.

SSS 2001 – Peng Hao 彭浩. *Zhangjiashan Han jian 'Suan shu shu' zhushi* 張家山漢簡算數書注釋 (Commentary and Interpretation of the Han [Dynasty Treatise] *Suan shu shu* [Written] on Bamboo Strips [and Found at] Zhangjiashan). Beijing: Kexue, 2001.

STS 1698 – Satō Shigeharu 佐藤茂春, *Sanpō tengen shinan* 算法天元指南 (Compass [for understanding the technique of] 'Celestial Element' [used in] computational methods) (preface 1698).

SWJZ – Xu Shen 許慎. *Shuowen jiezi* 說文解字 (Explanation of graphs and analysis of characters). *Qin ding Si ku quan shu hui yao* 欽定四庫全書薈要 edition, 1779.

Wang Wensu 王文素. 1524. *Suan xue bao jian* 算學寶鑑 (Precious Mirror of the Science of Computations). In Guo 1993, vol. 2, pp. 333–971.

Yang Hui 楊輝. 1274. *Cheng chu tong bian ben mo* 乘除通變本末 (Alpha and Omega of Continuation and Change in [methods] of Multiplication and Division). In Guo 1993, vol. 1, pp. 1047–1072.

YJ – *Yi jing tong zhu* 易經通注 (The Book of Changes with comprehensive commentaries), *Qin ding Si ku quan shu* 欽定四庫全書 edition, 1782.

YL – *Yi li ji shuo* 儀禮集說 (The *Rituals and Ceremonies* with Anthology of Explanations), *Qin ding Si ku quan shu hui yao* 欽定四庫全書薈要 edition, 1779.

Zhou Shuxue 周述學. 1558. *Shen dao da bian li zong suan hui* 神道大編曆宗算會. In: *Xuxiu Si ku quan shu* 續修四庫全書, vol. 1043, pp. 555–832. Shanghai: Shanghai guji chubanse.

ZBSJ 2001 – *Zhou bi suan jing* 周髀算經 (Computational treatise on the Gnomon of the Zhou [dynasty]). In *SJSS 2001*, pp. 29–65.

Part B: Translations of Chinese Texts

Berezkina, Él'vira I. [Березкина, Эльвира И.] (tr.). 1957. Математика в девяти книгах (Mathematics in nine books [= *Jiuzhang suanshu*], an annotated Russian translation). *Историко-математические исследования (Studies in the History of Mathematics)*, vol. 10, pp. 427–584.

Berezkina, Él'vira I. [Березкина, Эльвира И.] (tr.). 1963. О математическом труде Сунь-цзы. Сунь-цзы: Математический трактат. Примечания к трактату Сунь-цзы (On the mathematical work of Sun zi. Sun zi: *Mathematical treatise*. Annotations for Sun zi's treatise). *Из истории науки и техники в странах Востока (On the history of science and technology in the countries of the East)*, Moscow: Nauka, vol. 3, pp. 5–70.

Berezkina, Él'vira I. [Березкина, Эльвира И.]. 1969. О трактате Чжан Цю-Цзяня по математике (On Zhang Qiujian mathematical treatise). *Физико-математические науки в странах Востока (Physics and mathematics in the countries of the East)*, vol. 2, pp. 18–81 (The paper contains the full annotated translation of the treatise).

Berezkina, Él'vira I. [Березкина, Эльвира И.] (tr.). 1975. Ван Сяо-тун. Математический трактат о продолжении древних (методов) (Wang Xiaotong. *Mathematical treatise on the continuation of ancient (methods)*, an annotated Russian translation). *Историко-математические исследования (Studies in the History of Mathematics)*, vol. 20, pp. 329–371.

Berezkina, El'vira I. [Березкина, Эльвира И.] (tr.). 1985. Математический трактат Сяхоу Яна (Mathematical treatise of Xiahou Yang). *Историко-математические исследования (Studies in the History of Mathematics)*, vol. 28, pp. 293–337.

Chemla, Karine, Guo Shuchun. 2004. 九章算術。 *Les Neuf Chapitres: le classique mathématique de la Chine ancienne et ses commentaires*. Paris: Dunod.

Couvreur, Séraphin (tr.). 1951. *I-Li: Cérémonial*. Paris: Les Belles Lettres (Les Humanités d'Extrême-Orient, Cathasia, série culturelle des Hautes Études de Tien-Tsin). [Reprint; originally published in Tianjin in 1916].

Cullen, Christopher. 1996. *Astronomy and Mathematics in Ancient China: the* Zhou bi suan jing, (Needham Research Institute Studies, 1). Cambridge, Cambridge University Press.

Cullen, Christopher (tr.). 2004. *The* Suan shu shu *'Writing on reckoning': A Translation of a Chinese Mathematical Collection of the Second Century BC, With Explanatory Commentary.* Cambridge, Needham Research Institute (*Needham Research Institute working papers*, vol. 1).

Dauben, Joseph W. (tr.). 2008. 算數書 *Suan shu shu*: A Book on Numbers and Computations, *Archive for the History of Exact Sciences*, vol. 62, pp. 91–178.

Dauben, Joseph W. 道本周, Guo Shuchun 郭書春, and Xu Yibao 徐義保 (tr.). 2013. 九章筭术 *Nine Chapters on the Art of Mathematics.* Shenyang: Liaoning jiaoyu chubanshe.

Lam Lay Yong, and Ang Tian Se (tr.). 2004 [1992]. *Fleeting Footsteps. Tracing the Conception of Arithmetic and Algebra in Ancient China*, Singapore etc.: World Scientific.

Le Blanc, Charles, et Rémi Mathieu (tr.). 2003. *Huainan zi.* Paris: Gallimard.

Lim, Tina Su Lyn, and Donald B. Wagner. 2017. *The Continuation of Ancient Mathematics. Wang Xiaotong's Jigu suanjing, Algebra and Geometry in 7th-Century China.* Copenhagen: NIAS Press.

Schrimpf, Robert. 1963. *La collection mathématique "Souan king che chou": Contribtion à l'histoire des mathématiques chinoises des origins au VIIe siècle de notre ere.* (An unpublished PhD dissertation). Rennes: Universite de Rennes.

Steele, John (tr.). 1917. *The* I-Li, *or Book of Etiquette and Ceremonial.* London: Probsthain.

Part C: Secondary Works

Berezkina, Él'vira I. [Березкина, Эльвира И.]. 1980. *Matematika drevnego Kitaya* [Математика древнего Китая] (Mathematics of Ancient China, in Russian). Moscow: Nauka.

Boltz, William G. 1993. *I-Li* [*Yi li*, a bibliographical entry]. In: Michael Loewe (ed.), *Early Chinese Texts.* Berkeley: The Society for the Study of Early China and The Institute of East Asian Studies, University of California, pp. 234–243.

Bottéro, Françoise, and Christoph Harbsmeier. 2008. The *Shuowen Jiezi* Dictionary and the Human Sciences in China. *Asia Major*, vol. 21, pp. 249–271.

Brenier, Joël. 1994. Notation et optimization du calcul des grands nombres en Chine. Le cas de l'échiquier de *go* dans le *Mengqi bitan* de Shen Gua (1086). In: Isabelle Ang and Pierre-Étienne Will (eds.), *Nombres, Astres, Plantes et Viscères: Sept essays sur l'histoire des sciences et des techqniques en Asie Orientale.* Paris: Collège de France, Institut des Hautes Études Chinoises, pp. 89–111.

Chemla, Karine. 1982. *Etude du livre "Reflets des mesures du cercle sur la mer" de Li Ye.* (An unpublished PhD dissertation). Paris: Univeristy Paris-XIII.

Chemla, Karine. 1996. Positions et changements en mathématiques à partir de textes chinois des dynasties Han à Song-Yuan. Quelques remarques. In: Karine Chemla et Michael Lackner (eds.), *Disposer pour dire, placer pour penser, situer pour agir: Pratiques de la position en Chine* (*Extrême-Orient Extrême-Occident*, no. 18), Paris: PUV, pp. 116–147.

Chen Liang-Tso [Chen Liangzuo] 陳良佐. 1978. Xian Qin shuxue de fazhan ji qi yingxiang 先秦數學的發展及其影響 (The development and influence of pre-Qin mathematics). *Lishi yuyan yanjiusuo jikan* 歷史語言研究所集刊, vol. 49, pt. 2, pp. 263–320.

Chen Liang-Tso [Chen Liangzuo] 陳良佐. 1997. Woguo chousuan zhong de kongwei – ling – ji qi xiangguan de yixie wenti 我國籌算中的空位—零—及其相關的一些問題 (An empty position [used as] zero [when manipulating with] counting rods in our country [in the past] and some related problems). *Dalu zazhi* 大陸雜誌, vol. 54, no. 5, pp. 238–250.

Cheng, Anne. 1993. *Ch'un ch'iu, Kung yang, Ku liang* and *Tso chuan* [*Chunqiu, Gongyang, Guliang* and *Zuozhuan*, a bibliographical entry]. In: Michael Loewe (ed.), *Early Chinese Texts.* Berkeley: The Society for the Study of Early China and The Institute of East Asian Studies, University of California, pp. 67–76.

de la Grasserie, Raoul. 1896. Terrien de Lacouperie. *Bulletin de la Société de linguistique de Paris*, vol. 9, nos. 39–43, pp. lxxxi–lxxxvi.

des Rotours, Robert. 1932. *Le traité des examens, traduit de la Nouvelle histoire des T'ang (chap. 44–45)*. Paris: Librairie Ernest Leroux.

Djamouri, Redouane. 1994. L'emploi des signes numériques dans les inscriptions Shang. In: A. Volkov (ed.), *Sous les nombres, le monde: Matériaux pour l'histoire culturelle du nombre en Chine ancientne* (*Extrême-Orient, Extrême-Occident*, no. 16), pp. 12–42.

Han Wei 韩巍. 2012. Beida Qin jian zhong de shuxue wenxian 北大秦简中的数学文献 (Mathematical texts found among the [documents on] bamboo strips [preserved] in Peking University). *Wenwu* 文物 (Cultural relics), no. 6, pp. 85–89.

Hoe, John [=Jock]. 1977. *Les systèmes d'équations polynomes dans le* Siyuan Yujian *(1303)*. Paris: Collège de France.

Horiuchi, Annick. 1994. *Les mathématiques japonaises à l'époque d'Edo*. Paris: Vrin.

Hua Yinchun 華印椿 (ed.). 1987. *Zhongguo zhusuan shigao* 中國珠算史稿 (An unofficial history of "computation with beads" in China). Beijing: Zhongguo caizheng jingji chubanshe 中國財政經濟出版社.

Jami, Catherine. 1994. History of Mathematics in Mei Wending's (1633–1721) Work. *Historia Scientiarum*, vol. 4, no. 2, pp. 159–174.

Jami, Catherine. 1998. Abacus. In: Robert Bud and Deborah J. Warner (eds.), *Instrument of Science: an historical encyclopedia*, pp. 3–5. New York/London: Garland.

Kalinowski, Marc. 1994. La divination par les nombres dans les manuscrits de Dunhuang. In: Isabelle Ang and Pierre-Etienne Will (eds.), *Nombres, astres, plantes et viscères. Sept essais sur l'histoire des sciences et des techniques en Asie orientale.* (*Mémoires de l'Institut des Hautes Études chinoises*, vol. 35), Paris: Collège de France, pp. 37–88.

Kalinowski, Marc. 1996. Astrologie calendaire et calcul de position dans la Chine ancienne: Les mutations de l'hémérologie sexagésimale entre le IVe et IIe siècle avant notre ère. In: Karine Chemla et Michael Lackner (eds.), *Disposer pour dire, placer pour penser, situer pour agir: Pratiques de la position en Chine* (*Extrême-Orient Extrême-Occident*, no. 18), Paris: PUV, pp. 71–113.

Lam Lay Yong, and Ang Tian Se. 2004. *Fleeting Footsteps: Tracing the Conception of Arithmetic and Algebra in Ancient China*. Singapore: World Scientific.

Li Di 李迪. 1997. *Zhongguo shuxue tongshi. Shang gu dao Wudai juan.* 中國數學通史。上古到五代卷 (Comprehensive history of Chinese mathematics. Volume [on the period] from ancient times to 'Five Dynasties'). Nanjing: Jiangsu jiaoyu chubanshe.

Li Di 李迪 (volume editor). (1999). *Xi Xia, Jin, Yuan, Ming* 西夏金圓明 ([Mathematics of the state of] Xi Xia and of the Jin, Yuan and Ming [dynasties]). Volume 6 of Wu Wenjun 吳文俊 (general editor), *Zhongguo shuxue daxi* 中國數學史大系 (Encyclopedia of the history of Chinese mathematics). Beijing: Beijing Shifan daxue chubanshe.

Li Yan 李儼. 1931. *Zhongguo shuxue dagang* 中國數學大綱 (Outline of [the history of] Chinese mathematics).

Li Yan 李儼. 1955a [1935]. Chousuan zhidu kao 籌算制度考 (A study of the system of counting rods). In Li Yan 李儼, *Zhong suan shi luncong* 中算史論叢 (Collected works on the history of mathematics in China). Beijing: Kexue chubanshe, vol. 4, pp. 1–8.

Li Yan 李儼. 1955b [1954]. *Zhongguo gudai shuxue shiliao* 中國古代數學史料 (Materials for [study of the history of] ancient mathematics in China). Shanghai: Zhongguo kexue tushu yiqi.

Li Yan 李儼. 1983 [1937]. *Zhongguo suanxue shi* 中國算學史 (History of Chinese mathematics). Shanghai: Shangwu yinshuguan. [Reprinted *facsimile* by Taiwan shangwu yinshuguan (Taibei) in 1983].

Li Yan, and Du Shiran. 1987. *Chinese mathematics: A concise history*. (Translated by John N. Crossley and Anthony W.-C. Lun). Oxford: Clarendon Press.

Libbrecht, Ulrich. 1973. *Chinese Mathematics in the Thirteenth Century: The* Shu-shu chiu-chang *of Ch'in Chiu-shao*. Cambridge (Mass.) and London: MIT Press.

Libbrecht, Ulrich. 1982. Mathematical Manuscripts from Dunhuang Caves. In: Li Guohao, Zhang Mengwen, and Cao Tianqin (eds.), *Explorations in the History of Science and Technology in China*. Shanghai: Shanghai Publishing House, pp. 203–229.

Martzloff, Jean-Claude. 1997. *A history of Chinese mathematics*. Berlin etc.: Springer.

Mei Rongzhao. 1983. The decimal place-value numeration and the rod and bead arithmetic. In: *Ancient China's Technology and Science*, Peking: Foreign Language Press, pp. 57–65.

Mikami, Yoshio. 1911. The influence of abaci on Chinese and Japanese mathematics. *Jahresberichte der Deutschen Mathematiker*, vol. 20, pp. 380–393.

Mikami, Yoshio 三上義夫. 1926. Shina sūgaku-no tokushoku 支那數學の特色 (Special characteristics of Chinese mathematics). *Tōyō gakuhō* 東洋學報 (Reports of the Oriental Society of Tokyo), vol. 15, no. 4; vol. 16, no. 1.

Mikami Yoshio. 1934. *Zhongguo suanxue zhi tese* 支那數學之特色 (Chinese translation of Mikami (1926) by Lin Ketang 林科棠). Shanghai: Shangwu, 1934.

Mikami Yoshio. 1974 [1913]. *The development of mathematics in China and Japan*. Leipzig; reprinted by Chelsea Publishing Company (New York) in 1974 with an appendix authored by Fujisawa Rikitarō.

Needham, Joseph. 1959. *Science and Civilisation in China*. (Volume 3: *Mathematics and the Sciences of the Heavens and the Earth*.) Cambridge: Cambridge University Press.

Peterson, Willard J. 1982. Making connections: 'Commentary on the Attached Verbalizations' of the Book of Changes. *Harvard Journal of Asiatic Studies*, vol. 42, no. 1, pp. 67–116.

Siu Man-Keung, Volkov, Alexei. 1999. Official curriculum in traditional Chinese mathematics: How did candidates pass the examinations? *Historia Scientiarum*, vol. 9, no. 1, pp. 85–99.

Smith, David E. and Mikami Yoshio. 1914. *A History of Japanese Mathematics*. Chicago: The Open Court.

Terrien de Lacouperie, Albert Étienne Jean-Baptiste. 1883. The Old Numerals, the Counting-Rods and the Swan-Pan in China. *The Numismatic Chronicle and Journal of the Numismatic Society*, Third Series, vol. 3, pp. 297–340.

Vissière, A[rnold]. 1892. Recherches sur l'origine de l'abaque Chinois. *Bulletin de Geographie Historique et Descriptive*, [t. 28,] n. 1, pp. 54–80.

Vogel, Hans Ulrich. 1994. Aspects of metrosophy and metrology during the Han period. In: Alexei Volkov (Ed.), *Sous les nombres, le monde: Matériaux pour l'histoire culturelle du nombre en Chine ancienne* (*Extrême-Orient, Extrême-Occident*, no. 16), pp. 135–152.

Volkov, Alexei. 1994. Large numbers and counting rods. In: Alexei Volkov (Ed.), *Sous les nombres, le monde: Matériaux pour l'histoire culturelle du nombre en Chine ancienne.* (*Extrême-Orient Extrême-Occident*, no. 16), pp. 71–92.

Volkov, Alexei. 1997. The mathematical work of Zhao Youqin: remote surveying and the computation of *pi*. *Taiwanese Journal for Philosophy and History of Science*, no. 8 (1996–1997) (vol. 5, no. 1), pp. 129–189.

Volkov, Alexei. 1998. Counting rods. In: Robert Bud and Deborah J. Warner (eds.), *Instruments of Science: an Historical Encyclopedia*. London & New York: Garland, pp. 155–156.

Volkov, Alexei. 2000. From Numerology to Arithmetic: the Early Chinese Counting Devices. In: Zhang Jiafeng and Liu Juncan (Eds.), *Proceedings of the 1999 Conference on the History of Science (Taipei, Republic of China)*, Taibei: Academia Sinica, pp. 29–69.

Volkov, Alexeï. Le bacchette [Counting rods]. 2001. In: Sandro Petruccioli (Gen. Ed.), Karine Chemla, Francesca Bray, Daiwie Fu, Yilong Huang and Georges Métailié (Volume Eds.), *Storia Della Scienza* (Encyclopedia on History of science, in Italian), vol. 2: Cina, India, Americhe (China, India, America), pp. 481–485. Rome: Istituto della Enciclopedia Italiana.

Volkov, Alexei. 2014. History of mathematics education in Oriental Antiquity and Middle Ages. In: Alexander Karp and Gert Schubring (Eds.), *Handbook on the History of Mathematics Education*, New York etc: Springer, pp. 55–70, 79–82.

Volkov, Alexei. 2018. Visual representations of arithmetical operations performed with counting instruments in Chinese mathematical treatises. In: Fulvia Furinghetti and Alexander Karp (Eds.), *Researching the History of Mathematics Education: An International Overview* (ICME-13 Monographs). Cham: Springer, pp. 279–304.

Wang Ling. 1956. The *Chiu Chang Suan Shu* and the History of Chinese Mathematics during the Han Dynasty. (An unpublished PhD dissertation). [Cambridge:] Trinity College.

Wu Chengluo 吳承洛. 1937. *Zhongguo duliangheng shi* 中國度量衡史 (History of measures, volumes, and weights in China). Shanghai: Shangwu yinshuguan.

Yabuuti Kiyosi. 2000. *Une histoire des mathématiques chinoises.* Paris: Belin.

Yamazaki Yoemon. 1962. History of instrumental multiplication and division in China - from the reckoning blocks to the abacus. *Memoirs of the research department of the Toyo Bunko*, no. 21, pp. 125–148.

Yang Qiong-Ru 楊瓊茹. 2003. Ming dai lixuejia Zhou Shuxue ji qi suanxue yanjiu 明代曆算學家 周述學及其算學研究 (A study of Zhou Shuxue, an expert in calendrical science of the Ming dynasty, and of his mathematical methods). (An unpublished Master thesis). Taibei: National Taiwan Normal University.

Alexei Volkov is a professor of the Center for General Education and of the Graduate Institute of History of the National Tsing-Hua University (Hsinchu, Taiwan). His research focuses on the history of mathematics and mathematics education in East and Southeast Asia. He has published a number of papers and book chapters on these topics, including "Didactical dimensions of mathematical problems: 'weighted distribution' in a Vietnamese mathematical treatise", in Alain Bernard and Christine Proust (eds.), *Scientific Sources and Teaching Contexts Throughout History: Problems and Perspectives*, Dordrecht etc: Springer, 2014, pp. 247–272 and "Argumentation for state examinations: demonstration in traditional Chinese and Vietnamese mathematics", in Karine Chemla (ed.), *The History of Mathematical Proof in Ancient Traditions*, Cambridge University Press, 2012, pp. 509–551.

Interpreting Algorithms Written in Chinese and Attempting the Reconstitution of Tabular Setting: Some Elements of Comparative History

Charlotte-V. Pollet

Abstract The mathematics of 12th–14th-century China is known for its beautiful algebraic texts. Unfortunately, information concerning their context of transmission and instruction is scarce. One interesting pattern is that many of the texts share a predisposition for tabular setting and several of these texts refer to the same algebraic procedure named *tian yuan* 天元 (Celestial Source) used to set up polynomial equations. The setting of these equations on a counting surface is the result of a specificity of using counting rods for the algorithm of division. Precisely, the role of division for setting up and solving equations is fundamental to the algorithm. This chapter presents an excerpt borrowed from Li Ye's 李冶 *Yigu yanduan* 益古演段 (the Development of Pieces [of Area according to the Collection] Augmenting the Ancient [Knowledge], 1259). It presents first a basic example of the Celestial Source procedure, then attempts reconstitution of polynomials on the counting surface and ends with comparative observations related to the chapter on the *Bījagaṇitāvataṃsa* (BGA) written by Nārāyaṇa in 14th-century India. The description of the algorithm for setting up a quadratic equation is interesting from a comparative perspective. The way in which lists of operations are ordered shows that Indian and Chinese authors had different interests, addressed different difficulties and understood mathematical concepts differently, while referring to division, using tabular setting and "model" equations.

Keywords Algorithm · Division · Algebra · Song dynasty · China Li Ye

C.-V. Pollet (✉)
Center for General Education, National Chiao-Tung University, 1001 University Road, 30050 Hsinchu, Taiwan
e-mail: charlotte.pollet7@gmail.com

© Springer Nature Switzerland AG 2018   189
A. Volkov and V. Freiman (eds.), *Computations and Computing Devices in Mathematics Education Before the Advent of Electronic Calculators*, Mathematics Education in the Digital Era 11, https://doi.org/10.1007/978-3-319-73396-8_8

Introduction

Several high-level algebraic texts were produced in 12th–14th-century China. Yet, there is scarce information concerning their context of transmission and instruction. These texts were not part of any official curriculum and there is no information regarding their readers. It is known that from the middle of the first millennium onwards, a number of mathematical texts were used as textbooks in state educational institutions concerned mainly with administrative affairs and calendar making. However, during the Song (960–1279) and Yuan (1271–1368) dynasties, there existed mathematical texts unrelated to official education. For instance, Yang Hui 楊輝 (second half of the 13th century), Qin Jiushao 秦九韶 (1209–1261), Li Ye 李冶 (1192–1279) and Zhu Shijie 朱世傑 (late thirteenth century to early fourteenth century) composed texts that were not dealing with economical and astronomical matters and were never used for state examinations. These authors referred to texts that are more ancient and some of them claimed that they wrote their treatises for the purpose of transmission and clarification of ancient mathematical procedures. Unfortunately, their treatises deliver clues on neither their context of transmission nor their readership. There is almost no chance that these four mathematicians ever met. However, common technical points for some topics (algorithms, names of procedures and other elements of technical vocabulary) show that there was a circulation of knowledge.

For instance, the expression *tian yuan* (Celestial Source) appears in three of the works of the authors mentioned above. The term *tian yuan* first appeared in Qin Jiushao's *Shushu jiuzhang* 數書九章 (*Mathematical Treatise in Nine Chapters*, 1247), and its usage was related to what we identify as indeterminate analysis (*da yan* 大衍, translated as "great extension") (Libbrecht 1973, pp. 345–346). The works of Li Ye contain the earliest evidence of the procedure named *tian yuan shu* (天元術, literally "Celestial Source Procedure"), which is used to establish and solve algebraic equations. The method (*fa* 法) used by Li Ye was later titled "procedure" (*shu* 術) by the Qing dynasty (1644–1912) editor and mathematician Li Rui 李銳 (1768–1817).[1] The procedure was generalized independently by Zhu Shijie in the *Si yuan yu jian* 四元玉鑒 (*Precious Mirror of Four Sources*, 1303) to the "procedure of four sources," *si yuan shu* 四元術, which is generalized from a procedure of solution of a polynomial equation in one unknown to a procedure of solution of simultaneous polynomial equations in at most four unknowns.

Observation of algorithms presented in these mathematical works shows a predilection for tabular representation, use of certain types of equations as models and construction based on the algorithm of division. Arithmetical operations were already reconstituted by historians, and the relation between the algorithms of division and extraction of square roots in a tabular setting was already thoroughly studied; see, for example, Chemla (2006). Some manipulations with counting rods in the context of polynomial algebra are given in Hoe (2007), yet, we do not know precisely how arithmetical and algebraic operations were combined. For instance,

---

[1] Pollet and Ying (2017).

we will see that on the counting surface the initial step of the division of 3 by 4 and the polynomial $3x + 4x^2$ were represented in the same way. Neither do we know which part of the computation was mental and which was instrumental. It is possible to conceptualize these operations, or to symbolize them with modern notation, but it is not possible to reconstruct precisely their implementation on the counting device. This chapter explores the limits for the reconstitution of operations with counting rods for polynomial algebra.

An example is borrowed the *Yigu yanduan* 益古演段 (*Development of Pieces [of Area according to the Collection] Augmenting the Ancient [Knowledge]*)[2] and presents the algorithm later named "Procedure of Celestial Source" as it appeared in Li Ye's treatise written in 1259.[3] His literary name was Renqing 仁卿, and his appellation was Jingzhai 敬齋. He was born into a bureaucratic family in 1192. His first piece of work in mathematics was the *Ceyuan haijing* 測圓海鏡 (*Sea Mirror of Circle Measurement*) written in 1248, composed when he lived as a scholarly recluse in the Fenglong mountains 封龍山 in Hebei 河北, where he perhaps accepted students for instruction. In this environment, he produced the *Yigu yanduan* in 1259. He died in the Fenglong mountains in 1279. I present an excerpt of the first problem of the *Yigu yanduan*, which gives a clear and simple illustration of the Procedure of Celestial Source.

It is known that Chinese mathematicians were manipulating with counting rods on a countingsurface (Li and Du 1987, pp. 6–19; Martzloff 1988, pp. 194–196). We have no information concerning the counting surface and little information on the small bamboo rods. In August 1971, more than 30 rods of 140 mm long were unearthed from tombs from the time of Emperor Xuan (73–49 BC) of the western Han dynasty in Shanxi. In 1975, in Hubei, a bundle of rods was unearthed from the tombs of the reign of Emperor Wen (179–157 BC). The chapter titled "*Lü li zhi*" 律歷誌 ("Record on pitch-pipes and calendar") in the *Sui shu* 隋書 (*History of the Sui [Dynasty]*, compiled in the 7th century AD) states, "To calculate, one uses bamboo, two *fen* wide and three inches long" (Li and Du 1987, pp. 7–8). The counting rods gradually became shorter, but since no later artefacts have been discovered, the aspects of the rods in the Song–Yuan dynasties remain unknown.[4]

---

[2]"Development (演) of Pieces (段) [of Areas] [according to] [the collection] Augmenting (益) the Ancient (古) [Knowledge]." It is a presentation of a geometrical procedure named "Development by the Section of Pieces [of Areas]," based on an ancient treatise titled *Yiguji* 益古集 (*Collection Augmenting the Ancient [knowledge]*) of the 11th century (for a translation of terms see Pollet (2012) and Pollet and Ying (2017)).

[3]Biographies of Li Ye can be found in English in Mikami (1913, p. 80); Ho (1973, pp. 313–320); Lam (1984, pp. 237–239); Li and Du (1987, p. 114) and in Chinese in Mei (1966, p. 107). His life has been the object of several notices since the Yuan dynasty, with the first being written in 1370 in *Yuan Shi* 元史 (*Official History of the Yuan*, chapter 160) and the last being written in 1799 in *Chouren Zhuan* 疇人傳 (*Biographies of Astronomers [and Mathematicians]*) by Ruan Yuan 阮元 (1764–1849). I will not consider further this material in the present work.

[4]*See* Needham (1959, p. 365); Martzloff (1987, p. 194); Guo (1991, pp. 26–27); Li and Du (1987, pp. 6–24); Chemla (1982, Chap. 4, Sect. 3); Chemla (1996); Chemla and Guo (2004, pp. 15–20); Volkov (2001); Hoe (2007). *See also* the chapter authored by A. Volkov in the present volume.

There is even less information on the counting surface. We do not know if the surface was a board or a piece of cloth, or if any other support could be used as a counting surface. Strangely, the names of positions—"above" *shang* 上, "below" *xia* 下, "middle" *zhong* 中, "left" *zuo* 左, "right" *you* 右, "Celestial Source" *tian yuan* 天元, which appear in the example given in this chapter, coincide with the names of fixed positions on the table which is used to play *Weiqi* 圍棋 (better known in the West under its Japanese name *Go* 碁) with red and blue or white and black tokens (*chou* 籌) (He 2001). Martzloff (1998, p. 259) noticed that *tian yuan* is also the name of the central point of the grid on which the game is played. The *tian yuan* area on the grid is a circular area delimited at its north, south, east and west by intersections named *zhong fu* 中腹. The square surrounding the *tian yuan* circle is named "earth" (*di* 地). The same appellation is found in problems 19 and 43 of *Yigu yanduan*. There are 19 positions on the intersections of lines of the grid where the tokens are placed. In the paragraph of the *Jingzhai gujing tou* 敬齋古今黈 (*Commentary of Jingzhai on Things Old and New*) written by Li Ye, 19 positions were listed as positions on a counting surface. An identical table was also used for divination with sticks according to the method described in the *Yijing* (*Book of Changes*, 易經) since the Tang dynasty at least (Liu 2007). This board was a very common object for literati, and the ability to play *Weiqi* was listed among the basic skills that the noble people were supposed to master (Bottermans 2008). However, as we do not know what the counting rods were like in the Song–Yuan dynasties, it is difficult to know how wooden sticks would fit on the board and how they were manipulated.

If the reconstitution of manipulation on the counting surface has some limit, the description of the algorithm for setting up a quadratic equation becomes interesting from a comparative perspective. In this book (Chapter "Reading Algorithms in Sanskrit"), I already proposed an interpretation of an algorithm written in Sanskrit borrowed from the *Bījagaṇitāvataṃsa* (BGA) written by Nārāyaṇa in 14th-century India. Some curious points of resemblance deserve further investigation. The setting of a polynomial equation on a counting surface is the result of the specificity of using counting rods for the algorithm of division. Precisely, the role of division for setting up and solving an equation was also a fundamental step in the algorithm presented in Chapter "Reading Algorithms in Sanskrit" on the BGA. Both Indian and Chinese authors relied on tabular setting, division and, as we will see once again here, on model equations. That is to say, in both cases models of linear or quadratic equations were used as a frame to set up every equation. The role of division is pivotal in both algorithms. This chapter presents first a basic example of the Celestial Source Procedure, then an attempt of reconstitution of polynomials on the counting surface and ends with comparative observations related to Chapter "Reading Algorithms in Sanskrit" on the BGA.

Description of the Algebraic Procedure

The *Yigu yanduan* is composed of 64 problems solved first by the Procedure of the Celestial Source, then by a second procedure named "Section of Pieces (of Areas)" (*tiao duan* 條段).[5] The Celestial Source Procedure is considered a key element in the Chinese history of mathematics. The expression *tian yuan shu* was translated as "technique of the celestial element" by Li and Du (1987, p. 135) or as "heavenly element method" by Dauben (2007, p. 324). It is not uncommon to translate the character *shu* 術 as the technical mathematical term of "procedure," or to use its synonym "algorithm." The character y*uan* is sometimes translated as "element" and represents the unknown. The prefix *tian*—literally "sky" or "celestial"—indicates that it is the first unknown or the only one. Needham (1959, vol. 3, p. 129) pointed out the disadvantage of translating *yuan* as "element," because of the confusion with the elements of chemistry. Hoe (2007, p. 19) notices that in philosophical texts, *yuan* is "the source from which all the matter in the universe stems." *Yuan* means "the origin," "the beginning." However, he chose to translate it technically as "unknown." Our preference is to follow the ancient philosophers and Jock Hoe's observation and to translate *yuan* as "source."

Each of the statements proposed a problem of field surveying as a pretext for working on a quadratic equation. In the *Yigu yanduan*, the Celestial Source is a procedure for setting a quadratic equation in one unknown. In some cases (only two), linear equations are also solved. There is no case containing simultaneous equations in several unknowns. The procedure to solve the equation is never given because the reader is assumed to be familiar with it. Only one of the roots, always positive, is given by means of the solution of a polynomial equation. The simplicity of this systematic procedure is admirable.

First, one of the parameters is taken as unknown and then, based on the condition given in the statement of the problem, the equation that governs the unknown is found. The problems are solved after setting up an equation whose chosen quantity for the unknown is the root.

Each time, the statement of the problem presents two geometrical figures (i.e., a square field with an inner circular pond), gives one segment (i.e., the side of field) and asks for another segment (i.e., the diameter of the pond). For instance, a modern mathematical transcription of the problem that one may suggest is given in the following text (see the Appendix to this chapter for the Chinese text and for a translation of the problem and its solution).

Let a be the distance from the middle of the side of the square to the pond, 20 *bu*; let A be the area of the square field (S) less the area of the circular pond (C), 13 *mu* and 17.5 *fen*, or 3300 *bu*; and x be the diameter of the pond (Fig. 1).

---

[5]Pollet (2012, 2014).

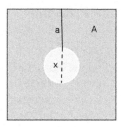

Fig. 1 Illustration of the statement of problem 1 of the *Yigu yanduan*

The solution of the problems with the Procedure of Celestial Source is composed of three steps made up of the nine repetitive sentences composing the mathematical discourse.[6] These sentences give a repetitive list of operations that rhetorically lead to the construction of mathematical expressions, and imply manipulations of rods performed on a counting surface. These manipulations are never described, so the reader is assumed to be acquainted with them. Three mains steps can be outlined:[7]

(1) The first mathematical expression corresponding to the area of one of the figures named in the statement of the problem is computed. This is used to express the first area in terms of a polynomial. This polynomial is to be placed on the top (*touwei* 頭位) of the surface.
For the problem above:

$$\text{Side of the square} = 2a + x = 40 + x$$
$$S = (2a + x)^2 = 4a^2 + 4ax + x^2 = 1600 + 80x + x^2.$$

(2) The second mathematical expression corresponding to the other figure is then computed. This gives a second polynomial. The second expression is subtracted from the first one or infrequently added depending on the segments asked in the problem (pb. 21; 23 to 30; 38; 43; 46 and 63). This polynomial is to be placed on the left.
For the problem above:
$C = 3/4x^2 = 0.75x^2$, since $\pi = 3$.[8]

---

[6]In the present case, the text is composed of a mathematical discourse made of sentences written with Chinese characters and tabular mathematical expressions, of diagrams with their legends and of commentaries. Here I discuss what I referred to as mathematical discourse. The two other items composing the text are the objects of other publications.

[7]There are many publications on this procedure. The reader can refer to Pollet (2012, 2014); Pollet and Ying (2017); Chemla and Guo (2004, pp. 314–322); Chemla (1982, Chap. 8); Li and Du (1987, pp. 17–19); Lam and Ang (2004, pp. 63–71); and Martzloff (1987, pp. 229–249) for publications in Western languages, among others. Here I just indicate the main thread of the procedure for the purpose of comparison with Indian documents discussed in Chapter "Reading Algorithms in Sanskrit" of the present volume.

[8]This approximates the value of π as 3. Arguably, this value does not mean that the author was not aware of the fact that π is not equal to three; rather, the value is the result of a process which

(3) The expression resulting from this operation is equal to the area given in the statement. The remainder or addendum of the two polynomials is equal to a constant, and the second is eliminated (*xiang xiao* 相消) from the former (or conversely) to give the equation.

For the problem above:

$$S - C = 4a^2 + 4ax + x^2 - 3/4x^2 = A$$
$$= 1600 + 80x + x^2 - 0.75x^2 = 1600 + 80x + 0.25x^2 = 3300bu.$$

The equation is:

$$A - \left(4a^2 + 4ax + x^2 - 3/4x^2\right) = 1700 - 80x - 0.25x^2 = 0.$$

There is no precise description of how two polynomials should be subtracted on the counting device.

Description of Tabular Setting

Quadratic polynomials and equations were represented on the counting surface as combinations of counting rods. Each polynomial was represented as a column of its coefficients placed vertically. The coefficients were represented as configurations of rods placed vertically and horizontally; when pictured in mathematical treatises, they were shown as configurations of horizontal and vertical lines.

Fig. 2 A polynomial in the *Yigu yanduan*

---

calculates the areas of circular figures from corresponding square figures. Thus, the four circular areas are equivalent to three squares with sides equal to their diameter. This arithmetical transformation is convenient for the geometrical transformation of areas. To represent four circles, three squares may be drawn instead. Liu Hui 劉徽 (*3rd century AD*) gives $\pi \approx 3.1416$ and Zu Chongzhi's 祖沖之 (AD 429–500) approximation was between 3.1415926 and 3.1415927.

For example, the tabular mathematical expression in Fig. 2 can be transcribed as

2700   *tai*

252

5.87

and denotes $2700 + 252x + 5.87x^2$. Each row indicates a term of consecutive degree. The upper row contains the constant term, the middle row contains the coefficient of the first degree of the unknown (say, x in modern algebraic notation) and the lower row contains the coefficient of the second degree of the unknown (x^2). The characters 太 *tai* or 元 *yuan* at the side of the array indicate the meaning of the numbers relative to the marked position. The character *tai* on the upper row indicates the constant term. *Yuan* is used when there is no constant term; it indicates the first-order term. There is no specific character to mark the terms corresponding to the degree of the unknown higher than the first.[9]

In problems 38, 44, 48, 56, 59 and 60, Li Ye refers to different rows as "what is below is the divisor, what is above is the dividend" (*xia fa shang shi* 下法上實). They are names borrowed from the algorithm of division. Another expression relative to division is used in problem 23: "to extract the square root by division" (*kai ping fang chu* 開平方除). The relation between division, root extraction and tabular settings, well known to historians of Chinese mathematics, is precisely the key to understanding Li Ye's concept of an equation (Wang and Needham 1955).

In Li Ye's work, tabular settings are not mere representations of the rods on the counting surface. In the available editions of *Yigu yanduan* (and in *Ceyuan haijing* as observed by Chemla), as Chemla (1996) shows, there were different practices of representation, and the transcription of tabular settings was not uniform in the Song–Yuan period. Li Ye distinguishes himself from contemporary mathematicians by the way in which he elaborates a transfer of the mathematical activity to paper. That is, he develops a way to represent polynomials and equations in written work evolving from the ancient practices of manipulations with counting rods. Li Ye's way of writing mathematical expressions was interpreted as a symbolization of the object he is treating. Chemla has shown that Chinese mathematical texts represented very different concepts of positional notation from the Han (206 BC–AD 220) to the Yuan (1279–1368) dynasties. Matrix arrays were always used, but there were variations in their transcriptions. This variation shows an evolution toward autonomy of work on paper. Despite these variations, the organization of the data is remarkably stable and the management of the operations on the support follows strict imperatives. This is a practice which testifies to a transition of mathematical activity from material support (that is, counting surface and counting rods) to paper-based work and of the development of symbolic representations of mathematical objects and operations with them.

The mathematical expressions are always written inside the space of the column containing the text, just like any part of a sentence. They are introduced by the character *de*, 得, "to yield" and after being interpreted, with the character *wei*, 為, "as." They are not represented like independent drawings and many times the

---

[9]For the signification of the absence or presence of the character *tai see* Pollet and Ying (2017).

Fig. 3 Tabular setting of problem 11 in the *Yigu yanduan*

author uses a pictogram representation titled *shi*, 式, "pattern," "configuration" (Fig. 3). There is continuity in the written text between the discourse and the configuration of numbers. Li Ye integrates the configuration to the written text as if it were a simple number, while the configuration itself extends the sentence as being inserted in the column. There are no such relations like picture/caption, therefore, a configuration cannot be considered an illustration.

To understand the peculiarity of Li Ye's ways of writing polynomials, one can refer to other contemporary mathematicians. For example, Qin Jiushao seems closer to a pictorial configuration. Contrary to Li Ye, Qin Jiushao inserts diverse states of support as illustrations (Fig. 4). The discourse and its illustrations are separated, the discourse being sometimes a caption to the illustration. We also notice that Qin Jiushao refers to the tabular setting by using the character *tu* 圖, "diagram."

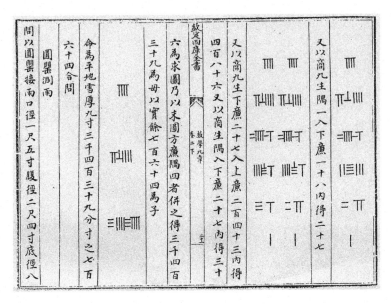

Fig. 4 The *Shu shu jiu zhang*, Chap. 2, p. 21

The configuration of numbers introduced by Li Ye portrays some states of the counting surface, yet we cannot consider this a pure transcript of the different steps of manipulating the rods. The configuration, as Li Ye presented it, is not a picture of the surface. It is a step in the process of creation of symbolic language.

Representation of Polynomials on the Counting Surface

Hoe (2007) described in great detail the operations of subtraction and addition that can be conducted with numbers as well as polynomials. The text written by Li Ye gives clues on how to situate polynomials on the counting surface simultaneously (see the Appendix to this chapter). On the basis of Li Ye's written discourse and polynomial representations, we can reconstruct the procedure as follows:

(1) The first polynomial is constructed. One rod is placed in the middle of the surface to stand for the unknown: "set up one Celestial Source as the diameter of the inside pond" (Fig. 5). I supposed that this place is the centre of the surface for the reason that the constant and other terms are to be placed around it. [10]

(2) Next, the constant given in the statement of the problem is added: "Adding twice *the reaching bu*," that is, 20 · 2 = 40. The "*reaching bu*" is the distance given in the statement of the problem which I named *a* in the mathematical

---

[10]I mark the center of the counting surface with a black dot. The representation of the surface is only partial for the reason of economy of space.

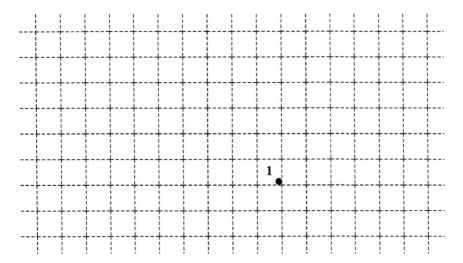

Fig. 5 Counting surface with one rod representing 1*x*

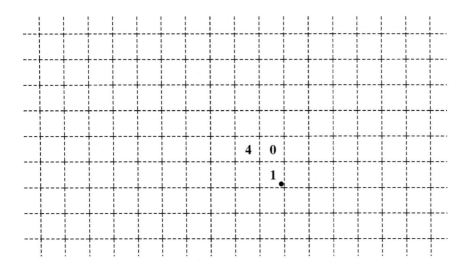

Fig. 6 Representation of 40 + *x* on the counting surface

description above. Following the standard for multiplication (Hoe 2007, pp. 51–55), the multiplicand should be placed above, the multiplier below and a place in between is left for the product, or, in the context of division, for the quotient. That is, 40 should be placed between 20 and 2. But, in the present case we have an arithmetical operation on constants 20 and 2 and an unknown added to it in the same setting. That is 20 · 2 + *x*. The result of the multiplication is placed above the rod for the unknown in the row of *tian yuan*. It is the row of *tai*, the constant. Thus we have 40 + *x* as the side of the square field. Yet

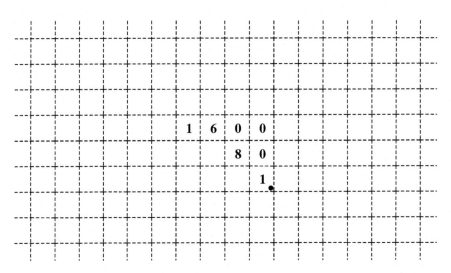

Fig. 7 Representation of $1600 + 80x + x^2$ on the counting surface

$20 \cdot 2$ is such a simple operation that it might not have been recorded. $40 + x$ is

recorded as [figure] . Figure 6 shows how it could be placed on the surface.

(3) This first expression is multiplied by itself: "Augmenting this by self-multiplying." The effect is to "augment" the configuration for one row/power. Figure 7 shows $(40 + x)(40 + x) = 1600 + 80x + x^2$, which is repre-

sented as [figure] , and is still placed in the center.

(4) This first polynomial is the area of the square (S) known from the statement of the problem. It has to be "sent on the top" to be kept for further operations as shown in Fig. 8.

(5) Next, the second polynomial is constructed. "Set up again one Celestial Source as the diameter of the inside pond" as before. In Fig. 9 the polynomial set in the central part is $1x$.

(6) The area of the circle (C) is to be computed as a function of the unknown. The unknown is first "self-multiplied," or squared, in modern terms. That is, it will be shifted down for one row to augment the power of x for one unit. Next, it is multiplied by 3, then divided by 4. Following the description of the algorithm

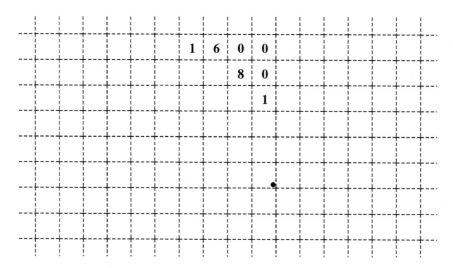

Fig. 8 The polynomial $1600 + 80x + x^2$ "sent to the top" of the counting surface

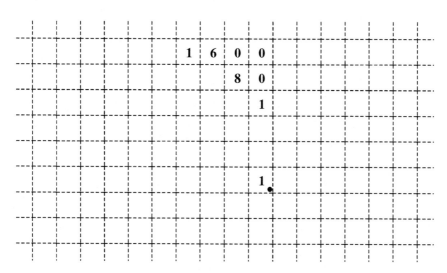

Fig. 9 The term $1x$ set in the center of the counting surface

of division, the dividend should be placed above and the divisor below (Fig. 10), thus expressing $3x^2/4$. The quotient, placed above the dividend

(Fig. 11) is $0.75x^2$, represented as .

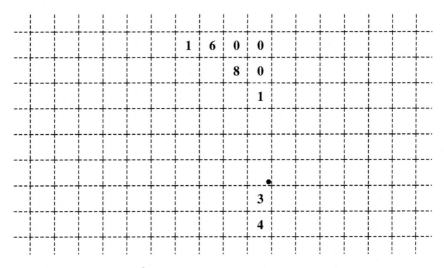

Fig. 10 Representation of $3x^2/4$ on the counting surface

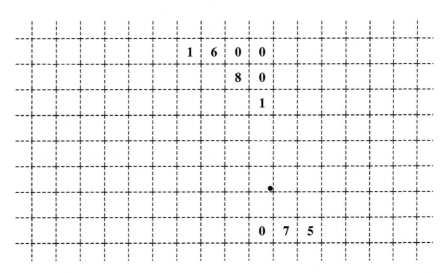

Fig. 11 Representation of $0.75x^2$ on the counting surface

This part of the problem poses a question about 0 (zero). While zero is clearly represented as a written object, there was no difference between a simple blank space and a blank space representing zero when the polynomials were represented with counting rods. This shows another limit of the reconstitution.

(7) This second polynomial is placed next to the first in order to be subtracted. There is no mention of place for the second polynomial. The latter is immediately used in subtraction or addition; therefore, one does not need to preserve

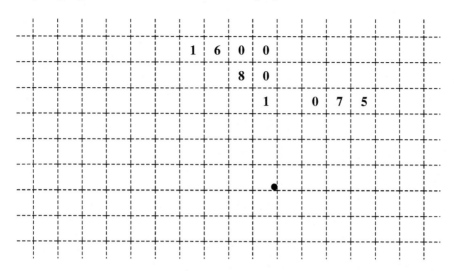

Fig. 12 Representation of the polynomial $1600 + 80x + x^2 - 0.75x^2$ on the counting surface

it. We can imagine that for addition and subtraction, the two expressions were represented on the counting instrument next to each other. Coefficients of the same degree of polynomials were on the same row facing each other, then added or subtracted as suggested by Hoe (2007, p. 73) and Chemla (1982, Chap. 7). This is suggested by three of the other problems of *Yigu yanduan*. In problems 21, 43 and 62, coefficients of three expressions have to be added together. The first one is placed above (*shang wei* 上位), the second on the next position (*ci wei,* 次位 in problems 21 and 62) or middle position (*zhong wei* 中位 in problem 43) and the third below (*xia wei* 下位) with the three positions (*san wei* 三位) being added. We can infer for the present case that the result of the subtraction is placed on the left. Figure 12 shows $1600 + 80x + x^2 - 0.75x^2$.

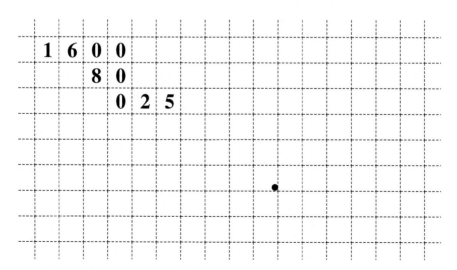

Fig. 13 Representation of the polynomial $1600 + 80x + 0.25x^2$ "sent to left."

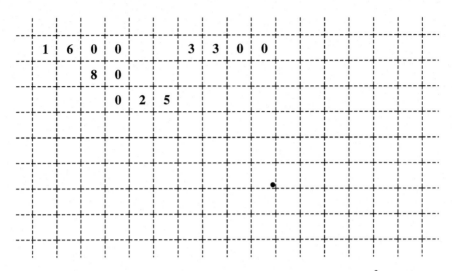

Fig. 14 Representation of the constant 3300 and polynomial $(1600 + 80x + 0.25x^2)$ placed on the counting surface

(8) The result of the subtraction, $1600 + 80x + 0.25x^2$, is "sent to the left" as

required by Li Ye, and represented as or as shown in Fig. 13.

(9) Next, "place the real area" which means that the area known from the statement is placed next to the result obtained at the previous step in order to prepare the latter to be "eliminated" (*xiang xiao*), that is, subtracted from it.[11] Figure 14 shows 3300 and $(1600 + 80x + 0.25x^2)$.

(10) "With what is on the left, eliminate from one another." The result of the elimination is $3300 - (1600 + 80x + 0.25x^2) = 1700 - 80x - 0.25x^2$ or

 . This setting is ready for the "extraction of the root," i.e., the

solution of the quadratic equation. The rows are not recorded as *tai* or *yuan* anymore. They are now "dividend" and "divisor." The "dividend" is on the top, the "divisor" on the second row. These are the names of rows used in the

---

[11]*See* Pollet and Ying (2017) for the distinctions between "to eliminate" (*xiang xiao*) and "to subtract" (*jian*).

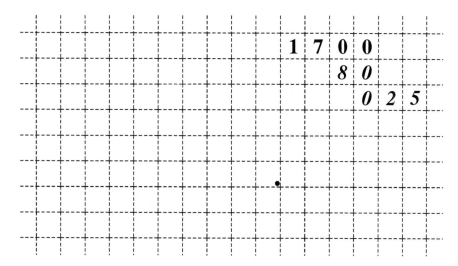

Fig. 15 Representation of the polynomial $1700 - 80x - 0.25x^2$ placed on the counting surface (the coefficients printed in italics are negative)

process of division, whose algorithm was developed into the method of extraction of square roots and solution of polynomial equations (see above). Figure 15 shows $1700 - 80x - 0.25x^2$.[12] The resulting equation contains two negative coefficients. They were marked by a diagonal stroke in the edition of the text by Li Rui. Li Ye does not mention the sign of coefficients. It is known that colors were used to differentiate the rods. According to the *Meng qi bi tan* 夢溪筆談 (*Dream Pool Essays*, Chap. 8, Sect. 2, p. 95) written by Shen Gua 沈括 (1031–1095), "in arithmetic, one uses red and black rods to differentiate the negative quantity from the positive one." And again, in the *Shu shu jiu zhang* by Qin Jiushao, in the diagram of the extraction of the root in the fourth chapter, we read: "the negative expressions are drawn in black, while positive expressions are drawn in vermilion." According to this, in this period, mathematical expressions were probably drawn in red or black in order to differentiate them. In Fig. 15 I represent the negative coefficients in italic.

This last setting is ready for the extraction of the root. Since I have used Arabic numerals and a matrix array here, it is not possible to represent the details of the operations. This point is interesting because the descriptions of some operations make sense only if actually performed on the device. This means that the counting device itself generates the concept of a polynomial equation here.

---

[12]The oblique lines at the coefficient mark the coefficient as negative. These were added by Qing dynasty editor, Li Rui. It is possible that in the original text the coefficients were not negative, but the disposition of rods was supposed to be interpreted as $0.25x^2 + 80x = 1700$. *See* Pollet (2014) and Pollet and Ying (2017).

Division and Equation

This setting for a polynomial and an equation finds its origin in the algorithm for root extraction, which itself finds its origin in the algorithm for division. Several studies have already thoroughly investigated the relation between these operations.[13] Lam and Ang (2004) summarized the elements based on simple examples provided in the *Sunzi suanjing* (孫子算經 *Computational Classic of Master Sun*).[14]

They first noted that the algorithm of division is the inverse of the one for multiplication, wherein the multiplier is placed in the upper position and the multiplicand in the lower position. The latter is moved to the left according to the number of digits of the former. The prescription of the *Sunzi suanjing* is to multiply the number placed below, digit after digit, by the greatest digit of the multiplier and to place the intermediate results in the middle row, where they are added progressively (Lam and Ang 2004). Similar to multiplication, where operations are based on the position of the multiplier relative to the multiplicand, operation of division is based on the placing of the divisor (*fa*) relative to the dividend (*shi*). The quotient (*shang*) is placed on top (Fig. 16). The initial position of the divisor relative to the dividend determines the position of the leftmost digit of the quotient. The characters *fa*, 法, and *shi*, 實, are standard names for the last two rows of the polynomial. The *Yigu yanduan* makes no exception.

The procedure of root extraction is similar to the procedures for division. Lam Lay Yong also explained the geometrical interpretation of the procedure, since, given an area of a rectangle and a side, finding the other side involves a division. Sunzi explains the method of extracting the square root with two examples (Chap. 2, Problems 19–20).[15] Parallelism appears with the following observations: the dividend is called *shi* and the number whose root is extracted is also called *shi*. This is the first number to be placed on the counting device and its digits define the positions of other operands set on the counting device (see chapter on Chinese counting rods in the present volume). The divisor termed *fa* is moved from right to left such that its first digit from the left is placed below the first or second digit of the dividend. The quantity set in the position of *fa* is used like a divisor and the quantity placed in the

---

[13]Different descriptions of the algorithms of division and root extraction from different sources are given in Western languages by Wang Ling and Needham (1955); Chemla and Guo (2004, pp. 314–322); Chemla (1982, Chap. 8); Li and Du (1987, pp. 7–19); Lam and Ang (2004, Chap. 3, Sect. 4); and Martzloff (1987, pp. 229–249), among others. We will not discuss the evolution and details of different algorithms and their historiography.

[14]*Sunzi suanjing* belongs to the *Shi bu suanjing* 十部算經 (*Ten Books of Mathematical Classic*) used in education during the Sui (581–618) and Tang (618–907) dynasties. Wang (1964) suggested that it was composed between 280 and 473. Its earliest surviving edition is dated from the Southern Song dynasty, 1127–1279 (Li and Du 1987, pp. 92–93; Qian 1963; Lam and Ang 2004, pp. 63–71).

[15]For other explanations on the extraction of square and cube roots *see* Qian (1963); Li and Du (1987, pp. 118–121); Chemla and Guo (2004, pp. 322–330); Lam and Ang (2004, Chap. 4).

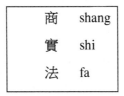

Fig. 16 Names of the rows in the operation of division

position of *shi* is treated like a dividend. The operation brings the extraction of the root back to that of division.[16]

A comparison of the algorithms of root extraction in Chinese mathematical treatises shows that there was an evolution of the algorithm, in which there appeared a place value notation for the equations associated with these extractions (Chemla 1994, 1996).[17] Chemla interprets the parallelism between the division and the extraction of a square root. Thus, terms such as "dividend" and "divisor" for positions corresponding to the successive steps of computation have two functions. First, they allow the same list of operations be performed in an iterative way. Second, they allow for the modeling of the extraction of the square root on the model of division (Chemla and Guo 2004, p. 327). The same position name is used and this position during the succession of operations is to be used in the same way. Names and the management of positions are the key indications of the correlation between the two procedures. In another publication, Chemla (1996) shows that there was a practice that consisted of the exploration of relations between the operations of root extraction and division; it resulted in a revision of the different ways of computing and naming positions. She concludes that the work presented in the Han dynasty classic *Jiu zhang suan shu* (九章算術 *Nine Chapters on Mathematical Procedures*) was in fact perpetuated in the 13th century, which was Li Ye's time.

The elaboration of the division procedure led to a general technique that mechanically extracts the square root of a number. This method is used as an algorithm and provides the basis for the further development of procedures for solving quadratic

---

[16]There is also a meaningful difference between algorithms. The root extraction uses a supplementary rod named the "borrowed rod" (*jie suan*). In her explanation of the algorithm of root extraction in the *Nine Chapters*, Chemla shows that the borrowed rod is first placed in the position of the units. This rod is later moved toward the left, from 10^{2n} to 10^{2n+2}, until it reaches the leftmost position under the dividend, that is, 10^{2n} if the first digit of the root is $a \cdot 10^n$, $1 \leq a \leq 9$. One deduces what is the power corresponding to the first digit of the root, named "the quotient" (Chemla and Shuchun 2004, p. 326). Since the algorithm of the *Nine Chapters* is slightly different from the one of the *Sunzi suanjing*, the interpretation of the role of the "borrowed rod," its position and modifications made in the latter text, require discussion. In her comparison of the algorithm of root extraction from the *Zhang Qiujian suanjing* and the one by Kushyar ibn Labban, Chemla explains the role of the "borrowed rod." She mentions that this rod has different roles in Chinese algorithms of root extraction (Chemla 1994, p. 17)

[17]For instance, the algorithm of root extraction in the *Nine Chapters of Mathematical Procedure* follows the same principles, but the setting is slightly different (Chemla 1982, Chap. 7, Sect. 7; Chemla and Guo 2004, pp. 324–326).

equations. The configuration necessary for conveying the meaning of these equations is inextricably expressed in the positions occupied by the rod numerals on the counting device. This justifies the use of the same terms to name positions on the counting device for division, extracting the root and, finally, the terms of the equation, as they are set in the same emplacement. Once the equation is set up on the counting device, one merely has to apply a well-known procedure to solve it. The development of the algorithm of root extraction led to the setting of the equation and to the solution.

Just as the development of the method of extraction of a square root was based on the algorithm for division, the concept of the polynomial equation was derived from the algorithm of the extraction of the square root. The form of the solution of quadratic equation is algorithmic; it is a series of operations with two different terms —a dividend and a divisor—which is solved by the algorithm of the extraction of the square root. There is also a geometric form of the equation.[18] The equality is expressed by the operation of "eliminating one another" (*xiang xiao* 相消). Two polynomials can be "eliminated" because their numerical values are equal. Thus, there is an equation in the sense that one must determine an unknown which satisfies an equality. In fact, what we identify as equations in the representation of tabular settings is an opposition between a "dividend" and "divisor." This peculiarity of the concept of an equation as an operation is due to the essential role played by the counting device and the way of articulating different algorithms together. This is an algorithmic conception of the equation based on a specific heuristic role of division. The algorithm of division plays an essential role in the elaboration of a quadratic equation. It is quite different from the treatment of equations seen in the BGA.

Comparative Perspective for Conclusion

In this chapter, and in the chapter dedicated to the BGA (i.e., Chapter "Reading Algorithms in Sanskrit" of the present volume), two procedures are given, referring to the same operation of division, but implying different understandings in China and India. That is to say they imply specific action and understanding. Moreover, even the same operation can imply different understandings. There are several known algorithms for the execution of division and we see here two different roles in different milieus for this operation. An interesting point is that in both examples, tabular settings for equations were modeled according to division. The Rule of Three contains a division used to set up and solve equations in India. The division is the framework in which algorithmic equations evolved in China.

The prescription of algorithmic operations delivers some clues as to what the authors expected their readers were supposed to do or to understand. The way in which lists of operations are ordered shows that Indian and Chinese authors had

---

[18]My forthcoming publication, "The Empty and the Full," is dedicated to it. *See also* Lam and Ang (2004).

different interests, addressed different difficulties and understood mathematical concepts differently. The algorithms used for the setting and solution of simple algebraic equations are particularly interesting in this respect. Both Indian and Chinese authors prescribed division and represented equations in a written tabular setting. Strangely, the authors of both countries relied on division to set up equations. Compared to mathematics nowadays, this seems an odd way to construct equations. No one uses division for this algebraic purpose. This coincidence of two unconnected milieus using division for a similar purpose deserves further investigation. Yet, naturally, the way division is prescribed in each algorithm and the way it interacts with the tabular arrays testify to very different practices in China and India.

If the diversity of objects and their corresponding relations were a preoccupation for the Indian author, it is not the case for Li Ye. Li Ye never discusses or re-interprets what the unknown is. Interestingly, the common point is somewhere else. The example of the Chinese algorithm provided here presents a case of a later algebraic procedure constructed as a set of other well-known former arithmetical procedures. There is the same purpose of grounding and verifying a later procedure and there is the same system of reference to a model equation. In China, the quadratic equation is a model. In India, the linear equation was the model. It is interesting to see the similar practice of using models.

Li Ye used the vocabulary of division to refer to each row of an equation, and the algorithm of division as a model to find a root of the equation. Also, root extraction was a model for setting up quadratic equations. The validity of the procedure is based on a well-known procedure where division (and its opposite, multiplication) grounds and verifies the algorithm used for setting and solving algebraic equations. The tabular setting on the counting device leads naturally to quadratic equations, where equations and operations are strongly interconnected.

In the BGA, the division gives a meaning and an origin to the terms of the equations and their roots, and this practice can be related to a specific practice of verification. There are traces of an investigation on mathematical objects (unknowns, constants, "variation" on unknowns and real objects like horses and monetary units) and the relation among objects is set according to their correspondences. Here the linear equation is the model and the equation is conceived as equality. In turn, in Li Ye's treatise, the tabular settings show an exploration of the relationships between operations of division, extraction of roots and composition of polynomials. That is, operations and equality matter more than the investigation of mathematical objects themselves. The method is not only an algorithm, but also a basis for further developments related to equations. Its function is heuristic.

From an historical point of view, it is interesting to notice that two mathematical milieus, apparently not connected, share a common focus on a specific operation in an algorithm used for algebra. Paradoxically, it is a common point! How many milieus rely on division to set up algebraic equations? Whether it is a matter of coincidence, historical transmission or cognitive evolution, this point should (and will) receive more investigation.

Appendix

Problem 1 of Chap. 1 of *Yigu yanduan*. The commentaries and the second procedure were removed from the text below. To read their translations, see Pollet and
Ying (2017). The present version of the Chinese text is based on the edition made
by the mathematician Li Rui (LR) in 1798. This edition was carefully compared to
two manuscripts of 1782 published in the imperial encyclopedia, the *Siku quanshu*
(*The Complete Library of the Four Treasuries* 四庫全書), namely the Wenjing
(WJG) and Wenyan (WYG) editions. See Pollet (2014) for details of the comparison. The numbering of the sentences is mine—(1.2) means sentence number 2
of problem 1—it is added in order to observe a correspondence between Chinese
text and translation.

(1.1) 第一問
今有方田一段, 內有圓池水占, 之外計地一十 三畝七分半. 竝不記內圓外方.
只云從外田楞 至內池楞, 四邊各二十步.
問內圓外方各多少?
荅曰: 外田方六十步, 內池徑二十步.

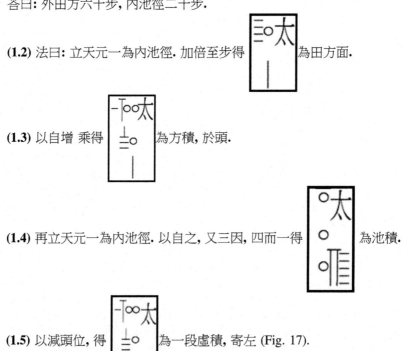

(1.2) 法曰: 立天元一為內池徑. 加倍至步得 [圖] 為田方面.

(1.3) 以自增 乘得 [圖] 為方積, 於頭.

(1.4) 再立天元一為內池徑. 以自之, 又三因, 四而一得 [圖] 為池積.

(1.5) 以減頭位, 得 [圖] 為一段虛積, 寄左 (Fig. 17).

(1.6) 然後列真積. 以畝法.通之, 得三千三百步. 與左相消

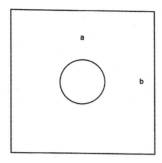

Fig. 17 The captions read: "(a) 至水二十步; (b) 方田六十步." For translation, see Fig. 18

(1.7) 得

開平方, 得二十步, 為圓池徑也. 倍至步, 加池徑, 即外方面也.

Translation

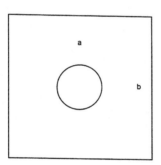

Fig. 18 The captions read: "(a) the *bu* reaching the water are 20; (b) the side of the field is 60 *bu*."

(1.1) Let us suppose there is one piece of square field, inside which there is a circular pond. Outside the [area] occupied by water, one counts thirteen *mu*, seven *fen* and a half [of *fen*] of land. Moreover, there is no record of the [dimensions] of the inner circle and the outer square. It is only said that [the distance] from the edge of the outer field to the edge of the inside pond on [all] four sides is twenty *bu*.

One asks how much are [the diameter of] the inner circle and [the side of] the outer square.

The answer says: the side of the outer field is sixty *bu*. The diameter of the inside pond is twenty *bu*.

(1.2) The method says: set up one Celestial Source as the diameter of the inside pond. Adding twice the reaching *bu* yields $\begin{matrix} 40 & tai \\ 1 \end{matrix}$ as the side of the field.

(1.3) Augmenting this by self-multiplying yields $\begin{matrix} 1600 & tai \\ 80 & \\ 1 & \end{matrix}$ as the area of the square, which is sent to the top.

(1.4) Set up again one Celestial Source as the diameter of the inside pond. This times itself and multiplied further by three then divided by four yields $\begin{matrix} 0 & tai \\ 0 & \\ 0.75 & \end{matrix}$ as the area of the pond.

(1.5) Subtracting this from the top position yields $\begin{matrix} 1600 & tai \\ 80 & \\ 0.25 & \end{matrix}$ as a piece of the empty area, which is sent to the left.

(1.6) Next, place the real area. With the divisor of *mu*, making this communicate yields three thousand and three hundred *bu*.

(1.7) With what is on the left, eliminating from one another yields $\begin{matrix} 1700 \\ -80 \\ -0.25 \end{matrix}$.

Opening the square yields twenty *bu* as diameter of the circular pond. Adding twice the reaching *bu* to the diameter of the pond gives the side of the outer square.

References

Primary Sources

Li Ye. 1248. *Ceyuan haijing* 測圓海鏡 (Sea Mirror of Circle Measurements).
Li Ye. 1259. *Yigu yanduan* 益古演段 (The Development of Pieces of Areas according to the Collection Augmenting the Ancient [Knowledge]).
[Both edited in the *Wen yuan ge Siku quanshu* 文淵閣四庫全書 (Complete library of the Four Treasuries, [copy preserved in] Wenyuan Pavilion), original edition of 1789 stored in National Palace Museum, Taiwan.]

Wen jin ge Siku quanshu 文津閣四庫全書 (Complete library of the Four Treasuries, [copy preserved in] Wenjin Pavilion), Beijing: Shangwu, 2005, vol. 799 [a reprint of the original edition of 1789].

Zhibuzu zhai congshu 知不足齋叢書 (Collected works of the Private Library of Knowing Our Own Insufficiencies), reprinted in Guo Shuchun 郭书春 (ed)., *Zhongguo kexue jishu dianji tong hui: Shuxue pian* 中國科學技術典籍通彙: 數學篇 (Source Materials of Ancient Chinese Science and Technology: Mathematics Section). *Henan jiaoyu chubanshe* 河南教育出版社 (Henan Education Press), 1993, vol. 1 [a reprint of the original edition of 1798].

Secondary Sources

Chemla, Karine. 1982. *Etude du livre "Reflets des mesures du cercle sur la mer" de Li Ye*. Thèse de doctorat de l'université Paris 13. Not published.

Chemla, Karine. 1994. Different concepts of equations in *The Nine Chapters on Mathematical Procedures* 九章算術 and in the Commentary on it by Liu Hui (3rd century). *Historia Scientiarum*. 4–2: 113–137.

Chemla, Karine. 1996. Positions et changements en mathématiques à partir des textes chinois des dynasties Han à Song-Yuan. Quelques remarques. *Extrême-Orient, Extrême-Occident* 18: 115–147.

Chemla, Karine, and Guo Shuchun. 2004. *Les neuf chapitres. Le classique mathématique de la Chine ancienne et ses commentaires*. Paris: Dunod.

Chemla, Karine. 2006. Artificial language in the mathematics of Ancient China. *Journal of Indian Philosophy* 34: 31–56.

Bottermans, Jack. 2008. *The Book of Games. Strategy, Tactics and History*. New York, NY: Sterling Publishing Company.

Guo Shuchun 郭書春. (1991). *Zhonguo gudai shuxue* 中國古代數學 (Mathematics in Ancient China). Jinan: *Shandong jiaoyu chubanshe* 山東教育出版社 (Shandong Education Press).

He Yunpo 何云波. (2001). *Weiqi yu Zhongguo wenhua* 圍棋與中國文化 (Weiqi and Chinese Culture). Beijing: Renmin Publishers.

Hoe, Jock. 2007. *The Jade mirror of the four unknowns by Zhu Shijie*. New Zealand: Mingming Bookroom.

Lam Lay Yong. 1984. Li Ye and his *Yi Gu Yan Duan* (Old Mathematics in Expanded Sections). *Archive for history of exact sciences*. 29: 237–266.

Lam Lay Yong, and Ang Tian Se. 2004. *Fleeting Footsteps, Tracing the Conception of Arithmetic and Algebra in Ancient China*. Singapore: World Scientific.

Libbrecht, Ulrich. 1973. *Chinese Mathematics in the Thirteen Century. The Shu-shu chiu-chang of Ch'in Chiu-shao*. Cambridge: The MIT Press.

Li Yan, and Du Shiran. 1987. *Chinese Mathematics, A Concise History*. Translated by John N. Crossley and Anthony W.C. Lun. Oxford: Clarendon Press.

Liu Shancheng 刘善承. (2007). *Zhongguo weiqi shi* 中国围棋史 (History of *weiqi* in China). Chengdu: Shidai 时代 Publishers.

Martzloff, Jean-Claude. 1988. *Histoire des mathématiques chinoises*. Paris: Masson. English edition: *A History of Chinese Mathematics*, Berlin etc.: Springer, 1997.

Needham, Joseph. 1959. *Science and Civilisation in China*, vol. 3. Cambridge: Cambridge University Press.

Pollet, Charlotte. 2012. Comparison of Algebraic Practices in Medieval China and India. PhD dissertation. National Taiwan Normal University/Université Paris 7. Not published.

Pollet, Charlotte. 2014. The influence of Qing dynasty editorial work on the modern interpretation of mathematical sources: the case of Li Rui's edition of Li Ye's mathematical treatises. *Science in Context*, 27(3), 385–422.

Pollet, Charlotte, and Ying Jia-Ming. 2017. One quadratic equation, different understandings: the 13th century interpretation by Li Ye and later commentators in the 18th and 19th centuries. *Journal for History of Mathematics. Korean Society for History of Mathematics*, 30 (3), 137–162.

Qian Baocong 錢寶琮. (1963). *Suanjing shishu* 算徑十書 (The Ten Classical Books on Mathematics). Beijing: Zhonghua shuju.

Volkov, Alexei. 2001. Le bacchette (Counting rods). In: Karine Chemla et al. (volume eds.), *Storia Della Scienza* (Encyclopedia on History of science, in Italian), vol. 2 *(Cina, India, Americhe)*. Rome: Istituto della Enciclopedia Italiana, 125–133.

Wang Ling. 1964. The date of the *Sun Tzu Suan Ching* and the Chinese remainder problem, *Actes du Xe Congrès International d'histoire des sciences*, Paris, pp. 489–492.

Wang Ling, and Joseph Needham. 1955. Horner's Method in Chinese Mathematics: Its Origins in the Root-Extraction Procedures of the Han Dynasty. *T'oung Pao. Second Series, 43*(5), 345–401.

Charlotte-V. Pollet is an assistant professor at National Chiao-Tung University (Hsinchu, Taiwan). She teaches philosophy and history of sciences. After studying philosophy, Sanskrit and Chinese languages in University Lille-3 (France), she obtained a dual Ph.D. in mathematics from National Taiwan Normal University and in history and philosophy of sciences from University Paris-7 in 2012. Her research interests are history of mathematics (algebra, combinatorics, and geometry), cross-cultural transmission of knowledge and the role of education in transmitting scientific knowledge, especially in Chinese and Indian worlds. She is the author of the forthcoming monograph *The Empty and the Full: Li Ye and the Way of Mathematics*. She also works as educator and has received several awards from France and Taiwan for her work in pedagogy of philosophy.

Same Rods, Same Calculation? Contextualizing Computations in Early Eighteenth-Century Korea

Young Sook Oh

Abstract The early eighteenth-century Korean mathematical sources testify that there were two types of authorship, the mathematical officials in the lower class and the literati in the upper class. This paper aims to show how each authorship, from dissimilar educational background, affected and transformed the algorithms and grounds of the computation differently in spite of the usage of the same computational tool, based on the analysis of two early eighteenth-century mathematical texts, the *Writings of Nine and One* (*Kuiljip* 九一集) by a skilled mathematical official, Hong Chŏng-ha 洪正夏 (1684–?), and the *Summary of Nine Numbers* (*Kusuryak* 九數略) by a renowned member of the literati, Ch'oe Sŏk-chŏng 崔錫鼎 (1646–1715). In their texts on the computational techniques using counting rods, Hong appraised the adeptness in handling counting rods and expanded the existing algorithms based on the real practice, while Ch'oe approved the algorithms in which he could find the meaning close to that conveyed by texts and images of Confucian philosophical tradition.

Keywords Counting rods · Early eighteenth-century Korea · Mathematical officials · Literati

Introduction

Counting rods had been the main computational tool in various countries in East Asia for more than one thousand years. Yet, the trajectory of their usage throughout the timeline can be sketched differently in the different social and cultural contexts. For example, in eighteenth-century China where calculation using counting rods had originally been invented and developed, it became obsolete in practice, whereas in Korea, it was still prevalent both in practice and

Y. S. Oh (✉)
Program of History and Philosophy of Science,
Seoul National University, Seoul, South Korea
e-mail: seyio@hotmail.com

© Springer Nature Switzerland AG 2018   215
A. Volkov and V. Freiman (eds.), *Computations and Computing Devices
in Mathematics Education Before the Advent of Electronic Calculators*,
Mathematics Education in the Digital Era 11, https://doi.org/10.1007/978-3-319-73396-8_9

in text (Needham 1959; Kim and Kim 1977; Li and Du 1987; Martzloff 1997; Kawahara 2010).[1] This kind of difference between China and Korea might already suggest a comparative history of mathematics with particular attention to computational tools, or a comparative history of "the contextuality of cognitive tools" in Netz's (2002, 344) terms.[2]

In this vein, another good case in a local context is that of various calculations contingent on the social ranks in the latter era of Chosŏn 朝鮮 dynasty (1392–1910) in Korea. In that period were produced two distinctive social strata among the literate: the *yangban* 兩班 literati with much more prestigious status, and *chungin* 中人 with lower and rather humble status, the latter usually being the term to designate an amalgam of the people who were illegitimate offspring of the *yangban* families and those who served as lower bureaucratic officials, including mathematical officials (Wagner 1987a, b; Chŏng 1993; Yi 1997; Han 1997). These two groups, basically having different mathematical education, adopted two divided stances in their approaches to computational tools and algorithms (Kim and Kim 1977). In particular, the period from the late seventeenth to the early eighteenth century witnessed this discrepancy in a clear and illuminating way. In the aftermath of consecutive wars against both Japan (1592–1593, 1597–1598) and China

---

[1]I am grateful for the helpful comments and kind interest of Lee Jongtae, Lee Jung, Jun Yong Hoon, and Jun's anonymous student. Special thanks go to C. and two anonymous commentators, who read drafts several times and gave precious comments and corrections, and two editors, who invited me to contribute this chapter.

For the romanization of Korean and Chinese, the McCune-Reischauer system and the Pinyin system, respectively, are used, with exception for the English articles in which authors' names and titles have already been romanized in other ways. Korean terms coined by Korean authors in Korean texts written in traditional Chinese will appear in the McCune-Reischauer system first, and then, the Pinyin system. Korean names, their titles of posts, and titles of texts will appear only in the McCune-Reischauer system. Korean and Chinese names are cited in their original order, family name first.

This paper does not explain the circumstantial differences between Korea and China, which eventually ended up with the different usage of the counting rods, but we could offer, until now, just a rough assumption that Korea didn't witness the rapid economic and commercial growth enough to cause the transformation of the tools from counting rods into abacus like China. This is a rich field that deserves much more detailed attention of the historians of mathematics and sciences. Also we can consider the long-lasting maintenance of the educational institution in firm support of the counting rods as one of the environmental factors that influenced their longevity for ages in Korea.

[2]Counting rods as material objects have been designated differently by diverse historians. For example, Li and Du (1987) called them "devices," while Martzloff (1997) put them into the "calculating instruments" section along with abacus. Needham (1959) also put them into the "mechanical aids" section with abacus that he referred to as "instruments." Meanwhile, what Netz (2002), as a cognitive historian, called "cognitive tools" included abacus as well as numerals. Even if there had been no sharp difference in the definitions of "tools" and "instruments," Taub (2011) acknowledged that the appellation "instrument" is now distinguished from the term "tool," as being used for more delicate work or for artistic or scientific purpose, and quoted an example of usage from the Oxford English Dictionary: "a workman or artisan has his tools, a draftsman, surgeon, dentist, astronomical observer, his instrument." In this chapter, I would like to use the word "tools" in order to avoid the image of "instruments" as being too sophisticated, or advanced.

(1627, 1636–1637), when the explosive need for the restoration of lost mathematical textbooks surged, a wide range of new mathematical texts were also authored by these two groups. And these texts reflected their allegedly dissimilar positions about computational algorithms using the seemingly identical tool, namely counting rods.

To articulate the dissimilarity, this paper will analyze two highly contrasting mathematical texts from this period, the *Writings of Nine and One* (*Kuiljip* 九一集), by a skilled mathematical official, Hong Chŏng-ha 洪正夏 (1684–?), and the *Summary of Nine Numbers* (*Kusuryak* 九數略), by a renowned member of the literati, Ch'oe Sŏk-chŏng 崔錫鼎 (1646–1715). The aim here is to show how the computations leading to the "same" numerical result were "differently" contextualized in terms of the socio-professional identity of the calculator, the tool he used, and the culture, particularly educational, he embodied.

Computation of Mathematical Officials in *Writings of Nine and One*

Most mathematical officials in Chosŏn society were low-level functionaries affiliated with the Ministry of Taxation (Kor. *hojo*, Chin. *hucao*, 戶曹), a government office in charge of the various tasks concerning government revenue, including household registration and taxation, supervisions on tribute products, labor, grains, and cash. Their recruiting system was different from that of the positions occupied by members of *yangban* literati. Officials worked for six months and then took six months off. They needed to pass the examination whenever they wanted to be posted again. Their positions were among the six lowest ranks of the eighteen that existed (Yi 1997). These governmental regulations, initially designed by the *yangban* literati to control the number and the quality of the lower officials, reflected the way the literati looked at the mathematical knowledge, or skill, which the mathematical officials had to offer. This knowledge was, of course, practical, but it was also seen as humble, or minor, and also required continuous exercise to be maintained at an acceptable level.

Under these restrictions, mathematical officials had created strong matrimonial relations among the families to build up exclusive lineages, which were in fact an efficient means of promoting their social interest (Hwang 1988, 1994). Their networks were often extended to the families of lower technical officials in other fields, including medical doctors, foreign language interpreters, astronomers, and lawyers. Ultimately, during the seventeenth century, mathematical officials, together with lower officials from other fields, came to form a distinctive social group, *chungin*, a social stratum between the elite *yangban* and the common, usually illiterate, people.

Many recent studies have pointed out that members of this new stratum strove to construct their own prestigious culture against the literati's. Since the late seventeenth century, many lower officials had formed poetry communities and published some collections of highly acclaimed poems (Kang 1997). Later in the nineteenth century, medical, interpretative, and mathematical officials made their own genealogical records, which had formerly been an exclusive practice of *yangban* families (Chŏng 1993; Yi 1997; Han 1997). In case of the mathematical officials in the late seventeenth century and early eighteenth century, as will be shown in the following, this effort was shrewdly manifested in the mathematical texts they published for themselves.

A sketch of their educational system can also help understand their intention to write mathematical texts. Throughout the entire Chosŏn dynasty, the educational system never changed. The candidates were selected at the early age to be trained in the governmental institution by a few teachers, who had worked as mathematical officials for decades; the list of three Chinese textbooks for the examination for selection of mathematical officials was never renewed (Kim and Kim 1977).[3] In this unaltered system, students learned mathematical practice in person from the teachers, who were already connected by matrimonial and consanguineous relationship. So it is safe to say that mathematical knowledge was not always transmitted by texts and that when some mathematical officials, usually the teachers, determined to publish their own mathematical texts, it might be for an extra purpose, in this case, to boost up their social status, more than just for the education of students.

First of all, they strategically chose, as a model for the content and the classi-fication of methods of their texts, the *Introduction to Mathematical Learning* (*Suanxue qimeng* 算學啓蒙, 1299) by Zhu Shijie 朱世傑 (fl. end of the 13th century), which was one of three Chinese textbooks used for the official exami-nation. This deliberate choice showed their awareness of the attention paid to this text by contemporary literati, compared to the other two texts.[4] For example, Kim Si-jin 金始振 (1618–1667), a ruling mayor, in charge of the postwar restoration of the lost mathematical textbooks, appraised that the methods of "the *Introduction to Mathematical Learning* are simplified (簡) and also preserve the essence (備實)," while describing the methods of other textbooks as too "shallow and vulgar (淺近)" or as "abandoning the easiness (舍易) and heading for the difficultness (趨難)" (Kim [1660] 1993). It was against this backdrop that Hong Chŏng-ha, a well-established mathematical official from a renowned lineage, decided to com-pose his mathematical text, *Writings of Nine and One*, with a similar structure of chapters and more or less the same version of the problems of *Introduction to Mathematical Learning* (Kawahara 2010, 104–113).

---

[3]The list of the three textbooks for the education of students was as follows: *Detailed Explanation of Mathematical Methods* (*Xiangming suanfa* 詳明算法, 1373) by An Zhizhai 安止齋 (fl. four-teenth century), *Yang Hui's Mathematical Methods* (*Yang Hui suanfa* 楊輝算法, ca. 1270) by Yang Hui 楊輝 (fl. end of the 13th century), and *Introduction to Mathematical Learning* (*Suanxue qimeng* 算學啓蒙, 1299) by Zhu Shijie 朱世傑 (fl. end of the 13th century).

[4]At least, two literati, Im Chun 任濬 (1608–?) and Pak Yul 朴繘 (1621–1668), were said to have written handbooks for this text.

However, two distinct changes could be noticeable: the reduction of the chapters on simple multiplication and division on the one hand and the expansion of the chapters on the "extraction of roots" (*kaifang* 開方) on the other hand.[5] These modifications probably implied that his mathematical text brought into focus not basic techniques for the beginner, but the more "expanded" algorithms for the advanced.

To make sense of the term "expanded," we have to explore the algorithms Hong Chŏng-ha adopted, but before that, it would be helpful to start with the problem 15 of the third chapter on the extraction of roots. This problem involved, in modern terms, setting and solving the cubic equation. In the former part, the equation was made by "the celestial element method" (*tianyuan shu* 天元術) and in the latter part, the equation was solved by "the method of the extraction of the cubic roots" (開立方法).

The problem reads as follows:

今有璞玉一塊形如鳥卵，內容方玉而空之，殼重二百六十五斤一十五兩五錢，只云殼厚四寸五分. 問玉方石徑各若干?

[Let us suppose that] now that there is a piece of unpolished jade, its shape is like a bird's egg. A [piece] of jade [having the form of] cube [was removed] from inside of it. The weight of [the remaining] crust [of unpolished jade] is 265 *jin* 斤 15 *liang* 兩 5 *qian* 錢. If it is only known that the depth of the crust [i.e., the distance from the face of the cube to the surface of the sphere] is 4 *cun* 寸 5 *fen* 分, then what are the edge of [the cube of] jade and the diameter of [the sphere of] stone [i.e., unpolished jade]? (Hong 1985, 599–600)

To solve this problem, the three things were assumed beforehand: the exact value of 1 *jin* as 16 *liang*, the rough density of the unpolished part of jade, which is 3, and the approximate estimation of "the cube inside the egg shape" as "the cube inscribed in the sphere." Hong let "the celestial element, *tianyuan* 天元," that is, the unknown x in modern terms, be the length of the edge of the cube, by which a solid sphere with an empty inscribed cube could be imagined as shown in Fig. 1.

Then he started the explanation of the way in which one is supposed to put and move the counting rods, which is reconstructed in Fig. 2. It should be noted that the top layer of the arrays of the counting rods represented the constant of the equation, and the degree of the unknown was ascending at each layer from the top downwards. Thus in the Step 1, the first array denoted x, and the second array, $x + 9$, which was the diameter of the sphere. Hong cubed the second formula, putting $(x + 9)^3$, i.e., $x^3 + 27x^2 + 243x + 729$ to the third array. And he mutiplied it by 9, making the fourth array $9x^3 + 243x^2 + 2187x + 6561$. As for the reason for the multiplication of the third array by 9, we should recognize that he used the traditional formula for the

---

[5]In the *Introduction to Mathematical Learning*, the problems on simple multiplication and division in the first five chapters made around 30% (79 problems out of total 259 problems) and the problems on the extraction of roots in the last chapter made up around 13% (34 problems out of 259 problems), whereas in *Writings of Nine and One*, the problems on simple multiplication and division in the first chapter made up around 4% (19 problems out of 473 problems; in this case the total sum of problems is calculated without the problems in the last "Miscellaneous Writing" chapter), and the problems on the extraction of roots in the last three chapters made up about 35% (166 problems out of 473 problems), the largest proportion of the entire text, as Kawahara (2010, 109–110) also pointed out.

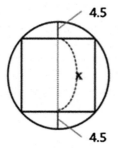
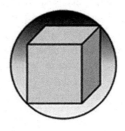

Fig. 1 Cutting section and the solid figure of the problem 15

calculation of the volume *V* of the sphere with a diameter *d*, $V = 9d^3/16$, which was provided in the *Nine Chapters on the Mathematical Procedures* (*Jiuzhang suanshu*, 九章算術, first century), one of the most influential mathematical texts in pre-modern East Asia.[6] So the last array, noted I in Fig. 2, was 16 times the volume of the sphere. Then he started another step and ended up with the final diagram, where array II was 16 times the volume of the inscribed cube. And in the Step 3, he performed a subtraction I–II, which resulted in 16 times the volume of the remaining unpolished part with the inside cube removed, represented with array III. In the Step 4, beside III, he put another array, IV, which featured the constant 22,696. This value was 16 times the volume of the remaining crust, calculated with the weight value, 265 *jin* 15 *liang* 5 *qian* (4255.5 *liang*) and the rough density, 3, all given in the problem and the prior assumptions. Now he got two arrays, III and IV, both of which had the same volume; but III was composed of the unknowns and numbers, while IV was composed of numbers alone. Finally he subtracted each layers of IV from III, as a result of which, he got V. From the Step 1 to the Step 4, what he had done was "the celestial element method," namely making the equation, V:

$$-7x^3 + 243x^2 + 2187x - 16135[= 0] \quad \text{(ibid., 600)}.$$

Afterward he just instructed "Extract the cube root," only providing the answer, *x* = 5, without any detail (ibid., 600). But if we consulted the other chapter on the explanation of the extraction of cube root in the same text, we could easily reconstruct the procedures as the Step 5, which basically iterated two operations: the multiplication of the guessed root 5 by the coefficient of the equation at the bottom layer and the addition of this to the next upper layer.[7]

---

[6]For the detailed explanation of this formula, see Martzloff (1997, 286).

[7]This kind of method for extracting the roots by iterated multiplication and addition was called "*zengcheng kaifangfa* 增乘開方法" and could be found in many Chinese mathematical texts from the twelfth century on, including, among others, Qin Jiushao's 秦九韶 (ca. 1202–1261) *Mathematical Treatise in Nine Chapters* (*Shushu jiuzhang* 數書九章, 1247). For more detailed study of this method, see Martzloff (1997, 231–249), Libbrecht (1973). Also, for the general explanation of the other operations with counting rods, including the method of the celestial element, see Marzloff (1997, 217–271).

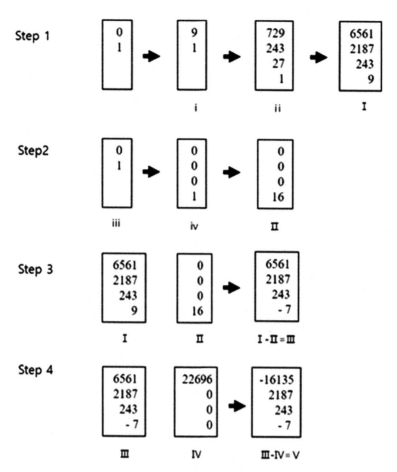

Fig. 2 Reconstruction of the algorithm of the problem 15 of the third chapter on the extraction of roots. In the Step 1, from the first to the last, each diagram represents x, $x+9$[i], $x^3 + 27x^2 + 243x + 729$ [ii], $9x^3 + 243x^2 + 2187x + 6561$ [I], where I is 16 times the volume of the sphere. In the Step 2, each diagram respectively represents x, x^3, $16x^3$[II], where II is 16 times the volume of the cube. In the Step 3, the final diagram, III, was obtained by I–II, $-7x^3 + 243x^2 + 2187x + 6561$, where III is 16 times the volume of the crust (i.e., the sphere with the inside cube removed). In the Step 4, IV was inserted, where 22,696 is 16 times the volume of the crust inferred from the given weight of the stone, and the final diagram, V, was obtained by III–IV, $-7x^3 + 243x^2 + 2187x - 16135$, where V is eventually equal to 0. Here the numbers represented by counting rods are changed into the corresponding Arabic numbers to help the modern readers' understanding

In the rest of the text, as in the most of Chinese and Korean mathematical texts, we can find this traditional type of explanation. All the layers of the coefficients were set up according to the ascending, or sometimes descending, degrees of the unknown x, and the elimination process, that is, the subtraction of two arrays to make one equation, like the Step 4 in Fig. 3, was essentially included at the last

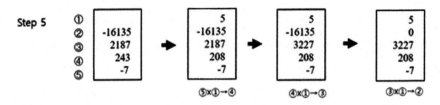

Fig. 3 Reconstruction of the algorithm of the solution of cubic equation, $-7x^3 + 243x^2 + 2187x - 16135 = 0$. This method is applied to deconstruct the equation in such a manner as $x\{x(-7x + 243) + 2187\} - 16135 = 0$, when $x = 5$; in the second diagram, 208 appeared from the calculation in the round brackets $(-7 \times 5 + 243)$; in the third diagram, 3227 appeared from the calculation in the curly brackets $\{5 \times 208 + 2187\}$; in the fourth diagram, 0 appeared from the addition of the original value –16,135 to the result of multiplication 5×3227

stage. However, in Hong's texts, a careful reader could find "expanded" version of the algorithm.

The first example is about how to expand the perfect expression of nth power with counting rods, in Fig. 2, from i to ii, which had never been offered in any other mathematical texts. Hong took the example of an expansion of $(x + 12)^3$ in the introductory part of *Writings of Nine and One*. He explained: put 1 in the lowest row and 12 in the uppermost row, and then multiply them and put the result into the second row from below; then again multiply the magnitudes in the second row from below and in the uppermost row, and put the result into the third row from below; this iterated procedure stops when some magnitude occupies the second row from the top. Then again he repeated this whole procedure with the number in the lowest row, still 1, and the one in the uppermost row, still 12, but this time he stopped when the third row from the top was occupied. This whole loop of algorithms stopped when every row under the uppermost one were occupied, and then, he concluded by removing 12 from the uppermost row. As shown in Fig. 4, the final array of the procedure attained the coefficients of the extension form of $(x + 12)^3$, that is, $x^3 + 36x^2 + 432x + 1728$ (ibid., 212). Note that this iteration itself was very similar to that of the solving the equation, in the former problem.

The second example is about the expansion of the polynomials of a hyperrectangle. Here is problem 34 in the third chapter on the extraction of roots, for instance, set to expand the polynomial, $x(x + 4)(x + 8)(x + 12)(x + 16)$, in modern terms:

今有面不等四乘方, 積二十萬八千八百四十五尺. 只云甲面多乙面四尺, 多丙面八尺, 多丁面一十二尺, 多戊面一十六尺. 問各面若于?

[Let us suppose that] now that there is five-dimensional [hyperrectangle] with unequal sides, (面不等四乘方) [whose] volume (積) is 208,845 *chi* 尺. If it is only known that [the length of] edge A is 4 *chi* more (多) than edge B, 8 *chi* more than edge C, 12 *chi* more than edge D, and 16 *chi* more than edge E, then what is [the magnitude of the length of] each edge? (ibid., 624–625).

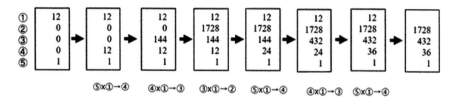

Fig. 4 Reconstruction of the sequential procedure for the expansion of $(x+12)^3$. The last diagram shows the column of the coefficients of $x^3 + 36x^2 + 432x + 1728$

Hong started with the instruction "Place in an array the differences [between the first edge and] all [other] edges" with the first diagram of Fig. 5 and promptly gave "what was obtained by the celestial element method (*tianyuan shu* 天元術)" with the last diagram of Fig. 5. Then in a commentary written in small size characters, he explained how to reach the last diagram: just with the same looping, but with the exception that 12 in the former algorithm now turned into four numbers in four layers, 4, 8, 12, 16, as shown in Fig. 5 (ibid., 624–625).

These two examples illustrated that his way of "expanding" the algorithm to similar problems was to stretch the layers of counting rods.[8] Note that, as Martzloff (1997, 230) has pointed out in his explanation of the extraction of roots, "in fact, Chinese authors took advantage of the resources provided them by the ancient terms although there was no longer any possibility of a geometrical interpretation." For example, the coinage of terms for the numerical values in the cubic equation, the geometrical origins could be easily detected. Where the term "*shi* 實" was for the dividend and the term "*fa* 法" was for the divisors, the geometrical terms like "*fang* 方" (literally "side") or "*cong fang* 從方" ("extension of the [coefficient] 'side'"), "*lian* 廉" ("edge") or "*cong lian* 從廉" ("extension of the [coefficient] 'edge'"), and "*yu* 隅" ("corner") were compounded as the terms for the coefficients as follows:[9]

$$yu[fa]x^3 + cong\,lian\,x^2 + cong\,fang\,x - shi = 0.$$

But in case of the higher dimensions, Hong had no obligation to interpret them geometrically, just as Martzloff said. He stretched the powers of numbers very freely, even saying the "volume (積)" of the five-dimensional hyperrectangle with unequal "sides (面)," without any consideration of the existence of the hyperrectangle in real three-dimensional space. However, he stretched the layers of the counting rods in an untraditional way. In Fig. 4, 12 at the top, and in Fig. 5, the four

---

[8]Hong's expansion started with the similar type of problems in the *Introduction to Mathematical Learning*, but while the *Introduction* dealt only with objects of no more than three dimensions that could be easily imagined in a real space, Hong expanded this version of problems up to ten dimensions.

[9]This is the expression with the coinage of the terms that Hong used mostly in his text. For other various expressions used, see Martzloff (1997, 229).

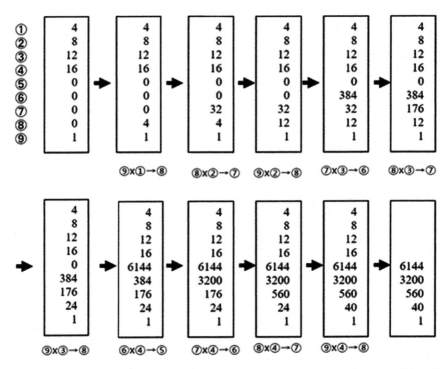

Fig. 5 Reconstruction of the sequential procedure for the expansion of $x(x+4)(x+8)$ $(x+12)(x+16)$. The last diagram shows the coefficients of $x^5 + 40x^4 + 560x^3 + 3200x^2 + 6144x$

numbers at the top, namely 4, 8, 12, 16, were put there only for the convenience of computing and then removed at the end. He ignored the traditional explanation: no strict frame of the layers according to ascending, or descending, degrees of the unknown; and no elimination process. So these examples might show what was casually done in real practice, or explain how what Kawahara (1998, 65) generally called "a big jump of the celestial element method" really occurred.[10]

Thus analysis of Hong's computation itself strongly suggests that this computation was derived from his actual practice with counting rods. Indeed, the first eight chapters of his book, full of the succinct prescriptions of the algorithms, seems like a manual without any second-order statements about mathematical calculation. But in the final ninth chapter, the reader was suddenly exposed to his strong opinion about mathematical calculation and keen recognition of social status. This chapter, called "Miscellaneous Writing" (雜錄 Kor. *Chapnok*, Chin. *zalu*), was a bricolage

---

[10]Kawahara provided a few reasons why he praised Hong's texts as "a big jump of the celestial element method": the expansion of the total numbers and the increasing percentage of the numbers of the problems on the extraction of roots, and the increasing portion of the problems using the celestial element method in those chapters. Kawahara (1998, 60–65).

of several different themes: a set of 16 problems and 5 diagrams related to astronomy, a basic description of how to calculate the lengths of the musical pipes in musical harmonics, one diagram of ten-by-ten magic squares, and an anecdote of a meeting with a distinguished Chinese astronomer who visited Chosŏn in 1713 (Hong 1985, 651–693).[11]

Prior to Hong's text, the first two subjects, astronomy and musical harmonics, had not been included in any mathematical text for or by officials in Chosŏn. The reason for his inclusion of these subjects seems to be crystal clear: to publicly display his mathematical knowledge, or skill, to be superior even to the literati's. All 16 problems related to astronomy were basically related with the passages in the *Book of Documents* (*Shujing* 書經), one of the most important texts for study by the literati.[12] Furthermore, in two problems, the doyen of the contemporary literati, Kim Chang-hŭp's 金昌翕 (1653–1722) entered the scene to ask about the basic meaning of the passages in the Classic:

三淵先生問曰，朞三百註九百四十分者，何也；又問曰，十九歲七閏則氣朔分齊者，何也.

(Problem 10) Kim Chang-hŭp asked "what [does it mean by] 'denominator 940' in the commentary on 'A Year with Three Hundred Days (*jisanbai* 朞三百)'?"

(Problem 11) Again he asked "what [does it mean by] that '7 leap months in 19 leap years, then surplus of tropical month (氣盈) and deficit of synodic month (朔虛) should be equally distributed'." (ibid., 659–660)

Significantly, a mathematical problem identical with problem 10 was already provided in problem 8, yet without mentioning Kim's name. This definitely reflects that Hong intentionally inserted the seemingly realistic situation in which he advised the influential member of the literati in order to make his mathematical knowledge stand out.

In fact, this strategy to set up the meeting scene with a member of the *yangban* literati had a precursor in a text from an earlier period written by another *chungin* mathematical official, Kyŏng Sŏn-jing 慶善徵 (1616–?). Kyŏng, the great-grandfather of Hong on his maternal side, with such mathematical expertise as to catch the contemporary literati's attention, had created a chapter called

---

[11]This anecdote was fully addressed in Horng's paper (2002b).

[12]For example, problems seem like the variations on the combinations of basic astronomical situations found in "A Year with Three Hundred Days" in the *Book of Documents*:

(Problem 2) [Let us suppose that] now there is a circle-shaped lake, whose circumference is 365 1/4 *chi* 尺. If it is only known that the large ant and the small one [started] together [to] walk, and that the small one goes 1 *chi* a day and the big one goes 13 7/19 *chi* a day, then how many days will pass before they meet [again]?

(Problem 3) [Let us suppose that] now there is heaven, whose circumference is 365 1/4 *du* 度, surrounding the earth and rotating clockwise. The sun and the moon [started to] go together with the heaven; the sun is 1 *du* slower than the heaven, and the moon is 13 7/19 *du* [slower than the heaven]. How many days will pass before the sun and the moon meet [again]? (Hong 1985, 651–653).

This kind of problems had been circulated in a wide range of areas for centuries. For more detailed discussion on these "pursuit problems," see Bréard (2002).

"Responses Exchanged" (*hwadap hohwan* 和答互換) in his text, *Mathematical Methods of the Writings of Muksa* (*Muksajip sanbŏp* 黙思集算法). In this chapter, he presented himself as an old man called "Muksa 黙思", literally meaning "contemplation," in the meetings where he exchanged eleven mathematical problems with local magistrates or with a schoolboy holding "books" under his arms, an unmistakable symbol of the boy's social status as *yangban* (Kyŏng 1985, 245–258). Here, again, the fact that most of these problems were presented in other chapters indicates that this chapter was not intended to provide more practical knowledge to the readers, but rather to present him as a superb mathematical expert.

Having developed this approach, Hong presented the anecdotes with real names like He Guozhu 何國柱 (fl. eighteenth century).[13] Hong first recorded the venues, dates, and names of the company at the meeting to give what Steven Shapin and Simon Schaffer have identified as "virtual witnessing" of the meeting to his readers (Shapin and Schaffer 1985). Hong continued to report a pedantic conversation on He Guozhu's several mathematical questions and his own answers. After a while, the Chinese envoy, Aqitu 阿齊圖 said, "He Guozhu's mathematical skill is the top fourth-ranked [among the people] all over the world; his stomach is completely full of mathematical methods, [which conventionally means that he has extensive knowledge of mathematical methods]. The bunch of people like you cannot beat [him]. He already asked a lot of questions, but you have not asked any question yet, so why don't you test his skill?" Instantly he gave the formerly mentioned egg-shaped amalgam of unpolished jade problem. According to Hong's record, "He Guozhu responded that the method for the solution [of the problem] was too difficult [so it cannot be solved] immediately and that he would solve it tomorrow. But after that, he never showed any answer." Then Hong proudly recounted his own algorithm and answer to the readers (Hong 1985, 676–693).

After continuing their dialogue about the other problems and solutions, at the end of this meeting, Hong recorded He Guozhu's admiration for his deft handling of the counting rods:

司曆曰, 算家諸術中方程正負之法, 極爲最難. 君能知之乎? 余曰, 方程之術, 卽中等之法, 何難之有? 余布算之際, 司曆曰, 中國無如此算子, 可得而誇中國乎? 余卽以與之, 則擇其中四十箇而去. 司曆曰, 君之姓名書去, 吾當以示大國. 因書吾與劉生姓名以去.

The [Chinese] astronomical official (He Guozhu) said, "Among the mathematicians' methods, the *fangcheng zhengfu* 方程正負 method is most difficult. Do you know this?" I said, "*fangcheng* method is the second-grade technique. How [can you say] it is difficult?" As I arranged the counting rods, the astronomical official said, "In China, we don't have this calculating equipment. Can I have it and boast of it in China?" Immediately I gave them [to him] and he picked up 40 [rods]. He said, "Write down your name and I will definitely inform the Great Country (大國, China) [of you]." Hence he wrote down my name and [the accompanying] Yu [Su-sŏk]'s 劉壽錫 name, and left (ibid., 692–693).

---

[13]He Guozhu had served as a staff member of the Office of Mathematics for Kangxi 康熙 (r. 1662–1722) in the Qing 淸 dynasty. For his general activities and missions, including his visit to Korea, see Jami (2012, 263–273, 277–279).

Here Hong's crafted way of reporting the dialogue hinted that the author aimed to let the reader witness his superiority over the "world-class" Chinese specialist, who was initially praised by Aqitu, but had no idea of the counting rods' calculation. Notably, Hong regarded this superiority as coming from his prowess in the manipulation of counting rods, which had been preserved in the education of the mathematical officials in Korea. For him, mathematical knowledge needed to be realized in the hands-on practice with counting rods, and only through that experience could mathematical officials lay claim to their positions sufficiently high to teach even the literati and Chinese scholars.

Computation of Literati in *Summary of Nine Numbers*

In contrast to the texts written by mathematical officials, recent historians have usually described the mathematical texts authored by the literati in the late seventeenth and early eighteenth century as composed "from the metaphysical viewpoint," or framed with "intellectual taste" or "Neo-Confucianism" (Kim and Kim 1977; Horng 2002a; Kawahara 2010). These modifiers derived from their observation that in mathematical texts, the literati usually borrowed vocabulary, rhetorics and literary styles from their philosophical counterparts in order to differentiate themselves from mathematical officials and to solidify their ruling social status. For instance, Cho Tae-gu 趙泰耈 (1660–1723), in his mathematical text, *Narrow View of Mathematical Text (Chusŏ kwan'gyŏn* 籌書管見), devoted the latter half to presenting himself as a teacher, appropriating the traditional Confucian question-and-answer literary style (Cho 1985, 125–195). Other people kept asking him about the justification of algorithms and Cho, as a teacher, answered them in a fluent and articulate way. Once he was asked about the foundation of the algorithm of the extraction of the root, he provided Liu Hui's 劉徽 (fl. end of the third century AD) long-established version of the explanation that typically regarded the "square" of a number as the area of the geometrical "square."[14] Remarkably, in his text, he was asked about what had never been asked in the mathematical officials' texts: if two-power was explained as a square, and three-power as a cube in space, for instance, what was the shape of four-power or five-power? In his answer, Cho claimed that those were of no shape, but only of numbers that was computable (ibid., 158–159). The very existence of this type of questions indicates the perspectival incongruities on the terms of mathematical problems between literati and mathematical officials, in that the latter ones could easily reduce them just to labels of magnitudes, as was shown in Hong's case.

---

[14]For the full commentary of Liu Hui, see Chemla and Guo (2004), and Guo et al. (2013).

The way of literati's acquisition of mathematical knowledge might be one of the crucial factors to give these features to literati's texts. Due to the lack of institutional education of mathematics for the literati, they had to achieve mathematical knowledge from other individual literati, or considerably by perusing and analyzing texts by themselves.[15] In this situation, some self-taught literati might well treat mathematical practice as one of their typical literary practice as imbuing texts with philosophical terms and passages.

Ch'oe Sŏk-chŏng, a former Chief State Councilor (*yŏngŭijŏng* 領議政), the highest-rank official and an influential scholar of great erudition, was one of those kind of the self-taught literati with deep mathematical interest. From the first pages of his mathematical text, the *Summary of Nine Number*, his presentation of the long, pretentious list of "Reference Texts," 70% of which were from the philosophical Confucius Classics, as shown in Table 1, seemed to warn the reader that his mathematics was different from that of the mathematical officials. The further the reader went into his mathematical text, the more often he would encounter the philosophical terms from the study in Neo-Confucianism, particularly in the Study of Image and Number (*Xiangshu* 象數), such as *yang* 陽, *yin* 陰, *taiyang* 太陽, *taiyin* 太陰, *shaoyang* 少陽, *shaoyin* 少陰, which could be, as Kawahara (1998, 51) aptly coined, "metaphysical embellishments."[16]

Yet these terms were not peppered over the text superficially but were systematically placed in his mature consideration. For the main part of the *Summary of Nine Number*, he executed a unique classification of the various algorithms into sixteen categories, according to what, he thought, the algorithms stood for, from the *sixiang* 四象, such as *taiyang*, *taiyin*, *shaoyang*, and *shaoyin*.

In his scheme, at the outset, he regarded all the known algorithms as ones with only four-row array of the counting rods, and attached the novel labels on every rows, like "the sun (*ri* 日)," "the moon (*yue* 月)," "the [five naked-eye] planets (*xing* 星)," and "the stars (*chen* 辰)," instead of the traditional terms that could tell the roles of the rows during the mathematical operation such as the dividend, the divisor, or the coefficients. These new labels were from the Shao Yong's 邵雍 (1011–1077) cosmological version of the Study of Image and Number, which regarded *taiyang* as "the sun," *taiyin* as "the moon," *shaoyang* as "the planets," and *shaoyin* as "the stars." In the first three rows, that is, "the sun," "the moon," and

---

[15]It was from the late eighteenth century on that it had been reported that literati had learned from, and sometimes co-worked with the *chungin* officials: at first, from astronomical officials, and then, in the nineteenth century, even from mathematical officials. By contrast, there were several anecdotes reporting that a member of literati, such as Hwang Yun-sŏk 黃胤錫 (1729–1791), hesitated to see any *chungin* officials to borrow their mathematical texts so that he could not have a chance to read the texts for years, or that another member of the literati, such as Hong Kil-ju 洪吉周 (1786–1841), learned mathematical methods from his mother when young, and soon was self-taught by mathematical texts when he grew up (Sun 2006; Koo 2012).

[16]The whole contents of *Summary of Nine Numbers* and the influence of the Study of Image and Number were thoroughly studied by Kawahara (1996). For the general relationship between number and the Study of Change (*yixue* 易學), or the general survey of the Study of Image and Number, see Ho (1995).

Table 1 Bibliography in *Summary of Nine and One* (Ch'oe 1985, 375–379)	Bibliography (引用書積)
	Book of Changes (Zhouyi 周易), *The [Book of] Songs in Mao [redaction] (Maoshi* 毛詩), *Book of Documents (Shangshu* 尙書), *Zuo's Commentary on the Spring and Autumn Annals (Chunqiu Zuoshizhuan* 春秋左氏傳), *Gong Yang Commentary (Gongyangzhuan* 公羊傳), *The Rites of Zhou (Zhouli* 周禮), *Book of Rites (Liji* 禮記), *Analects of Confucius (Lunyu* 論語), *Works of Mencius (Mengzi* 孟子), *Doctrine of Mean (Zhongyong* 中庸), *Great Learning (Daxue* 大學), *Approaching Correctness (Erya* 爾雅), *Works of Zhuangzi (Zhuangzi* 莊子), *Works of Xunzi (Xunzi* 荀子), *Works of Sunzi (Sunzi* 孫子), *The Masters of Huainan (Huainanzi* 淮南子), *Hidden Talisman Classic (Yinfujing* 陰符經), *Exemplary Sayings (Fayan* 法言), *Historical Records (Shiji* 史記), *Book of Han (Hanshu* 漢書), *Annotated Accounts (Gangmu* 綱目), *Selection of the Refined Literature (Wenxuan* 文選), *Works of Shao Yong (Shaozi quanshu* 邵子全書), *Complete Works of Master Zhu (Zhuzi daquan* 朱子大全), *Introduction to the Study of Changes (Yixue qimeng* 易學啓蒙), *Questions on Great Learning (Daxue jiwen* 大學幾問), *Nine Chapters of Mathematical Procedures (Jiuzhang suanjing* 九章算經), *Calculation of Seven Celestial Bodies (Qizheng suan* 七政算), *Introduction to Mathematical Learning (Suanxue qimeng* 算學啓蒙), *United Lineage of Mathematical Methods (Suanfa tongzong* 算法統宗), *Calculation of Multiplication and Division [by Yang Hui] (Chengchu suan* 乘除算), *Mathematical Methods for Elucidating the Strange Properties [of Numbers by Yang Hui] (Zhaiqi suanfa* 摘奇算法), *Classified Calculation of Surveying Fields [by Yang Hui] (Tianmu bilei* 田畝比類), *First Collection of Heavenly Learning (Tianxue chuhan* 天學初函), *Counting Rods [Napier's Bones] (Chousuan* 籌算), *Detailed Explanation of Mathematical Methods (Xiangming suanfa* 詳明算法), *Writings of Muksa (Muksajip* 默思集)
	−

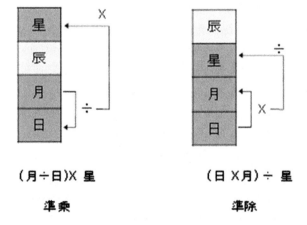

Fig. 6 Reconstruction of the procedure of *Chunsŭng* 準乘 and *Chunje* 準除

"the planets," the magnitudes were already given from the problem, and then, after undergoing some procedures, the result would be put into the empty row, "the stars." For example, "*Chunsŭng* (Chin. *Zhuncheng*) 準乘" and "*Chunje* (Chin. *Zhunchu*) 準除" were the methods following the calculating procedures shown in Fig. 6.

Ch'oe assigned each of the four basic arithmetical operation to one of the four symbols *sixiang*, and obtained a spatial scheme for the operations. Multiplication and addition typically implied *yang*, whose sign was ▬, and division and subtraction, *yin*, whose sign was ▬▬. Consequently, *Chunsŭng* belonged to *shaoyang*, whose sign was ▬▬, because it started with division (corresponding the broken line below) and ended with multiplication (corresponding with the complete line above), whereas *Chunje* belonged to *shaoyin*, whose sign was ▬▬, because it started with multiplication (corresponding to the complete line below) and ended with division (corresponding to the broken line above).[17]

This way of matching the philosophical terms with the operations performed with counting rods tells us that Ch'oe approached the mathematical calculation mainly through the spatial and visual images rather than the dexterity or speed of the operations.

Sometimes he made good use of the possibility to read some meaning off out of the images of the operations. With this reading, he went further to evaluate various algorithms, with the usual effect that he criticized the contemporary mathematical officials' practice. In the first quarter of the text, after presenting the step-by-step detailed prescriptions for the several algorithms for the simple multiplication and division with two magnitudes, he abruptly added his assessments:

> 故世之學算者, 率多舍此而取彼. 見今算法諸書, 俱以因歸爲主, 而絀步商. 正道之難明, 久矣.

> Therefore the contemporary people who study mathematics mostly avoid these [methods, *Bucheng* 步乘 and *Shangchu* 商除, due to their complexity and inconvenience] and follow those [methods, *Yin* 因, *Cheng* 乘, *Gui* 歸 and *Chu* 除, due to their simplicity and convenience]; having seen the contemporary mathematical texts, [they heavily] focus on the *Yin*, *Gui* methods, and abandon the *Bucheng*, *Shangchu*. [Hence] it has been long [time] since the proper way (*zhengdao*, 正道) [of calculation] was hard to shine. (Ch'oe 1985, 430)

Even though each set of the multiplication techniques, *Bucheng* 步乘, *Yin* 因, and *Cheng* 乘, and the division techniques, *Shangchu* 商除, *Gui* 歸, and *Chu* 除, should eventually reach the same numerical results, Ch'oe observed the mathematical officials' distinctive tendency in their choices toward the specific algorithms and pointed out an indelible difference between the algorithms that they usually employed, *Yin*, *Cheng*, *Gui*, and *Chu*, and those that they tended not to use, *Bucheng* and *Shangchu*. He continued to argue that the latter algorithms were "the exemplar (範驅)," whereas

---

[17]Take the other example, a procedure called "*Ch'ongsŭng* 總乘." The term was coined by Ch'oe and its meaning seemed the "total multiplication," because during the procedure, "the sun," "the moon," and "the planets" were multiplied all together. This procedure belonged to *taiyang*, whose sign was ▬, because it underwent multiplication below, and another multiplication above. For the other categories, see Kawahara (1996, 2010).

the former ones were just "deceitful expedients (詭遇)" for the sake of convenience or quickness (ibid., 430). Hence, for him, it was shameful that contemporary mathematical officials behaved like opportunists heavily reliant on the former ones.

On what grounds could he evaluate, with no hesitancy, some of the computational algorithms as the "right" or "proper way," and others not, if all of them might beget the same numerical results? As shown in Fig. 7, the highly appraised *Bucheng* and *Shangchu* were the algorithms with three-row arrays of the counting rods. During the procedure of the multiplication *Bucheng*, the digits of the product in the middle row started to appear from the right to the left, namely from the low place value to the high place value; in case of division *Shangchu*, the digits of the dividend in the middle row started to disappear from the left to the right. And in the *Bucheng*, the multiplier in the third row moved from the right to the left, and in the *Shangchu*, the divisor in the third row moved from the left to the right. According to him, these opposite movements of the numbers in the same row showed that two methods, *Bucheng* and *Shangchu*, acted as "spouses (配耦)."[18] However, in his opinion, the other algorithms, *Yin*, *Cheng*, *Gui*, and *Chu*, with two-row arrays of the counting rods, could not yield this oppositeness of the images during the computational operation, as the *Yin* and *Gui* methods shown in Fig. 8. These methods were only for the convenience in handling the counting rods, which might probably have improved the speed, but ultimately distorted the "proper way." In addition, Ch'oe kept rebuking the contemporary mathematicians for following mindlessly the different version of *Bucheng* in *Yang Hui's MathematicalMethods* (*Yang Hui suanfa* 楊輝算法, ca. 1270), "improperly" from the left to the right (ibid., 423–424).

His evaluations of other computational tools will also confirm that what was at stake was to make equivalence between the transformative images of the array of the counting rods during the computation and the philosophical concepts or the images in the Classics. In his last chapter called "Addendum," (*fulu* 附錄) he presented three computational tools other than counting rods and compared them: *gelosia* calculation (*wensuan* 文算 or *pudijin* 鋪地錦), abacus calculation (*zhusuan* 珠算), and Napier's bones calculation (*chousan* 籌算) (ibid., 641–666).[19]

Out of three tools, he manifested a clear endorsement of the *gelosia* calculation, saying "this is what a man of virtue who enjoys Arts (遊藝君子) ought to bear in mind," because of the following rationale:

加減數起於右而極於左, 乘除數起於右下而極於左上. 一則象易之重卦次序, 載乾於西, 而終坤於東也. 一則象易之重卦方圓, 位乾於西北, 而置坤於東南也.

[In the *gelosia* calculation during the procedures of] addition and subtraction, [the appearance of the digits of the resulting] number starts from the right and ends at the left, and [during the procedures of] multiplication and division, [the appearance of the digits of

---

[18]Ch'oe's interpretation of the multiplication and division as a pair was not at all his unique thought. For more detailed discussion, see Chemla (2010).

[19]For the detailed description of these computational tools, see Needham (1959), Martzloff (1997), and Li and Du (1987). For the Chinese sources that Ch'oe used, see Guo (2009).

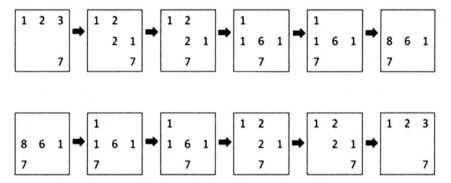

Fig. 7 Reconstruction of Ch'oe's idealized version of the procedures of 步乘 *Bucheng* (123 × 7), above, and *Shangchu* 商除 (861 ÷ 7), below

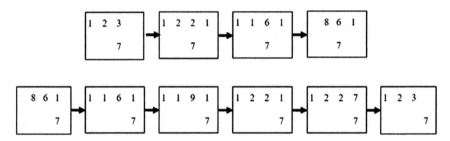

Fig. 8 Reconstruction of Ch'oe's version of the procedures of *Yin* 因 (123 × 7), above, and *Gui* 歸 (861 ÷ 7), below. Procedure *Yin* 因 starts from 3 × 7; one then removes 3 from the multiplicand and puts the result 21 into that space, instead. For the next step, one does 2 × 7 and then removes 2 from the multiplicand and puts the result 14 into that space. The same steps should be continued until the answer 861 could be obtained. To perform procedure *Gui* 歸, one needs the *jiugui* 九歸 rules; in the case of division by 7 the rules are as follows: *qi yi xia jia san* 七一下加三 ("Seven one, add three below"); *qi er xia jia liu* 七二下加六 ("Seven two, add six below"); *qi san sishi er* 七三四十二 ("Seven three, forty two"); *qi si wushi wu* 七四五十五 (Seven four, fifty five); *qi wu qishi yi* 七五七十一 (Seven five, seventy one); *qi liu bashi si* 七六八十四 (Seven six, eighty four); *feng qi jin cheng shi* 逢七進成十 ([When] seven is met, push ten forward). At first, one encountered 8, namely 7 and 1. According to the rule "[when] seven is met, push ten forward," one puts 1 in the next higher place, and with the remaining 1, according to "Seven one, add three below," one should add 3 into the next lower column, so 9 would appear. Then when one encounters 9, according to "Seven met, make ten forward," one puts 1 in the next higher column, and with the remaining 2, according to "Seven two, add six below," one should add 6 into the next column, thus the magnitude of the next column becomes 7. Now for the last step, one encounters 7, so according to "Seven met, make ten forward" one should put 1 into the next higher column, and then 123, the answer, appears. Note that while Ch'oe presented these procedures with the multiplier, 7, and the divisor, 7, fixed in their position, the usual prescriptions in other texts didn't suggest the physical existence of the multiplier and the divisor during the operations

the resulting] number starts from the bottom right and ends at the top left. The [former] ones emulate, [or are modeled on,] the Fuxi's Diagram of the Sequence of Sixty-Four Hexagrams (*Fuxi liushisi gua cixutu* 伏犧六十四卦次序圖), beginning in *qian* 乾 in the west and ending in *kun* 坤 in the east. The [latter] ones emulate the Fuxi's Diagram of Square and Circle of Sixty-Four Hexagrams (*Fuxi liushisi gua fangyuantu* 伏犧六十四卦方圓圖), placing *qian* in the northwest and placing *kun* in the southeast. (ibid., 651–652)

A hexagram is composed of six lines, each of which represents *yin* and *yang*, like each ideograms of *sixaing* is composed of two *yin/yang* lines, and among them, *qian* 乾 with all *yang* lines and *kun* 坤 with all *yin* lines are regarded as the most important ones, sometimes representing heaven and earth. Two Fuxi's Diagrams, the Fuxi's Diagram of the Sequence of Sixty-Four Hexagrams (*Fuxi liushisi gua cixutu* 伏犧六十四卦次序圖) and the Fuxi's Diagram of Square and Circle of Sixty-Four Hexagrams (*Fuxi liushisi gua fangyuantu* 伏犧六十四卦方圓圖), were known to show, respectively, how the whole set of sixty-four hexagrams had been sequentially generated or transformed, and how they were spatially arranged in the circle and square forms, as shown in Figs. 9 and 10.

What Ch'oe mentioned in this passage was Shao Yong's particular interpretation about the sequence of the hexagrams. In Fig. 9, every ideogram of Fuxi's Diagram of the Sequence on each layer had appeared, or transformed, from the right to the left. And if you apply this order to the ideograms of Fuxi's Diagram of Square and Circle in Fig. 10, every ideogram of the inside square shape had appeared from the bottom up, and from the right to the left.

He considered that these two types of the sequences of the hexagrams were embodied in the addition and multiplication of the *gelosia* calculation. During the procedure of the addition, the calculator started from the low place value so that the digits of the result appeared from the right to the left, just like the sequential appearance of the hexagrams in the Fuxi's Diagram of the Sequence, as shown in Fig. 9.

Similarly, as for the multiplication of the *gelosia* method, take 436×62 as an example. Firstly, the calculator should draw the grids and diagonals of each cell on the paper, then put 436 horizontally on the top, and 62 vertically on the left side like shown in Fig. 10.

The calculator then started by multiplying 6 and 2; he put the tens digit (1) of the product 12 on the left side of the diagonal in the lower rightmost cell, and units digit (2) on the right side. He proceeded in similar vein to fill all the cells of the table. After that, he added up the numbers slantwise to obtain the digits of the product. The digit "2" under the rightmost diagonal became the units digit of the result. To calculate the tens digit he added up 6, 1, and 6; the result was 13, so the tens digit of the result was "3" while "1" was carried to the next position. To calculate the hundreds digit he added 8, 8, 3, and the carried 1; the result was 20, so the hundreds digit was 0, and he carried 2 to the position of thousands. When he finished this procedure, he obtained 27,032. So the digits of the result appeared from the right bottom to the left top, viz. in sequence of 2, 3, 0, 7, 2, in the opposite direction of 27,032, which seemed to indicate the rough appearance of the hexagrams in the Fuxi's Diagram of Square and Circle, in which *qian* 乾 sits in the northwest (here, bottom right) and *kun* 坤 is placed in the southeast (here, upper left).

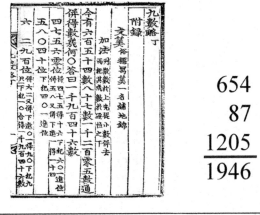

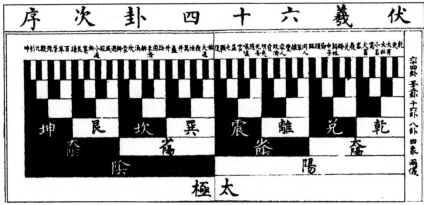

Fig. 9 Above: The problem of addition in the *gelosia* calculation in *Summary of Nine Numbers* and the reconstruction of addition 654 + 87 + 1205 in which the result 1946 is accumulated from the right to the left. Below: the *Fuxi liushisi gua cixutu* 伏犧六十四卦次序圖, "beginning in *qian* 乾 in the west (right) and ending in *kun* 坤 in the east (left)." (Above left: Ch'oe 1985, 641; below: *Zhu* 1985)

By contrast, according to him, "the Napier's bones calculation is limited and clumsy (偏而拙)," and the calculation using the abacus was "complicated, chaotic (煩亂)," "confined and limited (局滯)," and "could not reach the deep [level] of the counting rods' calculation." He didn't offer the apparent reasons for this allegation, but it is safe to say that he disapproved "the contemporary Chinese officials and [merchants in] the market, [as those who] all use the abacus, and abandon the counting rods, thus cannot be enlightened," because of the nonexistence of the similarity between the movements of beads during the calculating procedure and any philosophical concepts or images, like the dynamic relationships between the hexagrams described by the sage authors of antiquity (ibid., 659, 666).

When Ch'oe needed a criteria for acceptance or refusal of the new technique, but could not find any suitable images from the classics, he willingly expanded his credible sources to the traditional mathematical texts. In the survey problem with

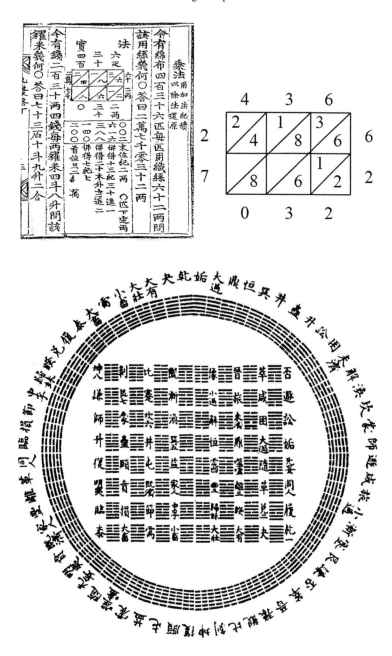

Fig. 10 Above: The problem of multiplication using *gelosia* calculation in the *Summary of Nine Numbers* and the reconstruction of addition 436 × 62 in which the result 27,032 is read clockwise from the right lower to the left upper position, viz. in order of 2, 3, 0, 7, 2. Below: the *Fuxi liushisi gua fangyuantu* 伏犧六十四卦方圓圖, "placing *qian* 乾 in the north west (bottom right), and placing *kun* 坤 in the south east (upper left)." (Above left: Ch'oe 1985, 645; below: *Zhu* 1985)

right-angled triangles, for instance, two algorithms were provided, one was a traditional one from the *Yang Hui's Mathematical Methods* and the other was a Western one from the newly imported *First Collection of Writings on Heavenly Learning* (*Tianxue chuhan* 天學初函, 1626), each of which respectively used the *Chunje* and *Chunsŭng* (see Fig. 6).

For a further illustration, let us consider a simplest survey problem to be solved, in modern terms, by right-angled triangles.

As shown in Fig. 11, when the length of a stick, b, the distance from the stick to observer's eye, a, and the distance from the observer's eye to the pagoda, c, were given, one had to find the height of the pagoda, x. According to the Western method, from the ratio of the similar triangles we can get the equation such as $b/a = x/c$; so the answer will be calculated by $x = (b \div a) \times c$, which involves a division followed by a multiplication; this order of operation corresponds to the *Chunsŭng* technique. On the other hand, according to the traditional method, the equality of the areas of the shaded rectangles yields the equation $ax' = bc'$, where $x' = x - b, c' = c - a$; so the answer will be obtained by $x = x' + b$, after calculating $x' = (b \times c') \div a$, which involves a multiplication followed by a division; this order of operations corresponds to the *Chunje* technique.

Ch'oe provided both algorithms in his text, with harsh criticism levelled at the Western one. Despite its strengths in the astronomical calculation, he claimed, the Western method applied to the survey problems "should" or "ought to" be replaced with the traditional one. And he significantly added:

禮曰, 其數易陳也, 其義難知也. 失其義陳其數, 祝史之事也. 余於算法亦云.

> The *Book of Rites* (禮記) says that the numbers are easily arrayed, but the meaning (義) is difficult to know. Losing the meaning, [just] deploying the numbers is the work of astrologers (祝史). I would like to say that so is the method of the calculation. (ibid., 542–543)

To him, the most important feature of the algorithms was not the result, but how one reaches the result, the way of computing, not least in order to let the "meaning (義)" be seen during the operations.[20] Here, the "meaning" was not directly bestowed from the passages or images in the Classics; it was only from the authority of the traditional algorithms. Each procedure, each order of the operations, was meant to, or rather had to, reveal its genesis, that is, its mathematical rationale that justified the procedure itself, as well as its affiliation to *sixiang* category. In a

---

[20]Here I translate "*yi* 義" as "meaning," but the term has already lots of other translation, such as "signification," "justice," and the like. In the field of history of ancient Chinese mathematics, as Chemla (2012, 481) pointed out, "the algorithms, together with the situations in relation to which they were introduced, provided means for determining the 'meaning' of an operation or a sequence of operations. This appears to a key act for proving the correctness of algorithms, and it is noteworthy that a term (*yi* 'meaning') seems to have been specialized to designate it in ancient China." If we accept this view, we might say that Ch'oe's usage of the term "meaning" expanded its "correctness" not only from the perspective of mathematics, but also from the perspective of philosophical behaviour.

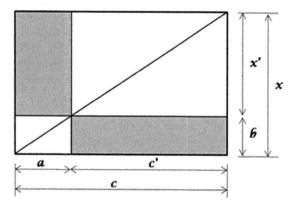

Fig. 11 Reconstruction for the simple survey problem

word, the procedures of survey problems had to show the "meaning," namely the genealogy and the proper *sixiang* category.

Note that the overall calculating performance without its visible "meaning" was likened to the practice of astrologers, who easily lost unconsciously, or hid consciously, the true "meaning." This simile might give the reader an inkling that he intended to degrade the mathematical officials who could deploy the counting rods very quickly without any "meaning."

However, we should bear in mind that his criticisms were only raised in case of the basic operations, not of the methods of the celestial element, which was the mathematical official Hong's self-assured strong point. In fact, Ch'oe's unnatural classification of mathematical algorithms with a fixed four-row format could not give any space to the algorithms that expanded multiple layers of rows. In *Summary of Nine Numbers*, Ch'oe did not, or rather could not mention the method of the celestial element at all. Another literati Cho Tae-gu was also unable to deal with it except for a general commentary that "the celestial element method of Zhu [Shijie] and the trigonometry of the Western scholars were excellent" (Cho 1985, 193). In this way, the literati with mathematical interest ignored the main computational algorithms of the mathematical officials, while disapproving their basic practices for their inability to reveal any "meaning" in Neo-Confucianism or in the traditional mathematical texts.

Conclusion

From the late seventeenth century to the early eighteenth century, the rivalry between *chungin* mathematical officials and the *yangban* literati became apparent in mathematical texts. Authors set out to present themselves as advisors or teachers with superior mathematical knowledge. Even though the mathematical knowledge of both groups was based on the calculation using same counting rods, each had distinctive features. On the one hand, Hong, as a mathematical official, advocated that adeptness in manipulating physical counting rods was essential. On the other hand, the mathematical knowledge of Ch'oe, as one of the Confucian literati, was mainly about seeing, or rather reading the algorithms, in order to find a meaning in the images of computation, and his choice of algorithms was a kind of normative judgment that came with responsibilities, as the terms "ought to," "should," and "a man of virtue" hinted at.

The discrepancy in their interpretation of the computation of the same counting rods, argued in this case study, mirrored the distance between the two educational cultures that the calculators embodied. Mathematical officials were educated and trained in a governmental institution with the unaltered curriculum. They had to stick to, and develop the algorithms of, calculation by handling physical counting rods. In contrast, literati usually learned mathematical knowledge by means of text. In their rendering, they tended to provide detailed explanation on more basic computation and wider exposure to new methods imported by text into the Chosŏn society. From the late eighteenth century on, the situation would start to change, when a small, but influential group of high official literati working under patronage of King Chŏngjo 正祖 (r. 1776–1800) promoted the written calculation (*bisuan* 筆算) introduced in the Chinese treatise *Essence of Numbers and their Principles* (*Shuli jingyun* 數理精蘊, 1722).[21] This promotion had been initiated from the need for more exact calculation of the calendar and consequently resulted in the reassignment of the textbooks featuring written calculations in the examinations conducted for selection of astronomical officials, whose positions had been more or less same as mathematical officials.

Admittedly, the textbooks for mathematical officials were still focusing on calculations with counting rods. In this situation, more officials passed both examinations for the astronomical and mathematical bureau, and more literati who were interested in Western, or new, mathematics and astronomy, were keen to co-work with them. Thus, this double interaction, firstly between calculations performed with the counting rods and written calculations, and secondly between mathematical officials and literati, caused another, much closer and mingled, perspectives of mathematical knowledge.[22] Yet until then, the discrepancy between the two social groups could not disappear.

---

[21]For the contents of *Shuli jingyun* and the written calculation in China, see Jami (2012).

[22]For the situation from the late eighteenth century, see Moon (2001), Lim (2009), Koo (2010, 2012).

References

Primary Sources

Cho Tae-gu 趙泰耉. [1718] 1985. *Chusǒ kwan'gyǒn* 籌書管見 (Narrow view of mathematical text). Reprint. In vol 2. of *Hankuk kwahakkisulsa charyo taeye: Suhakpyǒn* 韓國科學技術史資料大系: 數學篇 (Source Materials of the Korean science and technology: Mathematics). Seoul: Yǒgang ch'ulp'ansa 驪江出版社. 1–199.

Ch'oe Sǒk-chǒng 崔錫鼎. [ca. 1700] 1985. *Kusuryak* 九數略 (Summary of nine numbers). Reprint. In vol 1. of *Hankuk kwahakkisulsa charyo taeye: Suhakpyǒn* 韓國科學技術史資料大系: 數學篇 (Source Materials of the Korean science and technology: Mathematics). Seoul: Yǒgang ch'ulp'ansa 驪江出版社. 369–704.

Hong Chǒng-ha 洪正夏. [ca. 1724] 1985. *Kuiljip* 九一集 (Writings of nine and one). Reprint. In vol 2. of *Hankuk kwahakkisulsa charyo taeye: Suhakpyǒn* 韓國科學技術史資料大系: 數學篇 (Source Materials of the Korean science and technology: Mathematics). Seoul: Yǒgang ch'ulp'ansa 驪江出版社. 201–693.

Kim Si-jin 金始振. [1660] 1993. "Chunggan sanhak kyemong sǒ" 重刊算學啓蒙序 (Preface to the reprint of *Suanxue qimeng*). In vol. 1 of *Zhongguo kexue jishu dianji tonghui : Shuxue juan* 中國科學技術典籍通彙: 數學卷 (Source Materials of ancient Chinese science and technology: Mathematics section), ed. Guo Shuchun 郭書春, 5 vols. Zhengzhou: Henan jiaoyu chunbanshe 河南教育出版社.

Kyǒng Sǒn-jing 慶善徵. 1985. *Muksachip sanbǒp*. 默思集算法 (Mathematical methods of Muksa). Reprint. In vol 1. of *Hankuk kwahakkisulsa charyo taeye: Suhakpyǒn* 韓國科學技術史資料大系: 數學篇 (Source Materials of the Korean science and technology: Mathematics). Seoul: Yǒgang ch'ulp'ansa 驪江出版社. 1–368.

Zhu Xi 朱熹. 1985. *Zhouyi benyi* 周易本義. In vol. 201 of *Wenyuange Siku quanshu* 文淵閣四庫全書. Taipei 臺北: Shangwu yinshu guan 商務印書館.

Secondary Sources

Bréard, Andrea. 2002. Problems of pursuit: Recreational mathematics or astronomy? In *From China to Paris: 2000 years transmission of mathematical ideas*, ed. Yvonne Dold-Samplonius, Joseph W. Dauben, Menso Folkerts, and Benno van Dalen, 57–86. Stuttgart: Franz Steiner Verlag.

Chemla, Karine, and Guo Shuchun. 2004. *Les Neuf chapitres: le classique mathématique de la Chine ancienne et ses commentaris*. Paris: Dunod.

Chemla, Karine. 2010. Mathematics, nature and cosmological inquiry in traditional China. In *Concepts of nature: A Chinese–European cross-cultural perspective*, ed. Hans Ulrich Vogel and Günter Dux, 255–284. Leiden: Brill.

Chemla, Karine. 2012. Reading proofs in Chinese commentaries. In *The history of mathematical proof in ancient traditions*, ed. Karine Chemla, 423–486. Cambridge: Cambridge University Press.

Chǒng Ok-ja. 1993. Chosǒn sahoe-ǔi byǒnhwa-wa chunginkyech'ǔng-ǔi sǒngjang (The change of Chosǒn society and the development of chungin class). In *Chosǒn huki yǒksa-ǔi ihae* (The understanding of history of late Chosǒn), 165–172. Seoul: Ilchisa.

Guo Shuchun, Joseph W. Dauben, and Xu Yibao (eds.). 2013. *Nine chapters on the art of mathematics*. Shenyang, China: Liaoning Education Press.

Guo Shirong. 2009. *Zhongguo Shuxue Dianji Zaichaoxian bandaode Liuchuan Yingxiang* (Chinese mathematical texts that circulated in the Korean Peninsula and their influence). Shandong jiaoyu chunbanshe. 2009.

Han Yŏng-u. 1997. Chosŏn sitae chungin-ŭi sinbun, kyekŭp-chŏk sŏnggyŏk (The social status and class of chungin in Chosŏn period). In *Chosŏn sitae sinbunsa yŏn'gu* (History of Social Status in Chosŏn Period), 63-95. Seoul: Chimmundang.

Ho Peng Yoke. 1995. *Li, Qi and Shu: An introduction to science and civilization in China*. Seattle and London: University of Washington Press.

Horng Wann-Sheng. 2002a. Sino-Korean transmission of mathematical texts in the 19th century. *Historia Scientiarum* 12: 87–99.

Horng Wann-Sheng. 2002b. *Shiba shiji dongsuan yu zhongsuan de yi duan duihua: Hong Zhengxia vs. He Guozhu* (The eighteenth century dialogue of Eastern (Korean) mathematics and Chinese mathematics: Hong Chŏng-ha vs. He Guozhu). *Hanxue yanjiu* 20, no. 2: 57–80.

Hwang Chŏng-ha. 1988. Chosŏn Yŏngjo, Chŏngjo sitaeŭi sanwŏn yŏn'gu: *Chuhakipkyŏkan-ŭi* punsŏkŭl chungsimŭro (A study of mathematical officials in King Yongjo, King Chŏngjo period in Chosŏn: An analysis of *Chuhakipkyŏkan*). *Paeksanhakpo* 35: 219–258.

Hwang Chŏng-ha. 1994. Chosŏn huki sanwŏnjiban-ŭi hwaltong yŏn'gu: Kyŏngju Yisi ch'anggakyelŭl chungsimŭro (A study of activities of a family of mathematical officials: A case study of Kyŏngju Yi Family). *Chŏngjusarim* 6: 80–110.

Jami, Catherine. 2012. *The emperor's new mathematics: Western learning and imperial authority during the Kangxi Reign (1662–1722)*. Oxford: Oxford University Press.

Jun Yong Hoon. 2006. Mathematics in context: A case in early nineteenth-century Korea. *Science in Context* 19 (4): 475–512.

Kang Myŏng-gwan. 1997. *Chosŏn hugi yŏhang muhak yŏgu (A study of Yŏhang literature in late Chosŏn)*. Seoul: Ch'angjakkwa bip'yŏngsa.

Kawahara Hideki. 1996. Kusuryak: Sangaku to sisyō (Kusuryak: Arithmetics of four images). *Chōsen bunka kenkyū* 3: 77–93.

Kawahara Hideki. 1998. Tōzan to Tengenjutsu: 17seikichūki-18seikishoki no Chōsensūgaku (The eastern mathematics and the method of the celestial element). *Chōsengakuhō* 169: 35–71.

Kawahara Hideki. 2010. *Chōsen sūgakushi: Shushigakuteki na tenkai to sono shūen (History of mathematics in Chosŏn: Development and ending under the studies of Zhu Xi)*. Tokyo: University of Tokyo Press.

Kim Yong-un and Kim Yong-guk. 1977. *Han'guk suhaksa* (History of mathematics in Korea). Seoul: Kwanhak-kwa In'gansa. [Revised ed.: Yeolhwadang, 1982].

Koo Mhan-ock. 2010. Matteo Ricci ihu sŏyang suhak-e daehan Chosŏn chisigin-ŭi banŭng (The Chosŏn literati's response to the Western mathematics after the import of Matteo Ricci's mathematics). *Han'guk silhakyŏn'gu* 20: 301–355.

Koo Mhan-ock. 2012. How did a Confucian scholar in late Joseon Korea study mathematics? Hwang Yunseok 黃胤錫 and the mathematicians of late eighteenth-century Seoul (1729–1791). *The Korean Journal of the History of Science* 34: 227–256.

Libbrecht, Ulrich. 1973. *Chinese mathematics in the thirteenth century: The Shu-shu chiu-chang of Ch'in Chiu-shao*. Cambridge, Mass.: MIT Press.

Li Yan and Du Shiran. 1987. *Chinese mathematics, A concise history*. Trans. John. N. Crossley and Anthony W. C. Lun. Oxford: Clarendon Press.

Lim Jongtae. 2009. Emergence of literati mathematicians and the new culture of mathematics in late eighteenth-century Korea. *Hankuk-ŭi kirok munhwa-wa bŏpko-ch'angsin* (The Sourcebook of the 2nd international symposium of Kyujanggak Korean Studies).

Martzloff, Jean-Claude. 1997. *A history of Chinese mathematics*. Trans. Sthephen S. Wilson. Berlin: Springer.

Moon Joong-Yang. 2001. 18seki huban Chosŏn kwahakgisul ŭi Ch'ui wa Sŏnggykŭk (The scientific or ideological nature of royal astronomical projects and politics in the late eighteenth-century Korea). *Yŏksawa hyŭnsil* 39: 199–231.

Needham, Joseph. 1959. *Science and civilisation in China*, vol. 3. Cambridge: Cambridge University Press.

Netz, Reviel. 2002. Counter culture: Towards a history of Greek numeracy. *History of Science* 40: 321–352.

Shapin, Steven, and Simon Schaffer. 1985. *Leviathan and the air pump*. Princeton: Princeton University Press.

Taub, Lisa. 2011. Reengaging with instruments. *ISIS* 102: 689–696.

Wagner, Edward Willett. 1987a. The three hundred year history of the Haeju Kim *Chapkwa-Chungin* lineage. In *Song Chun-ho kyosu chŏngnyŏn kinyŏm nonch'ong pyŏlswae* [Essays in commemoration of Professor Song Chun-ho's retirement, offprint], Chŏnju, 1–22.

Wagner, Edward Willett. 1987b. An inquiry into the origin, development and fate of Chapwa-Chungin lineage. *Kuknaeoe e issŭsŏ Hangukhak ŭi hyŏnjae wa mirae*.

Yi Sŏng-mu. 1997. *Hankuk kwakŏ chedosa (History of civil service examination of Korea)*. Seoul: Minŭmsa.

Young Sook Oh is a Ph.D. candidate in the Program of History and Philosophy of Science, Seoul National University, Seoul, S. Korea. Her current research focuses on the history of mathematics and mathematical education in eighteenth- and nineteenth-century Korea.

The Education of Abacus Addition in China and Japan Prior to the Early 20th Century

Yifu Chen

Abstract Despite the voluminous prior research on the history of the abacus before the 20th century in East Asia, little has been studied on the educational ideas contained in the descriptions of manipulation of this counting device to perform elementary operations. The aim of this chapter is to explore the educational ideas related to the execution of the most elementary operation—addition—on abacus presented in the documents from the middle 15th century to early 20th century in China and Japan. The analysis focuses on different aspects related to the manipulation of this counting device in order to unveil their educational ideas. The results of this study show that the education of abacus addition experienced a historical change around the Meiji Restoration (1868), from an era of the use of verses to an era of the elaboration of "techniques of fingers," dealing with positions of fingers on instruments, and their movement. This study may be of importance in providing an understanding of the education of abacus addition in China and Japan before the early 20th century, as well as in proposing a method of analysis for exploring the educational ideas of other elementary operations performed on the abacus or other counting devices in the history of mathematical education.

Keywords Chinese abacus (*suanpan*) · Japanese abacus (*soroban*)
Counting instruments · Computing devices · Abacus education
Mathematics education

A large part of this chapter is based on my doctoral dissertation defended in the Paris Diderot University (University Paris 7) in 2013. I would like to thank the editors, especially A. Volkov, and two anonymous reviewers for their constructive comments and suggestions.

Y. Chen (✉)
Independent Scholar, Kaohsiung, Taiwan
e-mail: chen.yifu@gmail.com

© Springer Nature Switzerland AG 2018
A. Volkov and V. Freiman (eds.), *Computations and Computing Devices
in Mathematics Education Before the Advent of Electronic Calculators*,
Mathematics Education in the Digital Era 11, https://doi.org/10.1007/978-3-319-73396-8_10

243

Introduction

Addition is the most essential of the basic arithmetic operations performed on an abacus, and any addition is accomplished by executing a sequence of single-digit additions represented by the movement of beads. Moving correct beads to carry out single-digit additions therefore constitutes the base for further learning at the early stage of abacus education. Before the second half of the 19th century their executions in China and Japan were performed with the aid of a kind of mathematical verses, which appeared in all Japanese abacus manuals published prior to the Meiji period (1868–1912). These verses, related to the single-digit additions on the Japanese abacus, took at least three distinct forms, all of which allow us to investigate different ideas concerning the teaching of addition with the help of the Japanese abacus. The educational tradition changed in Japan after the decision to adopt Western methods for teaching arithmetic in the public education system, from the first years of the Meiji Restoration (1868). Newly compiled Japanese abacus textbooks under this policy reflected therefore some different ideas about Japanese abacus addition education—the subject of this chapter.

The chapter proposes a careful examination of how verses related to the single-digit additions on the Japanese abacus were formulated in the Chinese and Japanese abacus works published before the second half of the 19th century. The correspondence between single-digit additions, the movement of beads indicated by the verses, and the suggested order of movement of beads embedded in the verses are clues on which we can rely to investigate the ideas of education. In addition, analysis of the use of the verses in a particular exercise designed to teach addition presented in all ancient abacus documents permits the exploration of different modes of movement of beads, reflecting others aspects of abacus education as well, see Sect. "The Era of the Use of Verse". In the following, texts on the method of addition in two abacus textbooks published in Japan will be discussed. In each of them, the motion of moving beads and the fingering[1] played a role in both teaching and learning how to use the abacus in order to carry out operations of addition. These will be analyzed to note their implications in Sect. "Era of the Elaboration of Finger Techniques—Motion and Fingering". In the conclusion, I will try to show that abacus education in Japan, represented in this chapter with the example of teaching addition, experienced a historical change around the Meiji Restoration. This method of analysis may be applied to study other basic operations on the abacus, exploring the educational ideas of the authors of treatises in which these operations were described.

---

[1]Here and below I use the term "fingering" to refer to the method of using a particular finger to move a given bead or group of beads to perform an operation on an abacus.

The Abacus and Its Structure

The question of the origin of the Chinese abacus still remains open, and the sporadic evidence of its use starting in the 11th century through to the first half of the 16th century remains scattered here and there over different types of sources in China (for a detailed discussion of the origin of the Chinese abacus, *see* Hua (1987, pp. 26–66) and Volkov (2001, pp. 481–485)). According to extant sources, mathematical works entirely dedicated to computations on the abacus began appearing in China only in the late Ming (1368–1644) period, from the late 16th century through to the early 17th century. It is only during this period onwards that the use of the abacus diffused across Chinese borders and the instrument became known and used in Japan, Korea, and Vietnam. Even though its introduction into Japan cannot be dated with accuracy, the oldest documents testifying to its use date back to the end of the 16th century (Suzuki 1960, pp. 26–28; Horiuchi 1994, p. 33).

The Chinese abacus is a framed instrument made of a set of columns of beads divided by a transversal bar into an upper and lower compartment. Each column of beads represents one decimal position. Once a column is designated as the column of units by the operator, all other columns are assigned to the corresponding values of powers of ten relative to it. The beads count when they are pushed toward the dividing bar. The value of each of the lower beads is equal to the power of ten assigned to its column, and the value of an upper bead is equal to that of five lower beads of the same column. A number can be represented therefore by setting different combinations of beads on columns.

The earliest and most widely used type of abacus before the second half of the 19th century, in both China and Japan, contained two beads in its upper part and five in its lower, similar to the structure of the abacus presented graphically in two of the first published abacus books in the late Ming period, the *Avenue to Mathematics* (*Shuxue tonggui* 數學通軌, 1578) by Ke Shangqian 柯尚遷 (dates of life unknown) and the *Unified Lineage of Computational Methods* (*Suanfa tongzong* 算法統宗, 1592) by Cheng Dawei 程大位 (1533–1606) (Figs. 1 and 2).[2]

There existed other variants of the Chinese abacus (Jami 1998, p. 3), among them an abacus with one bead in its upper part and four in its lower one on each column which was most widely used in Japan, South Korea, and Taiwan in the 20th

---

[2]The illustrations of the abacus depicted in two other abacus books published in the same period—the *Computational Methods with the Beads in a Tray* (*Panzhu suanfa* 盤珠算法, 1573) by Xu Xinlu 徐心魯 (dates of life unknown) and the *Compass for the Computational Methods* (*Suanfa zhinan* 算法指南, 1604) by Huang Longyin 黃龍吟 (dates of life unknown)—have only one bead in the upper compartment and five in the lower, in each column. Since they were used to illustrate the results of operations, it is very doubtful that the illustrations correspond to the structure of the concrete abacus used by the authors. Besides, in the procedures of division, it is more reasonable to have two beads than one in the upper part from the point of view of necessity and convenience (Chen 2013, pp. 19–47). In brief, even if this 1 + 5 type instrument existed during that period, it seems that it was less used than the 2 + 5 types found in China and in Japan before the 20th century.

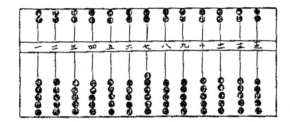

Fig. 1 An illustration of an abacus set to zero in the *Avenue to Mathematics* (Ke 1578, p. 1171) (The configuration of beads on the lower part of the middle column—six inactivated lower beads —should be corrected to five inactivated lower beads)

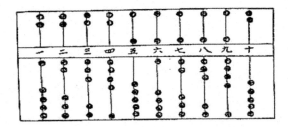

Fig. 2 An illustration of an abacus on which a number 1.23456789×10^N (column of units is not specified by the author) is represented in the *Avenue to Mathematics* (Ke 1578, p. 1172)

century. The idea behind its structure can be found in the *Primer for learning Computational Methods* (*Shogaku sanbō* 初学算法, 1781) by Nyūi Mitsugi 乳井 貢 (1712–1792) (Zenkoku shuzan kyōiku renmei 1971, p. 111). Since the 1890s, it was greatly recommended by specialists for the reason that this form allowed only one mode of representation of numbers on each single column, corresponding perfectly to the ten Arabic numerals used in written calculations, adopted in the public education system during this period.

The Era of the Use of Verse

Being the most basic operation on an abacus, addition has always been introduced before other operations. Its execution was proceeded with the aid of a kind of verse related to the single-digit additions presented in the documents published before the second half of 19th century in China and Japan. The most widely used form of the verse, the one appearing in the earliest extant Chinese abacus book—*Computational Methods with the Beads in a Tray* (*Panzhu suanfa* 盤珠算法, 1573) by Xu Xinlu 徐 心魯 (dates of life unknown)—will be studied carefully in the present chapter. It will be followed by a discussion of the other two forms of verse, concluding with the examination of an exercise of addition which is presented in almost all the documents. The aim is to unveil the educational ideas behind their use in abacus addition.

The verses related to the single-digit additions in *Computational Methods with the Beads in a Tray* are called "Lishou verses on mounting" (*Lishou shangjue* 隸首 上訣).[3] This title referred to nine sets of versified phrases, not necessarily grammatically correct, written in a few characters (Bréard 2014, p. 178), summarizing all the possible movements of beads on the abacus when one adds any of the nine single-digit numbers (1–9) to a single-digit number (0–9) or 10 represented by an upper bead and five lower beads, appearing on a single abacus column. It begins with the first set of three phrases summarizing the bead movements in the case of adding one, then those of adding two, and so on ... and ends with the ninth set of three phrases in the case of adding nine. Composed of 26 versified phrases, the "Lishou verses on mounting" indicate bead movements for the 99 possible single-digit additions (Table 1) which will be discussed later.

One of the immediate questions raised about the use of verse is what basis did the author use to formulate these versified phrases. It can be observed that these single-digit additions were classified into distinct groups according to whether the movement of the beads involved (1) 5-complement operating within a column or (2) carrying to the left adjacent column, and thereby the corresponding versified phrases were formulated for addition of numbers from 1 to 9 to every single-digit number. To illustrate this correspondence, I will take the first set of three phrases, "1 mount 1," "1 lower 5 remove 4," and "1 withdraw 1 advance 10," as examples to help explain the meanings and uses of these versified phrases.

The first phrase "1 mount 1" means that in order to add 1, one mounts (activates) a lower bead; the operations to which this verse applies could be: $0 + 1$, $1 + 1$, $2 + 1$, $3 + 1$, $4 + 1$, $5_U + 1$, $6 + 1$, $7 + 1$, $8 + 1$, and $9 + 1$. The execution of any one of the operations of this group can be carried out by moving up directly an inactivated lower bead, involving neither 5-complement operating nor carrying to the left adjacent column.

The second phrase, "1 lower 5 remove 4," means that to add 1, one lowers (activates) an upper bead and removes (deactivates) four activated lower beads. It applies to operations $4 + 1$ and $5_L + 1$, and the execution of the operations of this group involves 5-complement operating within a column. Take, for example, the addition of $5_L + 1$. Xu Xinlu explains to the reader the condition of use of the phrase of verse "1 lower 5 remove 4" as follows:

> If the five lower beads of the column are all at their activated places and [the operator] still wants to mount one, then there exists no one [bead] in the lower part that can be mounted. Therefore [the operator] lowers one [bead] from the upper part, that is worth five, then removes four [activated] beads from the lower part. [The operator] consequently mounts one.[4]

---

[3]Lishou is a legendary person who mastered arithmetic and was the historiographer of the mythical Chinese sovereign, the Yellow Emperor (*Huangdi* 黃帝), dated around 2600 B.C.

[4]Translated from Xu (1573, p. 1143). The original text reads (the chapter author's punctuation): 如本行下五子俱已在位。今又要上一。則下無一可上。故於上面下一。是五。復於下面去四。故上得一。．

Table 1 "Lishou verses on mounting" summarized

Single-digit additions, $number_{on\ the\ column} + number_{to\ add}$	Bead movements involved[a] 5-complement operating Carrying		Relevant verse[b]
$0+1, 1+1, 2+1, 3+1... 9+1$	No	No	1 mount 1
$4+1, 5_L+1$	Yes	No	1 lower 5 remove 4
$9+1, T+1$	No	Yes	1 withdraw 9
			advance 10
$0+2, 1+2, 2+2, 3+2... 8+2$	No	No	2 mount 2
$3+2, 4+2, 5_L+2$	Yes	No	2 lower 5 remove 3
$8+2, 9+2, T+2$	No	Yes	2 withdraw 8
			advance 10
...			...
$0+9, 1+9$	No	No	9 mount 9
$5_U+9,$	Yes	Yes	9 raise 4 unload 5
$1+9, 2+9, 3+9, 4+9, 5_L+9,$	No	Yes	advance 10
$6+9, 7+9, 8+9, 9+9, T+9$			9 withdraw 1
			advance 10

Notes: 5_L, five represented by five activated lower beads on a column; 5_U, five represented by one activated upper bead; T, ten represented by an upper bead and five lower beads on a column.[c]
[a]Since the value of an upper bead is equal to that of five lower beads on a column, there exists two groups of complementary numbers with respect to 5: 4 and 1, 3 and 2. A single-digit addition is classified into "5-complement operating," while the use of complementary numbers with respect to 5 is made in its execution on an abacus, for example, $4 + 2 = 4 + (5-(complement\ of\ 2)) = 4 + (5-3)$, which is indicated by the phrase of verse "2 lower 5 remove 3."
[b]Translated from the "Lishou verses on mounting" in Xu (1573, p. 1143); the original text reads 一上一。一下五除四。一退九進一十。二上二。二下五除三。二退八進一十。... 九上九。九上四去五進一十。九退一進一十。
[c]In some abacus works such as *Computational Methods with the Beads in a Tray*, the authors proposed to use the fifth, or the lowest, lower bead in a specific condition in addition, which will be discussed later in the chapter. In such cases, it happens to have the number 10 represented in a column by one activated upper bead and five activated lower beads. T will be used to indicate this representation of 10 in this chapter

That is, one carries out this operation $5_L + 1$ through adding 5 by pushing the upper bead down and subtracting the complement of the addend to 5, 4 in this case, by pushing down $(5 - number_{to\ add})$, here $(5 - 1)$, activated lower beads. Any operation $(number_{on\ the\ column} + number_{to\ add})$ of this kind, involving 5-complement operating within a column but without carrying to the left adjacent column, could be therefore translated to $(number_{on\ the\ column} + 5 - (5 - number_{to\ add}))$ on the abacus (Fig. 3).

The third phrase "1 withdraw 9 advance 10" means that to add 1, one removes the beads representing 9 and then adds the carry 1 to its left adjacent column.[5] The operations to which this phrase applies are $9 + 1$ and $T + 1$. The execution of the

---

[5]According to the extant Chinese abacus works published before the 20th century, the second upper, or the highest, bead was never used in the execution of an addition using the decimal system.

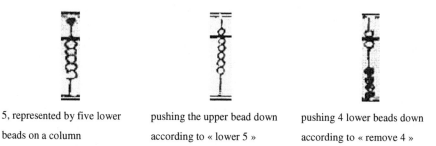

| 5, represented by five lower beads on a column | pushing the upper bead down according to « lower 5 » | pushing 4 lower beads down according to « remove 4 » |

Fig. 3 The execution of the operation that is 5 (represented by five lower beads) + 1 on an abacus according to the phrase of verse "1 lower 5 remove 4"

operations of this group involves no 5-complement operating within a column but does include carrying to the left adjacent column. Take the addition of $T + 1$ for example. Xu Xinlu explained the conditions of use of the verse:

> If all the beads of the column are at their activated positions and [the operator] still wants to add one, however there is no one [bead] that can be added. Therefore [the operator] withdraws the beads representing nine, then returns one bead in the high place [i.e., next left column] as ten [unit-]beads in the low place. It's thus exactly one [to be added].[6]

One carries out the operation $T + 1$ by subtracting the complement of the addend to 10, 9 in this example, through pushing away directly the activated beads representing $(10 - number_{to\ add})$ on the column, and then adding the carry 1 to the left adjacent column. Any operation $(number_{on\ the\ column} + number_{to\ add})$ of this kind, involving no 5-complement operating within a column but carrying to the left adjacent column, could be represented as $(number_{on\ the\ column} - (10 - number_{to\ add}) + 10)$.

In addition to the three groups classified on the basis of three distinct types of movement of beads studied above, the operations to which the following four versified phrases "6 raise 1 unload 5 advance 10," "7 raise 2 unload 5 advance 10," "8 raise 3 unload 5 advance 10," and "9 raise 4 unload 5 advance 10" apply could be classified into another group. The execution of any of the operations of this group involves both 5-complement operating within a column and carrying to the left adjacent column. Take, for example, the addition of $7 + 6$ to which "6 raise 1 unload 5 advance 10" applies. Since there are no remaining inactivated beads that can be set directly,[7] one has to subtract the complementary number of 6, that is 4, from 7 and add the carry 1 to the next left adjacent column, since $7 + 6 = 7 + (10 - 4) = (7 - 4) + 10$. As there are only two activated lower beads, in order to subtract 4, one adds 1, the complementary of 4 with respect to 5, by pushing one inactivated lower bead up and then subtracts 5 by pushing the activated upper bead up, as

---

[6]Translated from Xu (1573, p. 1143). The original text reads: 如本位子滿在位。又要加一。卻無一可加。故幾退去九子。卻于上位還一子。當下位十子。卻正一也。.

[7]It was mentioned earlier in the chapter that the second upper, or the highest, bead was never used in the execution of an addition using the decimal system. The result of the addition, 13, therefore cannot be set directly in the working column.

7 on a column	pushing 1 lower bead up according to « raise 1 »	pushing the upper bead up according to « unload 5 »	adding 1 to the next left adjacent column (supposed 1 on column) according to « advance 10 »

Fig. 4 The execution of the operation that is 7 + 6 on an abacus according to the phrase of verse "6 raise 1 unload 5 advance 10"

indicated by the words of the phrase "raise 1 unload 5." An operation (number$_{on\ the\ column}$ + number$_{to\ add}$) of this kind could be therefore translated to (number$_{on\ the\ column}$ + (5 − (10 − number$_{to\ add}$)) − 5 + 10) on an abacus (Fig. 4).

To sum up, the "Lishou verses on mounting," summarizing all the possible movements of beads for the 99 different possible single-digit additions, were formulated on the basis of the four types of movement of beads classified above. The correspondence between its 26 phrases and 99 single-digit additions is not one-to-one. Compared with another possible way of formulating phrases of verses, having 99 distinct indications of movement of beads, one for each of the 99 single-digit additions, which would permit an operator to carry out any single-digit addition mechanically without thinking, the "Lishou verses on mounting" require, to some degree, that the operators perform some reasoning to identify the movement of beads during an operation on an abacus, more specifically, to identify the proper verse that should be applied, as Xu Xinlu explained for "1 lower 5 remove 4" and "1 withdraw 9 advance 10" presented above.[8] In brief, the method for formulating the entire set of versified phrases might reflect an aspect of its author's thoughts on the early stages of abacus addition education, that is, the choice of a specific position between the two extremes, one being the idea of completely mechanical execution of operations without reasoning, the other being an execution of operations on the abacus on the basis of pure reasoning without any preliminary instructions or hints provided concerning their execution. This is key to my study of the various thoughts on addition found in other documents.

In addition to the information concerning the movement of beads to carry out an operation of addition, the author gave, through the versified phrases, an additional piece of information about the order of movements of the beads. It also reflects another aspect of the author's idea of abacus education, that is, the fluency of movement of beads. Let us examine two operations representing two groups of operations for which the movement of beads involves 5-complement operating.

---

[8]In addition to these two phrases, Xu Xinlu explained in the same way the conditions for the use of the two phrases, "2 lower 5 remove 3" and "2 withdraw 8 advance 10," which will not be presented here (Xu 1573, p. 1143).

Take again, for example, 7 + 6, to which the phrase "6 raise 1 unload 5 advance 10" will be applied (Fig. 4). As the phrase indicates, first one adds 1 by moving one inactivated lower bead up and then subtracts 5 by moving the activated upper bead up within the column where 7 was set. If the operator uses the same finger to move beads, he can execute it smoothly in one action, without change of direction, from the lower part to the upper part. If the order of movement of beads is reversed, that is, first unload 5 then raise 1, the operator will first move the activated upper bead up, and after this upward movement he will have to shift his finger down to the lower part of the instrument to move one inactivated lower bead up; precisely, to carry out the same operation the operator will make three actions—upward, downward, then upward. Even with two different fingers to move the beads, the order suggested by the author of the verse can be executed more smoothly than the reverse order. The same consideration remains valid in the case of the other example that is 4 + 2, to which the phrase "2 lower 5 remove 3" will be applied (Fig. 5): one downward action to carry out the operation according to the order suggested in the phrase vs. three actions—the movement of beads downward, upward, then downward in reverse order. In brief, the order of the movement of beads proposed by the author is smoother and simpler than the possible reverse order. Although fingering was not indicated in the texts, from the discussion above it is not unreasonable to assume that the author of the verse might have considered the fluency of movement of the beads as an important factor while formulating the phrases. Hence, the fluency of finger movement, moving beads within a column, is also a key to my study of abacus education (Chen 2013).

The other point that deserves special attention is related to the mode of movement of beads involving the use of the 5th, or the lowest, bead in the lower part. It reflects as well the author's idea about abacus education at its early stages. From the structure of the Chinese abacus, and the discussion about the correspondence between the 99 single-digit additions and the 26 versified phrases, it might be noticed that in some single-digit additions the operator would have two possible options of bead movement. These instances are for numbers whose sum is 5, 10, or 15—that is, the numbers whose sum can be represented in two different ways on an abacus. Take, for example, the addition of 4 + 1. An operator can perform it either

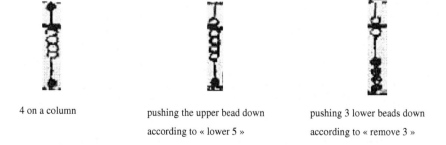

4 on a column   pushing the upper bead down   pushing 3 lower beads down
                         according to « lower 5 »   according to « remove 3 »

Fig. 5 The execution of the operation 4 + 2 on an abacus according to the phrase of verse "2 lower 5 remove 3"

by pushing one inactivated lower bead up or by pushing one inactivated upper bead down and pushing four activated lower beads down; the choice between these two options, therefore, is related to the choice of use of the lowest bead in a column. As far as this choice is concerned, it was demonstrated that there existed different modes of movement of beads in the ancient abacus documents published in China and Japan (Chen 2013, part I).

Some authors like Xu Xinlu, the author of *Computational Methods with the Beads in a Tray*, took a specific approach to dealing with the carry for the cases of $4 + 1_{carry}$ and $9 + 1_{carry}$: in these two cases, the lowest bead was activated; apart from these two cases, the lowest bead was never used in addition. According to my hypothesis concerning this specific rule, moving the lowest bead in these two operations makes the movement of the beads simpler and more convenient. Take, for example, $3,578 + 6,789$.[9] After the operation $3 + 6$, amounting to 9, is performed in the column of thousands, the operation of $5 + 7$ in the column of hundreds gives rise to a carry 1 to be placed in the left adjacent column, that is, in the column of thousands. At this stage, the operator has two options to perform $9 + 1_{carry}$ in the column of thousands. One option is that the operator simply pushes the lowest bead up without moving his hand too much from the column of hundreds and obtains the numeral ten represented by one upper bead and five lower beads in the column of thousands. This is the measure that Xu Xinlu proposed. The other possible option is that the operator can proceed by first deactivating all beads representing 9 in the column of thousands and then add the produced carry 1 to its left adjacent column, that is, the column of ten thousands. If this option is chosen, the operator not only has to take more actions than in the case when the choice of the former option is made, but also has to move his or her hand on the instrument with a greater amplitude in both directions, from right to left and from left to right, which may induce errors more readily. To be more precise, in this case the operator has to move his or her hand over two columns to the left, that is, from the column of hundreds to the column of ten thousands, and then move his or her hand over three columns to the right, that is, from the column of ten thousands to the column of tens, to perform the operation $7 + 8$ in the column of tens. In brief, by making use of the lowest bead to deal with the carry, Xu Xinlu proposed a specific mode of movement of beads for addition, which reflected an aspect of his ideas on abacus addition education[10]—the search for a particular pattern of movement of beads to achieve maximum efficiency of abacus computations.

---

[9]The example is extracted from the four rightmost places of the operation $246,913,578 + 123,456,789$, called Method of the third addition (*Disan shangfa* 第三上法) in the *Computational Methods with the Beads in a Tray* (Xu 1573, p. 1144).

[10]This specific mode of movement of beads might not have been created initially for didactical purpose, however, it is found in the very first page of Xu's book in an exercise of addition, where the illustrations of an abacus representing the results of the operations together with the versified rules of addition were provided to teach its readers how to move beads to perform certain operations on an abacus. This context suggests that the mode of movement of beads found in the book can be related to the educational ideas of the author.

Other authors like Huang Longyin 黃龍吟 (dates of life unknown), the author of the *Compass for the Computational Methods* (*Suanfa zhinan* 算法指南, 1604), did not use the 5th lower bead at all in the operation of addition in their works.[11] In other words, they placed more emphasis on the uniformity of representing numbers in one particular mode in terms of the movement of beads. To sum up, different styles of movement of beads reflect different approaches to abacus addition education—one being related to the search for maximum efficiency of computations through the design of a particular mode of movement of the beads, the other related to the search for a consistency in representing numbers in one particular way.

Let us now examine the other forms of verses related to addition, which reflected other ideas about abacus education. Prior to the "Lishou verses on mounting," presented in *Computational Methods with the Beads in a Tray*, there existed, in the *Great Compendium of the Comparison and Classification in Nine Chapters of Mathematical Methods* (*Jiuzhang suanfa bilei daquan* 九章算法比類大全, 1450) by Wu Jing 吳敬 (dates of life unknown) and in the *Precious Mirror of Mathematics* (*Suanxue baojian* 算學寶鑑, 1524)[12] by Wang Wensu 王文素 (1465–unknown), two verses related to the operations of addition.[13] They were

---

[11]Those who are not familiar with the arithmetic operations on a Chinese abacus may wonder why the 2nd upper bead and the 5th lower bead were still kept on the instrument if these two beads were not supposed to be used in addition. Actually, in the operations of multiplication and division performed according to the ancient methods used on the Chinese abacus, it happens that the number appearing in a column in the process of the operation may become larger than 10 (for example, 11 to 17 in the operation of division by a one-digit number (Chen 2013, Sect. 1.2.2.). If an abacus of type 1 + 4 was used, the operator from time to time had to keep in mind, temporarily, the numbers larger than 10 produced in the process of the operations. This would be very inconvenient and may have caused mistakes.

[12]Unlike *Computational Methods with the Beads in a Tray*, which was entirely dedicated to computation on an abacus, these two works presented computational methods performed on an abacus as well as counting rods, the principal mathematical instrument used in China before the second half of the 16th century. To be precise, it is considered by contemporary historians that the computational methods of the four elementary operations presented in these two works were performed on the abacus while the other more advanced mathematical methods, such as extraction of square roots, were performed using counting rods (for a more detailed discussion, *see* Hua (1987, pp. 66–73)).

[13]These two verses in the two books were presented together with the other two verses related to the operations of subtraction—"Verses on breaking five" (*Powu jue* 破五訣) and "Verses on breaking ten (*Poshi jue* 破十訣)." Since one of the four phrases of "Verses on breaking five," that is, "without 1 remove 5 return 4 below" (無一去五下還四), can be applied only to some operations performed on the abacus, these four verses, as, therefore, the entire collection of rhymed instructions, could not have been created at first for counting rods and then applied to the abacus, but must have been intentionally created for the abacus, even though the two verses related to the operations of addition can be applied correctly to both the abacus and counting rods (Chen 2013, p. 54).

entitled "Verses on erecting five" (*Qiwu jue* 起五訣) and "Verses on forming ten" (*Chengshi jue* 成十訣), and consisted of 13 versified phrases as follows:[14]

"Verses on erecting five"	"Verses on forming ten"
1 remove 4 make 5	1 remove 9 form 10, 2 remove 8 form 10, 3 remove 7 form 10
2 remove 3 make 5	4 remove 6 form 10, 5 remove 5 form 10, 6 remove 4 form 10
3 remove 2 make 5	7 remove 3 form 10, 8 remove 2 form 10, 9 remove 1 form 10
4 remove 1 make 5	

On what basis were these phrases classified into two groups and thereby formulated in this way? What aspect of the author's ideas concerning abacus education did they reflect? It could be observed that the classification of the two verses was based on the concept of complements of five and complements of ten, corresponding to the particularities of the structure of the Chinese abacus: the value of one upper bead is equal to that of five lower beads situated in a column, and the value of one lower bead in a column is equal to that of ten lower beads in its right adjacent column, respectively. The phrases of the "Verses on erecting five" were formulated under the form of (number$_{to\ add}$ remove (complement of number$_{to\ add}$ to five) make 5) to convey the concept of complements of five and to indicate partner numbers of complements of five. The same considerations apply to the "Verses on forming ten" under the form (number$_{to\ add}$ remove (complement of number$_{to\ add}$ to ten) form 10) to convey the concept of complements of ten and to indicate partner numbers which are complements of ten. Hence, it would be reasonable to say that the authors of these two verses stressed more the concepts of complement than the indications of movement of beads to carry out operations shown in the "Lishou verses on mounting." That is to say, the two verses demand a higher degree of reasoning about the movement of the beads for carrying out operations than the "Lishou verses on mounting." As for the fluency of the movement of beads in the two verses, it is not as well thought through as in the "Lishou verses on mounting."

Some abacus specialists might lay stress on both the concepts of complement and the practical indications of movement of beads, as Ke Shangqian did in his work *Avenue to Mathematics*. He presented both the "Verses on erecting five," "Verses on forming ten," and then listed the "Nine-nine statements on method of mounting" (*Jiujiu shangfa yu* 九九上法語), in which the phrases are identical to those of the "Lishou verses on mounting" (Ke 1578, p. 1171).

---

[14]Translated from Wu (1450, p. 16); Wang (1524, p. 358). The original text reads: 起五訣。一起四作五。二起三作五。三起二作五。四起一作五。成十訣。一起九成十。二起八成十。三起七成十。四起六成十。五起五成十。六起四成十。七起三成十。八起二成十。九起一成十。

The Chinese character *qi* 起 used in the title of the "Verses on erecting five" *qiwu jue* 起五訣, and in every phrase of these two verses, has various meanings in Chinese, I therefore translate it to different words in English according to its different contexts. For the name of the verse *qiwu jue*, I translate *qi*, whose meaning is rather close to "use" or "build" in this context, as "to erect;" the *qi* used in the phrases, whose meaning is rather close to "take off," I translate it as "to remove," such as "1 remove 4 make 5."

Li Gong 李塨 (1659–1733), the author of the *Course of Minor Learning* (*Xiaoxue Jiye* 小學稽業, 1705), took another approach. He tried to convey concurrently both the concepts of complement and indications of movement of the beads in the versified phrases under a specific form (see my earlier discussion about the phrases of the "Lishou verses on mounting.") The writing of the phrases took four different forms corresponding to four groups of single-digit additions: (1) "number$_{to\ add}$ mount number$_{to\ add}$" such as "1 mount 1;" (2) "number$_{to\ add}$ lower 5 remove (complement of the number$_{to\ add}$ to five)," for instance "1 lower 5 remove 4;" (3) "number$_{to\ add}$ withdraw (complement of the number$_{to\ add}$ to ten) advance 10," like "1 withdraw 9 advance 10;" and (4) "number$_{to\ add}$ raise (complement of (complement of the number$_{to\ add}$ to ten) to five) unload 5 advance 10," for example, "6 raise 1 unload 5 advance 10." It can be noticed that the concept of complements of five appears clearly in the phrases of the second form, since each of their corresponding single-digit additions involve a simple 5-complement operating within a column. The concept of complements of ten also appears clearly in the phrases of the third form. On the other hand, the concept of complements of ten does not appear directly in the phrases of the fourth form, since each of their corresponding single-digit additions involves a 10-complement operating with a carry in which 5-complement operating is needed. It is the phrases of this form that Li Gong modified in order to show clearly the concept of complements of ten. The modified phrases such as "6 remove 4 return 1 below form 10"[15] take a general form of "number$_{to\ add}$ remove (complement of the number$_{to\ add}$ to ten) return (complement of (complement of the number$_{to\ add}$ to ten) to five) below form 10."

Compared with the unmodified phrase "6 raise 1 unload 5 advance 10," the instruction "6 remove 4 return 1 below form 10" contains a new element "remove 4," and doesn't include the element "unload 5." The new element "remove 4," whose general form being "remove (complement of the number$_{to\ add}$ to ten)," was without doubt used to convey explicitly the concept of complements of ten, not to indicate the actual movement of beads. What can be said about the "unprescribed" indication "unload 5"? Suppose we put this element into the modified phrase, the hypothesized phrase will become "6 remove 4 return 1 below unload 5 form 10," containing 11 Chinese characters in total. It seems that 11 characters were somewhat too long for a phrase of verse, so Li Gong made a compromise between the length of the phrase and the information which he wanted to convey. Li Gong chose to omit the element "unload 5" probably because it remains unchanged in 5-complement operating and does not reflect the particularity of each of the phrases of this form (for a detailed discussion, *see* Chen (2013, pp. 104–106)).

In brief, Li Gong introduced a modified form of the versified phrases, different to those presented in the previously published Chinese abacus works, while trying to include concurrently both the concept of complement and the practical indications

---

[15]Translated from Li (1705, p. 134). The original text reads 六起四下還一成一十. This phrase is one of the three versified phrases for the cases of adding 6 to the column, the other two phrases are 六上六 and 六起四成一十.

of movement of beads to carry out an operation. It can be observed that compared with the "Lishou verses on mounting" the modified versified phrases demand a higher degree of reasoning about the movement of beads for carrying out the corresponding operations because the piece of information "unload 5" is not indicated.

My investigation conducted so far on abacus addition education from three viewpoints—the degree of reasoning about movement of beads, the fluency of finger movement when moving beads, and the modes of movement of beads—will be complemented by an examination of an exercise of addition. In the history of the Chinese abacus, there existed an exercise on addition so popular that it has been presented in almost all documents containing texts of calculations with abacus published in the Ming and Qing (1644–1912) dynasty. The execution of this exercise of addition begins by setting the number 123 456 789 on an abacus set to zero, and then proceeds with addition of the number 123 456 789 to the result of the previous operation successively eight times; thus 81 single-digit additions in total are executed. In all the texts containing the exercise, 81 applied versified phrases were presented to indicate the movement of beads to carry out the 81 single-digit additions; in some works like *Computational Methods with the Beads in a Tray*, nine illustrations of an abacus are provided, in which the results of the nine additions are represented. The question that might be raised immediately is one about the reason why this exercise was so popular. It can be noticed that the addend 123 456 789 is a number which is simple to memorize, with the final configuration of the beads indicating the result 1 111 111 101, represented by TTT TTT TT1, also being a configuration which is easy to identify. With such particularities, it was convenient for beginners to practice it even without the text in hand, as well as to check for themselves whether they made any errors during the execution of operation (Figs. 6 and 7).

Furthermore, since 25 of 26 versified phrases were applied in the exercise, the latter provided therefore good training to beginners so that they could acquaint themselves with the use of the versified phrases, that is, the correspondence between the single-digit additions and the movement of beads indicated by the phrases, not to mention the order of the movement of the beads specified in the phrases. In brief, this exercise served well as an educational vehicle for beginners learning how to perform the operation of addition on an abacus while using verse.

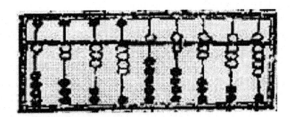

Fig. 6 Illustration of an abacus representing the result, 123456789, of the operation called "Method of the 1st addition" (*Diyi shangfa* 第一上法), 0 + 123456789, in *Computational Methods with the Beads in a Tray* (Xu 1573, p. 1143)

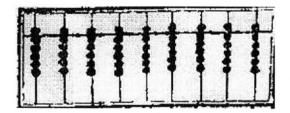

Fig. 7 Illustration of an abacus representing the result, TTT TTT TT1, of the operation called "Method of the 9th addition" (*Dijiu shangfa* 第九上法), 987654312 + 123456789, in *Computational Methods with the Beads in a Tray* (Xu 1573, p. 1147). (The configuration of beads on the ninth (the rightmost) column—one activated upper bead and five activated lower beads—should be corrected to one activated lower bead)

Era of the Elaboration of Finger Techniques—Motion and Fingering

To this point, my discussion about Japanese abacus addition education has been based on an analysis of the different forms of the verses and on a study of the different modes of movement of beads. For the rest of the chapter, we shall discuss an important historical change in Japanese abacus addition education that took place in the period of the Meiji Restoration in Japan in the second part of 19th century. In 1872, the Japanese Ministry of Education decided to adopt Western methods and to abandon completely the traditional Japanese methods and textbooks for teaching arithmetic in the public education system (Uegaki 1998). The *Elementary Textbook of Arithmetic* (*Shogaku sanjutsusho* 小學算術書, 1873), compiled by the Tokyo Normal School on the basis of four Western textbooks of arithmetic imported by Japan in that period, was published for this very purpose by the Ministry of Education (Uegaki 2001).[16] Nevertheless, because of a lack of teachers who could master Western methods, and the fact that the Japanese abacus was considered a good tool for helping understand written calculations, the Ministry of Education declared later a policy which was still based on Western methods, but which could be taught either by written calculation or by means of the Japanese abacus (Uegaki 2000). Since at that time there were no suitable Japanese abacus textbooks, a compilation of an Japanese abacus textbook based on Western methods became necessary in order to follow this policy. Published by the Tokyo Normal School in 1878, the *Textbook of the Art of Counting Beads* (*Sankajutsu jugyōsho* 算顆術授業書), was the first Japanese textbook in which its author Endō Toshisada 遠藤利貞[17] (1843–1915) developed content by combining the contents

---

[16]The four texts were H.N. Robinson's (1806–1867) *The Progressive Primary Arithmetic* (1862), H. N. Robinson's *First Lessons in Mental and Written Arithmetic* (1870), Ch. Davies' (1798–1876) *Primary Arithmetic* (1862) , and Davies' *Intellectual Arithmetic* (1858).

[17]Endō Toshisada was the author of *A History of Mathematics in Great Japan* (*Dai Nippon Sūgakushi* 大日本数学史, 1896, reprinted in 1918, 1960, and 1981), the first important book on

Fig. 8 The addition table (Endō, 1878, p. 19a)

加算九々之表

九	八	七	六	五	四	三	二	一	
十	九	八	七	六	五	四	三	二	一
十一	十	九	八	七	六	五	四	三	二
十二	十一	十	九	八	七	六	五	四	三
十三	十二	十一	十	九	八	七	六	五	四
十四	十三	十二	十一	十	九	八	七	六	五
十五	十四	十三	十二	十一	十	九	八	七	六
十六	十五	十四	十三	十二	十一	十	九	八	七
十七	十六	十五	十四	十三	十二	十一	十	九	八
十八	十七	十六	十五	十四	十三	十二	十一	十	九

presented in contemporary English and American textbooks of arithmetic with the methods and technical terms presented in the traditional Japanese abacus works (Kawamoto, 1968, p. 32; Suzuki 1960, p. 118); this textbook influenced greatly the pedagogy of calculations with the Japanese abacus in Japan.

Let us examine directly the part dedicated to the operation of addition in the *Textbook of the Art of Counting Beads*. It begins with a table of addition—numbers from 1 to 9 filled both in the top of the columns and in the first of the rows, and the sum of two numbers in every corresponding case (Fig. 8).

The author suggested that this table has to be remembered first and that "the most important are those whose sum equals ten …," which are related to the concept of complements of ten; "once the learner memorizes by heart the Nine-nine method of addition (*kasan kuku* 加算九九),[18] it would be appropriate for him to learn applying it to the Japanese abacus …."[19] This act of application was called "the employment of counting beads in addition." The author goes on to explain in

the history of traditional Japanese mathematics, *wasan* 和算. He was a practitioner of *wasan* and began to learn Western mathematics after the Meiji Restoration.

[18]It refers to the name of the table of addition shown in Fig. 8, which was never found in the older Japanese abacus books.

[19]Translated from Endō (1878, pp. 19a–b). "… 最モ緊要ナルハ相加ノ数正ニ十ニ満ツルモノトス… 學者既ニ加算九九ヲ諳記セハ宜ク之ヲ盤上ニ施スコトヲ習ハスヘシ…."

detail four ways of counting beads corresponding to four groups of operations of addition represented by four specific operations: 2 + 3, 3 + 8, 7 + 4, and 5 + 6.

For the operation that is 2 + 3, the way of counting beads which the author suggested was to move down an upper bead (+5) and then to move down two lower beads (−2) with the same finger.[20]

For the operation that is 3 + 8, the instructions were to move down two lower beads (−2) and with the same finger put the carry on the column of the position of tens (+10).

For the operation that is 7 + 4, the instruction was to move down a lower bead (−1), next move up an upper bead (−5), then with the same finger put the carry on the column of the position of tens (+10).

For the operation that is 5 + 6, the instruction was to move up a lower bead (+1), move up an upper bead (−5) with the same finger, then put the carry on the column of the position of tens (+10).

It can be observed that the classification was based on the finger movements when moving beads. Each of the operations, 2 + 3, 3 + 8, 7 + 4, and 5 + 6, actually represents a group of operations whose execution involve the same finger movements for moving beads suggested by the author. For example, 2 + 3 represents the group of operations (including 1 + 4, 2 + 3, 2 + 4, 3 + 2, 3 + 3, 3 + 4, 4 + 1, 4 + 2, 4 + 3, and 4 + 4) for which the finger movements for moving beads include "move down the upper bead" and then "move down lower bead(s)" with the same finger. In the same manner, we can list all the operations classified into 3 + 8, 7 + 4, and 5 + 6 groups, as shown in Table 2. From this table, it can be noticed that some operations are not classified into any one of the four groups. These represent operations, compared with the classified ones, which require much less effort to reason over the movement of beads, such as 1 + 1, 1 + 5, ..., etc. It is possible that the author did not classify them for this reason. The other point that needs to be brought to attention is that fingering was noted. Though the indications were not very detailed, to my knowledge the *Textbook of the Art of Counting Beads* is the first extant book containing such information.[21] Briefly, this first step of classifying single-digit additions in the new way, based on the finger movements for moving beads, accompanied with indications of fingering, turned over a new page in the pedagogy of calculations with an Japanese abacus in Japan.

Following the ideas introduced in the *Textbook of the Art of Counting Beads*, Japanese authors of Japanese abacus textbooks presented different classifications of the single-digit additions according to their own practices, and also provided exercises corresponding to different classified groups of operations. Among them, we choose to present the classification proposed byby Tamaoki Tetsuji 玉置哲二 and Ishibashi Umekichi 石橋梅吉 in their work *Essentials of Computation with the Beads* (*Shuzan seigi* 珠算精義, 1913), since this book permits us to observe more clearly the other point considered in the classification, namely, the fingering.

---

[20]In the book, the author did not inform the user which finger to use.

[21]This does not mean that the "finger movements" had not been taught by instructors beforehand.

Table 2 Classification of single-digit additions based on the finger movements for moving beads in the *Textbook of the Art of Counting Beads*

nbr1 / nbr2	1	2	3	4	5	6	7	8	9
1				↓					↑ ↑↓
2			↓	↓				↑ ↑↓	↑ ↑↓
3		↓	↓	↓			↑ ↑↓	↑ ↑↓	↑ ↑↓
4	↓	↓	↓	↓		↑ ↑↓	↑ ↑↓	↑ ↑↓	↑ ↑↓
5									
6				↑↓	↑ ↑↑	↑ ↑↑	↑ ↑↑	↑ ↑↑	↑↓
7			↑↓	↑↓	↑ ↑↑	↑ ↑↑	↑ ↑↑	↑↓	↑↓
8		↑↓	↑↓	↑↓	↑ ↑↑	↑ ↑↑	↑↓	↑↓	↑↓
9	↑↓	↑↓	↑↓	↑↓	↑ ↑↑	↑↓	↑↓	↑↓	↑↓

Notes

The grey tinted boxes represent operations 2 + 3, 3 + 8, 7 + 4, and 5 + 6, which the author Endō Toshisada took as representative for each group of operations

nbr1, number appearing on a column; *nbr2*, number to be added

Arrows in each case represent the finger movements for moving beads to be performed to execute nbr1 + nbr2. The significations of these arrows are as follows

↓ The high downward arrow represents moving down the upper bead, and the low downward
↓ arrow represents moving down the lower bead(s)

↑↓ The right downward arrow represents moving down the lower beads in the column of units, and the left upward arrow represents putting the carry on the column of tens

↑ ↑↓ The right lower downward arrow represents moving down the lower bead(s) in the column of units, the right higher upward arrow represents moving up the upper bead, the left lower upward arrow represents putting the carry on the column of tens

↑ ↑↑ The right lower upward arrow represents moving up the lower bead(s) in the column of units, the right higher upward arrow represents moving up the upper bead, the left lower upward arrow represents putting the carry on the column of tens

Table 3 Classification of single-digit additions based on the finger movements for moving beads in *Essential of Computation with the Beads*

nbr1 \ nbr2	1	2	3	4	5	6	7	8	9
1	↑	↑	↑	↓	↑	↑	↑	↑	↑↓
2	↑	↑	↓	↓	↑	↑	↑	↑↓	↑↓
3	↑	↓	↓	↓	↑	↑	↑↓	↑↓	↑↓
4	↓	↓	↓	↓	↑	↑↓	↑↓	↑↓	↑↓
5	↓	↓	↓	↓	↑	↑	↑	↑	↑
6	↓	↓	↓	↑↓	↑↑	↑↑	↑↑	↑↑	↑↓
7	↓	↑	↑↓	↑↓	↑↑	↑↑	↑↑	↑↓	↑↓
8	↑	↑↓	↑↓	↑↓	↑↑	↑↑	↑↓	↑↓	↑↓
9	↑↓	↑↓	↑↓	↑↓	↑↑	↑↓	↑↓	↑↓	↑↓

Note

The meanings of nbr1, nbr2, grey tinted boxes, and arrows are identical to those in Table 2

In the *Essentials of Computation with the Beads*, the operations are classified in two major groups: those of which the sum is lower than 10 and those of which the sum is equal or superior to 10. In the first group, the operations are classified further into four subgroups, represented by 5 + 3, 2 + 7, 3 + 4, and 2 + 3; in the second group, five subgroups are represented by 2 + 9, 7 + 5, 5 + 8, 8 + 4, and 7 + 3 (Tamaoki and Ushibashi 1913, pp. 8–13). We list all the operations classified into the nine different groups in Table 3.

It can be noticed in the table that the finger movements for moving beads for the operations 3 + 4 and 2 + 3 are the same, namely, "move down the upper bead" and "move down lower bead(s)," and these two operations are chosen to represent two subgroups. The nuance consists in the fingering. For the execution of 3 + 4, the authors suggested to move down the upper bead with the index finger and simultaneously to move down a lower bead with the thumb. On the other hand, for the execution of 2 + 3, the authors suggested to move down an upper bead and two lower beads with the index finger in one downward movement. The fingering is thus the choice of a particular organization of a suite of movements of the fingers in order to carry out an operation. Similar consideration was given to 8 + 4 and 7 + 3 which were classified as belonging to two different subgroups because of two different fingering routines. That is to say, the fingering played a role in the classification of single-digit addition. Compared with the fingering given in the *Textbook of the Art of Counting Beads*, the fingering presented in the *Essential of Computation with the Beads* was very precise.

To sum up, Japanese abacus education in respect to addition entered a new era in Japan from the publication of the *Textbook of the Art of Counting Beads*. The verses related to the single-digit additions used in the traditional Japanese abacus books were not included any more in the Japanese abacus textbooks of the new age.[22] Instead, a table of addition was introduced, the single-digit additions were classified according to the finger movements for moving beads in combination with the fingering, developed to articulate the movements in order to increase the speed at which the beads were moved. Besides this, the exercises of single-digit additions were designed according to different classified groups of single-digit additions. These designed exercises allowed learners to train themselves in movements and fingering. In this way the learner could establish and strengthen a direct link between a specific single-digit addition and specific finger movements for moving beads, along with specific fingering. A possible implication of this was the idea that the execution of addition could be performed on an Japanese abacus "automatically" after the direct link, mentioned above, had been mastered by the operator, being developed in Japan starting from the Meiji period more than ever through the precise description of the finger movements for moving beads and elaborated fingering.

Conclusions

This chapter provides a study of the education of Japanese abacus addition in China and Japan before the early 20th century. The results of the study show that teaching of addition with an Japanese abacus experienced a historical change in Japan around the Meiji Restoration, and passed from an era of using verses to an era of elaborated fingering techniques. In the era when verses were used, their various forms reflected different educational ideas concerning the degrees of reasoning about the movement of beads on an Japanese abacus and on the degree of fluency of finger movements for moving beads. Besides this, two distinct modes of bead movement reflected two different educational concerns, one related to the pursuit of the maximum efficiency of bead movement, the other related to the uniformity of the representation of numbers in one specific way. In the era that was marked by a focus on elaboration of fingering techniques, teaching of addition put special stress on the various ways in which the finger movements for moving beads were classified, in combination with developed fingering.

References

Bréard, Andrea. 2014. On the transmission of mathematical knowledge in versified form in China. In: *Scientific Sources and Teaching Contexts Throughout History: Problems and Perspectives*, Boston Studies in the Philosophy and History of Science 301, ed. A. Bernard and C. Proust, 155–185. New York etc.: Springer.
Chen Yifu. 2013. *L'étude des Différents Modes de Déplacement des Boules du Boulier et de l'invention de la Méthode de Multiplication Kongpan Qianchengfa et son Lien avec le Calcul Mental*. PhD dissertation. Paris: Université Paris Diderot-Paris 7.

---

[22]However, it cannot be concluded that the verses were not taught orally any more.

Endō Toshisada 遠藤利貞. 1878. *Sankajutsu jugyōsho* 算顆術授業書 [*Textbook of the Art of Counting beads*]. Tokyo: Tokyo shihangakkō 東京師範学校. National Diet Library Digitized Contents Database, http://dl.ndl.go.jp/info:ndljp/pid/829098.

Horiuchi, Annick. 1994. *Les mathématiques japonaises à l'époque d'Edo*. Paris: Librairie Philosophique J. Vrin. [For an English version of this work see Horiuchi 2000].

Horiuchi, Annick. 2000. *Japanese Mathematics in the Edo Period (1600–1868). A Study of the works of Seki Takakazu (?–1708) and Takebe Katahiro (1664–1739)*. Basel: Springer.

Hua Yinchun 華印椿 (ed.). 1987. *Zhongguo zhusuan shigao* 中國珠算史稿 [Sketch of a history of computation with beads in China]. Beijing: Zhongguo caizheng jingji chubanshe 中國財政經濟出版社.

Jami, Catherine. 1998. Abacus (Eastern). In *Instrument of science: an historical encyclopedia*, ed. R. Bud, and D. Warner, 3–5. New York/London: Garland.

Kawamoto Koji 川本亨二. 1968. Gakuseiki no shuzan kyōiku ni kansuru ichikōsatsu 学制期の珠算教育に関する一考察 [Survey on the Abacus Education in the Period from 1872 to 1879]. *Kyōikugaku zasshi* 教育学雑誌 [*Journal of Educational Research*] 2: 23–35.

Ke Shangqian 柯尚遷. 1578. *Shuxue Tonggui* 數學通軌 [Avenue to Mathematics]. Reprinted in Guo Shuchun 郭書春 ed. 1993. *Zhongguo kexue zhishu dianji tonghui* 中國科學技術典籍通彙. *Shuxue juan* 數學卷 vol. 2, 1167–1212. Zhengzhou: Henan jiaoyu chubanshe 河南教育出版社.

Li Gong 李塨. 1705. *Xiaoxue Jiye* 小學稽業 [Course of Minor Learning]. Beijing: Zhonghua shuju 中華書局, 1985. Reprint.

Suzuki Hisao 鈴木久男 and Toya Seiichi 戸谷清一 (ed.). 1960. *Soroban no rekishi* そろばんの歴史 [History of the *Soroban*]. Tokyo: Morikita Publishing Co., Ltd. 森北出版株式会社.

Tamaoki Tetsuji 玉置哲二 and Ishibashi Umekichi 石橋梅吉. 1913. *Shuzan seigi* 珠算精義 [Essential of Computation with the Beads]. Tokyo: Buneikan 文永館. National Diet Library Digitized Contents Database. Available at http://dl.ndl.go.jp/info:ndljp/pid/952188.

Uegaki Wataru 上垣渉. 1998. Gakuseiki ni okeru sanjutsu kyōkasho no taiyō 学制期における算術教科書の態様 [On a Phase of Textbooks for Arithmetic in the Period from 1872 to 1879]. *Nihon sūgaku kyōiku gakkaishi Sūgakukyōiku* 日本数学教育学会誌 数学教育 [*Journal of Japan Society of Mathematical Education, Mathematical Education*] 80(6): 89–96.

Uegaki Wataru 上垣渉. 2000. Meiji chūki ni okeru shuzan no fukkōundō ni kansuru ichikōshō 明治中期における珠算の復興運動に関する一考証 [A study on the Reconstructional Movement of the Calculation with Abacus (Soroban) in the Meiji Middle Period]. *Miedaigaku kyōikugakubu kenkyūkiyō Kyōiku kagaku* 三重大学教育学部研究紀要 教育科学 [Bulletin of the Faculty of Education, Mie University, Educational Science] 51: 1–20.

Uegaki Wataru 上垣渉. 2001. Shogaku sanjutsusho no tanehon ni kansuru saikôshô 小學算術書の種本に関する再考証 [Re-study on the Source Book of Shogaku sanjutsu-sho: Elementary text of arithmetic]. *Nihon sūgaku kyōiku gakkaishi* 日本数学教育学会誌 [*Journal of Japan Society of Mathematical Education*] 76: 3–16.

Volkov, Alexeï. 2001. L'abaco (Abacus). In *Storia Della Scienza* (Encyclopedia on History of science, in Italian), vol. 2 (Cina, India, Americhe), gen. ed. S. Petruccioli, volume eds. K. Chemla, F. Bray, D. Fu et al., 481–485. Rome: Istituto della Enciclopedia Italiana.

Wang Wensu 王文素. 1524. *Xinji tongzheng gujin suanxue baojian* 新集通証古今算學寶鑒 [Precious Mirror of Mathematicsm newly collected and provedly]. Reprinted in Guo Shuchun 郭書春 (ed.). 1993. *Zhongguo kexue zhishu dianji tonghui* 中國科學技術典籍通彙. *Shuxue juan* 數學卷 vol. 2, 337–971. Zhengzhou: Henan jiaoyu chubanshe 河南教育出版社.

Wu Jing 吳敬. 1450. *Jiuzhang Suanfa Bilei Daquan* 九章算法比類大全 [Great Compendium of the Comparison and Classification in Nine Chapters of Mathematical Methods]. Reprinted in Guo Shuchun 郭書春 (ed.). 1993. *Zhongguo kexue zhishu dianji tonghui* 中國科學技術典籍通彙. *Shuxue juan* 數學卷 vol. 2, 1–333. Zhengzhou: Henan jiaoyu chubanshe 河南教育出版社.

Xu Xinlu 徐心魯 (ed.). 1573. *Panzhu Suanfa* 盤珠算法 [Computational Methods with the Beads in a Tray]. Reprinted in Guo Shuchun 郭書春 (ed.). 1993. *Zhongguo kexue zhishu dianji tonghui* 中國科學技術典籍通彙. *Shuxue juan* 數學卷 vol. 2, 1143–1158. Zhengzhou: Henan jiaoyu chubanshe 河南教育出版社.

Zenkoku shuzan kyōiku renmei 全国珠算教育連盟 [League for Soroban Education of Japan]. 1971. *Nihon shuzanshi* 日本珠算史 [History of computation with beads in Japan]. Tokyo: Akatsushi shuppan Co., Ltd. 暁出版株式会社.

Chen Yifu defended his doctoral dissertation in University Paris-7 in 2013, his thesis titled "L'étude des différents modes de déplacement des boules du boulier et de l'invention de la méthode de multiplication *Kongpan qianchengfa* et son lien avec le calcul mental" was devoted to the history of computations with abacus in China and Japan.

Teaching Computation in 19th-Century Japan: The Transition from Individual Coaching on Traditional Devices at the End of the Edo Period (1600–1868) to Lectures on Western Mathematics During the Meiji Period (1868–1912)

Marion Cousin

Abstract The revolution that characterized the Meiji era (1868–1912) implied a complete revolution of teaching computation. Traditional individual teaching, based on the manipulations of the abacus and counting rods, had to be replaced by lecture-type teaching based on "paper computation," imported from the West. In this chapter, I will analyze the evolution of computation teaching during this period of transition, concentrating on the changes and continuities in the use of tools.

Keywords History of computation · History of education · Japan Edo period · Meiji period

The Meiji period (1868–1912) witnessed a nationwide modernization program in Japan, including sweeping reforms in the education system. More specifically, there was a profound transformation in the content and form of mathematics teaching. Formal education now introduced students to Western[1] mathematics, against the background of a strong popular tradition of mathematical practice (*wasan* 和算) based on Chinese tradition.[2]

---

[1]When I use the term "West" to refer to Europe and the United States, as the Japanese of the Meiji era did, it does not imply I consider that, from the scientific or cultural point of view, these nations form a uniform entity, or even a clearly delimited space. The actors of the modernization of Japan faced a heterogeneous ensemble of models, with several models for mathematical education. I use the term "West" to reflect the vision that Japanese people had of this geographic zone: a group of countries whose models had to be employed in order to occupy a strong position in the international context.

[2]On the *wasan* tradition, *see* for example Horiuchi (2010).

M. Cousin (✉)
Institut d'Asie Orientale, École normale supérieure de Lyon, Lyon, France
e-mail: cousin_marion@yahoo.fr

© Springer Nature Switzerland AG 2018
A. Volkov and V. Freiman (eds.), *Computations and Computing Devices in Mathematics Education Before the Advent of Electronic Calculators*, Mathematics Education in the Digital Era 11, https://doi.org/10.1007/978-3-319-73396-8_11

Before these reforms, the computation techniques taught in schools were based on these *wasan* methods from elementary level (in the *terakoya* 寺子屋)[3] right up to top-level private schools (*shijuku* 私塾). These methods privileged the use of the abacus (*soroban* 算盤) and counting rods (*sangi* 算木). Following the *Gakusei* 学制 (Decree on education, 1872), teachers were required to use new methods based on Western education, meaning that they had to integrate new types of mathematical texts and objects, and new pedagogic strategies into their teaching. Thus, they had to switch from individual teaching methods to "lecture-type" classes (or a "simultaneous teaching method"—*issei kyōjuhō* 斉教授法)[4] and had to abandon the abacus and counting rods for the exclusive use of computation on paper ("brush computation" *hissan* 筆算).[5]

In this chapter, I will study the evolution of mathematics teaching during this period of transformation from the middle of the 19th century to the beginning of the 20th century, focusing on certain computation methods. As the aim of this general work is to spotlight the use of computing devices for mathematical teaching, I will concentrate on elementary mathematics to examine the use of the abacus and algebraic methods with counting rods and I will determine what forms the teaching of computation methods took before and after the Meiji Restoration (1868).[6]

Computation Teaching Before the Meiji Era

After a long period of war, the peaceful Edo era (1600–1868) was characterized not only by (relative)[7] political, economic, and cultural isolation but also by economic growth and cultural development. Concerning education, several

---

[3]*Terakoya* were private elementary schools established in cities and in rural areas. Literally, *terakoya* means "temple school," coming from the fact that before the Edo era, these schools were established in temples.

[4]In every discipline, during the class, the teacher addressed the entire class rather than each student in turn. According to Galan, this pedagogical innovation is one of the first that was truly implemented in every classroom, *see* Galan (1999, p. 217).

[5]To implement the *Gakusei*, directives were promulgated to fix the curricula in elementary schools and in secondary schools. Both types of schools were ordered to base mathematical education on Western textbooks or on Japanese translations of Western works. Concerning teacher training in these new ways of teaching, some normal schools were established little by little but, in the beginning, teachers were from Edo schools. On the evolution of mathematics teaching during the Meiji era, *see* Cousin (2013, pp. 89–127) or Ogura (1974).

[6]The Meiji Restoration (*Meiji ishin* 明治維新) refers to the shift of power within the old ruling class: the powers were *restored* to the Emperor. More generally, "the larger process referred to as the Meiji Restoration brought an end to the ascendancy of the warrior class and replaced the decentralized structure of early modern feudalism with a central state under the aegis of the traditional sovereign, now transformed into a modern monarch" (Jansen 1989, p. 308). On the Meiji Restoration, *see* Jansen (1989).

[7]During the 16th century, the first contacts were established between Japan and Europe with the arrival of the Portuguese merchants as well as cultural exchanges with the Jesuits. But the activity

studies have shown that the school enrolment rate in Japan was high for a society where no national policy was established.[8] Galan underlined that: (1) there was a school network that covered the whole country; (2) a large part of the population agreed on the necessity of education; and (3) there were pedagogic practices and educational textbooks that had been used for several centuries.[9]

Another consequence of national development was the intensification of mathematical activity: new methods and results were established by Japanese traditional mathematicians (*wasanka* 和算家). By the end of the Edo period, computation executed with the abacus became very popular in all classes and regions of Japan (except in the samurai class), whereas the use of counting rods was reserved for specialists who also used rod representations as a highly developed notational system.

In China, counting rods[10] have been used since antiquity, for example, to execute all the procedures of the *Nine Chapters on the Art of Mathematics* (*Jiuzhang suanshu* 九章算術),[11] from elementary computation (such as division) to advanced classes of problems (such as the extraction of roots or the resolution of systems of

---

of Jesuits frightened the Japanese government and measures were taken to close the frontiers starting in the first half of the 17th century (the term *sakoku* 鎖国, literally "locked country," refers to this period). During the beginning of the 18th century, with the enlightened politics of the shogun Tokugawa Yoshimune 徳川吉宗 (1684–1751), the restrictions on European treatises were reduced and, in science, several studies on Western (especially Dutch) works (especially on medicine and on astronomy) were carried out during the Edo period. But, in mathematics, probably because of the success of the *wasan*, there were no translations of European works before the Meiji era. On the anti-Christian policy of the Tokugawa regime, *see* Elisonas (1991), and on the studies on European sciences during the Edo period, *see* Numata (1992).

[8]It is however necessary to note that a large gap existed between female and male education (in rural areas, only a few women attended *terakoya* before the Meiji era), and between education received in cities and the countryside (schooling was not well developed in the rural areas of the North and South), *see* Galan (1998, p. 7).

[9]On the state of Japanese education before the Meiji Restoration, *see* Galan (1998). For an English version, *see* Dore (1965) or Passin (1965), although one should be cautious concerning the quantitative results in these works, some of which have been discredited by recent studies. *See also* Rubinger (1982).

[10]The rods were small objects that could have various shapes (for example, cylindrical or prismatic) and be made of different materials (for example, wood, bone, or bamboo). *See* Martzloff (2006, p. 210) as well as the chapter 'Chinese Counting Rods' of the present volume.

[11]The *Nine Chapters on the Art of Mathematics*, often called *Nine Chapters*, is an ancient text compiled during the Han dynasty (206 BCE–220 CE) that can be compared to Euclid's *Elements* with respect to its impact on mathematical practices in China and in East Asia. For its translation, its presentation, and commentaries on its content, *see* Shen et al. (1999); Chemla and Guo (2004); or Guo et al. (2013).

linear equations).[12] It is unknown when the Chinese abacus (*suanpan* 算盤) first appeared but there is evidence that during the 14th century both devices existed, and, during the second half of the 16th century, the new device entered into common use.[13] The recent studies on the history of Chinese mathematics during the Ming dynasty (1368–1644) showed that the abacus progressively replaced the system of counting rods, as witnessed by the high number of mathematics works dedicated to this device published during the 16th century.[14] In Japan, the first documents that witnessed the use of the abacus (*soroban* 算盤 in Japanese), and that are available today, date from the end of 16th century.[15] The Chinese abacus has five beads in the lower part (each of them representing one unit) and two beads in the upper part (the "heaven" 天, Chinese *tian*, Japanese *ten*, where one bead represents five units), and it was mainly used during the Edo period. But, during this period, new types of abacus appeared. An abacus with five beads in the lower part but only one bead in the upper part (5 + 1 type)[16] was also used during the Edo period and it became more and more common after the 1880s. At the end of the 18th century, an abacus with four beads in the lower part and one bead in the upper part ("4 + 1" type) appeared and it was recommended by specialists after the 1890s. During almost half a century, both the 5 + 1 and 4 + 1 types were used and there were debates between specialists on the most appropriate type. However, the 4 + 1 type was definitively adopted in 1935, when the Ministry of Education decided to include it in the textbooks for elementary schools (for the various shapes of the abacus, *see* Fig. 1).[17] Learning how to use this device constituted one of the main elements of mathematical curricula of Japanese schools until the middle of the 17th century, when counting rods were also integrated into the advanced curricula in mathematics (whereas the abacus was still used for elementary teaching). Associated with the algebraic techniques studied in the *Suanxue qimeng* 算學啟蒙 (*Introduction to Computational Studies*, 1299) of Zhu Shijie 朱世傑, the counting rods generated many works in the *wasanka* community.

---

[12]The presence of the computation device is manifest in *Nine Chapters*, where counting rods are called *suan* 算 or *chou* 籌, *see* Chemla and Guo (2004, pp. 910–911, 988–989). There are several names for this device in Chinese tradition, *see* Martzloff (2006, p. 210), yet nothing is said about the counting board (a special surface where counting rods are manipulated) or about the counting rods' disposition; see the chapter 'Chinese Counting Rods' of the present volume especially devoted to the history of the instrument. However, historians have shown that the principles of the computation system (numeration system, surface management) were probably already the same as those developed for the device described in this chapter, *see* Chemla and Guo (2004, pp. 15–20).

[13]*See* Martzloff (2006, pp. 215–216).

[14]*See* Chen (2013, p. 3) and the chapter 'The Education of Abacus Addition in China and Japan Prior to the Early 20th Century' of the present volume authored by Yifu Chen.

[15]Chen (2013, p. 109).

[16]This notation is inspired by Chen (2013).

[17]On the evolution of the Japanese abacus and on the reasons for the changes of its shape, *see* Chen (2013, pp. 109–110) and the chapter 'The Education of Abacus Addition in China and Japan Prior to the Early 20th Century' of the present volume.

Fig. 1 The evolution of the abacus shape. The image on the left-hand side is copied from Cheng (1675, volume 2)

Teaching Elementary Operations with the Abacus: The Jinkōki 塵劫記 (Inexhaustible Treatise) by Yoshida Mitsuyoshi 吉田光由 (1598–1672), Published for the First Time in 1627

Yoshida Mitsuyoshi, a *wasanka* from the merchant class, wrote one of the earliest Japanese mathematical sources available today, the *Jinkōki*. This textbook was published several times during the Edo period and was also re-edited after the Meiji Restoration. Historical studies have shown that this textbook was widely used for mathematical teaching in *terakoya*, and it is therefore a reliable source concerning elementary teaching in computation.[18]

The *Jinkōki* gives an overview of the mathematical techniques useful in everyday life, and in several professions (merchant, carpenter, farmer, etc.) that developed during the Edo period. It is a sophisticated treatise whose contents bear witness to the maturity of the mathematical practices that must have developed during the late 16th century,[19] based on Chinese tradition and especially on the *Suanfa tongzong* 算法統宗 (*Unified Origins of Computational Methods*, 1592) written by Cheng Dawei 程大位 (1533–1606). Yoshida, who came from a family that played an important role in the development of sciences (and especially medicine) and international trade,[20] decided to write his textbook in Japanese and

---

[18]*See*, for example, Horiuchi (2014, p. 168).

[19]The content of the first Japanese treatises on mathematics, the *Warizansho* 割算書 (*Book on Division*, 1622) written by Mōri Shigeyoshi 毛利重能 and the *Shokanbumono* 諸勘分物 (*Estimations of Surfaces and Volumes*, 1622) by Momokawa Chihei 百川治兵衞, illustrates mathematical practices that probably developed during the second half of the 16th century, when the economic growth and development of cities established new requirements for mathematics education, *see* Horiuchi (1994, pp. 24–26).

[20]Yoshida was a descendant of the Suminokura 角倉 family, whose support was fundamental for the success of the *Jinkōki*. The Suminokura family included successful merchants, as well as reputed intellectuals. Suminokura Ryōi 角倉了以 (1554–1614) made the family famous by constructing fluvial routes that appeared to be crucial for Kyoto's economy. Yoshida was born surrounded by specialists of several cultures, he had access to Chinese sources and was helped by

introduce the basic knowledge of computation. As a consequence, this textbook was accessible to most of the population. To introduce the various mathematical methods he selected examples that might interest several occupations and classes that had been developing during the Edo period, so that the entire population could see the practical utility of mathematics.

Concerning the elementary operations, Yoshida decided to introduce them right after the presentation of "small numbers," "large numbers," and metrological units. On the one hand, addition and subtraction are not introduced and he only gives tables for multiplication ("Article 6: about [the table] 'nine-nine' *kuku* 九九,"[21] without using examples, probably because of the simplicity of the procedure). On the other hand, Yoshida emphasized division procedures. These procedures are introduced at the beginning of the textbook (they are therefore essential for the rest of the study), they are described in detail, with illustrations of the abacus,[22] with almost a third of the first volume of the *Jinkōki* devoted to this technique.

The rhymes for the division of numbers by a one-digit number, called the rhymes of the *hassan* 八算 ("the eight operations,"[23] the division by one not being expressed, *see* Fig. 2),[24] are given in the 7th Article of the *Jinkōki* and were to be memorized (in reality, the pupil was meant to sing the rhymes when he recited them).

These tables are very similar to those found in the *Suanfa tongzong*,[25] where they are presented in the section called *jiugui ge* 九歸歌 (the songs of the nine

---

Suminokura Soan 角倉素庵 (1571–1632), Ryōi's son, to study the Chinese treatises on computation. His high social rank also gave him access to the highest level teaching of the moment (for example, he followed Mōri's mathematical classes in Kyoto), as well as allowing him to print and distribute the *Jinkōki* (his family was known for the printing of sophisticated books). For more information on the Suminokura family and on the context in which Yoshida wrote his textbook, *see* Horiuchi (1994, pp. 33–40); WI (2000, pp. 2–50); and Fujiwara (1983, pp. 39–48, 190–216).

[21]In the Chinese tradition, the multiplication table is often called *jiu-jiu* 九九 (literally "nine [times] nine"). Yoshida also uses this title (*ku-ku* in Japanese) for Article 6, translated as "The Multiplication table" in WI (2000).

[22]Two operations are subject to a detailed presentation in the *Jinkōki*, division and the determination of the square root. Yoshida also uses *soroban* illustrations for particular examples. In the *Suanfa tongzong*, there are no explicative illustrations of the *soroban*, but Yoshida was probably inspired by one of Ming's textbooks where there are these kind of illustrations: three textbooks that have those illustrations are pointed out in Chen (2013), including the *Panzhu suanfa* 盤珠算法 (*Computation Methods with Beads in a Tray* [i.e., on an abacus]) written by Xu Xinlu 徐心魯 in 1573, that was reprinted in Japan according to Fujiwara (1983). These illustrations, being pedagogic tools, show that Yoshida wanted to make this text understandable by any reader.

[23]In the Chinese sources, the term usually employed is *jiugui* 九歸 (literally, "restitution by 9"). Yoshida chose to name these tables *hassan* 八算 (the eight operations), as his master Mōri did in the *Warizansho*.

[24]Figure 2 is extracted from Yoshida (1641, pp. 8b–9a). It was a sophisticated edition with color printing.

[25]There are only three differences between the table of the *Jinkōki* and those of the *Suanfa tongzong*, and Mōri made almost exactly the same modifications (there is only one difference between Mōri's and Yoshida's tables). *See* Cousin (2008, p. 38).

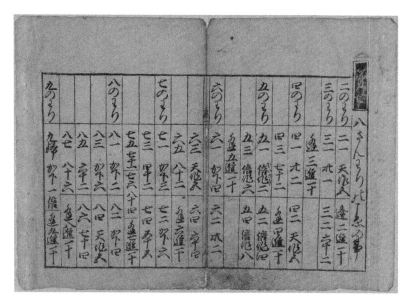

Fig. 2 The rhymes of the *hassan* in the *Jinkōki* (Yoshida 1641)

divisions). Created in China,[26] these "rules" make the learning of division easier because they make the division of any number by a one-digit number automatic: for each digit of the dividend, the pupil had to recite a rhymed formula corresponding to the configuration on the abacus. But, as revealed by the example below, while this technique is clearly well adapted to the abacus, it is not appropriate for written computation, as the digits of the dividend are substituted by the digits of the quotient. After having given these rhymes, extracted from the Chinese sources and presented in tabular form for each divisor from two to nine, Yoshida decided to repeat the song of the digit and to give a particular example (the division of 123,456,789 by this digit) with an illustration of the abacus. A representation of the abacus is also given for each digit between 2 and 9 (123,456,789 is represented on an 11-rod abacus) and, under each rod, the reader finds the rhymes corresponding to the successive configurations of beads,[27] and, over each rod, the configuration of the result. Figure 3 shows the page written by Yoshida for division by 4, along with its translation.

---

[26]According to Martzloff (2006), these rules were invented during the 11th century, originally for counting rods. In the *Suanxue qimeng*, Zhu Shijie commented on these rhymes as follows: "[T]he classical method (of division) is based on the *shangchu* [商除] technique (i.e., that which uses three superposed rows); but because it is difficult for beginners it has been replaced by the *jiugui* method which is not orthodox" (Martzloff 2006, p. 218, n. 1). For the description of the *shangchu* method, and for the methods of multiplication, division, and the manipulations on the *suanpan*, *see* Martzloff (2006, pp. 217–221).

[27]According to Chen (2013), the three textbooks mentioned in note 22 had these kinds of illustrations to describe elementary operations.

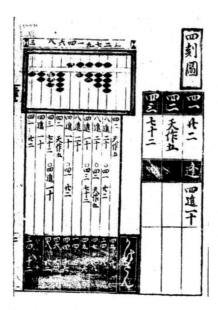

Fig. 3 The division by 4 in the *Jinkōki* (Yoshida 1641)

The text in this diagram reads as follows:
[On the right-hand side, from top to bottom] Diagram of the division by 4

4, 1	22	Encountering	4, move forward 10
4, 2	Make 5 in heaven[28]		
4, 3	72[29]		

[On the left-hand side, from top to bottom: example, where 123456789 set on the abacus is divided by 4.]

[Result:] 3086419725 [the digit 3 is placed above the rod where digit 1 of the dividend is set on the abacus.[30]]

[Recited rhymes. In this table, as the reader is supposed to start with the left rod, the first line of the table corresponds to the left column under the abacus.]

---

[28] As mentioned before, "heaven" means the upper part of the abacus.

[29] As shown in Fig. 3, the "rhyme for division by 4" is given in a table. This table is reproduced for each digit and progressively filled with the relevant rhymes. In the example of the division by 9, the table is full, as there are 9 rhymes in the song of the division by 9.

[30] As shown in Fig. 3, the result is given without any separator between integer and fractional part, and with a blank space used to represent the digit "0;" it looks as the sequence "3086419725." But, as explained below, the reader is supposed to know that, when the dividend (123,456,789) is displayed as shown in the picture, the result "appears" on the abacus, with the unit digit being displayed where the tens digit of the dividend (8) was displayed.

(Under 1)	4, 1, 22.	3, 4, 12.[31]
(Under 2)	4, move forward 10.[32]	
(Under 3)	4, 3, 72.○[33] 4, move forward 10.[34]	4, 8, 32.
(Under 4)	4, 2, make 5 in heaven.	4, 6, 24.
(Under 5)	4, move forward 10.○ 4, 1, 22.	4, 4, 16.
(Under 6)	8, move forward 20.	1, 4, 4.
(Under 7)	4, move forward 10.○ 4, 3, 72.	4, 9, 36.
(Under 8)	8, move forward 20.○ 4, 2, make 5 in heaven.	4, 7, 28.
(Under 9)	8, move forward 20.○ 4, 1, 22.	2, 4, 8.
(Under the empty rod)	4, 2, make 5 in heaven.	4, 5, 20.[35]

The numbers on the abacus are represented with decimal and positional notation: the position of the unit digit is chosen and then each digit of the dividend (on the left of the units, there are tens, hundreds, etc., and on the right-hand side, the decimal units are positioned) is entered by moving the beads to the horizontal bar,[36] as shown by Yoshida's text.[37] To execute the division algorithm, one should begin with the left-hand digit (here 1) and execute the movements corresponding to the rhyme: for example, with those of the division by 4, "4, 1, 22" means that, if there is only one bead on the rod, two beads (the quotient from division of 10 by 2) are activated, i.e. two beads are shifted to the bar instead of the original one bead, the dividend, and two beads (the remainder) are added to the column on its right, and then the next digit (on the right-hand side) is treated; "4, 2, make 5 in heaven" means that, when we encounter 2, the dividend 2 (interpreted as 20) is substituted by the quotient 5 (i.e., by one bead activated in the superior part, the "heaven"), and then the next digit is treated; similarly to the first rhyme, "4, 3, 72" means that when we encounter 3, 7 (the quotient of division of 30 by 4) replaces 3 (the dividend) and two beads (the remainder) are added to the column on its right, and then the next digit is treated; finally, "encountering 4, move forward 10" means that, if there are four beads, one bead (the quotient) is added to the column on its left while suppressing 4 beads on the column dealt with; in general, if the digit set in the position is larger than three,

---

[31]In this column (the last line under the *soroban* in the original page of the *Jinkōki*) the author gives verification that the result multiplied by 4 gives the original dividend.

[32]This line should be "Encountering 4, move forward 10." This rhyme is shortened in the example.

[33]This round symbol was used here to separate the several movements that were done on one rod.

[34]This last rhyme ("4, move forward 10") should be under the next rod.

[35]Yoshida (1641, volume 1, p. 9b). *See also* WI (2000 pp. 33–34).

[36]The beads of the lower part are initially moved down and the bead in the upper part (or both beads, if a Chinese-style $2 + 5$ abacus is used) is/are initially moved up. To represent n units, $0 \leq n \leq 4$, in a given decimal position, n beads from the lower part are moved up to the bar separating the upper and lower sections; to represent n units, $5 \leq n \leq 9$, 1 bead in the upper part is moved down to the bar and $(n - 5)$ are lifted up to the bar in the lower part.

[37]In the text, and in general in the *wasan* texts, only the beads that have been moved (to the horizontal bar) are represented. The beads that are down in the lower part or the beads that are up in the upper part are not drawn.

Table 1 The operations performed on the abacus to divide 123456789 by 4

R1	R2	R3	R4	R5	R6	R7	R8	R9	R10	Rhyme used
1	2	3	4	5	6	7	8	9	(0)	4, 1, 22
2	**4**	3	4	5	6	7	8	9	(0)	Encountering 4, move forward 10
3	0	**3**	4	5	6	7	8	9	(0)	4, 3, 72
3	0	7	**6**	5	6	7	8	9	(0)	Encountering 4, move forward 10
3	0	8	**2**	5	6	7	8	9	(0)	4, 2, make 5 in heaven
3	0	8	5	**5**	6	7	8	9	(0)	Encountering 4, move forward 10
3	0	8	6	**1**	6	7	8	9	(0)	4, 1, 22
3	0	8	6	2	**8**	7	8	9	(0)	Encountering 8, move forward 20
3	0	8	6	4	**0**	7	8	9	(0)	*(Empty position → go to next rod)*
3	0	8	6	4	0	**7**	8	9	(0)	Encountering 4, move forward 10
3	0	8	6	4	1	**3**	8	9	(0)	4, 3, 72
3	0	8	6	4	1	7	**10**	9	(0)	Encountering 8, move forward 20
3	0	8	6	4	1	9	**2**	9	(0)	4, 2, make 5 in heaven
3	0	8	6	4	1	9	5	**9**	(0)	Encountering 8, move forward 20
3	0	8	6	4	1	9	7	**1**	(0)	4, 1, 22
3	0	8	6	4	1	9	7	2	**2**	4, 2, make 5 in heaven
3	0	8	6	4	1	9	7	2	5	*(Result)*

one bead is added to the column on its left, four units are subtracted from the current position, and the process is repeated,[38] until the digit is inferior to 4; if it is positive, the appropriate rhyme is applied, and if there are no more beads, the next digit is treated. When the algorithm is finished, the quotient is represented on the abacus, displaced by one digit to the left: in the example, at the end of the algorithm, 3086419725 is represented on the abacus, with 2 (second to last digit) placed in the original position of units, so the result is 30,864,197.25.[39] A table that represents the successive configurations of the abacus is given above to help the reader (Table 1).[40]

---

[38]In fact, as we can see in the example, if the digit set in the position is equal or larger than eight, the rhyme is not repeated twice, the rhyme "Encountering eight, move forward 20" is used.

[39]For detailed explanations on division by other one-digit numbers, *see* WI (2000, pp. 31–35).

[40]In this table, on each line, the digit represented on the abacus is written in the 10 boxes on the left-hand side ("R1 stands for 'Rod 1'"), the number considered is typed in bold typeface, and the corresponding rhyme applied is given in the right-hand box. The units of the initial number (dividend) are placed in the position R9 and the units of the result appear in the position R8.

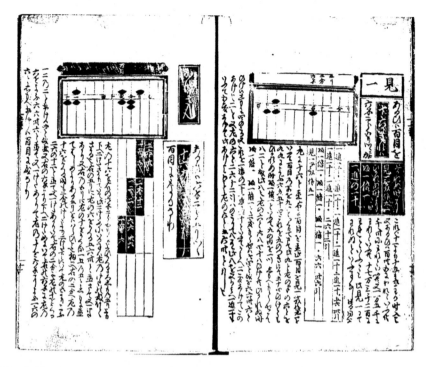

Fig. 4 The rhymes of the *ken-ichi* in the *Jinkōki* (Yoshida 1641)

In the 8th Article, named "Rhymes of the *ken-ichi* 見一" ("when seeing 1," title of the first of the rhymes), Yoshida introduced the procedure that allows the reader to make divisions by two-digit numbers. For each number *n* between 1 and 9, he enunciates the rhymes of the *ken-n* 見n, namely, the rhymes of *ken-ichi* 見一 ("when seeing 1," that allow to divide a number by any number between 11 and 19), the rhymes of *ken-ni* ("when seeing 2," that allow to divide a number by any integer between 21 and 29), and so on. As in the article about *hassan*, the author gives an example along with illustrations of the abacus for each number (Fig. 4).[41]

Concerning this part on division, Yoshida clearly took his inspiration from the work of his master, the *Warizansho* 割算書 (*Book on the Division* 1622) by Mōri Shigeyoshi 毛利重能 (dates of life unknown), who himself based his teaching on Chinese sources, especially the *Suanfa tongzong*. Imitating his Chinese sources,[42] Yoshida gave an important place to this elementary operation and added several illustrations of the abacus (probably being influenced by certain Chinese sources).[43]

---

[41]This portion is from Yoshida (1641, volume 1, pp. 14b–15a). For the *ken-ichi* procedure, and its text in the *Jinkōki, see* WI (2000, pp. 35–38).

[42]Yoshida is the first Japanese author to explicitly refer to the *Suanfa tongzong*.

[43]*See* note 22.

The *Jinkōki* shows that abacus computation represented the rudiments of mathematics learning during the Edo era, and it was the only arithmetic taught in traditional elementary schools. Students had to memorize the tables and practice computing on the abacus, but there was no attempt to explain to them why they had to perform operations this way. The abacus manipulations were taught dogmatically and procedures were applied automatically by students. Usually, professors did not give justifications for the procedures and neither did the textbooks (*see*, for example, the *Jinkōki*).[44] Even advanced students could not explain the processes, as Fujisawa Rikitarō 藤沢利喜太郎 (1861–1933) explained in his book on Japanese mathematical education (1912):

> Only a few gifted ones who are capable of learning without being taught, saw, as [if] it were from a distance, like looking through a mist gradually disappearing before the rising sun, the principle which underlies the operations; but even they were unable to explain it to others.[45]

Merchants and tradesmen had to master abacus computation and special stress was laid on the rapidity of the operations, which mainly depended on the movement of the fingertips. In fact, quick calculation became so popular that, among mercantile people, abacus calculation became a pastime and tournaments were often organized.[46]

During the Edo period, in private schools of mathematics, abacus computation was associated with basic learning (up to techniques like the extraction of the square root) but counting rods were used for more complex algorithms, especially those that would today be qualified as algebraic.[47] As a consequence, these two computing devices coexisted during the whole development of the *wasan* (*see* Fig. 5, an illustration taken from the *Sanpō benran* 算法便覧, written by Takeda Shingen 武田真元 in 1821, showing activities in the *wasan* school of Takeda, at the beginning of the 19th century: the master of the school is teaching advanced computation with counting rods and a young pupil is being helped to learn abacus computation by a more advanced pupil).

Computation with Counting Rods: A High-Level Algebraic Tool for Problem Solving

To calculate with this other device, the counting rods were placed on a counting board (Fig. 5).[48] When they were used for elementary calculation (in China, before the use of the abacus), the notation with counting rods was also decimal and

---

[44]There are few cases in which a justification was given. *See* Ueno (2012, p. 476).

[45]Fujisawa (1912, pp. 50–51).

[46]*See* Fujisawa (1912, pp. 22–23).

[47]For various examples of the use of rods for these types of problems, *see* Horiuchi (2010).

[48]A number of Chinese historians of mathematics claimed that such a board existed in China. *See* Martzloff (2006, p. 209) and the chapter "Chinese Counting Rods" of the present volume.

Fig. 5 Activities in the school of Takeda at the beginning of the 19th century

positional: a position was chosen for units, and then units, hundreds, etc., were represented vertically (as shown on the upper part of Table 2 below), while tens, thousands, etc., being represented horizontally (as shown in the lower part of the Table 2), and zero being represented by an empty space. For example 1023456789 would be represented as shown in Fig. 6.

However, starting in Chinese antiquity, counting rods were also used for more sophisticated problems, in particular for solving problems that would today be represented by simultaneous linear equations (*fangcheng* 方程 method).[49] In this case, on the calculation surface, the counting rods stand for the coefficients of the equations instead of the digits of a number. For example, they can be used to represent the system written with modern notation in Fig. 7, by the arrangement given in the same figure.[50]

---

[49]Historians have been unable to agree on how to translate the word *fangcheng* and that is why it is left in the *pinyin* transcription. Chemla has suggested the translation "squared measures," *see* Chemla and Guo (2004, pp. 599–602), whereas Guo, Dauben, and Xu chose "rectangular array," *see* Guo et al. (2013, p. 905). Various translations are discussed in Guo et al. (2013, pp. 905–907).

[50]The system and the rod configuration shown in Fig. 7 represent the data given in problem (8.1) of the *Nine Chapters*: "Suppose that 3 *bing* of high-quality grain, 2 *bing* of medium-quality grain and 1 *bing* of low-quality grain produce 39 *dou*; 2 *bing* of high-quality grain, 3 *bing* of medium-quality grain and 1 *bing* of low-quality grain produce 34 *dou*; 1 *bing* of high-quality grain, 2 *bing* of medium-quality grain and 3 *bing* of low-quality grain produce 26 *dou*. Then how much is produced respectively by one *bing* of high, medium and low-quality grain?", where *bing* 秉 is a bundle and *dou* 斗 is a unit of capacity. *See* Chemla and Guo (2004, pp. 617–621) and Guo et al. (2013, p. 907).

Table 2 Representation of digits on a counting surface

1	2	3	4	5	6	7	8	9
│	║	┃┃┃	┃┃┃┃	┃┃┃┃┃	⊤	⊤⊤	⊤⊤⊤	⊤⊤⊤⊤
1	**2**	**3**	**4**	**5**	**6**	**7**	**8**	**9**
—	═	≡	≣	≣	⊥	⊥	⊥	⊥

The configurations of counting rods in the upper row were used to represent digits corresponding to even powers of 10 (i.e., units, hundreds, etc.), while the configurations in the lower row were used for odd powers of 10 (dozens, thousands, etc.)

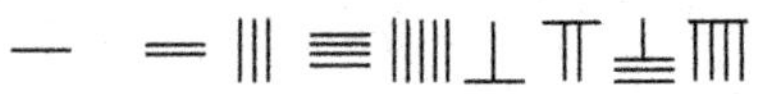

$$— \quad = \; ||| \; \equiv \; ||||\; \bot \; \top \; \underline{\bot} \; |||$$

Fig. 6 The representation of 1023456789 with rods

$$\begin{cases} 3x + 2y + z = 39 \\ 2x + 3y + z = 34 \\ x + 2y + 3z = 26 \end{cases}$$

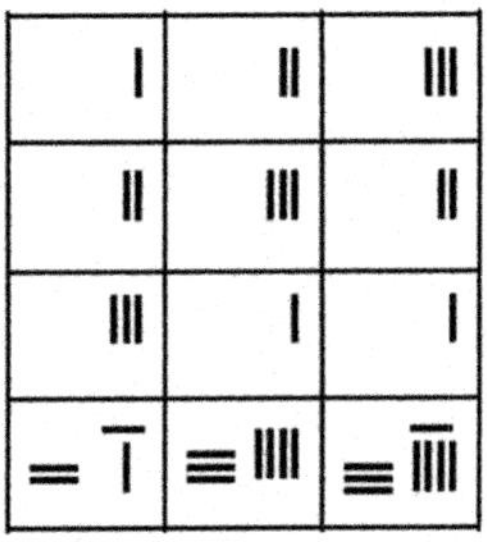

Fig. 7 The representation of linear systems of equations with rods

To solve the system, counting rods were moved as shown in Fig. 8,[51] following a method similar to modern row reduction. The solution is displayed in the last configuration and can be expressed, in modern terms, as $x = \frac{37}{4}, y = \frac{17}{4}, z = \frac{11}{4}$.

Moreover, during the Song (960–1279) and Yuan (1260–1368) dynasties, Chinese mathematicians developed the technique of the "celestial element" (天元 pronounced *tian yuan* in Chinese, *tengen* in Japanese), in order to determine what we would call today the roots of a single-variable polynomial.[52] In the middle of 17th century, while counting rods and the *tianyuan* method had disappeared from

---

[51]In Fig. 8, the successive configurations of the counting rods used to solve the problem are displayed, from left to right and from top to bottom. For the text of the procedure associated with this problem, *see* Chemla and Guo (2004, pp. 617–621). There are different interpretations for the last step of the procedure, Chemla and Guo's interpretation has been followed. For a detailed explanation of the procedure, *see* Chemla and Guo (2004, pp. 602–604).

[52]On these Chinese works, *see* Martzloff (2006, pp. 258–271). *See also* Libbrecht (1973) and Hoe (2007).

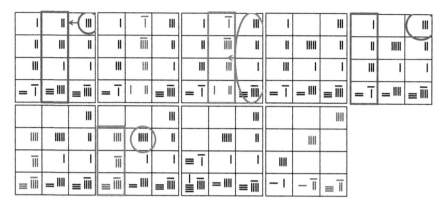

Fig. 8 Resolution of the system described in Fig. 7 with rods

the Chinese mathematical landscape because of what historians have called "the decline of Chinese mathematics," this technique was imported by Japan,[53] where research on "techniques for the resolution of problems" was becoming more and more important.[54] After the importation of the *Suanxue qimeng*, the *tianyuan* technique was integrated in these techniques and this domain became progressively one of the most important of the *wasan*. For example, the works of Seki Takakazu 関孝和 (ca. 1640–1708)[55] and Takebe Katahiro 建部賢弘 (1664–1739) established an evolved notational system based on rod representation that allowed high-level algebraic computation.[56]

In these practices, the counting rods represent the coefficients of a polynomial. The coefficients of the polynomial are represented one under the other, the uppermost coefficient being the constant coefficient. For example, the first image on the left of Fig. 9 represents the polynomial $1 - 4x^2 + 9x^4$ (the negative coefficient was marked with a slanted line and the null coefficient was represented with $\bigcirc$).[57]

This algebraic tool (the *tenzan* 点竄 algebra) was used throughout the Edo period for treating problems of several types, including geometrical problems (*see*

---

[53]The *Suanxue qimeng*, where the *tianyuan* method is presented, was published in Japan for the first time in 1658. *See* Horiuchi (1994, p. 91).

[54]The "techniques for the resolution of problems" (an expression used in Horiuchi (1994, p. 79)) were developed to calculate areas (or volumes) or to solve problems where, given a fixed figure whose area (or volume) is known, the measurements of the figure are asked. They consisted of "geometric techniques" (the translation of geometric properties into numeric relations between magnitudes, in other words, techniques for establishing equations) and the "computation techniques" (the isolation of the researched value, in other words, equation resolution). *See* Horiuchi (1994, pp. 79–80).

[55]Seki's date of birth is unknown. According to Majima Hideyuki, he was born between 1640 and 1645. See Majima (2013, pp. 6–7).

[56]For details on the techniques developed by Seki and Takebe, *see* Horiuchi (2010).

[57]Historians argued that the slanted "rod" was only used in texts. When computing with real counting rods, it seems that two systems of rods with different colors were used. *See* Horiuchi (1994, p. 97) and the chapter "Chinese Counting Rods" of the present volume.

Fig. 9 From left to right: (1) Representation of the polynomial $1 - 4x^2 + 9x^4$ with rods, (2) Extract from Hasegawa (1830), (3) Extract from Takebe (1685)

the image in the middle of Fig. 9).[58] Seki and Takebe developed a complex system of notation to write polynomials using symbols inspired by rod configurations (as in the Chinese tradition), and where coefficients could have nominal components with undetermined values[59] (which would be impossible to do using the original notation imported from China). For example, on the right-hand side of Fig. 9,[60] the polynomial represented is:

---

[58]The second image is taken from Hasegawa (1830, p. 82a). The aim is to solve the following problem: "Inside of the external circle, two circles *kō* 甲, two circles *otsu* 乙 and a circle *hei* 丙 are inserted as shown in the picture. The diameter of *kō* 甲 is worth 12 *sun* and the diameter of *hei*, 4 *sun*. We ask how much is the diameter of *otsu* 乙." The successive rod configurations that are used to resolve the problem and that are included inside the text represent the following polynomials (from right to left):$[x]$; $[12 + x]$; $[144 + 24x + x^2]$; $[20 - x]$; $[400 - 40x + x^2]$; $[464 - 40x + x^2]$; $[-320 + 64x]$ For a translation of the whole procedure along with its analysis, *see* Horiuchi (1996, pp. 251–253) and Cousin (2013, pp. 63–64).

[59]The (nominal) coefficients that are multiplied by the unknown are placed, next to each other, on the right-hand side of the counting rods that represent the numerical coefficients, and the (nominal) coefficients that divided the unknown were placed on the left-hand side of the counting rods representation.

[60]The last image is taken from Takebe (1685, volume 2, p. 9b).

$$A^9 - 9A^8x + 31A^7x^2 - 84A^6x^3 + 126A^5x^4 - 126A^4x^5 + 84A^3x^6 - 36A^2x^7 + 9Ax^8 - x^9,$$ where x is the unknown, the integer coefficients are represented with rod-type notations, A is the value designated by the *kanji*[61] *wa* 和, and the powers of A are also represented with *kanji*.[62] The calculus on these "polynomials" was first done using counting rods, but their notation on paper progressively made them detached from the device, and these written configurations (called *shiki* 式 by Takebe) became more and more like modern "formulas."[63]

The teaching associated with these traditional devices was individual: students followed the instructions of their teacher, or of a more advanced student, and trained by executing new computations using their devices. Elementary operations executed on the abacus were taught by a *terakoya* teacher to the children in non-elite elementary schools. Meanwhile, in the *shijuku* dedicated to *wasan* learning, higher level computation on abacus and sophisticated techniques executed using counting rods (or representations of counting rods on paper) were studied, following the methods taught by the instructor.

Revolutionizing the Teaching of Computation During the Meiji Era

During the Edo period, several Japanese scholars became interested in European scientific theories, especially in medicine, astronomy, and botany.[64] Nevertheless, considering the strong mathematical tradition that developed during the Edo period, little Western mathematical knowledge interested Japanese scholars before the arrival of Commodore Perry's ships (1853).[65] Trigonometric tables and logarithms were included in the *wasan* curricula and Japanese scholars agreed on the practical

---

[61]The *kanji* 漢字 are the Chinese ideograms that have several pronunciations in Japanese; besides, there are two types of phonetic characters, the syllabary *hiragana* ひらがな (used for Japanese terms), and the syllabary *katakana* カタカナ (used for foreign terms).

[62]For a detailed explanation of these notations, *see* Horiuchi (1994, pp. 174–175).

[63]See Horiuchi (1994, p. 175).

[64]After the first contact with Jesuit missionaries during the 16th century, several Chinese works on European sciences were imported following the efforts of the shogun Tokugawa Yoshimune 徳川吉宗 (1684–1751) to introduce Western learning into Japan at the beginning of the 18th century. But Japanese scholars really became interested in European theories and techniques only at the end of the 18th century, with the *rangaku* 蘭学 (Dutch learning) movement, at first led by physicians. *See* Numata (1992).

[65]In 1852, the American Matthew C. Perry (1794–1858) was sent to Japan to persuade the government to open their ports. The signature of the Treaty of Kanagawa (1854) marked the beginning of the Japanese modernization movement that characterized the Meiji period.

efficiency of Western mathematical works in astronomy and navigation but *wasan* studies did not change in appearance following contact with these theories imported from Europe.[66]

Nevertheless, before the Restoration (1868), the authorities became aware of the urgent need to import Western sciences and techniques in order to attain a strong position on the international stage.[67] In 1872, shortly after the Restoration, the *Gakusei* (Decree on education, originally based on the American system) was promulgated to establish a modern national educational system with Western scientific subjects taught from primary schools to universities. At first, schools and teachers from the old feudal system were used to carry out the reform, but normal schools (*shihan gakkō* 師範学校) were gradually set up to train a renewed educational staff.

The First Textbooks of Arithmetic According to Western Methods: The Life and Death of the Traditional Devices

The first Japanese treatises on *yōsan* 洋算 (western computation) appeared during the 1850s, when methods of computation on paper, imported from Europe and the United States, were taught in technical and military schools as part of the Western scientific curriculum. The *Yōsan yōhō* 洋算用法 (*Rules for the Use of Western Computation*), written by Yanagawa Shunsan 柳河春三 (1832–1870) in 1857, was prepared in order to allow Japanese students of the *Bansho shirabesho* 蕃書調所 (Institute for the Study of Barbarian Books) and the *Nagasaki kaigun denshūjo* 長崎海軍伝習所 (Nagasaki center for naval training)[68] to learn mathematical computation so that they could assimilate Western technical sources.

The content of the *Yōsan yōhō* is very original in the context of the beginning of the Meiji era, as it is one of the first textbooks written in Japanese to give a proper presentation of Western mathematical elementary methods of computation (in particular, with notations employing Arabic numerals and Western symbolism).[69]

---

[66]Until the middle of the 19th century, the *wasan* schools were still very popular in Japan and their curricula were devoted to the tradition developed during the Edo period and based on Chinese works. Trigonometric and logarithmic tables were integrated in *wasan* curricula, but the mathematics imported by the Jesuits had less success than in China. For example, the axiomatic and deductive reasoning that characterized Euclidean geometry was never studied by mathematicians during the Edo period. Nevertheless, after the first contact with Jesuit science, Japanese scholars were convinced that mathematics had a decisive utilitarian and practical role in the development of Western science. *See* Horiuchi (1996, pp. 256–258), and Cousin (2013, 2017).

[67]*See* Numata (1992, pp. 147–160).

[68]These institutions were established to fulfil the military and diplomatic needs of the government facing the international configuration of the middle of the 19th century.

[69]To compare, in the same year, Fukuda Riken 福田理軒 (1815–1889) wrote a book on "Western computation," *Seizan sokuchi* 西算速知 (*Rapid Learning of the Western Computation*), which is in fact a treatise on the "brush computation" based on the Chinese works of Mei Wending 梅文鼎 (1633–1721), where, for example, Arabic numbers are not used. *See* Horiuchi (1996, p. 260).

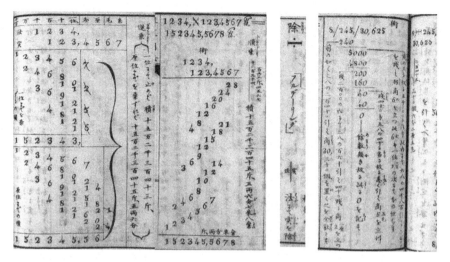

Fig. 10 Extracts from Yanagawa (1857)

It was also one of the only books of the Meiji period to present mathematical content using a comparative approach putting Japanese and Western mathematics side by side. Yanagawa's aim, as he noted in his preface, was to show that "Western arithmetic (*sanjutsu* 算術) is not that different from our rule of the *tenzan*."[70]

In the case of the elementary operations, although it is a book whose content is completely devoted to Western mathematics, the tradition of the *wasan* is clearly present in the text, as one can see in the first extract (on the left-hand side of Fig. 10:[71] the multiplication of 1234 by 123.4567 *kin, kin* being a unit of weight). First, the operation is translated using *tenzan* notation: there is a vertical line with 1234 and 123.4567 on its right written in Japanese, one next to the other: *sen ni-hyaku san-jū yon* 千二百三十四 and *hyaku ni-jū san kin, shi-go-roku-shichi* 百二十三斤、四五六七 (this representation is given at the beginning of the text, that is to say on the right-hand side of the extract). Next, the written computation is presented. One can also note that, to give detailed explanations about the results obtained when neglecting the decimal digits or when considering only the first two decimals of 123.4567, Yanagawa uses a configuration that reminds us of the illustrations of the positional representations on the abacus or on the calculation surface for counting rods. The pupil had to execute the same process but instead of positioning the numbers on the abacus, it had to be done on paper.

At the beginning of the textbook, division is defined as shown on the second extract of Fig. 10:[72] the Japanese *kanji* that signifies "division," *jo* 除, is followed by the sign "÷," its pronunciation in Dutch "verdeling," transcribed into

[70]Yanagawa (1857, p. 7a).

[71]This extract is taken from Yanagawa (1857, p. 37b).

[72]Yanagawa (1857, p. 13a).

katakana.[73] There then follows its transcription using *tenzan* notation: a representation of a vertical line, with the kanji *jitsu* 実 (operand, dividend) on its right, and the *kanji hō* 法 (operator, divisor) on its left. Finally, the sentence "*hō* divides *jitsu*" is given as an explanation for the process. In other words, this "definition" is a complete translation of the Dutch terms and symbols into Japanese, employing the traditional *wasan* notation.

The process of division on paper is explained later in the book, employing several examples. One can see in the presentation of the division of 245 by 8 that the division, written with Arabic numbers and Western symbolism is given with its result, and then translated into Japanese, using *tenzan* notation. The computation leading to the result is then described in detail as shown in Fig. 10 (image on the right-hand side), and, although Arabic numerals are used in the execution of the computation, the traditional Japanese characters are mobilized for the explanation. This approach to division reflects the whole tone of the book: Japanese traditional terminology and notation are used to introduce readers who already have a basic formation in *wasan* (they had to know the rod notational system) to the computation techniques found in Dutch sources.

To finish, although there is no algebraic or high-level computation (the textbook only goes up to proportions), Yanagawa offers a comparison between *wasan* and the Western tradition concerning the conventions of naming mathematical objects, so as to introduce the symbolism that would allow students to read books about these computations as well as other Western technical or scientific sources:

> And there is also this way of proceeding where one uses the letter *a, b, c, d,* etc. as temporal signs for given numbers. It is exactly like when our mathematicians use the ideograms *kō* 甲, *otsu* 乙, *hei* 丙, *tei* 丁, etc. or the ideograms *ne* 子, *ushi* 丑, *tora* 寅, *u* 卯, etc. Except for the fact that the letters *a, b, c* and *d* are often used for the numbers 'that we now have,' whereas *x, y* and *z* that are at the end of the alphabet designate numbers that are requested.[74]

In mathematics, although some books were published about computation on paper, there was still no movement for the translation of Western mathematical treatises at the end of the Edo period, as had been the case for medicine or astronomy. This all changed in the 1870s, following the promulgation of the *Gakusei* in 1872 and the end of *wasan* teaching throughout the school system. At this point, several authors tried to offer an appropriate textbook for teaching this new arithmetic in elementary schools. This situation led to an active but poorly structured rush to translate Western (and especially American)[75] textbooks. The

---

[73]See note 61.

[74]Yanagawa (1857, p. 8b). Translated to French in Horiuchi (1996, p. 260).

[75]Theoretically, the Japanese scholars could have used any model, from Europe or the United States, but the political and cultural context had a great impact on the choices made by Japanese scholars. Today, the image of the "rational shopper" (Westney 1987) who would thoughtfully and rationally choose the best model for Japanese modernization is discredited by several works. Historians now emphasize the complex processes that were involved in these importations of Western models. *See*, for example, Westney (1987) and Duke (2009), or in terms of the history of mathematics, Ogura (1974), Horiuchi (1996), and Cousin (2013).

Japanese authors also naturally turned to Chinese works on Western mathematics written during the 19th century (China being the oldest interlocutor with Japan), which were the results of collaboration between Protestant missionaries and Chinese mathematicians.[76]

One of the first books to be recommended by the government for teaching mathematics in elementary schools was *Hissan kunmō* 筆算訓蒙 (*Introduction to Brush Calculation* 1869), written by Tsukamoto Neikai 塚本寧海 (1833–1885). In this treatise, only the Western "brush-calculus" was described. "Western numbers" (*yōji* 洋字) are introduced at the beginning of the book and 69 pages are devoted to elementary operations (addition, subtraction, multiplication, and division are described and illustrated with examples), with no reference to the *wasan* tradition. The general aim of the government is clearly visible in this source: the total elimination of *wasan* teaching.

In the elementary schools provided by the *Gakusei*, "arithmetic with Western methods" and its basic computations were now taught starting from the lowest grades. The "Teaching directives for elementary schools" (*Shōgaku kyōsoku* 小学 教則, 1872) accompanying the *Gakusei*, for lower elementary schools, indicated that:

> With *Hissan kunmō* 筆算訓蒙 (*Elementary Learning of Brush Calculation*) and *Yōsan hayamanabi* 洋算早学 (*Elementary Study of Western Mathematics*), Western numeric characters are taught one by one, by writing them on the blackboard and making pupils copy them on paper, and so are [all the other lessons] from the numbers to operations of addition and multiplication and right up to the recitation of multiplication tables. For addition and subtraction, first the methods are taught, and then exercises are done on the blackboard. As for paper computations (written computations) and mental computations, they [the pupils] do exercises on them every two days.[77]

There were 8 grades (4 years) in lower elementary school and the learning of numbers along with the elementary operations and their applications were supposed to be taught between the first year and the middle of the third year. Subsequently, operations with fractions and proportions were to be studied until the end of the 4th year. From the beginning of higher elementary school, geometry was also taught, but algebra teaching started only after entry into middle school. One of the most notable features of this teaching was that all the students in the class had to learn the same subject at any given time, leaving less freedom for teachers than in the case of traditional individual teaching.[78] During the Edo period, mathematics teachers had abacus, counting rods, and books (Japanese textbooks and imported Chinese treatises) as tools to offer some personal training to their students, but, during the Meiji era, the same teachers had to use blackboards, the *kakezu* 掛図 (educational wall charts inspired by American originals), and textbooks recommended by the Ministry.

---

[76]*See*, for example, Martzloff (2006, pp. 111–122).

[77]Extracted from the "Teaching directives for elementary schools" of 1872 as reproduced in Ogura (1974, pp. 231–232).

[78]Ueno (2012, pp. 477–478).

Shortly after the promulgation of the *Gakusei*, the *Shōgaku sanjutsusho* 小学算
術書 (*Arithmetic Book for Elementary Schools*, 1873) was published by the
Ministry of Education (*Monbusho* 文部省) and widely distributed to teach numbers
and elementary operations in Meiji schools. It was compiled under the supervision
of the American education advisor, Marion McCarrel Scott (1843–1922), whose
pedagogy was clearly influenced by Johann H. Pestalozzi (1746–1827): the
mathematical textbook he compiled was based on the *First Lessons in Arithmetic on
the Plan of Pestalozzi* written by Warren Colburn (1793–1833). According to
Scott's directives, children should not sit passively and memorize facts and figures,
but they should discover new concepts through everyday life situations. Moreover,
the children should also play and enjoy themselves while learning, as excessive
learning harms children's sensitivity. They should also discover new concepts
gradually, spending a long time on basic techniques.[79]

As a result, the *Shōgaku sanjutsusho* gave a presentation of Western elementary
computation, using many examples extracted from Japanese everyday life (Fig. 11).
It is composed of 5 volumes, each one of which describes one of the elementary
operations: addition, subtraction, multiplication, division, and operations on fractions. Following Colburn's point of view, the textbook gives a table for each
operation with a one-digit number along with a collection of practical questions
related to the table. In Colburn's mind, children should have an inductive approach
to elementary arithmetic: they should not discover numbers through theory learning, but by doing practical exercises and by solving problems. The techniques for
computation on paper were not introduced and pupils performed examples "in the
mind, or by means of sensible objects, such as beans, nuts, etc."[80]

The most innovative tools available in classes were the *kakezu*, illustrated large
(56 cm × 74 cm) wall charts printed for the first time by Tōkyō Normal School in
1873 to support the achievements of the new "simultaneous teaching method"
(associated to "lecture-type" classes). The first set of wall charts were translated and
adapted from the *School and Family Charts; Accompanied by a Manual of Object
Lessons and Elementary Instruction* (1870 or 1871) compiled by Marcius Willson
(1813–1905) and Norman A. Calkins (1822–1895) and imported into Japan in
1872. Out of the 28 charts that were printed and disseminated into Japanese
schools, 7 concerned mathematics and 2 were designed to teach elementary operations: addition and multiplication tables (*Kasan kukuzu* 加算九九図 and *Jōsan
kukuzu* 乗算九九図).[81]

Concerning computing devices, we have seen that, according to the *Gakusei*,
only Western computation on paper was to be taught in elementary schools. But in
the first decade of the Meiji era, the teachers of mathematics in elementary schools

---

[79]See Duke (2009, pp. 125–126), Matsubara (1982, pp. 187–238), and Cousin (2013, p. 101).
[80]Colburn (1825, p. v).
[81]The other mathematical wall charts were the following: the "Lines and angles chart" (*Sen oyobi
do zu* 線及度図), the "Figures and solids chart" (*Katachi oyobi tai zu* 形及体図), the "Numerals
chart" (*Sūjizu* 数字図), the "Arabic numerals chart" (*San-yō sūjizu* 算用数字図), and the "Roman
numerals chart" (*Rōma sūjizu* 羅馬数字図). *See* Galan (1999, pp. 230–233).

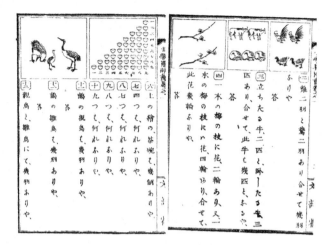

Fig. 11 Extracts from the *Shōgaku sanjutsusho* (1873)

were reconverted *terakoya* teachers, or *wasanka*, with no training in this kind of computation, which meant that the prohibition ran up against the practical need to teach useful techniques that teachers could master themselves. As it was a device that had been used for several centuries, people kept on using the abacus and there was an important debate among mathematicians about using the ancient device in schools. In elementary schools, abacus computation was often reinstated but, as Fujisawa underlined, "it lost its character of being a part of the antiquated mathematics, and metamorphosed itself into an essential constituent of new arithmetic."[82] Teachers used traditional techniques to teach elementary everyday computation on the abacus, and the *Jinkōki* was re-edited in modern versions so as to provide an appropriate support to this teaching. For example, there were editions of the *Jinkōki* that introduced abacus computation as in the original version of the traditional Japanese book with a parallel introduction of Western computation on paper (Fig. 12),[83] or re-editions that concentrated on the methods taught in elementary schools, with problems of arithmetic (and a little geometry) that were to be resolved using the abacus, eliminating all the problems of the original *Jinkōki* that had nothing to do with the official curriculum.[84]

---

[82]Fujisawa (1912, p. 25).

[83]This extract is a presentation of the division of 123456789 by 5 in *Seiyō Jinkōki* 西洋塵劫記, written by Matsui Koretoshi 松井惟利 in 1872, with the traditional computation at the top and the imported computation method at the bottom.

[84]For example, *Shōgaku Jinkōki* 小学塵劫記, written by Fujitsuka Tadaichi 藤塚唯一 and Itō Yūrin 伊藤有隣 in 1880, introduced the elementary operations methods in the same way as in the original *Jinkōki*, and then suggested problems that had to be solved with the computation device. Additionally, an important part of the original *Jinkōki* was suppressed to present only the methods that had to be taught in elementary schools.

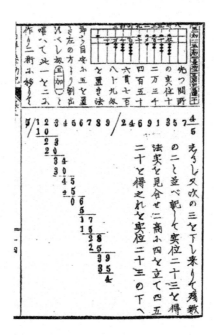

Fig. 12 Extracts from Matsui (1872)

Concerning counting rods, following the end of *wasan* teaching, the use of this ancient computation device (and the notations associated with it) completely disappeared from education. For example, in one of the first Japanese textbooks on algebra, *Tenzan mondai shū* 点竄問題集 (*Collection of Questions About the Tenzan,* 1872),[85] all the equations are of Western type and there is no mention of the *tenzan* tradition, even though it is mentioned in the title. Contrary to the case of the abacus about which there were many debates, the abandonment of counting rods was immediate and very few documents can be found where both traditional and Western methods are presented.

Even though the practices in relation to *tenzan* were completely abandoned in education, the parallel between Western algebra and Japanese *tenzan* was easy to make, as Yanagawa showed in his introduction (see the excerpt quoted earlier) and as shown in Fig. 13, where, in a manuscript written by Satō Noriyoshi 佐藤則義 (1794–1844), a geometrical problem typical of the *wasan* is first solved using a *tenzan* technique, and then with a Western approach. As one can see in the picture, the symbolism used in Western algebra was compared to rod-type notation. In fact, this parallel led to a fierce debate between the *wasanka* and the mathematicians who defended a rapid and radical importation of Western mathematics (trying to put an

---

[85]This source is a compilation of problems extracted from the works of Horatio N. Robinson (1806–1867) and Isaac Todhunter (1820–1884).

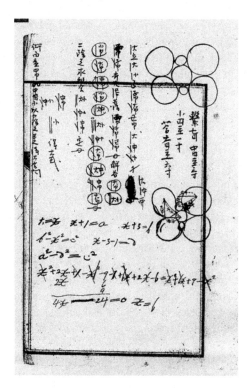

Fig. 13 Manuscript written by Satō Noriyoshi

end to *wasan* completely).[86] But, contrary to what had been done in China,[87] Japanese mathematicians did not try to transform Western symbolism in order to give it a "Japanese look." Although there were debates about the qualities of both traditions, Japanese scientists were convinced that assimilating Western mathematics was the only way to access Western sciences and techniques, and for the same reason, although *tenzan* symbolism inspired from rod configurations was effective, Western symbolism was considered essential for gaining access to other scientific works.[88]

During the 1870s, the first textbooks on mathematics were mainly translations of their American counterparts,[89] but, at the beginning of the 1880s, compilations of

[86]See Horiuchi (1996, pp. 260–262).

[87]On the transformation of Western symbolism in Chinese writings, *see* Martzloff (2006, pp. 119–122).

[88]*See* Horiuchi (2004) and the studies of Cousin (2013) about the integration of the symbolism in geometrical textbooks.

[89]The books translated are mainly works of Horatio N. Robinson and Charles Davies (1798–1876). However, there are also some translations of the textbooks by the English author Isaac Todhunter. *See* Ogura (1974, pp. 238–243) and Neoi (1997).

foreign books based on a deeper reflection on mathematical teaching were produced and, starting at the end of the 1880s, original works began to appear.[90] But, in algebra, until the end of the 1890s, the American sources were still the most common references. In particular, the works of Charles Smith (1844–1916) were widely used in middle schools.[91]

The Contributions of Fujisawa

With his textbooks of algebra and arithmetic, the Japanese mathematician Fujisawa Rikitarō played an important role in stabilizing "Westernized teaching" imposed by the Meiji authorities. Having graduated from the recently established Imperial University of Tōkyō in 1882, he was sent to Europe to study mathematics (London University, Berlin University, and Strasburg University),[92] and later played an important part in the modernization of Japan.[93] Upon his return to Japan, in 1887, he contributed to the establishment of a modern curriculum of mathematics at the Imperial University. Then, he prepared textbooks of arithmetic and algebra that would, along with the geometry textbooks of Kikuchi Dairoku 菊池大麓 (1855–1917), establish the new modern teaching of mathematics in elementary and middle schools.

After having provided a *syllabus* for the teaching of arithmetic in elementary schools,[94] in which he recommended classes oriented toward the daily use of mathematics, Fujisawa wrote *Sanjutsu kyōkasho* 算術教科書 (*Textbook of Arithmetic*) in 1896. Following Fujisawa's ideas on education, mathematics taught in elementary schools was completely different from other mathematics (taught starting in middle school), as they were culture dependent. As a consequence, it was

---

[90]At the beginning of the 1880s, Tanaka Naonori 田中矢徳 (1853–?), who knew Western mathematics better than the authors of the 1870s, designed a set of textbooks for teaching in middle schools. In geometry, Tanaka kept his distance from the American (and Todhunter's) works, as he wrote a compilation based also on the works of James M. Wilson (1836–1931) and William Chauvenet (1820–1870). In arithmetic, he used the textbooks of William Chambers (1800–1883) and Robert Chambers (1802–1871) as well. However, in algebra, his treatise was based only on Robinson's and Todhunter's books. Moreover, at the end of the 1880s, in arithmetic and in geometry, authors that studied abroad produced their own original works. *See* Cousin (2013, pp. 114–118, 123–124).

[91]*See* Neoi (1997, pp. 36–40).

[92]Fujisawa spent a little time at London University in 1883, then moved to Berlin, where he followed lectures given by Leopold Kronecker (1823–1891) and Karl Weierstrass (1815–1897). In 1884, he entered a doctoral program in Strasburg University, and got his degree in 1887, with a dissertation on Fourier analysis.

[93]On Fujisawa's education and career, *see*, for example, Sasaki (1994, p. 183) or Sasaki (1999, p. 30).

[94]In 1895, Fujisawa wrote the *Sanjutsu jōmoku oyobi kyōjuhō* 算術条目及教授法, where he suggested a curriculum for the teaching of arithmetic in elementary schools.

decided that theory should not be studied during these elementary courses, and instead the Japanese tradition of problem solving should be used to teach mathematics.[95]

Concerning computing devices, starting with the implementation of the *Gakusei*, the use of the abacus became more and more important. On the mathematical wall charts published in 1879,[96] elements of computation were directly linked to the abacus.[97] In 1905, the government ordered the preparation of national textbooks that followed Fujisawa's views. For each year of elementary school, two textbooks were published (one for the pupils, one for the teachers). They presented Western methods of computing in a "Japanese way," that is to say using several problem-solving exercises based on the daily use of mathematics, and they continued to be used until 1934 (with four revisions). In 1907, the Ministry of Education also published a teacher's textbook for the course about abacus computation *Shōgaku sanjutsusho - Shuzan - Kyōshiyō* 小学算術書•珠算•教師用 (*Book of Arithmetic for Elementary Teaching—[Computations with] the Abacus—For the Use of Teachers*, 1907), so that there was an official pedagogic support for the courses on the traditional device.[98]

In algebra, Fujisawa's textbook, *Shotō daisūgaku kyōkasho* 初等代数学教科書 (*Textbook of Elementary Algebra*, 1900), presented the rudiments of algebra for middle school teaching, using only Western methods. In his opinion, arithmetic should be taught following Japanese tradition but "Algebra cannot possibly differ much in different countries except in regard to the scope allotted to the particular kind of schools."[99] As a consequence, *wasan* methods in algebra were not mentioned in the textbook and, as one can see with the resolution of a linear system of equations below, there is no trace of the ancient device:

For the solution, let x, y, z be three unknown numbers that verify the alliance of three linear equations, the 1st and the 2nd equations are multiplied by the appropriate numbers, thus, in the two [new] equations obtained, the coefficients of a certain unknown number, for example z, are equal to each other. Then, the former one is subtracted from the latter one and z is eliminated. Furthermore, the 1st and the 3rd equations take the place of the 2nd and the 3rd equations and z is eliminated in the same way as before. We should obtain two unknown numbers that verify two linear equations, we resolve this and we determine the values of x and y. We enter these values of x and y in one of the equations and we should determine the value of z.

---

[95]Nevertheless, Fujisawa criticized the tradition of posing too many problems and underlined the fact that the elimination of theory teaching did not exclude an analysis of the meaning of problems. *See* Ueno (2012, p. 480).

[96]These new wall charts, called *Shōgaku shikyōzu* 小学指教図 (Pictures for the teaching in elementary schools), measured 72 cm × 50 cm.

[97]See Galan (1999, pp. 237–238).

[98]See Matsubara (1983, pp. 616–624).

[99]Fujisawa (1912, p. 140).

Example: Resolve the following system of equations:

$$7x + 3y - 2z = 16 \tag{1}$$

$$2x + 5y + 3z = 39 \tag{2}$$

$$5x - y + 5z = 31 \tag{3}$$

(1) is multiplied by 3: $21x + 9y - 6z = 48$

(2) is multiplied by 2: $4x + 10y + 6z = 78$

We add (1) and (2) to obtain

$$25x + 19y = 126 \tag{4}$$

(1) is multiplied by 5: $35x + 15y - 10z = 80$

(3) is multiplied by 2: $10x - 2y + 10z = 62$

We add (1) and (3) to obtain

$$45x + 13y = 142 \tag{5}$$

Then we should determine the value of x and y with (4) and (5):

(4) is multiplied by 9: $225x + 171y = 1134$

(5) is multiplied by 5: $225x + 65y = 710$

We subtract (5) from (4) to obtain $106y = 424$

Then $y = 4$. This value of y in entered in (4) and we obtain $25x + 76 = 126$, from that we obtain $x = 2$.

Then, we put $x = 2, y = 4$ in (1) and find that $14 + 12 - 2z = 16$, and from that we obtain $z = 5$. Therefore, we obtain the following answer: $x = 2, y = 4, z = 5$"[100]

Contrary to previous authors, Fujisawa had precise ideas about the teaching of algebra and algebraic computation that he developed during his stay in Europe. For example, he rejected the ideas of Smith[101] (that had great success in Japan), but his works for the teaching of algebra were influenced by Isaac Todhunter (1820–1884), George Peacock (1791–1858), and Augustus De Morgan (1806–1871).[102] Considering his mathematical curriculum and his great knowledge of the Japanese context, Fujisawa was able to provide materials that managed to stabilize the teaching of algebraic computation in Meiji Japan.[103]

---

[100]Translation of Fujisawa (1900, pp. 187–188).

[101]Nevertheless, before he published his textbook and rejected Smith, Fujisawa participated in the translation of a textbook by this author in 1897. Moreover, his presentation of the method for solving linear systems of equations and the exercises associated to it are very similar to the ones found in Smith's textbook, *see* Smith (1886, pp. 78–80).

[102]On Fujisawa's ideas about algebra teaching, *see* Ogura (1974, pp. 269–271).

[103]His textbook was used until the end of the Meiji era. *See* Neoi (1997, pp. 35–40).

Conclusion

The Meiji era witnessed a complete reorganization of Japanese society, including an educational revolution concerning computation teaching. As part of a plan to westernize scientific education, the devices used in the *wasan* tradition were abolished by the authorities. Nevertheless, the abacus, especially criticized by modernizers for its dogmatic learning, remained in use by the population, and the authorities decided to keep its teaching in elementary schools. On the other hand, counting rods, a sophisticated tool, especially in algebraic problem solving, were completely abandoned in education with the introduction of Western mathematics. These cases show the heterogeneities in the integration of Western knowledge in Japanese curricula.

Fujisawa, the mathematician whose textbooks stabilized arithmetical and algebraic teaching, defended the use of the abacus and the traditional way of teaching mathematics for elementary schools. In his mind, arithmetic needed to be taught through problem solving, following Japanese cultural heritage, but this teaching did not have to be completely dogmatic. In algebra, he defended the complete abandonment of traditional practices and introduced a modern vision of this domain based on a deep reflection on mathematical language and the new English school of algebra.

Bibliography

Chemla, Karine, and Guo, Shuchun. 2004. *Les neuf chapitres. Le Classique mathématique de la Chine ancienne et ses commentaries*. Paris: Dunod.

Cheng Dawei. 1675. *Shinpen chokushi sanpō tōsō* 新編直指算法統宗 (New edition of the *Suanfa tongzong*), 17 volumes, Japan.

Chen, Yifu. 2013. *L'étude des différents modes de déplacement des boules du boulier et de l'invention de la méthode de multiplication Kongpan Qianchengfa et son lien avec le calcul mental*. Université Paris VII, Ph.D. thesis.

Colburn, Warren. 1825. *First lessons in arithmetic, on the plan of Pestalozzi with improvements*. Boston: Cummings, Hilliard, and Co.

Cousin, Marion. 2008. *Les premiers visages du wasan. Etudes du Jinkôki (1627) de Yoshida Mitsuyoshi et du Jugairoku (1639) d'Imamura Tomoaki*. Université Lyon I, Master thesis.

Cousin, Marion. 2013. *La " révolution" de l'enseignement de la géométrie dans le Japon de l'ère Meiji (1868–1912): Une étude de l'évolution des manuels de géométrie élémentaire*. Université Lyon I, Ph.D. thesis.

Cousin, Marion. 2017. Sur la création d'une nouvelle langue mathématique japonaise pour l'enseignement de la géométrie élémentaire durant l'ère Meiji (1868–1912). *Revue d'histoire des mathématiques, 23*(1), 5–70.

Dore, Ronald P. 1965. *Education in Tokugawa Japan*. New York: Routledge and Kegan Paul.

Duke, Benjamin. 2009. *The history of modern Japanese education. Constructing the National School System, 1872–1890*. New Bunswick–New Jersey–London: Rutgers University Press.

Elisonas, Jurgis. 1991. Christianity and the daimyo. In *The Cambridge history of Japan* (Vol. 4: *Early Modern Japan*, Chap. 7, pp. 301–372). New York: Cambridge University Press.

Fujisawa Rikitarō. 藤沢利喜太郎. 1895. *Sanjutsu jōmoku oyobi kyōjuhō* 算術条目及教授法 (Syllabus and teaching methods for arithmetics). Tokyo: Fujisawa Rikitarō.

Fujisawa Rikitarō. 藤沢利喜太郎. 1896. *Sanjutsu kyōkasho* 算術教科書 (Textbook of arithmetic). Tokyo: Dainihon tosho.

Fujisawa Rikitarō. 藤沢利喜太郎. 1900. *Shotō daisūgaku kyōkasho* 初等代数学教科書 (Textbook of elementary algebra). Tokyo: Dainihon tosho.

Fujisawa Rikitarō. 1912. *Summary report on the teaching of mathematics in Japan*. Tokyo: Sanshūsha.

Fujitsuka Tadaichi. 藤塚唯一, Itō Yūrin 伊藤有隣. 1880. *Shōgaku jinkōki* 小学塵劫記 (Jinkōki for elementary schools), Tochigishi: Shūeidō.

Fujiwara Matsusaburō. 藤原松三郎. 1983. *Meijizen nihon sūgakushi* 明治前日本数学史 (A history of Japanese mathematics prior to Meiji era) (Vol. 1). Tokyo: Nihon Gakushiin.

Galan, Christian. 1998. Le paysage scolaire à la veille de la restauration de Meiji: écoles et manuels. *Ebisu, 17,* 5–47.

Galan, Christian. 1999. Les manuels de langue au lendemain de la Restauration de Meiji. Les innovations de la période du décret sur l'éducation. *Cipango: Cahiers d'études japonaises, 8,* 215–257.

Guo Shuchun (modern Chinese translation), Joseph W. Dauben, and Xu Yibao (English translation). 2013. 九章筭ʼ, *Nine chapters on the art of mathematics. A critical edition and English translation* (3 Vols.). Shenyang: Liaoning Education Press.

Hasegawa Hiroshi. 長谷川寛. 1830. *Sanpō shinsho* 算法新書 (New book on mathematics), Tokyo: Sūgaku dōjō zōhan.

Hoe, Jock. 2007. *A study by J. Hoe of the fourteenth-century manual on polynomial equations* The jade mirror of the four unknowns *by Zhu Shijie*. Christchurch (New Zealand): Mingming Bookroom.

Horiuchi, Annick. 1994. *Les mathématiques japonaises à l'époque d'Edo (1600–1868). Une étude des travaux de Seki Takakazu (?–1708) et de Takebe Katahiro (1664–1739)*, Paris: Vrin. English edition: Horiuchi, Annick. (2010). *Japanese mathematics in the Edo period (1600–1868). A study of the works of Seki Takakazu (?–1708) and Takebe Katahiro (1664–1739)* (trans.: S. Wimmer-Zagier). Basel: Birkhäuser.

Horiuchi, Annick. 1996. Sur la recomposition du paysage mathématique japonais au début de l'époque Meiji. In Catherine Goldstein, Jeremy Gray, & Jim Ritter (Eds.), *L'Europe mathématique: histoires, mythes, identities* (pp. 247–268). Paris: Les editions de la MSH.

Horiuchi, Annick. 2004. Langues mathématiques de Meiji: à la recherche du consensus? In Pascal Crozet & Annick Horiuchi (Eds.), *Traduire, transposer, naturaliser: la formation d'une langue scientifique hors des frontières de l'Europe au XIXe siècle* (pp. 43–70). Paris: L'Harmattan.

Horiuchi, Annick. 2014. History of mathematics education in Japan. In Alexander Karp & Gert Schubring (Eds.), *Handbook on the history of mathematics education* (pp. 166–174). New York: Springer.

Jansen, Marius B. 1989. The Meiji restoration. In *The Cambridge history of Japan* (Vol. 5: *The nineteenth century*, Chap. 5, pp. 308–366). New York: Cambridge University Press.

Libbrecht, Ulrich. 1973. *Chinese mathematics in the thirteenth century: The* Shu-shu chiu-chang *of Ch'in Chiu-shao* [Qin Jiushao]. Cambridge (Mass.)-London: The MIT Press.

Majima, Hideyuki. 真島秀行. 2013. Seki Takakazu, his life and bibliography. In Eberhard Knobloch, Komatsu Hikosaburo & Liu Dun (Eds.), *Seki, founder of modern mathematics in Japan: a commemoration on his tercentenary* (pp. 3–20). Tokyo & New York: Springer.

Martzloff, Jean-Claude. 2006. *A history of Chinese mathematics*. Berlin, Heidelberg: Springer.

Matsubara Gen'ichi. 松原元一. 1982. *Nihon sūgaku kyōikushi* 日本数学教育史 (History of mathematical teaching in Japan) (Vol. 1). Tokyo: Kazama shobō.

Matsubara Gen'ichi. 松原元一. 1983. *Nihon sūgaku kyōikushi* 日本数学教育史 (History of mathematical teaching in Japan) (Vol. 2). Tokyo: Kazama shobō.

Matsui Koretoshi. 松井惟利. 1872. *Seiyō Jinkōki* 西洋塵劫記 (Western Jinkōki). Tokyo: Aoyamu seikichi.

Monbushō 文部省 (Ministry of Education). 1907. *Shōgaku sanjutsusho. Shuzan. Kyōshiyō.* 小学算術書・珠算・教師用 (Book of arithmetic for elementary teaching—(Computation with) the abacus—For the use of teachers). Tokyo: Nihon shoseki.

Momokawa Chihei. 百川治兵衞. 1622. *Shokanbumono* 諸勘分物 (Estimations of surfaces and volumes), Japan.

Mōri Shigeyoshi. 毛利重能. 1622. *Warizansho* 割算書 (Book on division). Tokyo: Nihon shuzan renmei.

Neoi Makoto. 根生誠. 1997. Meiji ki chūtō gakkō no sūgaku kyōkasho ni tsuite 「明治期中等学校の数学教科書について」 (On the mathematical textbooks used in the middle schools of the Meiji era). *Sūgakushi kenkyū* 数学史研究, *152*, 26–48.

Numata, Jirō. 1992. *Western learning. A short history of the study of western science in early modern Japan.* Tokyo: The Japan–Netherlands Institute.

Ogura Kinnosuke. 小倉金之助. 1974. Sūgaku kyōiku no rekishi 「数学教育の歴史」 (History of mathematical education). In *Ogura Kinnosuke chosakushū.* 小倉金之助著作集 (Selected works of Ogura Kinnosuke.) (Vol. 6). Tokyo: Keisō shobō.

Passin, Herbert. 1965. *Society and education in Japan.* Tokyo-New York: Kōdansha International.

Rubinger, Richard. 1982. *Private academies of Tokugawa Japan.* Princeton: Princeton University Press.

Sasaki, Chikara. 1994. The adoption of western mathematics in Meiji Japan. In Chikara Sasaki, Mitsuo Sugiura, & Joseph W. Dauben (Eds.), *The intersection of history and mathematics* (pp. 165–186). Basel–Boston–Berlin: Birkhäuser Verlag.

Sasaki, Chikara. 1999. The development of mathematical research in twentieth-century Japan. In Chikara Sasaki (Ed.), *The introduction of western mathematics in modern Japan: collected papers* (pp. 29–44). Tokyo: University of Tokyo.

Sekiguchi, Hiraki. 関口開. 1872. *Tenzan mondai shū* 点竄問題集 (Collection of questions about the *tenzan*). Kanazawa: Garyōbō.

Shen, Kangsheng, John N. Crossley, and Anthony W.-C. Lun. 1999. *The nine chapters on the mathematical art: Companion and commentary.* Oxford–Beijing: Oxford University Press.

Shihan Gakkō (SG). 師範学校 (Normal school, ed.). 1873. *Shōgaku sanjutsusho* 小学算術書 (Arithmetic book for elementary schools). Tokyo: Monbushō.

Smith, Charles. 1886. *Elementary algebra.* London: Macmillan and Co.

Takebe Katahiro. 建部賢弘. 1685. *Hatsubi sanpō endan genkai* 発微算法演段諺解 (Commentaries in vernacular of the *endan* of the *Hatsubi sanpō*). Kyoto: Hishiya.

Tsukamoto Neikai. 塚本寧海. 1869. *Hissan kunmō* 筆算訓蒙 (Introduction to the brush calculation). Numazu: Numazu gakkō.

Ueno, Kenji. 2012. Mathematics teaching before and after the Meiji Restoration. *ZDM: The International Journal on Mathematics Education, 44*(4): 473–481.

Wasan Institute (WI). 2000. *Jinkōki.* (English translation). Tokyo: Tokyo Shoseki Printing Co.

Westney, Eleanor. 1987. *Imitation and innovation: The transfer of western organizational patterns to Meiji Japan.* Cambridge (Mass.)–London: Harvard University Press.

Yanagawa Shunsan. 柳河春三. 1857. *Yōsan yōhō* 洋算用法 (Rules for the use of western computation). Tokyo: Yamatoya

Yoshida Mitsuyoshi. 吉田光由. 1641 (first edition: 1627). *Jinkōki* 塵劫記 (Inalterable treatise). Japan, reproduced in WI (2000).

Part IV
Early Modern Europe and Russia

Counting Devices in Russia

Alexei Volkov

Abstract No definitive history of the Russian abacus *schyoty* (счёты) has ever been written; the hypothesis about its Asian origin stated by a number of Western authors was strongly opposed by I. G. Spasskiĭ (1904–1990) in his book-length article of 1952. The chapter is devoted to two topics: (1) the extant sources relevant to the history of the Russian instrument and (2) the operations with common fractions performed with it as described in Russian arithmetical manuals of the seventeenth century. The author concludes that the theory suggesting that the instrument was imported to Russia from the Golden Horde is worth to be explored, and offers an interpretation of the descriptions of operations with common fractions performed with the instrument.

Keywords Russian schyoty (счёты) · Chinese abacus (*suanpan* 算盤)
Golden Horde · *Schyotnaya mudrost'* (*Счётная Мудрость*)

A. Volkov (✉)
National Tsing-Hua University, Hsinchu, Taiwan
e-mail: alexei.volkov@gmail.com

© Springer Nature Switzerland AG 2018

299

A. Volkov and V. Freiman (eds.), *Computations and Computing Devices in Mathematics Education Before the Advent of Electronic Calculators*, Mathematics Education in the Digital Era 11, https://doi.org/10.1007/978-3-319-73396-8_12

Introduction

The counting instrument called *schyoty* (счёты), that is, literally, "counts/reckonings" (in plural),[1] was widely used in Russia in shops, in accountant offices, and in classrooms until the late twentieth century. The instrument used in the nineteenth and twentieth centuries consisted of a wooden frame whose width and length varied, approximately, from 20 to 35 cm, and from 30 to 50 cm, respectively; there existed even larger and smaller specimens (especially those used for educational purposes). The longer sides of the frame were connected with bars usually made of metal; on each bar, there were nine or ten sliding beads (except one or two bars with only four beads). The most common version of the instrument used in the twentieth century had bars with ten beads. The numbers of bars may have varied. The picture of the instrument shown in Fig. 1 is reproduced from a manual of 1925; it shows an instrument with 15 bars. Thirteen of them (namely 11 bars in the upper section and two bars in the lower sections) have ten beads each, while the first and the fourth bars from below have only four beads. It is generally believed that the fourth bar from below was used for counting of quarters of (monetary unit) ruble, the third bar for dozens of (monetary units) kopecks, the second bar for kopecks, and the first bar for quarters of kopecks.[2]

One (in the case of nine beads on one bar) or two beads (in the case of four or 10 beads on one bar) in the middle of each bar often had a color different from that of the other beads on the same bar; this helped the operator visually identify the number of beads shifted to left (i.e., activated) on each bar. The beads of the large-size instruments used in accounting offices and shops in the twentieth century were often made of wood, while beads of some earlier extant specimens were made of colored glass. When working with the instrument, the operator placed it on a flat or slightly inclined surface (e.g., a desk) facing the short side of the frame. The beads shifted to the left side of the frame were considered "active" and represented digits of the number set on the instrument, while the beads that stayed shifted to the right side were "inactive" and were not taken into account.

---

[1]This name of the instrument is relatively recent. Its depiction in the mathematical manuscripts of the seventeenth century are accompanied with the caption "дщица щетная (= счетная")" [*dshchitsa shchetnaya*], that is, "counting board"; see Spasskiĭ (1952, pp. 322–327, Figs. 15–20). In sources dated of the seventeenth century the instrument was also referred to as "счет" [*schyot*] or "дощаный счет" [*doschchanyĭ schyot*], that is, "counting" or "board counting," and this word was used in singular even when the instrument most probably contained two counting surfaces joined together; see Spasskiĭ (1952, pp. 314–320). Spasskiĭ also reports that the term "доска" [*doska*] ("board") applied to the instrument was still in use as late as the first half of the nineteenth century (1952, p. 318); by that time the term "*schyoty*" (i.e., literally "counts" or "reckonings," in plural) was already widely used; see, for example, Orlitskiĭ (1830a, b) and Tikhomirov (1830).

[2]The number of bars in the upper part of the instrument varied considerably. For example, the *schyoty* made in 1830s or 1840s shown in (Spasskiĭ 1952, p. 392, Fig. 32) had only six bars in the upper part, while the lower part was the same as that of the instrument depicted in Kiryushin (1925) and reproduced in Fig. 1, i.e., had four bars with 4, 10, 10, and 4 beads, counting from the bottom. There existed modifications of the *schyoty* with 20 bars and 10 beads on each bar (Spasskiĭ 1952, p. 396, Fig. 33), 11 bars with 10 beads on each bar (ibid., p. 400, Fig. 36), and so on.

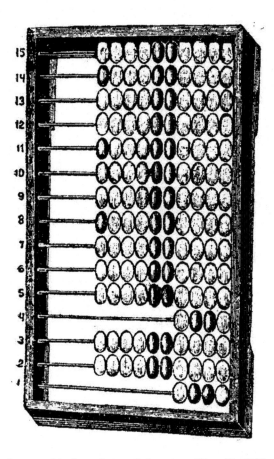

Fig. 1 Russian *schyoty* used in the early twentieth century (Kiryushin 1925, p. 12)

Numerous questions concerning the instrument and the history of its early use in Russia remain unanswered until now. One of the most intriguing problems is that of the origin of the instrument. The earliest accounts of the use of *schyoty* in Russia were produced starting from the mid-seventeenth century by travelers from Western Europe; these descriptions together with one extant specimen presumably made prior to 1618 (see Fig. 2) suggest that the instrument existed in the early seventeenth century and, possibly, even in the late sixteenth century. The construction of the instrument described by the travelers, namely a rectangular frame with beads sliding on bars parallel to the shorter sides of the frame, to some extent resembled that of Chinese *suanpan* 算盤 and Japanese *soroban*, and thus allowed a number of Western authors make conjectures about its Asian (in particular, Chinese) origin; given that the instrument was used in Russia but remained unknown in the countries to the West of it, the theory of Oriental (more specifically, Chinese) origin of the instrument apparently looked convincing enough to these authors, despite obvious differences between the constructions of the instruments and techniques of

Fig. 2 Russian abacus from Tradescant collection (reproduced from Ryan (1972))

manipulation with them.[3] The exact time of its hypothetical transmission from China to Russia remains unknown, yet some authors suggested that it may have happened in the late fourteenth century; as we shall see later, the earliest relevant sources actually did not mention the Chinese origin and were misinterpreted by these authors.[4] Interestingly enough, the instrument was not mentioned in the first Russian-printed arithmetical manuals, those of I. F. Kopievskiĭ [И.Ф. Копиевский] (or Kopievich [Копиевич], also known as Elias Kopijewitz, 1651–1714) and of L. F. Magnitskiĭ [Л.Ф. Магницкий] (1669–1739). This omission can be tentatively explained by the fact that both manuals were largely based upon Western mathematical textbooks of the sixteenth and seventeenth centuries in which the Russian abacus was never mentioned.[5]

---

[3]The structure of the Russian instrument is different from that of the Chinese one: the numbers of beads on the bars are not the same, and the Russian instrument never had two sections used for beads of values 1 and 5, as did the Chinese one (even though the early specimens of the Russian instrument did have a vertical bar separating beads into two groups having different values; see below). Moreover, the positions in which the instruments were set on a flat surface for use were different, the Russian one was placed "vertically" (that is, the operator faced the short side of the frame) while the Chinese one was always set "horizontally." The structure of the instrument and its orientation relative to the operator thus determined different procedures for performance of arithmetical operations.

[4]Spasskiĭ (1952) advanced several arguments against this hypothesis; he claimed that the instrument had no connection with the Chinese *suanpan* and was invented in Russia independently. Unfortunately, his work remained virtually unknown outside USSR/Russia; for rare exceptions, see Ryan (1972), Ryan (1991, esp. see pp. 373–374), and Burnett and Ryan (1998). On Spasskiĭ and his study of the *schyoty* see below.

[5]On Kopievskiĭ/Kopievich see the chapter in the present volume authored by A. Karp; on Magnitskiĭ's textbook and on the Western textbooks used by him see Freiman and Volkov (2012, 2015).

In this chapter, I will introduce the earliest extant sources containing mentions or descriptions of the instrument and then pass to a discussion of the hypotheses concerning its origin, in particular, to its connections with the archaic Russian abacus which, according to R. A. Simonov, existed in Russia as early as the eleventh century. The second part is devoted to the use of the instrument, especially to the operations with common fractions.

The Counting Instruments Used in Russia Prior to the Early Eighteenth Century

In his book-length paper on the history of Russian *schyoty*, Ivan G. Spasskiĭ (Иван Георгиевич Спасский, 1904–1990) explored seven Russian manuscript arithmetical manuals bearing the generic title *Schyotnaya mudrost'* [Счётная мудрость] (Counting Wisdom) and mentioning the instrument (1952, pp. 274, 279). One of the manuscripts contained a record dated of 1642, and another one, reprinted in 1879, was originally produced in 1691.[6] Spasskiĭ, as a number of earlier authors, believed that the manuscript copies of the manuals that remained undated when he was writing his work were also produced in the seventeenth century.[7] The manuals studied by Spasskiĭ mentioned two counting instruments: One of them was the *schyoty* discussed in the present chapter, and the other was the Western-style abacus using counting (metallic) round flat tokens placed on a flat surface with a drawn rectilinear grid. The latter instrument was referred to as "[instrument for] counting with seeds" (*schyot kost'mi* [счёт костьми]; hereafter "seeds counting"). Spasskiĭ suggested that there existed a third instrument, originally also called "seeds counting"; according to him, this instrument considerably differed from the Western-style abacus but was later replaced by it and entirely forgotten. No specimens of this third instrument had been found by the moment when he was writing his paper, and no extant sources contained descriptions or depictions of this instrument. Spasskiĭ's hypothesis about the existence of the third kind of counting instrument remained unnoticed for more than 30 years after the publication of his essay; for instance, one of the leading experts in the history of the Russian mathematics, A. P. Yushkevich (1906–1993) in his monograph on the history of mathematics in Russia (1968) mentioned "seeds counting" as identical with the Western-style computation on counting surface (pp. 27–28), thus tacitly rejecting Spasskiĭ's theory about the archaic Russian abacus that was originally called "seeds counting." However, some

---

[6][Anon.] (1879). On Russian mathematics prior to the eighteenth century see Bobynin (1884, 1886).

[7]Spasskiĭ (1952, p. 275, esp. see footnote 2).

time later R. A. Simonov positively estimated Spasskiĭ's hypothesis and provided a reconstruction of this counting instrument used in Russia, according to him, as early as the eleventh century.[8]

The Early History of the Russian *Shyoty*: The Extant Evidence

The Russian abacus (*schyoty*) was for the first time very briefly mentioned by Jacob (Iacobus) Reutenfels in his *De Rebus Moschoviticis* published in 1680:[9]

> Supputandi artem lapillis, seu corallis ferreo filo trāsmissis, ac in duas tabellas despensatis, Tatarorum, & Chinensium ritu tractant. (Reutenfels 1680, p. 204)

The Russian translator of the early twentieth century rendered this statement as follows:

> They [=the Russians] count with the help of stones and corals threaded on metallic wires and arranged in two rows, similarly to the Tatars[10] and Chinese.[11]

Interestingly enough, Reutenfels described the counters disposed on *two* tables/ tablets (*tabellas*), while the Russian translator of the twentieth century speaks about "two rows" ("два ряда"). It is possible that the translator did not know how to interpret the words "*tabellas*" and tried to find a solution by equating the Russian instrument with the Chinese counting device *suanpan* (which was relatively well known in Russia by that time) and interpreting the words about two *tabellas* as referring to the two sections (upper and lower) of the *suanpan*. However, he did not notice that the resulting translation thus became incorrect, since the Russian *shyoty* did not have the bar that separated the entire instrument into two parts, as it was the case of the Chinese abacus. It would be plausible to suggest that by two *tabellas* the author meant two counting fields (two tablets joint together) attested in the extant

---

[8]Simonov (1975, 1988, 1990, 1993a, b, 2001b [1987], 2001c [1997]), 2015. For more details see below.

[9]On the book and its author see the introductory note of the Russian translator (A. Stankevich) in Reutenfels (1905, pp. I–X). The dates of life of the author are unknown. Stankevich conjectures that Jacob Reutenfels stayed in Russia in 1670–1672 and suggests that the original text of Reutenfels may have been written in an unidentified language and translated into Latin by a German scholar in the late 1670s; the Latin translation was published in Padova in 1680, while the original text was lost.

[10]It is not clear who are the "Ta(r)tars" mentioned here. Apparently, the author distinguishes them from the "Chinese", yet the term still may have been used to refer to Manchus who at that time governed China. There is, however, another possibility, namely, that the author referred to the Tatars inhabiting the territory of the present-day Russia, esp. the basin of lower Volga and areas to the East of it (including Ural mountains and Siberia).

[11]"Считают они посредством камешков и кораллов, нанизанных на проволоку и расположенных в два ряда, на подобие Татар и Китайцев." (Reutenfels 1906, pp. 158–159).

Russian descriptions of the *schyoty* of the seventeenth century (see Fig. 2); the instrument of this kind disappeared well before the beginning of the twentieth century, and the translator who lived at that time most likely was not aware of it. If this interpretation is correct, the mention of similarity with the "Ta(r)tar or Chinese" abacus made by Reutenfels should be understood as a general remark about the similarity of the two instruments (viz. using sliding beads on bars) and not a statement about the identity of their construction.[12]

The next description of the instrument was provided by Nicolaas (or Nicolaes) Witsen (1641–1717)[13] who travelled in Russia prior to Reutenfels, in 1664, but published his book titled *Noord en Oost Tartarye* [*North and East Tartary*] only in 1692. It contains a very brief mention of the *schyoty* made in the context of the legend of the origin of the Stroganov family who became members of Russian nobility in the eighteenth century due to their highly successful business activities. The relevant excerpt reads as follows:[14]

> De afkomste van de ryke boeren in Ruschlant, *Stroganovi* genoemt, is aldus: haer voor-vader is uit het lantschap *Solota*, of *Solitaja Orda*, niet ver van *Astracan* gelegen, afkomstigh, en was aldaer des konings zoon; welke lust bekomende tot de Kristelyke religie, begaf zich naer Ruschlant, en liet zich aldaer op de Grieksche wyze doopen; en gaf den Zaer van Ruschlant hem zyn eigen dochter tot een vrouwe. [...] [D]ezen ouden *Stroganof* heeft in Ruschlant, zo gezeght wert, de cyfer of rekenkonst gebracht, welk zy noch gebruiken, zynde met beene koraeltjens op yzere roostertjens geregen.[15]

> The rich peasants named Stroganovs appeared in Russia as follows: their ancestor originated from the country of Gold, or Golden Horde,[16] located not far from Astrakhan, and was a son of the king of that [country]. He had desire to become Christian, and went to Russia where he was baptized according to Greek tradition. The Czar [=King] of Russia gave him his daughter for marriage. [...] This old [=the elder] Stroganov, people say, brought *schyoty*, or arithmetic, to Russia, and they [=the Russians] still use it until now. This [instrument consisted of] bone beads sliding on iron sticks.[17]

---

[12]Spasskiĭ (1952, p. 411) made the following comment on Reutenfels' description: "Reutenfels only briefly remarked that [the Russian *shyoty*] are designed 'similarly to the Chinese or Tatar [ones]'". It is possible that Spasskiĭ did not have access to the original text of the Reutenfels and used the translation of A. Stankevich (Reutenfels 1905–1906) instead. It is also interesting that in his quote Spasskiĭ (most probably, intentionally) changed the order and the connective found in the original ("Chinese *or* Ta(r)tar" instead of "Ta(r)tar *and* Chinese").

[13]On Witsen's life and work, see the monograph of Peters (2010); see also Peters (1994) and Hoving (2012). On Witsen's travels to Russia, his personal connections there (among his Russian acquaintances was, for instance, Peter the Great) and his work see the introductory article of the recent three-volume annotated Russian translation of Witsen's work (2010).

[14]I would like to express my gratitude to Professor Jan van Maanen who kindly helped me obtain access to the 1692 edition of Witsen's book preserved in the Library of Utrecht University.

[15]Witsen (1692, p. 472).

[16]Золотая Орда [Zolotaya Orda] in Russian, hence "Solitaja Orda" of Witsen.

[17]The English translation is mine; it is based on a Russian translation published in Witsen 2010. Its last two sentences read as follows: "[...] Этот старый Строганов привез в Россию, как говорят, счеты, или арифметику, которые они еще употребляют до сего дня. Это костяные бусинки, нанизанные на железные прутики." The modern translator thus interpreted the Dutch word *cyfer* (i.e., literally, digit, figure, numeral) as "schyoty".

Spasskiĭ (1952, p. 414), provides his own translation of the last part of this description:

> Вот тот-то старший Строганов, как говорят, принес в Россию счет или счетное искусство, употребляемое еще и теперь, которое заключается в том, что костяные корольки передвигаются на железных решетках.

> It was this elder Stroganov, people say, who brought to Russia counting or counting art, which is still in use now, and which consists of bone beads (*korol'ki*) moving on iron grids.[18]

It is worth noting that Spasskiĭ did not mention that the legendary "elder Stroganov," according to the legend, came from the Golden Horde (1224–1502) whose capitals, Old Saraĭ (Saraĭ Batu) and New Saraĭ (Saraĭ Berke or Saraĭ al-Jedid), were located in the Southern part of Russia on Volga River, and instead claimed that Witsen suggested a *Chinese* origin of the Russian *schyoty*.[19] Apparently Spasskiĭ equated the Golden Horde with the entire Mongolian Empire which included China till 1368. However, the event described by Witsen, if it ever took place, occurred in the late fourteenth century, as suggests a detailed analysis of this episode provided by Miller (1750, pp. 68–69);[20] by that time the unified Mongolian Empire that included a large portion of the European part of present-day Russia and the entire territory of China did not exist anymore, and its part, the Golden Horde, which occupied the area within the present-day Russia, was *de facto* independent from the Mongolian Empire since 1269. Miller suggested that the legendary Tartar ancestor of the Stroganovs, a baptized Ta(r)tar nobleman (whose Christian name, Miller claims, was Spiridon), lived during the reign of the Great Prince Dmitriĭ Donskoĭ (1350–1389, r. 1359–1389);[21] he also briefly mentioned the story of the introduction of the "Tartar *shyoty*" by this first member of Stroganov family.[22] Apparently, at the time when Miller was writing his book (1750) it still would not be too much surprising for the reader to find a mention of Ta(r)tar origin

---

[18]Spasskiĭ used the edition of 1785 (1952, p. 349) as did Ryan (1972, p. 85, n. 7); Ustryalov in his book (1842, p. 1, n. 2) quotes the legend about Spiridon from 1705 edition.

[19]The majority of researchers agree that the "Old Saraĭ" was located near the present-day Selitrennoe Gorodishche (Селитренное городище) in Astrakhan region, while the location of the "New Saraĭ" remains a matter of controversy. Some authors suggest that it was located near the present-day Russian city of Volgograd (better known in Western Europe as Stalingrad), while others argue against this theory; for more details see (Egorov 1985; Il'ina 2005; Pachkalov 2009a, b; Zaĭtsev 2006).

[20]Gerhard Friedrich Müller, 1705–1783; his name was traditionally transliterated in Russian as "Герард Фредерик Миллер" (hence "Miller").

[21]Spasskiĭ (1952, p. 416) mentions a genealogical tree of Stronganovs found in a document of the eighteenth century; according to it, Spiridon was born in 1362. In turn, the genealogical tree reproduced in (Ustryalov 1842) suggests that the first Stroganov passed away in 1395. These dates of life of Spiridon are consistent with the claim stating that he lived during the reign of Dmitriĭ Donskoĭ (1359–1389).

[22]The legend of a nobleman escaping from the Golden Horde in the mid-fourteenth century to be baptized as Eastern Orthodox Christian can be to some extent corroborated by the fact that Islam became the state religion of the Golden Horde in 1320s, during the rule of Uzbek-Khan (r. 1313–

of the well-known Russian instrument. In 1792, the story about the connection between the Stroganov family and the *schyoty* was reproduced by J. B. Scherer (1741–1824) (on his stay in Russia see Spasskiĭ (1952, p. 367, n. 2)) who, while preserving the part about the "Ta(r)tar" origin of the first Stroganov, already spoke about this Russian instrument as imported from *China*:

> C'est elle [= la famille des Stroganov] qui a introduit en Russie cette machine arithmétique des Chinois aussi simple qu'ingénieuse, et dont les plus grands mathématiciens se servent avec tant d'avantage pour abréger et vérifier toutes leurs opérations de calcul.[23]

> It was [the family of Stroganovs] that introduced to Russia this arithmetical machine of the Chinese, as simple as ingenious, and which the most famous mathematicians use with so much advantage to simplify and check all their computational operations.

Scherer's description of the *schyoty* suggests that he had an opportunity to explore an actual specimen of the instrument; it apparently contained only nine beads on each wire:

> C'est une petite table garnie de fils de fer ou de laiton parallèles, et qui correspondent à ce qu'on appelle colonnes en arithmétique. Ces fils de fer portent chacun neuf perles qui glissent d'un bout à l'autre et représentent les neuf chiffres.[24]

> It is a small tablet with parallel bars made of iron or brass, which correspond to what we call "columns" in arithmetic. These bars have nine pearls each that slide from one side to another and represent the nine digits.

The instruments of this kind having only one counting field indeed became widely spread in Russia after the Petrine monetary reforms of 1700–1704 (Spasskiĭ 1952, p. 375).

Apparently, for Scherer, as for a number of researchers later (including Spasskiĭ), the ethnonym "Ta(r)tars" equaled to "Chinese," and the instrument mentioned in the earliest extant accounts as used in the Southern outskirts of Grand Duchy of Moscow (1263–1547) along Volga River thus became an arithmetical machine imported from far-away China. This theory became widespread rather quickly and was reiterated by a number of later authors;[25] however, it did not look satisfactory to Spasskiĭ who in his work claimed that Miller's mention of the introduction of the *schyoty* by the first Stroganov is but an additional remark stemming from an oral tradition not based on any written documents, and, as he made it understood, not deserving serious attention (1952, p. 415). As for Scherer's

---

1341), and therefore in the mid-fourteenth century some Tartar Christians indeed may have had strong reasons to move to the Grand Duchy of Moscow.

[23]Scherer (1792, v. 1, p. 153).

[24]Scherer (1792, v. 1, pp. 153–154, note 1).

[25]For example, Kiryushin (1925) in the very beginning of his introductory chapter writes "It is not known when exactly did the Russian *schyoty* appear. There is a hypothesis that the idea of the design [of the instrument] was borrowed from the Chinese counting device *Suan-pan* in the 16th century by the merchants Stroganovs who lived at the border of Siberia" (p. 9). Interestingly enough, this author does not notice that the lands possessed by the Stroganovs were located near the *Western*, and not *Eastern*, border of Siberia, that is, quite far from China.

account, Spasskiĭ states that it was full of "hostile statements about Russia and Russians" (1952, p. 367, n. 2) and, one can understand, it does not deserve the reader's trust either. Spasskiĭ's claims are apparently related to his explicitly expressed conviction that the *schyoty* is a "purely Russian" instrument developed independently from Chinese *suanpan*:

> Основываясь исключительно на формальном сходстве русских счетов с китайским счетным прибором (суан-пан), западные ученые обычно принимают за истину высказанную в XVII веке догадку Н. Витзена о китайском происхождении счетов, заимствованных якобы от золотоордынских татар в XIV веке. (Spasskiĭ 1952, p. 272)

> On the sole basis of the formal similarity between the Russian *schyoty* and the Chinese counting device (*suanpan*), Western scholars usually take as true the guess of N. Witsen made in the 17th century about the Chinese origin of the Russian *schyoty* borrowed, [he] believed, from the Tatars of the Golden Horde in the 14th century.

Spasskiĭ does not explain why he interpreted the text of Witsen as suggesting that the hypothetical Tartar instrument presumably used in the fourteenth century in Golden Horde, that is, on the territory of the present-day European part of Russia, must have been "of Chinese origin"; it is possible that Spasskiĭ's vision was strongly influenced by his conviction that the conventional Chinese *suanpan* was invented in the time of "high antiquity,"[26] and therefore the instrument used in Golden Horde (which, he apparently believed, had strong ties with China) necessarily must have originated from China. Spassskiĭ did not notice that this claim contradicts his own theory of independent invention of similar counting instruments (or rather instruments based on similar principles) in different cultural traditions (pp. 289–296). He is right, however, when saying that the Western scholars (and later their Russian counterparts; see for instance, Kiryushin's work quoted above) conjectured about the Chinese origin of the Russian instrument without sufficient grounds.

The story told by Witsen may provide an additional hint concerning the origin of the Russian *schyoty*: As it becomes clear from the materials provided by Spasskiĭ, the earliest mentions of the instrument are dated of the seventeenth century, and the earliest extant specimen (the so-called Russian abacus from Tradescant collection, see below) was produced in the late sixteenth or early seventeenth century. Moreover, it was suggested that prior to the mid-seventeenth century another

---

[26]"Китайский *суан-пан*, возникновение которого относится к глубокой древности ..." (Chinese *suanpan*, whose origin is dated of the high antiquity...), see Spasskiĭ (1952, p. 291). Apparently, Spasskiĭ was not aware that the time of the creation of the Chinese *suanpan* is a matter of controversy, and the earliest evidence of its use of China is a depiction of an object visually resembling the modern *suanpan* and found in a painting produced in the 12th century. A description of a counting instrument similar to the *Roman* abacus is found in the Chinese mathematical treatise *Shu shu ji yi* 數術記遺 (Procedures of "numbering" recorded to be preserved [for posterity]), compiled by Xu Yue 徐岳 (b. ca. 185 – d. ca. 227 AD), with commentaries of Zhen Luan 甄鸞 (b. ca. 500 – d. after 573 AD) (Xu 1993); this device is discussed below. However, dating the treatise of Xu Yue is a quite complex matter; for more details, see, for example, Volkov (1994).

counting device, a counting board with counters of two sizes (for units and fives), was used in Russia (see below). If the instrument later known as "Russian *schyoty*" (or its early version) was indeed used in the Golden Horde, it would seem rather natural if it did become popular in Russia after the fall of the Golden Horde in 1502 and inclusion of a part of its territory into the Grand Duchy of Moscow, or even earlier, during the time of coexistence of the two states.

Spasskiĭ (1952, pp. 356–357, 367–368) quotes mentions of the Russian *schyoty* found in the works of other travelers to Russia (Korb 1700, p. 205; Perry 1716; 1717, p. 202) and Western diplomats (Scherer 1792). J. Korb reports that the instrument allows to obtain correct results with amazing speed ("… quorum usu, mirabili celeritate ad summum, certúmque numerum deveniunt"), while J. Perry, on the contrary, claims that the computation with the instrument is "très ennuyeuse et sujette à de grandes erreurs" (very boring and subject of large errors). The reason for this discrepancy in evaluation of the instrument is unclear; probably, Korb and Perry encountered local experts having different levels of skillfulness. Interestingly enough, Perry informs the reader that the instrument was used in all the governmental offices of the Czar; this statement, duly translated by Spasskiĭ but left without commentaries, is highly important, given that Perry describes the use of *schyoty* by Russian functionaries of highest level that was going on even *after* the first educational institution for teaching Western mathematical sciences, School of Navigation, was established in Moscow;[27] a description of this school and of difficulties encountered by foreign instructors is also found in his book (Perry 1717, pp. 202–205). Later, the instrument drew attention of A. von Humboldt who in 1829 attempted to provide a comparison of systems of numeration developed in various cultural contexts.[28]

"Russian Abacus" of John Tradescant

The arguably earliest specimen of the Russian *schyoty* is the one preserved in the Founder's Room of the Ashmolean Museum, Oxford; it remains unknown whether the founder of the collection of curiosities, John Tradescant the Elder (b. 1570?—d. 1638?) purchased the instrument during his travel to Russia in 1618, or it was added to the collection in the mid-seventeenth century.[29] For some time, the instrument stayed in the collection identified as "numerical table of the Chinese" (in a

---

[27]On this school see Freiman and Volkov (2015) and literature cited there.

[28]Spasskiĭ 1952, p. 368, n. 3.

[29]In his paper of 1972, William F. Ryan stated that "The evidence… points to a strong probability [that the instrument was brought from Russia in 1618 by John Tradescant the Elder]… [but one] cannot definitely exclude possible alternative sources [of the instrument]" (pp. 85–86); for instance, it is possible that the instrument was purchased in 1650s from an individual working for Muscovy Company (p. 86).

description of 1836), while in an even earlier descriptions produced in 1656 it was mentioned as an "Indian" instrument.[30]

The instrument is designed as two shallow boxes measuring 65 × 97 mm joined together at their longer sides (Fig. 2). Each box has six metallic bars with spherical metallic beads; the left box contains five bars with 10 beads on each and one bar with only four beads; the right box contains four bars with 10 beads, one bar with four beads, and one bar with two beads. Since the beads are made of metal and could not be easily broken, the small amounts of beads on three bars most likely did not result from loss of beads but were part of the original design. Indeed, numerous instruments with small numbers of beads in lower part are depicted in the Russian mathematical manuscripts of the seventeenth century studied by Spasskiĭ (1952, Figs. 8–20); on the use of such instruments see below.

Peder Von Haven's Description of Operations with the *Schyoty*

Peder von Haven (1715–1757) published his *Travel to Russia* in Danish in 1743. A German translation of his book printed in 1744 contained an appendix titled "Anhange, darin das Chinesische und ietzo in Russland gebröuliche Rechen-Brett beschrieben und erkläret wird" (pp. 513–570) not included in the Danish edition. This appendix contained a history of the Russian *schyoty* based on the author's conviction of its Chinese origin (pp. 516–521), followed by descriptions of operations that can be performed with the instrument including the method of representation of numbers (p. 534), addition (p. 536–545), subtraction (pp. 545–548), multiplication (pp. 548–552), division (552–554), raising to second, third, and higher powers (pp. 555–558), extraction of square (pp. 558–559), and cube roots (pp. 559–560); two more sections are devoted to operations with fractions (pp. 561–565). All the operations were supposed to be performed on an instrument having nine beads on each bar.

Given that von Haven travelled to Russia and thus may have observed how local experts operated with the instrument, it was usually assumed that he describes the arithmetical operations as they were actually performed with the *schyoty* in Russia in the first half of the eighteenth century. However, as Spasskiĭ promptly noticed (1952, p. 363), the representation of numbers described by von Haven *differed* from the conventional one considerably. Namely, according to the latter's description, the units were supposed to be set on the uppermost bar, the dozens on the bar below it, and so on.[31] For instance, the result of subtraction of 7594 from 9246, 1652, was shown as in Fig. 3.

This system of representation thus differed radically from the conventional one known from all the other extant sources and used until today: The units are set on

---

[30]Ryan (1972, pp. 83, 85).

[31]This system of representation of numbers is described not only in the appendix of 1744, but in the book itself (p. 119).

Fig. 3 Configuration representing 1652, according to von Haven (1744, p. 546)

the lowest bar, the dozens above them, etc. Spasskiĭ suggests that von Haven "either did not notice or intentionally modified" this system (1952, p. 363). The latter conjecture seems more plausible, especially if one assumes that von Haven indeed based his description on the observations that he presumably conducted in Russia during his stay there in 1736–1739, and it would be impossible for him not to notice that the units were placed at the bottom of the instrument. Spasskiĭ suggested that this "deviation from truth" should make the reader cautious about "other details in [von Haven's] description"—apparently, the latter's theory of the Chinese origin of the Russian instrument which Spasskiĭ strongly opposed (1952, pp. 363–364).[32]

Von Haven's description contained one more peculiarity that remained unnoticed by Spasskiĭ. The table added by the Danish author at the end of his description of the Russian abacus contained pictures of three instruments: the Roman beads abacus, Chinese *suanpan*, and Russian *schyoty*. The picture of the Chinese instrument contained names of decimal positions assigned to the bars (Fig. 4).

Von Haven most probably was not familiar with Chinese language, and the terms written on his drawing contain some irregularities; however, it is possible to identify them.[33] These terms are the units of weight *hao* 毫, *li* 厘[釐] = 10 *hao*, *fen* 分 = 10 *li*, *qian* 錢 = 10 *fen*, and *liang* 兩 = 10 *qian*.[34] The next measure in this system is *jin* 斤 equal to 16 *liang*, and von Haven's diagram does not show it; instead, the caption of the bar next to *liang* reads "*Chelearg*," that is, most likely, *shi liang* 十兩, "dozens of *liang*."

Von Haven presents *suanpan* as disposed vertically, while in China the instrument was always placed horizontally when used for calculations. Moreover, even when turned counterclockwise for 90°, the picture of von Haven would still depict *suanpan* incorrectly: It would show the consecutive powers of 10 represented by the bars from the left side of the instrument to its right side, while in Chinese instrument the increasing powers of 10 were always ordered from right to left; that

---

[32]Interesitngly enough, Snelling (1769, p. 16) also reports that the units occupy the uppermost position, the dozens are set below them, etc. Moreover, he believed that only the beads pushed to the right side are counted. He suggests that this disposition can be modified: the units can be set below, the dozens above them, etc. (yet still only rightmost beads are counted); the picture at the end of his paper shows the latter disposition. Each bar has nine beads.

[33]See, for example, the Chinese mathematical treatise *Pan zhu suan fa* 盤珠算法 (Computational methods for abacus [lit. "pearls in tray"]) of 1573 (Xu 1993 [1573], *juan* 1, p. 20b).

[34]These terms are transcribed by von Haven as *Hao*, *Li*, *Füen*, *Tsïen*, and *Learg*, respectively.

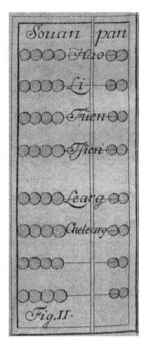

Fig. 4 Chinese *suanpan*, according to von Haven (1744, Fig. 2)

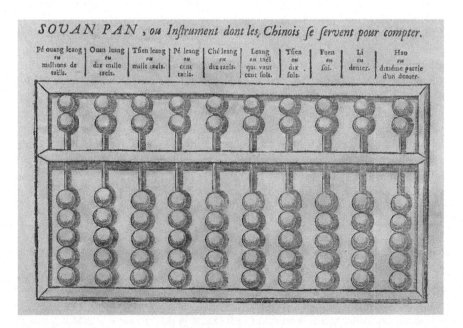

Fig. 5 Chinese *suanpan* of the eighteenth century (Du Halde 1735, Vol. 3, p. 267)

is, the rightmost bar was used for representation of the smallest power of 10. This feature of Chinese abacus was well known in Europe by the time of publication of von Haven's work; see, for example, the picture shown in Fig. 5 found in the book Du Halde (1735) mentioned by von Haven. It is likely that von Haven used the work of Du Halde (1735) for his description of the Chinese instrument, as suggest almost completely identical transcriptions of the names of monetary units (*Hao, Li, Fuen, Tsien, Leang,* and *Ché leang*) accompanying the picture of the instrument in Du Halde's work.

The fact that von Haven provided a depiction of the Chinese abacus which was deliberately modified, most likely, in order to fit into his vision of the "correct" disposition of the powers of 10 may suggests that his description of the Russian instrument was also modified for the same purpose and in the same way. Moreover, given that von Haven described not only four arithmetical operations, but also raising to powers, extraction of square and cube roots, operations with fractions and problems on proportions, and taking into account the oddities of his description, it would be reasonable to suggest that at least in some cases he did not record the operations with the *schyoty* as they were actually performed in Russia in 1730s, but presented operations designed by himself or borrowed from a work of an unidentified author. The latter hypothesis can be indirectly supported by a mistake found in von Haven's description of division. On p. 549, he discusses the procedure of multiplication and takes as example the multiplication of 85213 and 76253; he calculates correctly the product, 6497746889. On page 552, he discusses the procedure of division, and to provide an example he describes the division of the same product, 6497746889, by one of the two multiplicands, 76253. He suggests to find first digit of the result, 8, and to subtract $8 \times 76253 \times 10000 = 6100240000$ from the dividend. If one follows this instruction, the difference should be equal to $6497746889 - 6100240000 = 397506889$, yet the book of van Haven contains 39*8*506889 instead of 397506889. Apparently, if one continues division with this wrong value, the correct answer can never be obtained because of the mistake made at the first step. Interestingly, von Haven does not provide the values supposed to be obtained at the consecutive steps of division, and unless we deal with a misprint here, this mistake suggests that van Haven did not actually perform this division by himself but copied his example from some unidentified work.

In his book, von Haven mentions two types of *schyoty*: a "common" instrument with 10 to 12 bars with nine beads on each bar, and an "improved" one, with 24 bars and two counting fields (Fig. 6). The bars formed four groups with six bars in each, and the instrument contained two sections, similar to the one preserved in Ashmolean Museum.

Von Haven claims that the beads on the six bars in each of the four sections had six different colors: white, red, blue, yellow, green, and black, if listed from top to bottom. He also stated that there existed an even larger instrument, with six vertical sections, each with six groups of six bars. No extant specimens of these large-size instruments have ever been found in Russia or elsewhere.

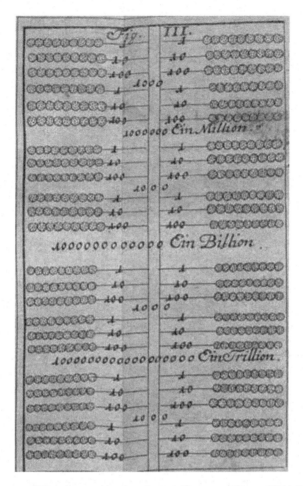

Fig. 6 An "improved" Russian counting board, according to von Haven (1744, Fig. 3)

The Origin of the Russian *Schyoty*: Spasskiĭ's and Simonov's Hypothesis

In his seminal work on the history of the Russian *schyoty*, Spasskiĭ advanced a hypothesis about the counting instrument mentioned in the mathematical manuals of the seventeenth century as "seeds counting." He conjectured that this term referred to a particular counting device mentioned by two travelers from Western Europe: Heinrich von Staden (1542—d. after 1579), who stayed in Russia in 1564–1576, and Adam Olearius (1599–1671) who passed by Russia on his way to Persia in 1630s. Von Staden mentioned that the Russians performed computations with the help of cherry and plum seeds,[35] while Olearius mentioned only plum seeds but also claimed that all the "secretaries and copyists [of the six Governmental Offices] ... do Arithmetic well in their own way, by using the plum seeds which they carry in little bags."[36] The details of the manipulations with the seeds remained unknown; Spasskiĭ conjectured that these descriptions mentioned an original counting device used in Russia for relatively long period of time prior to the seventeenth-century translation of Western methods of manipulation with counting tokens (*jetons*) found in the extant editions of the *Counting wisdom* and misleadingly also referred to as "seeds counting." In other words, the "seeds counting" described in the Russian manuscripts bearing the generic name *Counting wisdom* as computations with the Western-style counting board and metallic tokens was *not* the original Russian computational device whose name the anonymous Russian compilers of the seventeenth century borrowed to describe the Western device. What was the original Russian way of counting remained unclear; moreover, it remained unknown when it started to be used, when it disappeared, and whether there were any connections between this device and the *schyoty*.

The hypothesis of Spasskiĭ did not have any supporting evidence except the two aforementioned short and relatively obscure mentions of von Staden and Olearius. This probably was the reason why it remained only an unconfirmed hypothesis for quite long time after publication of his paper (1952); this hypothesis was not even mentioned in A. P. Yushkevich's monograph on the history of mathematics in Russia (1968). However, R. Simonov in his book of 1977 supported Spasskiĭ's claim (pp. 46–51) and analyzed a series of arithmetical problems found in the additional chapters of the *Russkaya Pravda* (Russian Justice), a compendium of legal regulations and related materials originally compiled in the eleventh century (1977, pp. 51–58). Even though the chapters with the arithmetical problems are found in its editions of the fifteenth century while earlier editions are no longer extant, the monetary system mentioned in the problems was not used after the early twelfth century, which suggests that these problems can be dated of the eleventh century (Simonov 2001b [1987], p. 30). Simonov conjectured that the numerical data of these problems indicate that they may have been represented with a counting device designed as a flat surface with a series of drawn columns and counters of two types (Simonov 1977, pp. 49–60). To support his claim, he provided a picture showing configurations of dots separated by vertical lines found in the fifteenth century manuscript copy of the *Radzivill Chronicle* compiled in the early thirteenth

---

[35]"В приказах были [...] сливяные и вишневые косточки, при помощи которых производился счет" (In the offices there were seeds of plum and cherries used for computations) (Shtaden [Staden] 1925, p. 83); "Счет ведут при помощи сливяных косточек" ([They] conduct counting with the help of plum seeds) (Shtaden [Staden] 1925, p. 123).

[36]Spasskiĭ (1952, p. 300); for details of the edition of 1647 that he used see footnote 2 on that page; see also Olearius (1656, p. 132); for a Russian translation see Oleariĭ (1906 [2003], p. 250).

Fig. 7 Page 228r of the *Radzivill Chronicle* (fragment). Note the configurations of dots in the lower right corner identified by Simonov as depictions of counters

century (Simonov 1977, p. 62), see Fig. 7; according to Simonov, the picture shows counters on a counting instrument.[37]

In a later publication (2001c [1997], pp. 17–18), Simonov reproduced a drawing of the thirteenth century which, in his opinion, represented positions of counters on a counting surface. His reconstruction is shown in Fig. 8a; unfortunately, Simonov did not provide the original picture. In his description, he mentions the elements shown in the four cells of the grid as "points," yet in his diagram they are represented as relatively large circles. Moreover, if the diagram found by Simonov represents the counters (seeds) on an actual abacus, operations with such an instrument would have required seeds of only one type, and not of two types (plum and cherry seeds) mentioned by the authors of the sixteenth and seventeenth centuries.

The counting device reconstructed by Simonov thus bears a certain structural similarity with the Greek abacus "with decimal columns" ("à colonnes décimales")

---

[37]It should be noted, however, that the diagram of this kind is found in only one picture of this richly illustrated chronicle; moreover, Simonov himself failed to provide an interpretation of the configurations of dots shown in the diagram (in his monograph of 1977 he suggests that these configuration "may have represented results of a divination," p. 63).

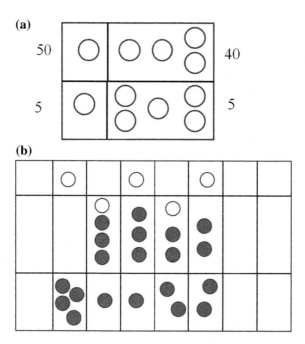

Fig. 8 a Hypothetical representation of number 100 on the Russian abacus of the thirteenth century (Simonov 2001b [1997], p. 18, Fig. 2). **b** Chinese counting device "counting with pearls" (*zhu suan* 珠算) reconstructed on the basis of description found in the *Shu shu ji yi;* the represented number is 8372

reconstructed by Schärlig (2001)[38] as well as with the Roman, Chinese and Japanese *abaci*.

If such an abacus indeed existed in Russia, Simonov argues, then it would be possible to explain a mysterious error in the aforementioned *Russkaya Pravda* where the result of division of 256 by two was (incorrectly) recorded as 124. Simonov suggests that in this case the division was performed on the abacus shown in Fig. 8a and the correct result (128) was obtained; however, when writing the result down, the scribe misinterpreted one token in position of fives and three tokens in position of units (together representing eight) as four tokens in the position of units and thus wrote it down as digit "4". This reconstruction seems plausible; it confirms that (1) the counters of only one kind and not of two kinds (viz. plum and cherry seeds) were used for computations, and (2) the decimal positions on the instrument were not physically subdivided into sections for units and fives, and counters from one section may have been moved to the other by chance.

In the same paper, Simonov mentions a report of archeological excavations of a tomb of a young man buried in the eleventh century conducted in 1985 near the

---

[38]Schärlig (2001) mentions two representations of the instrument of this type found in Greece, the first of them is found in a depiction dated of ca. 280 BC (pp. 88–89), and the second one is drawn on a piece of marble dated of the 1st century AD (pp. 90–91).

town of Суздаль [Suzdal'] located about 200 km to the Northeast of Moscow. In the tomb, a purse was found; it contained an iron weight, a silver coin, and plum and cherry seeds (2001b [1997], pp. 26–27). This archeological find, according to Simonov, strongly suggests that the abacus with counters of two types (plum and cherry seeds) mentioned (yet not described) in the seventeenth century *Schyotnaya mudrost'* existed as early as the eleventh century.[39]

If the hypothesis of Spasskiĭ and Simonov is correct, an interesting parallel can be drawn with an instrument described in the Chinese mathematical treatise *Shu shu ji yi* 數術記遺 (Records of the procedures of numbering left behind for posterity) compiled by Xu Yue 徐岳 (b. before 185–d. after 227) and commented by Zhen Luan 甄鸞 (fl. ca. AD 570). The treatise briefly describes an instrument called "counting with pearls" (*zhu suan* 珠筭); it consists of a board with three (horizontal) sections; the central section is subdivided into decimal positions, and spherical tokens (referred to as "pearls") of two different colors are placed in the upper and in the lower sections of the board. The value assigned to the tokens in the upper section is five, while the value assigned to the tokens in the lower section is one. To set a digit n from one to four, the operator has to place in the central section n tokens taken from the lower section, and to set a digit n from five to nine, the operator should place in the central section one token from the upper section and $(n - 5)$ tokens from the lower section (see Fig. 8b).[40]

The representation of decimal figures with counters of two types (spherical or round tokens of different colors) used in this instrument is surprisingly similar to the use of seeds of two kinds in the Russian archaic abacus reconstructed by Simonov, but it would be probably premature to make any conclusions concerning the possible transmission of mathematical expertise (from China to Russia?) on the sole basis of the similarity of these two instruments.[41]

Russian Abacus of the Seventeenth Century

Detailed description of the instrument and of operations with it can be found in the aforementioned paper of Spasskiĭ (1952) dealing with manuscript arithmetical anthologies bearing the generic title *Счётная мудрость* [*Schyotnaya mudrost'*] (*Counting Wisdom*) compiled in the seventeenth century.[42] The instructions concerning operations with the abacus are found in its section titled *Книга сошному и*

---

[39]On the archaic Russian abacus see also Vilenchik (1984).

[40]Guo and Liu (2001, p. 450).

[41]The problem is rather complex given that the dates of compilation of the Chinese treatise (the 2nd century) and of the commentary in which a detailed description of the instrument is provided (the 6th century) are not completely certain, since the original treatise was lost by the late first millennium AD and the version available nowadays is based on a manuscript found in the late 12th century. For more details concerning this treatise see, for example, Volkov (1994).

[42]For a bibliography of the extant Russian mathematical manuscripts see Shvetsov (1955).

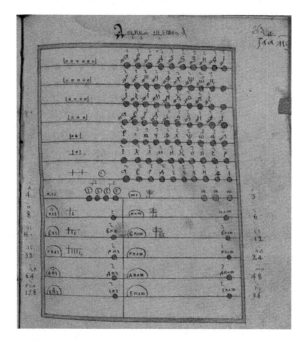

Fig. 9 A diagram of "Counting board" for surface units (Anon. 1865, p. 96r)

вытному письму [Kniga soshnomu i vytnomu pis'mu] (Book of *sokha* and *vyt'*
script) accompanied by numerous illustrations; some of them were reproduced in
Spasskiĭ (1952) and in Tsaĭger (2010).[43] Figures 9 and 10 present drawing of the
instrument found in one of the manuscript copies of the *Книга сошному и
вытному письму* [Kniga soshnomu i vytnomu pis'mu].

According to Spasskiĭ, diagrams of this type are found in a number of Russian
mathematical manuscripts of the seventeenth century; the boards for land measures
and monetary units usually are drawn next to each other which may reflect the
actual construction of the instrument containing two joined counting fields as
shown in Fig. 2 above. The wires in the lower part of the instrument (viz. wires 1–5
from the bottom) were used for operations with fractional amounts of the basic land
measure *sokha* and of the monetary unit *denga*, respectively. The beads in the left
section of each lower part represented fractions with denominators 2^n, $n = 2,\ldots,7$,
while the beads in the right section represented fractions with denominators $3 \cdot 2^n$,
$n = 0,\ldots,5$. According to Spasskiĭ's study, the design of the instrument was closely
related to the system of land units and used for calculation of land taxes. In the
1670s–1680s, the old system of land taxation started being considerably modified
and eventually was abolished; the related computations performed with the *shyoty*
thus were no longer useful. Similarly, the monetary reform of 1700–1704 made

---

[43]On "sokha script" see Shvetsov (1966).

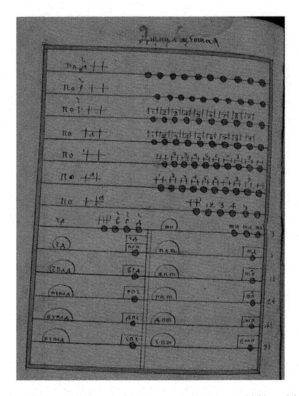

Fig. 10 A diagram of "Counting board" for monetary units (Anon. 1865, p. 96v)

obsolete the computations with fractional amounts of the basic unit *denga* performed on the *schyoty* of the seventeenth century. Spasskiĭ (1952, p. 375) suggests that these reforms were the main reason of the transition from the counting instruments with two or even four counting fields to the instrument with only one field that took place in the early eighteenth century.

Conversion of Fractions

The aforementioned manuscript manuals *Schyotnaya mudrost'* contain a table providing information concerning the conversion of fractions with denominators 2^n and $3 \cdot 2^n$ represented with the *shyoty*. This table reproduced in the manuscript edition (Anon. 1865) is shown in Fig. 11; its slightly different version is found in the edition of the manual studied by Spasskiĭ (Fig. 12).[44]

---

[44]Spasskiĭ (1952, p. 338, Fig. 21).

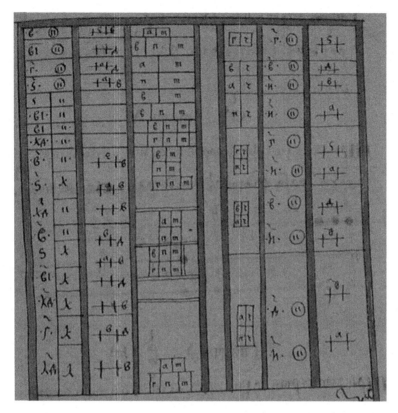

Fig. 11 Table for reduction of fractions (Anon. 1865, p. 98r)

In modern transcription the table shown in Fig. 12 looks as follows (Table 1).[45]

Monetary units mentioned in the table are *sokha, altyn,* and *denga*; 1 *sokha* = 8 *altyns,* 1 *altyn* = 6 *dengas,* so 1 *sokha* = 48 *dengas.*

Line 1 of the table provides the following information. According to Spasskiĭ (1952) and Tsaĭger (2010),[46] column 1 contains the number of *космu* [*kosti*] ("*seeds*") corresponding to the amounts of money in the second column; in modern terms, the number in this column is the denominator n of the fraction $^1/_n$ expressing the amount in the second column in terms of *sokhas*. The symbol used in columns 1 and 5 is the letter "K" in a circle; both Spasskiĭ and Tsaĭger interpret it as an abbreviation of the Russian word *кость* [*kost'*] ("seed"). For example, 4 *dengas* (in second column) equals to $^1/_{12}$ of 1 *sokha*, and this is why the first column contains the number 12. The numbers of *seeds* are set according to the convention explained

---

[45]In this table "[A]" means that the term A should be added, and "<B>" means that the term B should be removed or replaced.

[46]In his work (2010) M. Tsaĭger reconstructs and explains two upper lines of the table shown in Fig. 12 (pp. 53–56). On Tsaĭger's book, see Simonov (2010, 2011).

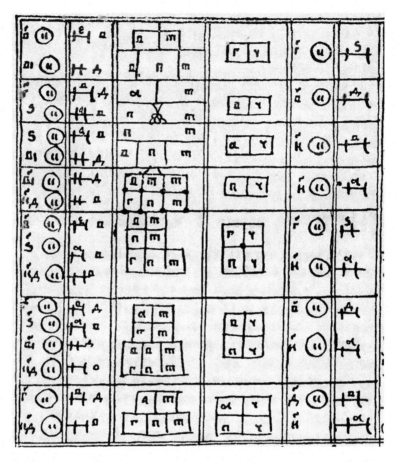

Fig. 12 Table for reduction of fractions from the Hermitage MS (Spasskiĭ, p. 338, Fig. 21)

in another table provided in the treatise before the one shown in Fig. 11 (Anon. 1865, p. 97v): Two-third of *sokha* corresponded to two *seeds*, three-fourth of *sokha* corresponded to three *seeds*, and two-fourth (i.e., one-half) of *sokha* corresponded to two *seeds*.[47] In all other cases, the numbers of *seeds* listed in the table are equal to the denominators of the fractions listed in the third and the fourth columns. Column 2 contains two amounts of money: $M_1 = 5$ *altyns* 2 *dengas* (i.e., 32 *dengas*) and $N_1 = 4$ *dengas*; these amounts are expressed with "grid symbols." Column 3 contains expressions of these two amounts represented as fractions of *sokha*

---

[47]Tsaĭger suggests that in the Table shown in Fig. 12 the scribe forgot to mention that in the first line the "price" of one *seed* in the first column equals to $^1/_3$ (Tsaĭger 2010, p. 54). However, the "price" of one *seed* mentioned in the fifth column of the same line equals to $^1/_4$, and most likely we are dealing here not with an omission of a scribe but with a particular convention used only in the cases of fractions with denominators 3 and 4 and numerators larger than 1.

Table 1 Conversion of fractions with denominators 2^n and $3 \cdot 2^n$

2 K 12 K	5 altyn 2 denga 4 denga	Two-third [of sokha] Double-half of third [of sokha]	Three-fourth [of sokha]	3 K	6 altyn
3 K 6 K	2 altyn 4 denga 1 altyn 2 denga	One-third [of sokha] One-half of third [of sokha]	Two-fourth [of sokha]	2 K	4 altyn
6 K 12 K	1 altyn 2 denga 4 denga	One-half of third [of sokha] Double-half of third [of sokha]	One-fourth [of sokha]	<8> [4] K	Double altyn
12 K 24 K	4 denga 2 denga	Double- <third> [half] of third [of sokha] Triple-half of third [of sokha]	Half of fourth [of sokha]	8 K	1 altyn
2 K 6 K 24 K	5 altyn 2 denga 1 altyn 2 denga 2 denga	Two-third [of sokha] One-half of third [of sokha] Triple-half of third [of sokha]	Three-fourth [of sokha] Half of fourth [of sokha]	3 K 8 K	6 altyn 1 altyn
3 K 6 K 12 K 24 K	2 altyn 4 denga 1 altyn 2 denga 4 denga 2 denga	One-third [of sokha] One-half of third [of sokha] Double-half of third [of sokha] Triple-half of third [of sokha]	Two-fourth [of sokha] Half of fourth [of sokha]	2 K 8 K	4 altyn 1 altyn
3 K 24 K	2 altyn 4 denga 2 denga	One-third [of sokha] Triple-half of third [of sokha]	One-fourth [of sokha] Half of fourth [of sokha]	4 K 8 [K]	Double altyn 1 altyn

(equal to 8 *altyns* and to 48 *dengas*; hereafter [S]): $M_1 = {}^2/_3$ [S], $N_1 = \frac{1}{2} \, \frac{1}{2} \, {}^1/_3$ [S] $= {}^1/_{12}$ [S]. Column 4 contains the sum $M_1 + N_1$ (equal to 6 *altyns*) expressed as a fraction of *sokha*, namely ¾ [S]. Column 5 contains the number of *seeds* here equal to 3, that is, the numerator of the fraction $^3/_4$. Column 6 contains the sum of the amounts in column 2 expressed in *altyns*: $[M_1 + N_1] = 6$ *altyns*.

Line 2 is to be interpreted as follows: Column 1 contains the number of *kosti* (*seeds*), 3 and 6, corresponding to the amounts of money in the second column. Column 2 contains two amounts of money: $M_2 = 2$ *altyns* 4 *dengas* (i.e., 16 *dengas* equal to $^1/_3$ of *sokha*) and $N_2 = 1$ *altyn* 2 *dengas* (= 8 *dengas* equal to $^1/_6$ of *sokha*) expressed with "grid symbols." Interestingly, $M_2 = \frac{1}{2} M_1$ and $N_2 = 2N_1$. Column 3 contains expressions of these two amounts represented as fractions of *sokha*: $M_2 = {}^1/_3$ [S], $N_2 = \frac{1}{2} \, {}^1/_3$ [S] $= {}^1/_6$ [S]. Column 4 contains the sum $M_2 + N_2$ expressed as a fraction of *sokha*, namely $^2/_4$ [S]. Column 5 contains the number of *seeds* here equal to 2. Column 6 contains the sum of the amounts in column 2 expressed in *altyns*: $[M_2 + N_2] = 4$ *altyns*.

Line 3 reads as follows: Column 1 contains the number of *seeds*, 6 and 12, corresponding to the amounts of money in the second column. Column 2 contains two amounts of money: $M_3 = 1$ *altyn* 2 *dengas* (i.e., 8 *dengas*) and $N_3 = 4$ *dengas* expressed with "grid symbols." Note that $M_3 = M_2/2 = N_2$ and $N_3 = N_1$. Column 3 contains expressions of these two amounts represented as fractions of *sokha*: $M_3 = \frac{1}{2} \frac{1}{3}$ [S] $= \frac{1}{6}$ [S], $N_3 = \frac{1}{2} \frac{1}{2} \frac{1}{3}$ [S] $= \frac{1}{12}$ [S]. Column 4 contains the sum $M_3 + N_3$ expressed as a fraction of *sokha*: $\frac{1}{4}$ [S]. Column 5 contains the number of *seeds* specified as 8 *seeds*. This number is wrong and it should be corrected to 4, since it is supposed to represent $\frac{1}{4}$ of a *sokha* (compared with the number of *seeds* in the last line); the copyist most probably erroneously copied in this cell the number 8 from the cell below. Column 6 contains the sum of the amounts in column 2 expressed in *altyns*: $[M_3 + N_3] = 2$ *altyns*.

Line 4 reads as follows: Column 1 contains the number of *seeds*, 12 and 24, corresponding to the amounts of money in the second column. Column 2 contains two amounts of money: $M_4 = 4$ *dengas* and $N_4 = 2$ *dengas* expressed with "grid symbols." Note that $M_4 = \frac{1}{2} M_3 = N_3$ and $N_4 = \frac{1}{2} N_3$. Column 3 contains expressions of these two amounts represented as fractions of *sokha*: $M_4 = \frac{1}{2} \frac{1}{2}\frac{1}{3}$ [S] $= \frac{1}{12}$ [S], $N_4 = \frac{1}{2} \frac{1}{2} \frac{1}{2}\frac{1}{3}$ [S] $= \frac{1}{24}$ [S]. Column 4 contains the sum $M_4 + N_4$ expressed as a fraction of *sokha*: $\frac{1}{2}\frac{1}{4}$ [S] $= \frac{1}{8}$ [S]. Column 5 contains the number of *seeds* (equal to 8 *seeds*). Column 6 contains the sum of the amounts in column 2 expressed in *altyns*: $[M_4 + N_4] = 1$ *altyn*.

As for line 5, its first column contains the number of *seeds*, 12, 6, and 24, corresponding to the amounts of money in the second column. Column 2 contains three amounts of money: $L_5 = 5$ *altyns* 2 *dengas*, $M_5 = 1$ *altyn* 2 *dengas,* and $N_5 = 2$ *dengas* expressed with "grid symbols." Note that $L_5 = M_1$, $M_5 = N_2 = M_3$, $N_5 = N_4$. Column 3 contains expressions of these three amounts represented as fractions of *sokha*: $L_5 = \frac{2}{3}$ [S], $M_5 = \frac{1}{2} \frac{1}{3}$ [S] $= \frac{1}{6}$ [S], $N_5 = \frac{1}{2} \frac{1}{2} \frac{1}{2}\frac{1}{3}$ [S] $= \frac{1}{24}$ [S]. Column 4 contains the sum $L_5 + M_5 + N_5$ expressed as fractions of *sokha*: $\frac{3}{4}$ [S] $+ \frac{1}{8}$ [S]. Column 5 contains the number of *seeds*: 3 *seeds* and 8 *seeds*. Column 6 contains the sum of the amounts in column 2 expressed in *altyns*: $[L_5 + M_5 + N_5] = 6$ *altyns* $+1$ *altyn* $[= 7$ *altyns*$]$.

Line 6 is to be read as follows: Column 1 contains the number of *kosti* (*seeds*), 3, 6, 12, and 24, corresponding to the amounts of money in the second column. Column 2 contains four amounts of money: $K_6 = 2$ *altyns* 4 *dengas*, $L_6 = 1$ *altyn* 2 *dengas*, $M_6 = 4$ *dengas* and $N_6 = 2$ *dengas* expressed with "grid symbols." Note that $K_6 = M_2$, $L_6 = M_5$, $M_6 = N_1 = N_3$, $N_6 = N_5$. Column 3 contains expressions of these four amounts represented as fractions of *sokha*: $K_6 = \frac{1}{3}$ [S], $L_6 = \frac{1}{2} \frac{1}{3}$ [S] $= \frac{1}{6}$ [S], $M_6 = \frac{1}{2} \frac{1}{2} \frac{1}{3}$ [S] $= \frac{1}{12}$ [S], $N_6 = \frac{1}{2} \frac{1}{2} \frac{1}{2}\frac{1}{3}$ [S] $= \frac{1}{24}$ [S]. Column 4 contains the sum $K_6 + L_6 + M_6 + N_6$ expressed as fractions of *sokha*: $\frac{2}{4}$ [S] $+ \frac{1}{8}$ [S]. Column 5 contains the corresponding numbers of "seeds," that is, 2 seeds and 8 seeds. Column 6 contains the sum of the amounts in column 2 expressed in *altyns*: $[K_6 + L_6 + M_6 + N_6] = 4$ *altyns* $+1$ *altyn* $[= 5$ *altyns*$]$.

The last line contains the following information: Column 1 contains the number of *kosti* (seeds), 3 and 24, corresponding to the amounts of money in the second column. Column 2 contains two amounts of money: $M_7 = 2$ *altyns* 4 *dengas* and $N_7 = 2$ *dengas* expressed with "grid symbols." Note that $M_7 = K_6$ and $N_7 = N_6$. Column 3 contains expressions of these two amounts represented as fractions of *sokha*: $M_7 = {}^1/_3$ [S], $N_7 = {}^1/_2 \; {}^1/_2 \; {}^1/_2{}^1/_3$ [S] $= {}^1/_{24}$ [S]. Column 4 contains the sum $M_7 + N_7$ expressed as a sum of fractions of *sokha*: ${}^1/_4$ [S] $+ {}^1/_8$ [S]. Column 5 contains the corresponding numbers of *seeds*: 4 *seeds* and 8 *seeds*. Column 6 contains the sum of the amounts in column 2 expressed in *altyns*: $[M_7 + N_7] = 2$ *altyns* +1 *altyn* [= 3 *altyns*].

As far as operations with fractions are concerned, the table provides the following seven identities:

I1: ${}^2/_3 + {}^1/_{12} = {}^3/_4$;
I2: ${}^1/_3 + {}^1/_6 = {}^1/_2$;
I3: ${}^1/_6 + {}^1/_{12} = {}^1/_4$;
I4: ${}^1/_{12} + {}^1/_{24} = {}^1/_8$;
I5: ${}^2/_3 + {}^1/_6 + {}^1/_{24} = {}^3/_4 + {}^1/_8$;
I6: ${}^1/_3 + {}^1/_6 + {}^1/_{12} + {}^1/_{24} = {}^2/_4 + {}^1/_8$;
I7: ${}^1/_3 + {}^1/_{24} = {}^1/_4 + {}^1/_8$.[48]

One can conjecture that these identities were used to transform combinations of fractions with denominators $3 \cdot 2^n$ to fractions with denominators 2^m; apparently, they may have also been used in the opposite direction, that is, to represent fractions with denominators 2^m as combinations of fractions with denominators $3 \cdot 2^n$.

Tsaĭger (2010, pp. 56–57) noticed that the third identity can be derived from the second one if one divides all the terms of I2 by 2, and the fourth identity can be derived from the third one in the same way. Apparently, this operation can be reiterated, and new identities can be obtained; for instance, if one divides both sides of I6 by 2, the result will be ${}^1/_6 + {}^1/_{12} + {}^1/_{24} + {}^1/_{48} = {}^1/_4 + {}^1/_{16}$. There are other relationships between the identities: I5 results from addition of I1 and I4, and I6 can be obtained if one adds I2 and I4, while I7 can be obtained as the difference of I6 and I3 (or, in other words, to obtain I6 one has to add I3 and I7). I3 also can be obtained if we add I1 and I2, and subtract 1 from both sides of the identity.

When the abacus was used, these identities allowed the operator to transform fractions with denominators $3 \cdot 2^n$ into fractions with denominators 2^m (and vice versa, if necessary). The table is followed by a list of 48 identities that allowed the operator reduce combinations of fractions of these two kinds to entire *sokhas* or relatively simple fractions of *sokha*, or a difference between one *sokha* and a unit fraction of *sokha*; each identity is accompanied with a picture representing the configuration of beads on the *schyoty* (see examples below). These identities are in turn subdivided into two groups: The first group contains identities representing reduction of sums of fractions of *sokha* to an entire *sokha* or to a relatively "simple" fraction of *sokha*; the second group contains identities to be used to reduce groups of fractions of *sokha* to one *sokha* without a unit fraction of it. The first group contains 28 identities, and the second, 20. Below I list the identities of both groups.

---

[48]This list of seven identities was for the first time reconstructed by Tsaiger (2010, p. 56, Fig. 24).

Group A

Subgroup 1 contains identities A1–A7 dealing with "reduction of fractions of *sokha* to entire *sokha*."

[A1] $^3/_4 + {}^1/_6 + {}^1/_{12}$ [=1];

[A2] $^3/_4 + {}^1/_8 + {}^1/_{12} + {}^1/_{24}$ [=1];[49]

[A3] $^1/_4 + {}^2/_3 + {}^1/_{12}$ [=1];[50]

[A4] $^1/_8 + {}^2/_3 + {}^1/_6 + {}^1/_{24}$ [=1];

[A5] $^2/_4 + {}^1/_3 + {}^1/_6$ [=1];

[A6] $^2/_4 + {}^1/_8 + {}^1/_3 + {}^1/_{24}$ [=1];[51]

[A7] $^1/_4 + {}^1/_8 + {}^1/_3 + {}^1/_6 + {}^1/_{12} + {}^1/_{24}$ [= 1].[52]

The configurations of beads representing these identities were reproduced in Spasskiĭ (1952, pp. 341–344) from the "Hermitage manuscript" copy of the *Schyotnaya Mudrost'*. The drawings from the edition (Anon. 1865) of the treatise presenting configurations of beads corresponding to identities A2 and A3 are shown in Fig. 13.

Identities of this group reveal strong connections with Table 1 and identities I1-I7 listed in it. For instance, to obtain A1 one can use I3 ($^1/_6 + {}^1/_{12} = {}^1/_4$); to obtain A2, identity I4 can be used; for A3, one can use I1; to obtain A4, the identity I5 can be used; for A5 one can use I2; to prove A6 one can use I7, and A7 follows from I6.

Subgroup 2 contains five identities presenting reduction of fractions of *sokha* to three quarters of *sokha*:

[A8] $^2/_3 + {}^1/_{12}$ [=$^3/_4$];

[A9] $^2/_4 + {}^1/_6 + {}^1/_{12}$ [=$^3/_4$];

[A10] $^2/_4 + {}^1/_8 + {}^1/_{12} + {}^1/_{24}$ [=$^3/_4$];

[A11] $^1/_4 + {}^1/_3$ [+ $^1/_6$] [=$^3/_4$];[53]

[A12] $^1/_4$ [+ $^1/_8$] + $^1/_3 + {}^1/_{24}$ [=$^3/_4$].[54]

Below I will explain how these identities may have been obtained.

Subgroup 3 contains four identities presenting reduction of fractions of *sokha* to two-third of *sokha*.

---

[49]The picture of the *schyoty* from Anon. (1865) reproduced in Fig. 13 erroneously shows two copies of position of 1/12.

[50]The picture of the *schyoty* from Anon. 1865 reproduced in Fig. 13 erroneously shows three beads in the position of 1/3; the picture reproduced in Spasskiĭ (1952, p. 342, Fig. 23) correctly shows only two beads in this position.

[51]Picture for this identity is not found in Anon. (1865).

[52]The picture of the *schyoty* provided in Anon. 1865 for this identity does not contain a bead in the position of 1/24; the picture reproduced in Spasskiĭ (1952, p. 344, Fig. 25) correctly shows a bead in this position.

[53]The manuscript Anon. (1865) erroneously misses a bead in the position of 1/6.

[54]The manuscript Anon. (1865) erroneously misses a bead in the position of 1/8.

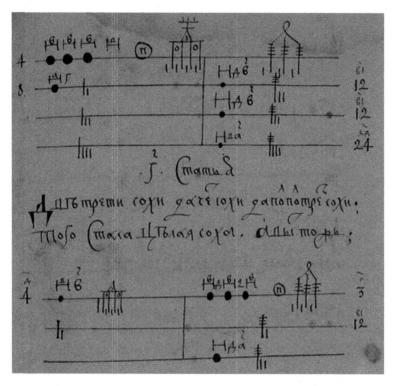

Fig. 13 Page 93 r from Anon. 1865, p. 99 (fragment): configurations of beads for identities A2 and A3

[A13] $^2/_4 + ^1/_6 [=^2/_3]$;
[A14] $^2/_4 + ^1/_8 + ^1/_{24} [=^2/_3]$;
[A15] $^1/_4 + ^1/_3 + ^1/_{12} [=^2/_3]$;
[A16] $^1/_3 [+ ^1/_8] + ^1/_6 + ^1/_{24} [=^2/_3]$.[55]

Subgroup 4 contains four identities presenting reduction of fractions of *sokha* to one-half of *sokha*.

[A17] $^1/_3 + ^1/_6 [=^1/_2]$;
[A18] $^1/_8 + ^1/_3 + ^1/_{24} [=^1/_2]$;
[A19] $^1/_4 + ^1/_6 + ^1/_{12} [=^1/_2]$;
[A20] $^1/_4 + ^1/_8 + ^1/_{12} + ^1/_{24} [=^1/_2]$.

---

[55]The manuscript Anon. (1865) erroneously misses a bead in the position of 1/8.

Subgroup 5 contains two identities presenting reduction of fractions of *sokha* to one-quarter of *sokha*.

[A21] $\frac{1}{6} + \frac{1}{12}$ [$=\frac{1}{4}$];
[A22] $\frac{1}{8} + \frac{1}{12} + \frac{1}{24}$ [$=\frac{1}{4}$].

Subgroup 6 contains two identities presenting reduction of fractions of *sokha* to one-third of *sokha*.

[A23] $\frac{1}{4} + \frac{1}{12}$ [$=\frac{1}{3}$];
[A24] $\frac{1}{8} + \frac{1}{6} + \frac{1}{24}$ [$=\frac{1}{3}$].

The last, seventh subgroup of group A contains four identities:

[A25] $\frac{1}{8} + \frac{1}{24}$ [$=\frac{1}{6}$];
[A26] $\frac{1}{12} + \frac{1}{24}$ [$=\frac{1}{8}$];
[A27] $\frac{1}{24} + \frac{1}{48}$ [$=\frac{1}{16}$];
[A28] $\frac{1}{32} + \frac{1}{48} + \frac{1}{96}$ [$=\frac{1}{16}$].

Group B

All the 20 identities in group B share the same pattern: They express a sum of fractions as a difference between a unit and a fraction $1/(3 \cdot 2^n)$ or $1/2^m$. Their list runs as follows:

[B1] $\frac{3}{4} + \frac{1}{6} = 1 - \frac{1}{12}$;
[B2] $\frac{3}{4} + \frac{1}{8} + \frac{1}{12} = 1 - \frac{1}{24}$;
[B3] $\frac{3}{4} + \frac{1}{12} = 1 - \frac{1}{6}$;
[B4] $\frac{3}{4} + \frac{1}{12} + \frac{1}{24} = 1 - \frac{1}{8}$;
[B5] $\frac{3}{4} + \frac{1}{8} + \frac{1}{16} + \frac{1}{32} = 1 - \frac{1}{32}$;
[B6] $\frac{3}{4} + \frac{1}{8} + \frac{1}{16} = 1 - \frac{1}{16}$;
[B7] $\frac{3}{4} + \frac{1}{8} = 1 - \frac{1}{8}$;
[B8] $\frac{1}{4} + \frac{2}{3} = 1 - \frac{1}{12}$;
[B9] $\frac{2}{3} + \frac{1}{12} = 1 - \frac{1}{4}$;
[B10] $\frac{2}{3} + \frac{1}{6} = 1 - \frac{1}{6}$;
[B11] $\frac{2}{3} + \frac{1}{8} + \frac{1}{24} = 1 - \frac{1}{6}$;
[B12] $\frac{2}{3} + \frac{1}{6} + \frac{1}{24} = 1 - \frac{1}{8}$;
[B13] $\frac{2}{4} + \frac{1}{3} = 1 - \frac{1}{6}$;
[B14] $\frac{2}{4} + \frac{1}{6} = 1 - \frac{1}{3}$;
[B15] $\frac{2}{4} + \frac{1}{8} + \frac{1}{24} = 1 - \frac{1}{3}$;
[B16] $\frac{1}{4} + \frac{1}{8} + \frac{1}{32} + \frac{1}{3} + \frac{1}{6} + \frac{1}{24} + \frac{1}{48} = 1 - \frac{1}{32}$;
[B17] $\frac{1}{4} + \frac{1}{8} + \frac{1}{3} + \frac{1}{6} = 1 - \frac{1}{8}$;
[B18] $\frac{1}{4} + \frac{1}{8} + \frac{1}{16} + \frac{1}{32} + \frac{1}{6} + \frac{1}{12} + \frac{1}{48} + \frac{1}{96} = 1 - \frac{1}{4}$;
[B19] $\frac{3}{4} + \frac{1}{8} + \frac{1}{16} + \frac{1}{32} + \frac{1}{64} + \frac{1}{128} = 1 - \frac{1}{128}$;
[B20] $\frac{2}{3} + \frac{1}{6} + \frac{1}{12} + \frac{1}{24} + \frac{1}{48} + \frac{1}{96} = 1 - \frac{1}{96}$.

Figure 14 shows a page from Anon. (1865) with the identities B17–B18 represented on the counting device.

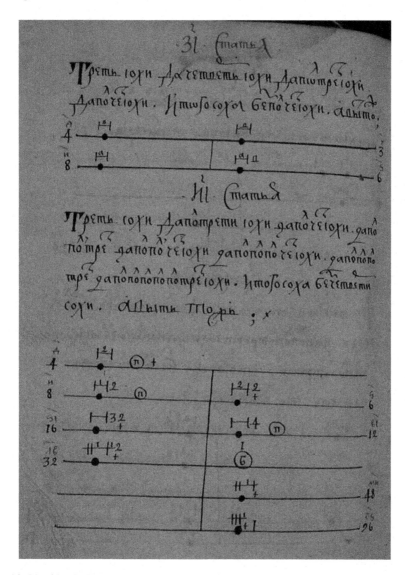

Fig. 14 Identities B17 and B18. From Anon. 1865, p. 110v

How these identities were obtained? The treatise does not contain any information to this effect; however, some conjectures can be made. Above I mentioned obvious links between the identities A1–A7 and I1–I7. Similarly, connections between other identities of groups A and B can be reconstructed; they are as follows.

In subgroup 2, A8 is identical with I1, A9 follows from I3, A10 can be deduced from I4, A11 is based on I2, and A12 directly follows from I7. In subgroup 3, A13 follows from A5, A14 from A6, A15 from A3, and A16 can be easily deduced on the basis of I5. In subgroup 4, A17 is identical with I2, A18 directly follows from I7, A19 results from division of all the terms of A5 by 2 or can be directly obtained from I3, and A20 can be obtained on the basis of I3 (all the terms of I3 should be divided by 2). In subgroup 5, A21 is the result of division of all the terms of A17 by 2, and A22 can be obtained as the sum of $^1/_8$ and of the result of division of the table value I2 $(^1/_3 + ^1/_6)$ by four. In subgroup 6, A23 can be easily deduced if $^1/_4$ is represented as $^1/_6 + ^1/_{12}$ according to I3, and A24 directly follows from A4. Finally, A25 results from division by 2 of all the terms of A23, A26 can similarly be obtained from A21, A27 results from division by 2 of all the terms of A26, and A28 can be obtained if one divides all the terms of A22 by 4.

As far as the group B is concerned, they were most likely obtained on the basis of identities $a + b + \ldots + f = 1$ including a unit fraction f and transformed into $a + b + \ldots = 1 - f$. Some hypotheses concerning production of these identities look plausible; they are as follows. B1 and B3 are based upon A1; B2 and B4 on A2; B5–B7 are based on the identities $1 = \frac{3}{4} + ^1/_4 = \frac{3}{4} + ^1/_8 + ^1/_8 = \frac{3}{4} + ^1/_8 + ^1/_{16} + ^1/_{16} = \frac{3}{4} + ^1/_8 + ^1/_{16} + ^1/_{32} + ^1/_{32}$. Identities B8 and B9 are based on A3. B10 directly follows from $1 = ^2/_3 + ^1/_3 = ^2/_3 + ^1/_6 + ^1/_6$. B11 and B12 result from $^2/_3 + ^1/_6 + ^1/_8 + ^1/_{24} = 1$ which in turn results from addition of 1/8 to both sides of I5. B13 and B14 stem from A5, while B15 can be easily obtained on the basis of I7. In turn, B16 can be obtained on the basis of A27 and I2, while B17 is apparently based on $1 = \frac{1}{2} + ^1/_3 + ^1/_6 = ^1/_3 + ^1/_6 + ^1/_4 + ^1/_4 = ^1/_3 + ^1/_6 + ^1/_4 + ^1/_8 + ^1/_8$. B18 follows from A1 and A28, while B19 can be obtained from the following series of identities: $1 = \frac{3}{4} + ^1/_4 = \frac{3}{4} + ^1/_8 + ^1/_8 = \frac{3}{4} + ^1/_8 + ^1/_{16} + ^1/_{16} = \frac{3}{4} + ^1/_8 + ^1/_{16} + ^1/_{32} + ^1/_{32} = \frac{3}{4} + ^1/_8 + ^1/_{16} + ^1/_{32} + ^1/_{64} + ^1/_{64} = \frac{3}{4} + ^1/_8 + ^1/_{16} + ^1/_{32} + ^1/_{64} + ^1/_{128} + ^1/_{128}$. Finally, B20 can be obtained in a similar way: $1 = ^2/_3 + ^1/_3 = ^2/_3 + ^1/_6 + ^1/_6 = ^2/_3 + ^1/_6 + ^1/_{12} + ^1/_{12} = ^2/_3 + ^1/_6 + ^1/_{12} + ^1/_{24} + ^1/_{24} = ^2/_3 + ^1/_6 + ^1/_{12} + ^1/_{24} + ^1/_{48} + ^1/_{48} = ^2/_3 + ^1/_6 + ^1/_{12} + ^1/_{24} + ^1/_{48} + ^1/_{96} + ^1/_{96}$.

To conclude this section, several remarks are certainly due. First, if a configuration of beads set on the instrument represented a number N, then "shifting up" and "shifting down" this configuration resulted in valid representations of the numbers $2N$ and $^N/_2$, respectively.[56] Examples of such "shifting" are found in Table 1: Configurations for I3 and I4 result from "shifting down" configurations for I2 and I3, respectively. Another example is identity A27 resulting from "shifting down" identity A26 (identical with I4). Second, the listed configurations may have been used even when the beads referred to in the rules of reduction constituted a proper subset of the beads "activated" on the device. For instance, if the beads representing $^1/_4$, $^2/_3$, $^1/_8$ and $^1/_{12}$ are activated, then the operator may have used the table entry A3 to reduce $^1/_4$, $^2/_3$, and $^1/_{12}$ to one *sokha* and operate with the obtained amount $1\,^1/_8$

---

[56]By "shiftning up" I mean a transformation of a given configuration consisting in placing active beads in n-th wire (counted from the top) on $(n - 1)$-th wire; "shifting down" is moving active beads from n-th wire to $(n + 1)$-th wire.

sokha. Third, the identities listed above, apparently, may have been used in both ways, that is, to simplify configurations on the counting instrument (e.g., to replace $^2/_3 + {}^1/_{12}$ by ¾, according to A8) or, conversely, to represent one fraction as a combination of several smaller fractions, for instance, to replace ¾ by $^2/_3 + {}^1/_{12}$, to use the same example.

Identities in the Book of *sokha* script of the year 7137 [=1629]

A list of identities similar to those discussed above can also be found in the manual *Kniga soshnogo pis'ma 7137 goda* (Book of *sokha* script of the year 7137 [=1629]) (Anon. 1629 [1853]). Unlike the identities analyzed in the previous section, they are not grouped together and are mixed up with other data. The groups of identities listed in this text are as follows:

List C (Anon. 1629 [1853], pp. 33–34):

[C1] $^1/_3 + {}^1/_6 = \frac{1}{2}$;
[C2] $^1/_3 + {}^1/_8 + {}^1/_{24} = \frac{1}{2}$;
[C3] $^1/_6 + {}^1/_{12} = {}^1/_4$;
[C4] $^1/_8 + {}^1/_{12} + {}^1/_{24} = {}^1/_4$;
[C5] $^1/_{12} + {}^1/_{24} = {}^1/_8$;
[C6] $^1/_4 + {}^1/_{12} = {}^1/_3$;
[C7] $^1/_6 + {}^1/_8 + {}^1/_{24} = {}^1/_3$;
[C8] $^1/_8 + {}^1/_{24} = {}^1/_6$;
[C9] $^1/_{16} + {}^1/_{24} + {}^1/_{32} = {}^1/_6 - {}^1/_{32}$;
[C10] $^1/_6 + {}^1/_{24} [+ {}^1/_{48}] = {}^1/_4 - {}^1/_{48}$;
[C11] $^1/_8 + {}^1/_{12} = {}^1/_4 - {}^1/_{24}$;
[C12] $^1/_6 + {}^1/_{16} + {}^1/_{24} + {}^1/_{32} = {}^1/_3 - {}^1/_{32}$;
[C13] $^1/_3 + {}^1/_{12} + {}^1/_{24} = \frac{1}{2} - {}^1/_{24}$;
[C14] $^1/_3 + {}^1/_8 = \frac{1}{2} - {}^1/_{24}$;
[C15] $^1/_4 + {}^1/_8 + {}^1/_{12} [+ {}^1/_{96}] = \frac{1}{2} - {}^1/_{32}$;
[C16] $^1/_4 + {}^1/_{12} + {}^1/_{16} + {}^1/_{24} + {}^1/_{32} = \frac{1}{2} - {}^1/_{32}$.

Some identities in this list look familiar: C1 is the same as I2 (or A17), C2 is identical with A18, and C3 is the same as I3 and also can be obtained from C1. C4 is the same as A22, C5 is identical with I4, and C6 is the same as A23. C7 is identical with A24, and it can be used to obtain C8 (which is identical with A25) if we subtract $^1/_6$ from both parts; C8 also results from division of both parts of C6 by 2 (i.e., by "shifting down" the configuration representing it on the abacus). C8, in turn, can be used to obtain C9: To do so, one can apply the standard method of "splitting" the fraction $^1/_8$ into two equal parts: $^1/_8 = {}^1/_{16} + {}^1/_{16} = {}^1/_{16} + {}^1/_{32} + {}^1/_{32}$. C9 opens a series of identities in which the right part is a difference of two fractions; even though these identities remind us of the group B1–B20, they are actually different, because none of them has the expression $(1 - {}^1/_p)$ on the right side. C10 contains an obvious mistake of the copyist; this identity can be obtained from C3: ${}^1/_4 = {}^1/_6 + {}^1/_{12} = {}^1/_6 + {}^1/_{24} + {}^1/_{24} = {}^1/_6 + {}^1/_{24} + {}^1/_{48} + {}^1/_{48}$. C11 directly follows from C4 (or, in other words, from A22), and C12 can be obtained from C7 if we "split" ${}^1/_8$ twice. C13 can be obtained from C1 in a similar way, and C14 directly follows from C2. To obtain C15 (in which the term ${}^1/_{96}$ is apparently missing on the left side), one can divide both parts of C8 by 4 (that it, "shift it down" on the abacus twice) to obtain ${}^1/_{32} + {}^1/_{96} = {}^1/_{24}$; then $\frac{1}{2} = {}^1/_4 + {}^1/_4 = {}^1/_4 + {}^1/_8 + {}^1/_8$ and since, according to C5, ${}^1/_8$ equals to ${}^1/_{12} + {}^1/_{24}$, we have $\frac{1}{2} = {}^1/_4 + {}^1/_8 + {}^1/_{12} + {}^1/_{24} = {}^1/_4 + {}^1/_8 + {}^1/_{12} + {}^1/_{32} + {}^1/_{96}$, hence the result. Finally, to obtain C16, one can again use $\frac{1}{2} = {}^1/_4 + {}^1/_8 + {}^1/_{12} + {}^1/_{24}$ and "split" ${}^1/_8$ as ${}^1/_{16} + {}^1/_{32} + {}^1/_{32}$.

Another group of identities is found on p. 35 of the same manual. It reads as follows:

[D1] ${}^1/_3 + {}^1/_6 = \frac{1}{2}$;
[D2] ${}^1/_4 + {}^1/_6 + {}^1/_{12} = \frac{1}{2}$;
[D3] ${}^1/_4 + {}^1/_8 + {}^1/_{12} + {}^1/_{24} = \frac{1}{2}$;
[D4] ${}^1/_4 + {}^1/_8 + {}^1/_{16} + {}^1/_{32} = \frac{1}{2} - {}^1/_{32}$;
[D5] ${}^1/_6 + {}^1/_{12} = {}^1/_4$;
[D6] ${}^1/_8 + {}^1/_{12} + {}^1/_{24} = {}^1/_4$.

In this group, D1 is identical with I2 and A17, D2 with A19, and D3 with A20. D4 apparently follows from "splitting" $\frac{1}{2}$ $({}^1/_2 = {}^1/_4 + {}^1/_4 = {}^1/_4 + {}^1/_8 + {}^1/_8 = \cdots)$. D5 and D6 can be obtained via "shifting down" D1 and D2, respectively.

Group E (p. 36) contains only two identities:

[E1] ${}^1/_8 + {}^1/_9 = {}^1/_4 - {}^1/_{32}$;
[E2] ${}^2/_9 = {}^1/_4 - {}^1/_{32}$.

Both of them are wrong; I will not discuss here their possible origin.
Group F (p. 41) contains four identities:

[F1] ${}^1/_3 + {}^1/_6 = \frac{1}{2}$;
[F2] ${}^1/_6 + {}^1/_{12} = {}^1/_4$;
[F3] ${}^1/_{12} + {}^1/_{24} = {}^1/_8$;
[F4] ${}^1/_{12} + {}^1/_{24} = {}^1/_{16}$.

F4 is an obvious mistake of the copyist, it should have been ${}^1/_{24} + {}^1/_{48} = {}^1/_{16}$. Here, F1 is I2 (or A17), and F2–F4 are obtained from it via "shifting down," or, in other words, via division of both parts of the previous identity by 2 (note also that F2 is identical with I3 and C3, F3 is identical with A26, and corrected F4, with A27).

The last group of identities found on p. 42 of the same publication is as follows:

[G1] ${}^1/_3 + {}^1/_6 = \frac{1}{2}$;
[G2] ${}^1/_{12} + {}^1/_6 = {}^1/_4$;
[G3] ${}^1/_{24} + {}^1/_{12} = {}^1/_3$;
[G4] ${}^1/_8 + {}^1/_{12} = {}^1/_6$.

Here, G1 is again the familiar identity I2 (A17), G2 is obtained from it via "shifting down" and is the same as I3 (or C3), G3 is apparently a miswritten A23 (in this case G3 is obtained from G2 via one more "shifting down"), and the final identity G4 is miswritten again; most likely, it is a distorted C8 (i.e., A25) in which one prefix *pol-* ("half") was erroneously omitted and $1/_{24}$ thus became $1/_{12}$.

Conclusions

A summary of this brief presentation of the history of counting instruments in Russia may look as follows. The earliest instrument referred to as "seeds counting," as R. Simonov claims, existed in the early second millennium AD; interestingly enough, this instrument bears certain similarity with a device described in a Chinese treatise of uncertain date (which is conventionally dated of the early first millennium AD, but may as well be a spurious text of the early second millennium). Another Russian instrument appeared no later than the early seventeenth century; it construction was similar to that of the Russian *schyoty* of the nineteenth and twentieth centuries yet included two counting fields each having two additional sections used for representation of natural fractions and for operations with them. As I. G. Spasskiĭ suggested (1952, p. 374), the sections for fractions became obsolete after the land tax reforms of 1680s and monetary reforms of the early eighteenth century, and the rules of operation with them fell into oblivion soon after; as for the instruments with two counting fields, they still existed in the early eighteenth century, but were soon replaced by the well-known nowadays simplified instrument with only one counting field (ibid., p. 375).

The arithmetical manuals of the seventeenth century featured sections on addition of fractions to be performed with the *schyoty*; these sections contained lists of fractions supposed to be summed up accompanied by diagrams showing representations of the fractions on the counting device. The diagrams were apparently introduced with the didactical purpose, yet the treatises did not explain the rules according to which the lists of identities were formed; most likely, the learners were supposed to memorize these rules, test them, and even try to create their own identities. The lists of identities thus could have been used in at least two ways: (1) as reference materials applied when actual operations with the *shyoty* were performed, and (2) as a model for learners studying operations with fractions. In other words, the lists of identities were indispensable companions of the counting device and, at the same time, played the role of didactical materials.[57]

---

[57]One can only regret that the traditional Russian *schyoty* with sections for fractions disappeared by the early 18th century and thus remained unknown to the Western educators of the 19th century who argued for the use of counting devices (and, in particular, for the use of the Russian *schyoty*) in school classroom. It would be interesting to see what results might have been obtained if teaching common fractions were performed with the help of the original instrument.

References

A. Russian Mathematical Textbooks of the 17th Century

[Anon.] 1629 [1853]. Книга сошного письма 7137 года [Kniga soshnogo pis'ma 7137 goda] (Book of *sokha* script of the year 7137 [=1629]). Transcribed by G. Trekhletov. Временник Императорскаго Московскаго общества истории и древностей российских [*Vremennik Imperatorskago Moskovskago obshchestva istorii i drevnostei rossiĭskikh*], Book 17, pp. 33–65.

[Anon.] 1865. *Книга сошному и вытному письму* [*Kniga soshnomu i vytnomu pis'mu*] (Book of *sokha* and *vyt'* script). A manuscript acquired by the library of the Troitsk-Sergiev Monastery in 1865.

[Anon.] 1879. *Счётная мудрость* [Schyotnaya mudrost'] (*Counting Wisdom*). Sankt-Peterburg: Obshchestvo lyubitelei drevnei pis'mennosti.

B. Primary Works in Other Languages

Du Halde, Jean-Baptiste. 1735. *Description Geographique, Historique, Chronologique et Physique de l'Empire de la Chine et de la Tartarie Chinoise*. Paris: Le Mercier.

Guo Shuchun 郭書春 (ed.) 1993. *Zhongguo kexue zhishu dianji tonghui* 中國科學技術典籍通彙 (Comprehensive compendium of classical texts related to science and technology in China). *Shuxue juan* 數學卷 (Mathematical section). Zhengzhou: Henan jiaoyu chubanshe 河南教育出版社.

Guo Shuchun 郭書春, and Liu Dun 劉鈍 (eds.) 2001. *Suanjing shi shu* 算經十書 (*Ten mathematical treatises*). Taibei: Jiuzhang Publishers.

von Haven, Peder von. 1743. *Reise udi Rusland*. Kjøbenhavn: Christoph Georg Glasing.

von Haven, Peter von. 1744. *Reise in Rußland* [*German translation of von Haven (1743)*]. Coppenhagen: Gabriel Christian Rothe.

Korb, Joanne Gergio [=Johann Georg]. 1700. *Diarium itineris in Moscoviam*. Viennae: Leopoldi Voigt.

Miller, Gerard Frederik [Миллер, Герард Фредерик] [=Müller, Gerhard Friedrich]. 1750. *Opisanie Sibirskago Tsarstva i vseh proizoshedshih v nem del ot nachala, a osoblivo ot pokoreniya ego Rossiĭskoi Derzhave, po sii vremena* [Описание Сибирскаго Царства и всех произошедших в нем дел от начала, а особливо от покорения его Российской Державе, по сии времена] (*Description of the Siberian Kingdom and all the events that happened in it from the very beginning, and especially from the subjugation of it by the Russian State, until the present times*). Sankt-Peterburg: Academy of Sciences.

Olearius, Adam. 1656. *Relation du voyage de Moscovie, Tartarie, et Perse*. Paris: Francois Clouzier.

Oleariĭ [Олеарий] [=Olearius], Adam. 1906. *Opisanie puteshestviya v Moskoviyu i cherez Moskoviyu v Persiyu i obratno* [Описание путешествия в Московию и через Московию в Персию и обратно] (*Description of the travel to Muscovy and through Muscovy to Persia and back*). Translated into Russian by A. M. Lovyagin. Sankt-Peterburg. [Reprint: Smolensk: Rusich, 2003].

Perry, John. 1716. *The state of Russia under the present Czar*. London: B. Tooke.

Perry, Jean [=John]. 1717. *État présent de la Grande-Russie*. La Haye: Henry Dusauzet.

Reutenfels, Iacobus. 1680. *De Rebus Moschoviticis ad Serenissimum Magnum Hetruriae Ducem Cosmum Tertium*. Patauii: Petri Mariae Frambotti (The name of the author is not found on the front page; it is mentioned in the *Admonitio* section in the beginning of the book.).

[Reutenfels, Iacobus] Yakov Rejtenfel's [Яков Рейтенфельс]. 1905–1906. Skazaniya svetleĭshemu gertsogu Toskanskomu Ko'zme tret'emu o Moskovii (Paduya, 1680 g.) [Сказания светлейшему герцогу Тосканскому Козьме третьему о Московии (Падуя, 1680 г.)] (Accounts [presented] to the illustrious Duke of Tuscany, Cosimo the Third, about the Muscovy (Padua, anno 1680)). [Russian translation of Reutenfels 1680 by Aleksei I. Stankevich.] *Чтения в Императорском обществе Истории и Древностей Российских при Московском Университете.* Москва: Типография об-ва распространения полезных книг. [Part 1:] 1905, Book 3, Sect. II, pp. I–X, 1–128; [Part 2:] 1906, Book 3, Sect. III, pp. 129–228.

Scherer, Jean-Benoît [Johann Benedikt]. 1792. *Anecdotes intéressantes et secrètes de la Cour de Russie, tirées de ses archives.* Londre et Paris: Buisson.

Snelling, Thomas. 1769. *A view of the origin, nature, and use of jettons or counters.* London.

Shtaden, Genrikh [Штаден, Генрих = Heinrich von Staden]. 1925. *О Москве Ивана Грозного: Записки немца опричника* [O Moskve Ivana Groznogo: Zapiski nemtsa *oprichnika*] (About Moscow of Ivan the Terrible: Notes of a German oprichnik). Moscow: Sabashnikovs Publishers [an annotated Russian translation of a manuscript copy of the 16th century preserved in a German archive].

Witsen, Nicolaes. 1692. *Noord en Oost Tartarye, Ofte Bondigh Ontwerp Van Eenige dier landen, en volken, zo als voormaels bekent zyn geweest. […] Zedert nauwkeurigh onderzoek van veele jaren, en eigen ondervindinge beschreven, getekent, en in 't licht gegeven, door Nicolaes Witsen.* Amsterdam. [Second edition (under slightly different title): Amsterdam: M. Schalekamp, 1705; reprint 1785].

Witsen, Nicolaas (Витсен, Николаас). 2010. Северная и восточная Тартария: включающая области, расположенные в северной и восточной частях Европы и Азии [Severnaĭa i vostochnaĭa Tartarīĭa: vklĭuchaĭushchaĭa oblasti, raspolozhennye v severnoĭ i vostochnoĭ chasfīakh Evropy i Azii]. Amsterdam: Pegasus Publishers. [An annotated Russian translation of Witsen 1785].

Xu, Xinlu 徐心魯. 1993 [1573]. *Pan zhu suan fa* 盤珠算法 (Computational methods for abacus [lit. "pearls in tray"]). In Guo (1993), Vol. 2, pp. 1143–1164.

Xu, Yue 徐岳. 1993. *Shu shu ji yi* 數術記遺 (Procedures of "numbering" recorded to be preserved [for posterity]). In Guo (1993), Vol. 1, pp. 347–351; Guo and Liu 2001, pp. 445–452.

C. Russian and Soviet School Manuals Related to *Schyoty*

Kiryushin, E. D. [Кирюшин, Ефим Данилович]. 1925. *Вычисления на счётах* [Vychisleniya na schyotakh] (*Calculations with the schyoty*), 6th ed. Moscow: Kooperativnoe izdatel'stvo.

Orlitskiĭ, D. [Орлицкий, Д.] 1830a. Умножение и деление чисел на счетах, с присовокуплением особых правил сложения и вычитания [Umnozhenie i delenie chisel na schyotakh, s prisovokupleniem osobykh pravil slozheniya i vychitaniya] (Multiplication and division of numbers on the *schyoty*, with additional special methods of addition and subtraction). Санкт-Петербург: Греч.

Orlitskiĭ, D. [Орлицкий, Д.] 1830b. Умножение и деление чисел на счётах [Umnozhenie i delenie chisel na schyotakh] (Multiplication and division of numbers on the *schyoty*). *Syn Otechestva* [Сын Отечества], 131: 297–307.

Tikhomirov, Petr [Тихомиров, Петр Васильевич]. 1830. Арифметика на счётах или легчайший способ производить все арифметические действия на счётах, усовершенствованных генерал-майором г. Свободским [Arifmetika na schyotah ili legchaishiĭ sposob proizvodit' vse arifmeticheskie deĭstviya na schyotakh, usovershenstvovannykh general-maĭorom g. Svobodskim] (Arithmetic on the *schyoty*, or the easiest way to produce all the arithmetical operations of the *schyoty* improved by general-major Mr Svobodskoi). Санкт-Петербург: Морская типография.

D. Secondary Works

Bobynin, Viktor [Бобынин, Виктор Викторович]. 1884. Состояние математических знаний в России до XVI века [Sostoyanie matematicheskikh znaniĭ v Rossii do XVI veka] (The state of mathematical knowledge in Russia prior to the 16th century; in Russian). *Журнал Министерства Народнаго Просвещения* [Zhurnal Ministerstva Narodnago Prosveshcheniya]. *Journal of the Ministry of Education*, 132(2): 183–209.

Bobynin, Viktor [Бобынин, Виктор Викторович]. 1886. *Очерки история развития физико-математических знаний в России. XVII столетие* [Ocherki istoriya razvitiya fiziko-matematicheskikh znaniĭ v Rossii. XVII stoletie] (Outlines of the development of physics and mathematics in Russia. The 17th century; in Russian). Moscow: Mamontov.

Burnett, Charles, and William F. Ryan. 1998. Abacus (Western). In *Instruments of science: An historical encyclopedia*, ed. Robert Bud, Deborah J. Warner, 5–7. London and New York: Garland.

Egorov, Vadim L. [Егоров, Вадим Леонидович]. 1985. *Историческая география Золотой Орды в XIII-XIV в.в.* [Istoricheskaya geografiya Zolotoĭ Ordy v XIII-XIV vv.] (Historical geography of the Golden Horde in the 13th–14th century; in Russian). Moscow: Nauka.

Freiman, Viktor, and Alexei Volkov. 2012. Common fractions in L. F. Magnitskiĭ's Arithmetic (1703): Interplay of tradition and didactical innovation. *Paper delivered at the 12th International Congress on Mathematics Education*, Seoul, South Korea, July 8–15, 2012.

Freiman, Viktor, and Alexei Volkov. 2015. Didactical innovations of L. F. Magnitskiĭ: setting up a research agenda. *International Journal for the History of Mathematics Education*, 10(1): 1–23.

Hoving, Ab J. 2012. *Nicolaes Witsen and shipbuilding in the Dutch Golden Age*. Translated by Alan Lemmers; foreword by Andre Wegener Sleeswyk; with an appendix by Diederick Wildeman. College Station: Texas A&M University Press.

Il'ina, Ol'ga A. [Ильина, Ольга Алексеевна]. 2005. Историческая топография и локализация золотоордынских городов Нижнего Поволжья [Istoricheskaya topografiya i lokalizaciya zolotoordynskih gorodov Nizhnego Povolzh'ya] (Historical topography and localization of towns of the Golden Horde in the lower basin of Volga; in Russian). Unpublished Ph.D. dissertation. Volgograd.

Pachkalov, Alexander [Пачкалов, Александр Владимирович]. 2009a. Золотоордынские города Нижнего Поволжья в конце XIV в. [Zolotoordynskie goroda Nizhnego Povolzh'ya v konce XIV v.] (Towns of the Golden Horde of the lower Volga in the late 14th century; in Russian). In: Золотоордынское наследие [Zolotoordynskoe nasledie] (Heritage of the Golden Horde). *Proceedings of the international conference "Political and socio-economical history of the Golden Horde* (13th–15th centuries)". Issue 1. Kazan', FEN Publishers, Academy of Sciences of the Republic of Tatarstan, pp. 210–218.

Pachkalov, Alexander [Пачкалов, Александр Владимирович]. 2009b. Очерк по истории Старого и Нового Сараев – столиц Золотой Орды [Ocherk po istorii Starogo i Novogo Saraev—stolits Zolotoi Ordy] (Outline of the history of the Old and New Saraĭs, the capitals of the Godlen Horde; in Russian). *Azerbaijan and Azerbaijanis* (Baku), Vol. 103–104, No. 1–2, pp. 122–128.

Peters, Marion. 1994. From the study of Nicolaes Witsen (1641–1717). His life with books and manuscripts. *LIAS. Sources and documents relating to the early modern history of ideas*. Vol. 21, No. 1, pp. 1–49.

Peters, Marion. 2010. *De wijze koopman. Het wereldwijde onderzoek van Nicolaes Witsen (1641–1717), burgemeester en VOC-bewindhebber van Amsterdam* [Mercator sapiens (Wise Merchant). The Worldwide Research of Nicolaes Witsen (1641–1717), Mayor and EIC-boardmember of Amsterdam]. Amsterdam: Uitgeverij Bert Bakker.

Ryan, William F. 1972. John tradescant's Russian Abacus. *Oxford Slavonic Papers*, 5: 83–88.

Ryan, William F. 1991. Scientific instruments in Russia from the middle ages to Peter the Great. *Annals of Science*, 48(4): 367–384.

Schärlig, Alain. 2001. *Compter avec des cailloux. Le calcul élémentaire sur l'abaque chez les anciens Grecs.* Lausanne: Presses polytechniques et universitaires romandes.

Shvetsov, Konstantin [Швецов, Константин Иванович]. 1955. Бібліографія староруських математичних рукописів [Bibliografiya starorus'kikh matematichnikh rukopisiv] (Bibliography of old Russian mathematical manuscripts; in Ukrainian). *Станіславський Державний Педагогічний Институт. Наукові Записки. Фізико-математична серія* [Stanislavs'kiĭ Derzhavniĭ Pedagogichniĭ Institut. Naukovi Zapiski. Fiziko-matematichna seriya] (Scientific Notes of Stanislavskiĭ State pedagogical institute, Series of Physics and Mathematics), issue 1, pp. 49–103.

Shvetsov, Konstantin [Швецов, Константин Иванович]. 1966. Математика сошного письма [Matematika soshnogo pis'ma] (Mathematics of *sokha* script). In I. Z. Shtokalo [Штокало, Иосиф Захарович] (ed.), *История отечественной математики* [Istoriya otechestvennoĭ matematiki] (*History of mathematics of the fatherland; in Russian*), Kiev: Naukova Dumka, Vol. 1, pp. 79–82.

Simonov, Rem A. [Симонов, Рэм Александрович]. 1975. О проблеме наглядно-инструментального счета в средневековой Руси [O probleme naglyadno-instrumental'nogo scheta v srednevekovoĭ Rusi] (On the problem of visual-instrumental counting in medieval Russia; in Russian). *Sovetskaja arkheologiya*, vol. 19, no. 3, pp. 82–93.

Simonov, Rem A. [Симонов, Рэм Александрович]. 1977. *Математическая мысль древней Руси* [Matematicheskaya mysl' drevneĭ Rusi] (*Mathematical thought of Ancient Russia; in Russian*). Moskva: Nauka.

Simonov, Rem A. [Симонов, Рэм Александрович]. 1988. Учебные задачи для абака по пересчету натуры на деньги Русской Правды [Uchebnye zadachi dlia abaka po pereschetu natury na den'gi *Russkoi Pravdy*] (Problems from the Russian Justice used for teaching the methods of computations on abacus of amounts of money corresponding to goods; in Russian). In *Древности славян и Руси* [Drevnosti slavyan i Rusi] (*Antiquities of Slavs and of Russia*), ed. B. A. Timoshchuk [Б. А. Тимощук], pp. 279–286. Moscow: Nauka.

Simonov, Rem A. [Симонов, Рэм Александрович]. 1990. Древнерусский абак для пересчета натуры на деньги [Drevnerusskij abak dlya perescheta natury na den'gi] (Ancient Russian abacus used to calculate the amounts of money corresponding to goods; in Russian). *Вопросы истории естествознания и техники* (Voprosy istorii estestvoznaniya i tekhniki) (Problems of the history of science and technology), No. 3, pp. 90–93.

Simonov, Rem A. [Симонов, Рэм Александрович]. 1993a. 900-летний возраст древнерусского абака («счёта костьми») [900-letniĭ vozrast drevnerusskogo abaka ("schyota kost'mi")] (900 years of ancient Russian abacus ("counting with seeds")). *Вопросы истории естествознания и техники* (*Voprosy istorii estestvoznaniya i tekhniki*) (Problems of the history of science and technology), No. 1, pp. 96–98.

Simonov, Rem A. [Симонов, Рэм Александрович]. 1993b. Die Archäologie bestätigt 900jähriges Alter des altrussischen Abakus (des „Zählens mit Kernen"). *Jahrbücher für Geschichte Osteuropas*, Neue Folge, Bd. 41, H. 1, pp. 111–113.

Simonov, Rem A. [Симонов, Рэм Александрович]. 2001a. Естественнонаучная Мысль Древней Руси: Избранные Труды [*Estestvennonauchnaya Mysl' Drevneĭ Rusi: Izbrannye Trudy*] (*Scientific thought of Ancient Russia: Selected Works; in Russian*). Moscow: Moscow State University of Printing Arts.

Simonov, Rem A. [Симонов, Рэм Александрович]. 2001b [1987]. Древнерусский абак (по данным моделирования математической основы денежных систем) [Drevnerusskiĭ abak (po dannym modelirovaniya matematicheskoĭ osnovy denezhnykh system)] (Ancient Russian abacus (according to the data obtained via emulation of the mathematical foundation of monetary systems); in Russian). [Originally published in 1987]. Reprinted in: Simonov (2001a), pp. 29–37.

Simonov, Rem A. [Симонов, Рэм Александрович]. 2001c [1997]. Археологическое подтверждение использования на Руси в XI в. архаического абака ("счета костьми") [Arkheologicheskoe podtverzhdenie ispol'zovaniya na Rusi v XI v. arhaicheskogo abaka ("scheta kost'mi")] (Archaeological proof of the use of the archaic abacus ("counting with seeds") in Russia in the 11th century; in Russian). Originally published in *Истоки Русской культуры (археология и лингвистика)* [Istoki Russkoĭ kul'tury (arkheologiya i lingvistika)] (Origins of Russian culture (archeology and linguistics)), Moscow: Institute of Archeology, 1997, pp. 178–196; reprinted in Simonov (2001a), pp. 12–29.

Simonov, Rem A. [Симонов, Рэм Александрович]. 2010. К истории счета в допетровской Руси [K istorii scheta v dopetrovskoj Rusi] (On the history of counting in pre-Petrine Russia [a review of Tsaĭger 2010]; in Russian). *Математика в высшем образовании* [*Matematika v vysshem obrazovanii*] (Mathematics in higher education), No. 8, pp. 135–142.

Simonov, Rem A. [Симонов, Рэм Александрович]. 2011. О книге М.А. Цайгера по истории русской науки [O knige M.A. Tsaĭgera po istorii russkoĭ nauki] (About the book by M. A. Tsaĭger on the history of Russian science; in Russian) [a review of Tsaĭger 2010]. *Древняя Русь. Вопросы медиевистики* [Drevnyaya Rus'. Voprosy medievistiki] (*Ancient Russia. Problems of Medieval Studies*, in Russian), Vol. 12, No. 1, pp. 122–126.

Simonov, Rem A. [Симонов, Рэм Александрович]. 2015. Изучение творчества Кирика Новгородца за рубежом [Izuchenie tvorchestva Kirika Novgorodtsa za rubezhom] (Studies of Kirik from Novgorod abroad). *Математика в высшем образовании* [Matematika v vysshem obrazovanii] (Mathematics in higher education), No. 13, pp. 125–142.

Spasskiĭ, Ivan G. [Спасский, Иван Георгиевич]. 1952. Происхождение и история Русских счетов [Proiskhozhdenie i istoriya Russkih schetov] (Origin and history of the Russian schyoty; in Russian). *Историко-математические исследования* [Istoriko-matematicheskie issledovaniya] (*Studies in History of Mathematics*), Vol. 5, pp. 269–420.

Tsaĭger, Mark A. [Цайгер, Марк Аркадьевич]. 2010. *Арифметика в Московском государстве XVI века* [Arifmetika v Moskovskom gosudarstve XVI veka] (*Arithmetic in Moscow state of the 16th century; in Russian*]. Be'er-Sheva: Berill.

Ustryalov, Nikolai G. [Устрялов, Николай Герасимович]. 1842. *Именитые люди Строгановы* [Imenitye lyudi Stroganovy] (*The honorable Stroganovs*; in Russian), Sankt-Peterburg: Publishers of the headquarter of military educational institutions.

Vilenchik, B. [oris (?)] Ya. [Виленчик, Б(орис? Я(ковлевич?)]. 1984. Новые доказательства существования русского архаического абака [Novye dokazatel'stva sushchestvovaniya russkogo arhaicheskogo abaka] (New supportive evidence for [the theory of] existence of the Russian archaic abacus; in Russian). *Советская Археология* [Sovetskaya Arkheologiya] (*Soviet archaeology*), Vol. 28, No. 3, pp. 59–65.

Volkov, Alexei. 1994. "Large numbers and counting rods". In: A.Volkov (Ed.), *Sous les nombres, le monde.* (*Extrême-Orient Extrême-Occident*, no. 16), pp. 71–92. Paris: PUV.

Yushkevich, Adol'f P. [Юшкевич, Адольф Павлович]. 1968. *История математики в России до 1917 года* [Istoriya matematiki v Rossii do 1917 goda] (*History of mathematics in Russia until 1917*; in Russian). Moscow: Nauka.

Zaĭtsev, Il'ya V. [Зайцев, Илья Владимирович]. 2006. *Астраханское ханство* [Astrakhanskoe khanstvo] (*The Astrakhan Khanate*; in Russian). Moscow: Vostochnaya Literatura.

Alexei Volkov is a Professor of the Center for General Education and of the Graduate Institute of History of the National Tsing-Hua University (Hsinchu, Taiwan). His research focuses on the history of mathematics and mathematics education in East and Southeast Asia. He has published a number of papers and chapters on these topics, including "Didactical dimensions of mathematical problems: 'weighted distribution' in a Vietnamese mathematical treatise," in Alain Bernard and Christine Proust (eds.), *Scientific Sources and Teaching Contexts Throughout History: Problems and Perspectives*, Dordrecht etc.: Springer, 2014, pp. 247–272, and "Argumentation for state examinations: demonstration in traditional Chinese and Vietnamese mathematics," in Karine Chemla (ed.), *The History of Mathematical Proof in Ancient Traditions*, Cambridge University Press, 2012, pp. 509–551.

A Short History of Computing Devices from Schickard to de Colmar: Emergence and Evolution of Ingenious Ideas and Technologies as Precursors of Modern Computer Technology

Viktor Freiman and Xavier Robichaud

> On a toujours cherché les moyens de diminuer la fatigue d'esprit et d'abréger le temps qu'entrainent les opérations arithmétiques (People always looked for ways to make the spirit less tired and to reduce the time taken by arithmetic operations).
> Louis Thomas (Chevalier de Colmar) 1852

Abstract Even the brief analysis of the history of the invention of mechanical computing devices from the 17th to early 20th centuries that we provide in this chapter demonstrates the rich potential available to mathematics educators today. The first devices designed by Schickard and Pascal showed a technological complexity when dealing with the issue of representing even simple arithmetic operations, such as addition, by means of gears and wheels. The knowledge they developed, along with the ideas they did not succeed to put in practice, inspired further generations of inventors who not only pursued the search for better aids for calculation practices but also envisioned novel mathematical structures allowing for a more universal approach to computing which, along with technological know-how, eventually led to the modern era of electronic computers, the Internet, and other digital tools and technology-rich environments. At the end of the chapter, we explore the educational potential of this historical development.

Keywords History of computing devices · Mechanical calculators Arithmometers · Educational implications

V. Freiman (✉)
Université de Moncton, Moncton, NB, Canada
e-mail: viktor.freiman@umoncton.ca

X. Robichaud
Université de Moncton, Shippagan, NB, Canada
e-mail: xavier.robichaud@umoncton.ca

© Springer Nature Switzerland AG 2018
A. Volkov and V. Freiman (eds.), *Computations and Computing Devices in Mathematics Education Before the Advent of Electronic Calculators*, Mathematics Education in the Digital Era 11, https://doi.org/10.1007/978-3-319-73396-8_13

Introduction

One of the reform-based documents sharing a futurist vision of mathematics in the 21st century, dated back to 1980s, "Project 2061: Science for all Americans" (1989), underlines the double-edged nature of mathematics, being beauty and intellectual challenge, on the one side, and a tool for useful applications in a multitude of practical domains, on the other. These two, at first glance contrasting approaches to mathematics have a long and rich history, for example, in the case of mathematics in Renaissance Europe in the 15th and 16th centuries. By quoting several sources, Johnston (1996) noticed:

> [M]athematics could be a spiritual discipline, as a guide to meditation on the divine; alternatively, it could be acquired as a vocational resource by merchants, developing their bookkeeping skills. Mathematics was both a preliminary to natural philosophy in university arts courses and, through studies of fortification, artillery and the ordering of troops, an element of informal aristocratic military education. The subjects encompassed within the cycle of mathematical arts might range from arithmetic to perspective, mechanics to surveying, and navigation to astronomy, each topic with canonical texts, instruments and practices at elementary and advanced level. (p. 93)

By picturing a complex portrayal of mathematical practice and mathematical practitioners in 16th-century England, the author reveals its public character as being visible and accessible, as opposed to mathematics at courts and at academies.

From an educational point of view, in his analysis of 16th-century arithmetic, Jackson (1906) considers two types of the so-called "subject-matter" of arithmetic, one as a basis for the "needs of the trader" (it includes simple operations with comparatively small integers and fractions, as well as more complex denominate numbers) and the other, for the "needs of the scholar" (covering a larger domain, with emphasis on more theoretic topics, like proportions, progressions, roots, along with extended arithmetic, to the solution of equations) (p. 186). The former (the needs of the trader) inspires "commercial arithmetic;" the later (the needs of the scholar) a "disciplinary school arithmetic." Indeed, according to the author, "when a nation passes from pastoral and agricultural pursuits to those of manufacture and trade, the transition seems to be reflected in the subject-matter of its arithmetic" (Jackson 1906, p. 187).

Regarding the tools for calculations, the author notices another modification of the subject-matter during the transition from "counting on lines" (medieval abacus) to "counting with numerals" (algorism); this observation is particularly interesting in terms of the development of computing devices, since this modification is based on using a "better method" for solving business problems (algorism), whereas a modern (19th century) tendency rather goes in the opposite direction, where the change is directed from the "head work" (written calculations) to automatic machinery (mechanical calculator). Both trends have, however, a common purpose: to make easier the "method of practical calculation" (Jackson 1906, p. 189).

More recently, when introducing their book about technology, mathematics, and secondary schools, from the historical perspective based on the materials from the United Kingdom, Johnston-Wilder and Pimm (2004) considered the original aim behind the creation of devices, such as Babylonian tables, Napier rods, slide rules, mechanical and electronic calculators, and finally, modern computers in order to "assist with the doing of mathematics, especially for the carrying out of algorithmic computations and the coming up with (and holding by means of tables) the values of particular functions" (p. 3). The use of these devices could be purely technical (enhanced or substituted performance), but sometimes also have "greater or lesser pedagogical intent," or even be specifically designed for mathematics classrooms, such as geo-boards, Dienes multibase blocks, and, more recently, interactive geometry software (Johnston-Wilder and Pimm 2004, p. 3).

A very recent book by Jones (2016) provides a detailed account of the development of reckoning, invention of calculating machines, and other innovations from Pascal to Babbage. The author mentions, among others, that relatively few people in early modern Europe had the skills necessary to perform addition, even when using an abacus. With numeracy unevenly distributed among European countries there was a need for more people to be capable of doing sums (Jones 2016). Moreover, even better educated people were having difficulties and needed help in performing simple multiplication, and even more so when calculating interest. At the same time, from the more elitist point of view, reckoning was viewed as a merely mechanical activity compared to the new (at that time) geometry and algebra which were considered as higher level mathematics. In a similar vein, logical reasoning, even in its most formal way, was considered "greater than the debased mechanical procedures of commercial arithmetic with symbolic reasoning yet to be developed" (Jones 2016, p. 4).

When looking into the development of mechanical computing devices between the 16th and 19th centuries, we aim to get additional insight into particular contexts which required increasing numbers of calculations. Examples include tasks which people had to solve but felt the capacity of existing methods (using written arithmetic and some manipulative tools) was too limited, which led to the search for new means, including more automatic approaches; ideas pushing beyond the mere mechanization of computational procedures toward a more philosophical view of what a computing device of the future could be like; the struggles and failures that accompanied the technological process of the creation of machines; and the reflective attempt to understand what did and did not work and how it could be improved, both in terms of new engineering solutions and ideas.

While becoming a science in its own right, "computing" science has closely interacted with mathematics since very ancient times; pebbles, counters, rods, abaci, etc., are often mentioned as the first calculators (or even computers). The whole history of this interaction that has developed since antiquity still remains to be written; our contribution will focus on one particular aspect of it, the history of mechanical computing devices that appeared at this particularly important period of development.

Initial Steps Toward Automatization of Calculations in the 17th Century

As European life became, toward the 16–17th centuries, more complex and diversified in terms of societal needs and issues, mathematicians became increasingly involved in the (technological) process of (better) responding to practical needs, that implied an increasing role and complexity of computations and of the apparatus aiding such calculations, leading to the creation of computing machines.

The technological revolution we are experiencing today has radically changed the way we do things, in our case, calculations, and has been developed over a long time by pioneers like Wilhelm Schickard (1592–1635), Blaise Pascal (1623–1662), Gottfried Wilhelm Leibniz (1646–1716), Samuel Morland (1625–1695), Thomas de Colmar (1785–1870), and Charles Babbage (1791–1871), to mention but a few of the most important figures we find in the literature on the history of computing and computer science. Nordhaus (2007) called them "mathematically inclined inventors" (p. 130). From the technological point of view, Cortada (1993) identifies two phases in the history of the development of mechanical calculators, one in the 17th century when the demand was rather limited and the capability of tools was inadequate; the second in the 19th century when the technology was significantly better and opened doors for many possibilities which led not only to the invention of more reliable arithmometers but also to their eventual commercialization and increasing use in many spheres of the economy and science (Cortada 1993).

At the beginning of the 17th century, it became necessary to simplify the methods of calculation for astronomy and algebra, which had made great progress, but also for commerce, banking, and the development of public finance (most commercial or financial calculations were still done using tokens that were moved on a special table). For instance, John Napier (1550–1617) brought important improvements to the technique of calculation and introduced logarithms (Nepero 1614), followed by an invention of particular rods (now known as Napier bones) to help with multiplication and division (1617). Ten years later, Edmund Gunter (1581–1626), William Oughtred (1574–1660), and Edmund Wingate (1596–1656) realized the first slide rules by combining two slats with equal logarithmic scales, intended to further facilitate calculations conducted, among others, by astronomers (Taton 1963).[1]

Regarding the origins of the so-called mechanical (automatic) computing devices, for a relatively long period of time, Pascal was believed to be the first inventor of such a machine. In the second part of the 20th century, this view was challenged in the debate about the possible invention of a certain calculating mechanism by Leonardo Da Vinci (1452–1519), near the end of the 15th century and beginning of the 16th century (see, for example, Belanger and Stein 2005), based on the discovery of Da Vinci's drawings in the late 1960s. According to Nordhaus (2007), Leonardo could

---

[1]For more details about the complex history of this invention, see, for example, Cajory (1909).

have possibly sketched, around 1502, a mechanical adding machine which, however, "was never built and probably would not have worked" (p. 130).

Also, later in the 20th century, a drawing found in Joannes Kepler's (1571–1630) archive suggested that Schickard was the first scientist to design an arithmetic machine and also to build a mechanized hand planetarium. A Professor of oriental languages in Tubingen (Germany), Schickard was well known in the scientific community as an astronomer, mathematician, and cartographer. In his design of the machine, according to Graf (1995), he integrated the technique of Napier's bones together with the technological inventions of clockmakers, this later fact acknowledged in the labeling of Schickard's device as a "calculating clock" (in German, Rechenuhr). He wrote to Kepler in 1623 and 1624 to announce that he had realized an arithmetic machine with cogwheels which "calculates from given numbers in an instantaneous and automatic manner, it adds and subtracts, multiplies and divides," and also made a sketch of it (Taton 1963, p. 143).

The cogwheel was already known for a long time as a principle of automatic clockwork and as a mechanism of operation of astrological instruments such as Schickard's hand planetarium, which presented the movement of the planets. In fact, the history of mechanical planetariums and astronomical devices goes back to antiquity and is, per se, a complex field of study which is beyond the scope of this chapter. As astronomical calculations often involved large numbers, so the idea of an automated arithmetic machine was more likely to occur in the astronomy domain. For instance, a Greek calculating mechanism, Antikythera, dated 60 BC, which has been extensively studied during the past two decades, could predict, for many years ahead, solar eclipses, produce a Metonic calendar, and calculate the timing of the Olympic Games, among many other functions (Freeth 2002; Freeth et al. 2008; Freeth 2014).

Nevertheless, going back to Schickard's invention, there is also evidence that the model that Schickard had built was destroyed in a fire, and that Schickard's subsequent correspondence no longer contained records related to this machine. Schickard seemed to have abandoned his research on the mechanization of arithmetic calculation (Taton 1963). Moreover, Kepler did not seem to have recognized the importance of this realization. We do not have any contemporary testimony on Schickard's machine, and it seems that Pascal designed his arithmetic machine independently. Yet, an interesting discussion could emerge from Cole's (1995) analysis of the possible sources of ideas for calculating machines that could have inspired Pascal. For instance, Pascal's father, a mathematician himself, had noticed his son's mathematical talents at a very young age and sent him to M. Mersenne (1588–1648), another famous mathematician who might have heard (from Kepler?) about Schickard's work and could have shared it with Pascal's father, who, in turn could have shared it with his son, encouraging him to start work on a computing device.

Whatever is the truth in this story, when beginning his work in 1640, Pascal wanted to create a machine that would perform the four elementary arithmetic operations to facilitate his father's routine work which required a lot of calculations, and which turned, in fact, into an attempt to resolve a more general problem of the mechanization of arithmetic calculation. Consequently, he succeeded to create, in 1642, a prototype of an "arithmetic machine" named *Pascaline*.

In 1645, Pascal presented his machine to the public with an *Advis nécessaire à ceux qui auront curiosité de voir la Machine Arithmétique et de s'en servir.*[2] Reading this document, one notices that Pascal does not only merely present his machine; he explains, in fact, its functioning, but he also examines different problems he had to overcome and explains how he succeeded in this process. In short, he uses a "didactical" discourse rather than one of the inventor, by sharing his thinking from a larger, mathematical, as well as technical, perspective, which is illustrated, for example, in the following quotation:

> I do not doubt that, after having seen it, it falls at first in your mind that I must have explained it in writing, and its construction and use; and that, in order to render this discourse intelligible, I was even obliged, according to the method of geometers, to represent by figures the dimensions, the arrangement, and the ratio of all the parts, and how each one must be placed to compose the instrument and put its movement into perfection [...]. (Pascal 1645, p. 9)

In short, he sees his machine as an "execution of his thought," which gives access to a "method" and explains the simplicity of the movement of parts, their convenience, and speed. However, he avoids the difficulty of expressing in writing, "the measures, the forms, the proportions, and the properties of so many different pieces," considering that this can only be taught orally, written discourse being "useless and embarrassing" in this matter (p. 10). Denis Diderot (1713–1784) nevertheless did not shy away from this difficulty, and completed the didactic effort of Pascal, providing a detailed description of the arithmetic machine in the first volume of *Encyclopaedia* (Diderot and D'Alembert 1751). The detailed description by Diderot allowed the potential user of the machine to understand how addition, subtraction, multiplication, and division worked using the *Pascaline*. Producing this type of document formed part of the effort to make the affordance of the machine more explicit and therefore to make it more accessible to the public.

The artifacts used by Pascal, such as wheels, gears, number discs, that already existed, provided him with affordances to create something new, in his case, the arithmetic machine. But Pascal also encountered great difficulties in the process of creation, however, one such difficulty regarding carrying units he could resolve. According to Temam (2009), when there is no need to carry units in a process of calculation, the mechanism of addition is quite simple, as all corresponding units can be added independently. It is when the sum of two digits exceeds nine, that it becomes necessary to have a mechanism that will turn the wheel of the next decimal position by one unit. Moreover, this operation must be done in a cascading system, for example, when the operation reaches 999 and one must add a unit to get 1000. Pascal's ingenious solution brought the idea of a system of cascading jumpers ("sautoires en cascade") which apparently survived up until the appearance of electronic calculators, with the same principle eventually used in mechanical odometers in cars (Temam 2009).

---

[2]Notice necessary to those who will have the curiosity to see the arithmetical machine and to use it.

Even if these ideas allowed Pascal to deal with additions and subtractions, multiplications and divisions were, however, more laborious with this approach. Overall, Pascal evokes the complexity of his machine: "You will be able to notice a kind of paradox, to make the movement of the operation simpler, the machine had to be constructed with more compound motion" (Pascal 1645 p. 14). The machine needed to be more and more complex, in order to make operations with it simple enough for the end user. Pascal wanted his machine to be reliable and robust, but the novelty of this creation did not allow the workers, who were not ready for the task and lacked knowledge in both domain and tools, to construct it in all its complexity (Pascal 1645, pp. 14–15). In the 19th century, Lardner (1834, cited in Walford 1871) mentioned that Pascal's machine could do only restricted arithmetical operations and still required manual work which could lead to errors in calculations, and overall, did not do better than a human calculator using a pen. It is also worth mentioning the numerous technical difficulties and higher costs that prevented Pascal's machine from wide distribution. All these challenges were eventually addressed by Leibniz in his later attempt to build his machine known as *Step Reckoner* that we analyze in the next section.

Calculators of Morland and Leibniz: When Practical Needs, Philosophy, Mathematics, and Technological Creativity Push the Boundaries of Automatic Computing

While Leibniz's name comes often next to Pascal's in the chronology of the historical development of calculating machines, several authors point at the important work of the English "courtier-inventor," Morland, who invented three types of calculators, one for addition and subtraction, one for multiplication and division, and one for trigonometry, which were manufactured by the then famous Sutton workshop (a trigonometric device called *Maccina Cyclologica Trigonometrica* was invented in 1663; two other machines are dated 1666, see Ratcliff 2007). These are the oldest surviving English calculating machines that were still being sold by London instrument makers up to around 1710. Called by Whitelocke[3] "ingenious mechanist" (cited, for example, in Halliwell 1838), Morland was essentially looking for providing better aid in calculations for people who had not very much experience with arithmetic, thus focusing on the routine tasks to facilitate calculation, to make it more affordable, and eliminate some rote learning required for those who wanted to perform them with exactitude (Ratcliff 2007).

From the technological perspective, these machines were essentially "gear versions" of some of the commonly used mathematical instruments of the time (among them, abacus, counting boards, and Napier's bones) (Ratcliff 2007). In his more

---

[3]Sir Bulstrode Whitelocke (1605–1675) was Head of the English diplomatic mission in Sweden in 1653; Morland took part in that mission.

pragmatic approach, different from Pascal's, Morland dropped the idea that gear design could work automatically (as in the case of *Pascaline*), and therefore escaped some technical challenges (like an automatization of the carrying algorithm). Hence, his first machine could only perform addition and subtraction, without the function of carrying units to higher positions which still had to be done by pencil and paper.

However, in his device, Morland succeeded to integrate a mechanism for calculations with English monetary units, which consisted of a complex combination of three different number systems, with base 4, base 12, and base 20, and this invention was recognized as innovative (Tent 2012, p. 87). This device, as well as Morland's second machine which was able to do multiplications and divisions was seen by Leibniz, who, according to Tent (2012), had a meeting with Morland, saw his machines, compared them to those that he had been building himself, and made some critical comments, the first regarding their inability to carry units to higher positions when performing addition, and the second, relating to multiplication, consisted of the observation that the machine was but a mechanical version of Napier's bones.

On his part, Leibniz, a Doctor of Law who practiced philosophy and mathematics in Germany, was convinced that he would be able to build a machine for the four mathematical operations, even better than that of Pascal. The idea of having a "suitably arranged" machine which could accomplish "not only counting but also addition and subtraction, multiplication and division easily, promptly, and with sure results" came to Leibniz when he saw an instrument capable of recording automatically a number of steps made by a pedestrian (Leibniz 1685, p. 173). Having designed a prototype of the calculator, he traveled to London and Paris to share his ideas with a circle of renowned philosophers and inventors.

During his stay in Paris in 1672, Leibniz also studied the writings of Christian Huygens (1629–1695), an astronomer, mathematician, and physicist, including his work on clocks. Having a chance to meet with Huygens, Leibniz was inspired by the ideas to study geometry and the mathematics of motion. Huygens also advised him to study Pascal's work *Pensées* (Pascal 1670, cited in Bloch 2016) about his *Pascaline*. While praising Pascal's machine as an example of the "most fortunate genius," Leibniz recognized its limits to facilitate only additions and subtractions, while the operations most needed for "practical use to people engaged in business affairs" are multiplications and divisions which were still cumbersome with Pascal's device (Leibniz 1685, p. 180).

Unlike Pascal, who had to innovate from scratch, Leibniz used the affordances (the mechanisms of the *Pascaline*) already created by Pascal. He wanted to improve the arithmetic machine, trying to correct its defects, especially regarding multiplications and divisions. In 1673 he succeeded in building his calculating machine: the *Step Reckoner*. Leibniz designed the concept of a mechanism where the upper row is the *Pascaline*, and the lower row is used by the machine for multiplications. The central idea was to imitate the order in which the mathematical operations are carried out on paper. For example, he describes that in order to multiply 365 by 124, the first number (365) needs to be multiplied by 4, at the first step (Fig. 1).

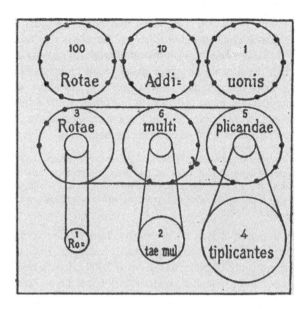

Fig. 1 Drawing by Leibniz showing how his calculator could use Pascal's calculator. This was the first description of a pinwheel design (1685)

Here is the algorithm of the operation described by Leibniz (which we cite here grouped in four stages to help the reader follow it):

(1) Turn the multiplier-wheel 4 by hand once; at the same time, the corresponding pulley will turn four times (being as many times smaller) and with it the wheel of the multiplicand 5, to which it is attached, will also turn four times.

(2) Since the wheel 5 has five teeth protruding, at every turn. 5 teeth of the corresponding wheel of addition will turn once and hence in the addition box there will be produced four times 5 or 20 units.

(3) The multiplicand-wheel 6 is connected with the multiplicand-wheel 5 by another cord or chain and the multiplicand-wheel 3 is connected with wheel 6.

(4) As they are equal, whenever wheel 5 turns four times, at the same time, wheel 6, by turning four times, will give 24 tens (it namely catches the decadic addition-wheel 10) and wheel 3, catching the addition-wheel 100, will give twelve hundred so that the sum of 1460 [365 × 4 = 1200 + 240 + 20. – V.F. & X.R.] will be produced (Leibniz 1685, p. 176).

Next, to multiply 365 by 2 (or rather by 20) it is "necessary to move the entire adding machine by one step so to say, so that the multiplicand-wheel 5 and the multiplier-wheel 4 are under addition-wheel 10, while they were previously under 1, and in the same manner 6 and 2 under 100 and also 3 and 1 under 1000" (Leibniz 1685, pp. 176–177).

Overall, Leibniz saw his device being "desirable to all who are engaged in computations which, it is well known, are the managers of financial affairs, the administrators of others' estates, merchants, surveyors, geographers, navigators, astronomers, and [those connected with] any of the crafts that use mathematics" (Leibniz 1685, p. 180). Despite the long period of work he undertook (from 1672 to his death in 1716), despite his promises that even a child could do everything without knowing about calculations, and despite that, according to later authors, Leibniz's machine inspired many calculating machines of the 19th and 20th centuries, he did not succeed in making the multiplication mechanism work, that is, he did not solve the problem that he faced from the beginning:

> Multiplication was, after all, an iterated addition in that the same number had to be added to itself as many times as the multiplier required, and division was the reverse of this process. But how to find a way to store the number to be repeated in the addition/subtraction mechanism, how to count the number of repetitions and how to input easily any desired number? (Morar 2015, p. 126)

While aiming to improve Pascal's invention, Leibniz had in mind more than building a device to save time and reduce errors in calculations. In fact, it was noticed, among others by Lardner (1843, cited in Lister et al. 2009), that the real significance of his invention was to demonstrate that reasoning (which means all kinds of reasoning: moral, legal, philosophical, and scientific, all of which follow rules, as do calculations) can be mechanized not by means of arithmetic but using some kind of formal logic ("language of thought") to represent something "vague and general" (Lister et al. 2009, p. 365). This would mean that every "reasonable man, faced with a difficult question of philosophy or policy would express the question in a precise language and use rules of calculation to carry out precise reasoning" (Beeson 2004, p. 82). Wiener (1988, cited in Sawday 2007, p. 241) argued that Leibniz's computing machines were "offshoots of his interest in a computing language, a reasoning calculus which again was in his mind … ." In this sense, according to Lister et al. (2009, p. 365), the calculator itself was a "by-product of this larger project."

Leibniz continued to work on his machine until the end of his life, but despite this, none of the models of calculators developed during this period worked. Leibniz died in 1716 without completing the construction of the third model of the calculator. In fact, technological difficulties seemed to prevent his machine from doing accurate computations. Horsburgh (1914) points that, "this was long before the days of accurate machine tools" (p. 122). Leibniz himself blamed for this failure to "properly execute what was so well designed" the artisans he employed. In his defense, one can quote Huygens who affirmed in 1675 that "these things are not meant for demonstrating utility, they are more for demonstrating the force of the human mental capacities" (Morar 2015, p. 133). Indeed, history has retained two important Leibniz's inventions: the so-called *Leibniz Wheel* with variably extractable teeth and movable carriage, which remained the basic components of most calculators for a long time after his death, as well as the idea of using a binary system for computations which continues to be the mainstream of computer science nowadays.

Indeed, in his paper written in 1703 (in French), he introduces the table of numbers 1–32 written using only digits of 0 and 1, as well as basic operations in this number system (Fig. 2).

TABLE 86 MEMOIRES DE L'ACADEMIE ROYALE

DES
NOMBRES.

100	4
10	2
1	1
111	7

1000	8
100	4
1	1
1101	13

0	0	0	0	0	0	0
0	0	0	0	0	0	1
0	0	0	0	0	1	0
0	0	0	0	0	1	1
0	0	0	0	1	0	0
0	0	0	0	1	0	1
0	0	0	0	1	1	0
0	0	0	0	1	1	1

bres entiers au-deſſous du double du plus haut degré. Car ici, c'eſt comme ſi on diſoit, par exemple, que 111 ou 7 eſt la ſomme de quatre, de deux & d'un?
0
1
2 Et que 1101 ou 13 eſt la ſomme de huit, quatre & un. Cette propriété ſert aux Eſſayeurs pour
3 peſer toutes ſortes de maſſes avec peu de poids,
4 & pourroit ſervir dans les monnoyes pour don-
5 ner pluſieurs valeurs avec peu de piéces.
6 Cette expreſſion des Nombres étant établie, ſert à faire
7 très-facilement toutes ſortes d'opérations.

Fig. 2 Table des nombres (table of numbers) (Leibniz 1703, p. 86)

The explanations given by Leibniz provide additional insight into the origins of his idea and its rationale. Namely, he claims that the system of grouping numbers 2 by 2 instead of 10 by 10 (which is the decimal system) adopted in binary arithmetic is the easiest compared to all other bases, and serves to "the perfection of the Science of Numbers". He notices, among other reasons the easiness in continuation of his conversion table representing numbers beyond 32. Also, he argues that for the execution of arithmetic operations, there is no more need to proceed by trial and error, or to learn tables by heart (Leibniz 1703). From the point of view of the science of numbers, other observations can be made from the binary tables which lead to the discovery of exponents (square, cube, etc.), as well as geometric numbers (triangular, pyramidal, etc.).

The idea of the binary system was not new at the time of Leibniz. Bacon (1623, pp. 277–284; 1640, pp. 264–271; see also Lions 1991) had mentioned that this (i.e., the binary) way of writing numbers opens the door for humans to "expresse and signifie the intentions of his minde, at any distance of place, by objects which may be presented to the eye, and accommodated to the eare; provided those objects be capable of a twofold difference only" (Bacon 1640, p. 266; see also Lions 1991, p. 228).[4] Moreover, Leibniz tracked the origins of the binary system to the very ancient time (some 4000 years before him) and credited it to the authorship of the Chinese king and philosopher Fohy 伏羲 (Fuxi in modern Chinese *pinyin* transliteration), who is believed to be the founder of the Chinese Empire and science, and the inventor of eight divinatory symbols composed of a combination of continuous and broken lines known as Eight Trigrammes or Eight Coua (八卦 *bāgùa* in modern Chinese) (Fig. 3).

While the meaning of these signs seems to be lost in the past, tables provided by Leibniz could, according to him, be a key to the mystery of binary numbers from Fohy's time. While it was left to the modern authors to discuss Leibniz's hypothesis of the origins of the binary system, it seems plausible to suggest that his interest in Chinese history and philosophy was a part of his bigger project which was the search for a universal language of science (Ryan 1996). While the detailed analysis

---

[4]For a detailed historical outline of the binary system, in general, and prior to Leibniz, in particular, see Glaser (1981), Chap. 2.

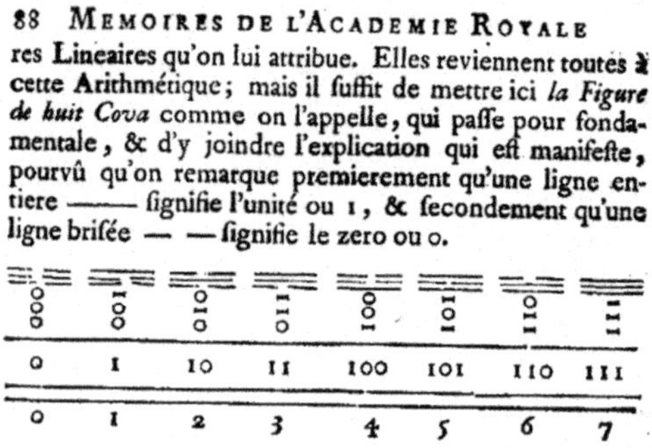

88 Memoires de l'Academie Royale
res Lineaires qu'on lui attribue. Elles reviennent toutes à cette Arithmétique; mais il suffit de mettre ici *la Figure de huit Cova* comme on l'appelle, qui passe pour fondamentale, & d'y joindre l'explication qui est manifeste, pourvû qu'on remarque premierement qu'une ligne entiere ——— signifie l'unité ou 1, & secondement qu'une ligne brisée — — signifie le zero ou o.

Fig. 3 La Figure de huit Cova (the figure of eight Cova [= Coua, *gua*]) (Leibniz 1703, p. 88)

of this part of Leibniz's work lies beyond the scope of our chapter, there is a strong claim in modern literature that his idea could have influenced the development of computer science in the 20th century and the choice of the binary system as the base for representation of information in electronic devices (Davis 2000).

However, in the 18th century, according to Lions (1991, p. 228), an important development in the field of representing information by means of punch cards led to the invention of a device that could organize storing, and the automatic treatment of, a large number of punch cards (used by Joseph Marie Jacquard, 1752–1834, who invented a loom programmed with perforated cards which later inspired Charles Babbage, 1791–1871, in using perforated cards in his analytical engine). Along with Pascal's and Leibniz's work on calculation machines, as well as the ideas of systematic organization of calculations with operation systems allowing one to separate the control of the (automatic) process from its realization, it influenced further significant improvement in mechanical computing devices in the 19th century which we will describe in the next section of this chapter.

Getting Closer to Having Computations Done Automatically by Machines: The 19th-Century Breakthrough

In the 18th century, many inventors continued working hard on building and improving computing devices. Yet, they seemed to remain within the limits set by Pascal, Morland, and Leibniz. It was only during the 19th century that real progress toward the better automation of calculations was made, due to the inventions of Charles Babbage in England and of Thomas de Colmar in France, among others. In this section, we will analyze their work in more detail.

In the first half of the 19th century Charles Babbage, who is now considered the "father" of the modern computer, started designing his *Difference Engine* (1822) and *Analytical Engine* (1834). His idea was to create an automatic and mechanical calculator with an external program, capable of operating with both numbers and symbols (Durand-Richard 2010).

According to Green (2005), the idea to design a machine that calculates and prints numbers automatically, with no mistakes in computation or copying, came to Babbage after his discovery of errors in the Board of Longitude's tables. As the mathematical principle for his design, Babbage used a mathematical technique, called the "method of finite differences" for constructing polynomials to generate the values that could be used to approximate other functions.

Babbage began working on the *Difference Engine* around 1822. Originally, he was mainly focused on astronomical calculations (and their applications, for instance, in nautical almanacs). The *Difference Engine* of Babbage allowed the computation of successive values of a function, obtained by the method of finite differences. Mathematically, it is a question of calculating successive values of a function knowing its initial value and those of its successive differences, for example[5]:

$$\Delta_n f(x) = \Delta_{n-1} f(x+1) - \Delta_{n-1} f(x)$$

In a more simplistic way, as was explained by Babbage himself, a simple table of squares can be built, on one side, as a number multiplied by itself (like 5 times 5 giving 25), or it could be obtained by means of successive additions of differences of the first and the second order (illustrated in Table 1).

Table 1 Table of successive additions of differences (Menabrea 1842, p. 355)

A Colonne des nombres carrés.	B Différences premières.	C Différences deuxièmes.
1		
	3	
4		2 *b*
	5	
a 9		2 *d*
	7	
c 16		2
	9	
25		2
	11	
36		

[5]Durand-Richard (2010, p. 293).

One can notice that each next term in column A of the table is the result of the previous one (like 1, for example) added to the first difference (3), so 1 + 3 gives 4. The next square (9) is the result of addition of the previous one (4) and the next first difference (5), so 4 + 5 = 9. This pattern works for all square numbers; it works because the second difference remains the same (2) for all steps. According to Babbage, "for the first series of differences may be formed by repeatedly adding the constant difference (2) to (3) the first number of in the column B, and we then have series of numbers 3, 5, 7, etc.: and again, by successively adding the each of these to the first number (1) of the table, we produce square numbers" (Babbage 1961, p. 319).

The *Difference Engine* marks a triple break with the line of Pascal's arithmetic machines: as a printer (never completed), as a tabulator, and in its wide possibilities for calculation (Mosconi 1983). These provisions "suggested to the inventor by the inherent nature of the mechanism confer possibilities which he had never foreseen" (p. 79). This machine was no longer "a docile instrument of the calculations; it astonishes its creator and will even enlighten him" (Babbage 1961, p. 79).

The different stages of the process are materialized by a part of the machinery. This is why historians consider this "automatic mechanical calculator with external program" to foreshadow the von Neumann computer architecture: input–output devices, memory unit, control unit, arithmetic unit (Cragon 2000, p. 5), and the use of different punched cards of Jacquard's type (Chazal 2000, p. 197). However, it is foremost the principle of the division of labor that results from this organization of computational work (Mosconi 1983). In his own account of this principle, Babbage recalls the development of mathematics at the time of the French Revolution which gave a boost to the production of (new at this time) mathematical tables "on the most extensive scale" to "facilitate the application of the decimal system which they had so recently adopted."

With reference to de Prony (Gaspard Clair François Marie Riche de Prony, 1755–1839), Babbage cites a joint enterprise of English and French governments that hired a large number of individuals who performed calculations to accomplish a very ambitious task of compiling logarithmic and trigonometric tables:

> M. de Prony engaged himself, with the governmental committees, to create logarithmic and trigonometric tables for centesimal division of the circle which would not only be perfect as far as precision is concerned, but also would form the largest and most impressive monument of calculation which had never been executed, or even conceived, before.[6]

In his explanation of de Prony's ideas, Babbage analyzes the organization of computations in three sections. The first section, formed by five or six of the "most eminent mathematicians" of France had to investigate, among the various analytical expressions, the ones that could be employed in a single function. Furthermore, this function needed to be adapted to simple numerical calculation by many individuals

---

[6]"M. de Prony s'était engagé avec les comités de gouvernement, à composer pour la division centésimale du cercle, des tables logarithmiques et trigonométriques, qui, non-seulement ne laissassent rien à désirer quant à l'exactitude, mais qui formassent le monument de calcul le plus vaste et le plus imposant qui eût jamais été exécuté, ou même conçu" (Anon. 1820, p. 7, cited by Babbage 1832, p. 155; Babbage 1961, p. 316).

employed at the same time. The second section consisted of seven or eight "persons with considerable acquaintance with mathematics" whose duty was to convert into numbers the function put in their hands by the first section. Those numbers were delivered to the third section consisting of 60–80 people who had almost no knowledge beyond two basic operations, addition and subtraction, but who mastered them to the point of making far fewer mistakes than the more knowledgeable individuals grouped in sections one and two. It is exactly as a substitute for this third category of calculators that the calculating-engine was to be produced (Babbage 1961, p. 317) (see also Campbell-Kelly et al. 2013, pp. 108–110 for a more detailed explanation of de Prony's principle).

Babbage had to abandon his *Difference Engine* in 1842 due to a lack of governmental support, but that meant that he could focus on the *Analytical Engine*, another device he had already began to work on in the mid to late 1830s in order to materialize the ideas of Symbolic Algebra (this machine was never finished either). Symbolic Algebra was an attempt to constitute algebra as science and aimed, above all, to explain the foundations of an algorithmic conception of computation. Babbage recognized the *Analytical Engine* as a universal machine capable of calculating values of any function (Durand-Richard 2010, p. 296).

According to Green (2005), "these ideas were critical to the new machine, comparing to the *Difference Engine*, thus representing one of the highest forms of rational thought known" (p. 38). As also mentioned by Green (2005), Jacquard's automated loom was used by Babbage to get cards lashed "together end to end and punch holes in them that could be 'read' by a number of movable pins in the machine" (p. 38). This system ensured, according to the author, the process of "encoding" information, along with the possibility of repeating the same set of cards ("looping") and making decisions about which cards to execute based on intermediate results ("conditional branching")—all processes still central to modern computing theory (Green 2005).

But besides this technical invention, the ideas of a universal machine explored by Babbage were connected to more general ones related to the ability of a machine to think widely as discussed by Ada Lovelace (1815–1852) and Luigi Frederico Menabrea (1809–1896). Menabrea (1842, cited in Green 2005), for instance, argued that the tasks of mathematics may be divided into two parts, the mechanical one and the other "demanding the intervention of reason, [which] belongs more specially to the domain of the understanding" (p. 39). The authors believed that machines could be employed for executing the mechanical part of mathematics (and thus eventually replace the earlier mentioned third section of human calculators used in the division of computational labor). Lovelace (also cited in Green 2005) added notes to Menabrea's text mentioning the *Analytical Engine* as a "uniting link" between mind and matter which she believed could be a way toward the eventual design of the "thinking machine." Menabrea himself added that the "*Analytical Engine* would be able to replace the second section as well (those who plug the numbers into the formulas produced by the expert mathematicians in the first section)" (Green 2005, pp. 40–41).

Having much more limited and rather pragmatic ambitions then those of his predecessors (Pascal and Leibniz), as well as one of his contemporaries, i.e.,

Fig. 4 The front panel ("platine") of the 1822 *Arithmomètre* (La companie d'assurance "Le Soleil," 1929)

Babbage, who faced much more difficulties in creating a machine that would meet more universal needs and which would adapt to several domains, Thomas de Colmar, an insurance agent, the founder of the *Compagnie du Phénix*, focused on the task of creating a machine that would meet the needs of a specific domain (bookkeeping). With access to much better technology available at that time (compared to the time of Pascal and of Leibniz), Thomas de Colmar used a stepped drum with a set of teeth of incremental lengths (the so-called "Leibniz wheel") which he further improved with his own inventions. As a result, he succeeded in producing the first arithmetic machine that could work fully automatically and provide stable and exact results for basic operations. Moreover, his *Arithmomètre* was commercialized, manufactured on a larger scale, and used in different industries for several decades (Fig. 4).

For more than 60 years after de Colmar's invention, many improvements were made to his *Arithmomètre*, its practical qualities were quickly recognized, and it was reliable (it could be used every day for many years without any need for repair) (Marguin 1997). It was expensive, reserved for large companies, and not at all for schools. Despite some enthusiasm from French mathematics university teachers, the *Arithmomètre* was never considered at the time to be used as an educational tool (Ageron 2016). However, it was very effective as an aid for increasingly compli-cated calculations. De Colmar's *Arithmomètre* made it clear for the first time that a "steel brain" could be faster and more reliable than a human brain.

Educational Implications of the *Arithmometer* from the Late 19th Century to the First Half of the 20th Century

As work on the improvement of mechanical calculators, and making them more automatic, continued through the 19th century, along with their increased use in science, commerce, banking, as well as other areas of everyday life, schools relied on more "old fashioned" tools as aids for calculations, such as various objects (sticks and pebbles used as counters), tables, drawings, or bead *abaci*. The first part of the 19th century had already seen rapid growth in the pedagogical movement in Europe, the United States, and in Russia, based on the use of different manipulatives including bead abacus (also known in different parts as "numeral frame," "ball-frame," "schyoty," "Kugel Rechner," or "boulier") to support the early learning of arithmetic. One of the typical books of this period, Mayo and Mayo's (1837) *Practical Remarks on Infant Education* advises:

> The children's first lessons upon arithmetic should be with visible objects, that through the medium of their senses they may obtain accurate ideas of number before they attempt their combinations. To facilitate this, the frame of balls called the arithmometer, is very useful; it presents the children with an actual representation of number and enables them to compare and combine for themselves. When they have acquired the knowledge and names of numbers, they must apply it to all they see around them, and find examples in their own persons or in the room. They should next be exercised in mental arithmetic, using the arithmometer to correct any error they may make, or to help them through any difficult solution. (p. 78)

From the beginning of the 19th century, at least two different aspects that make the computing device (here, "frame of balls") useful in teaching and learning were underscored in the didactical texts: visualizing numbers and helping with correct mental calculations. This double functionality of the tool remains an important didactical issue in today's era of electronic devices. Almost two centuries ago, according to yet another book on the education of young children in the United Kingdom published by the Central Society of Education, in more "advanced schools," "the use of the abacus, arithmometer, or bead-table" was viewed as an important help for explaining addition, subtraction, multiplication, division, reduction, and proportion (Baker 1839, p. 35). It is interesting to notice that in the educational context, compared to the mechanical calculators, the term "arithmometer" was applied to a larger spectrum of computational devices. For instance, toward the end of the century, in his arithmometer for schools, Sonnenschein used several types of blocks to represent 10-base numbers and operations with them, as shown in Fig. 5. The teacher helped students learn about conversion of units (cubes) into tens (staves) and tens into hundreds (plates) while doing addition (Sonnenschein 1879).

Regarding the possible educational use of Thomas's mechanical arithmometer, Ageron (2016) mentions the issues of fabrication (it was a very complex instrument) and high cost which made its usage limited to management, banking, or insurance, with limited possibilities for educational institutions (lyceums and universities). Yet, the author cited public demonstrations of it, one made in 1867 by Rivière, professor of physics from the Imperial Lyceum in Rouen, and another, made by Edouard Lucas (1842–1891), professor of mathematics in Paris. However, the use of the arithmometer was limited to demonstrations, and not to teaching practices. Another pedagogical artefact analyzed by Ageron (2016) was a user's manual (*L'Instruction pour se servir de l'arithmomètre*) by an unknown author which had several editions from 1850 to 1915. The modifications brought to the manual over time mostly reflected technological improvements of the arithmometer instead of pedagogical ones (Ageron 2016). One of the examples, however, is of

Examples.

Addition.   *Problem:* Add : 352, 165, 226, 104.

Solution: Let the pupil place on the blue part of the black-board the four given quantities, and at the same time write under the three headings "plates," "staves," "cubes," the numbers expressing the respective quantities, and the board will look thus :

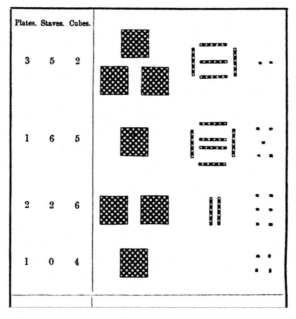

Fig. 5 Example of the use of Sonnenschien's apparatus (1879)

certain pedagogical interest as it illustrates the modifications of the traditional procedure to make the extraction of square roots better adaptable to the functioning of the arithmometer. The following example illustrates it for the root of the number of four digits:

> In general form, the digit of tens for the square root of A is calculated as follows:
>
> $R = A - (1 + 3 + 5 - \ldots + (2b_1 - 1)) \, 10^2 \geq 0$, where b_1 is the largest integer satisfying the inequality.
>
> Then, we look for the largest integer b_0 to satisfy another inequality:
>
> $R - ((20b_1 + 1) + (20b_1 + 3) + (20b_1 + 5) + \ldots + (20b_1 + 2b_0 - 1) \geq 0$ which gives the digit of units for the square root.

As a concrete example of the application of the algorithm, Ageron (2016) cited Reuleaux[7] (1861): "to get square root of 2209, we can write $(1 + 3 + 5 + 7)$ $10^2 + (81 + 83 + 85 + 87 + 89 + 91 + 93)$, which gives 47 as final result."

---

[7]Reuleaux, Franz (1829–1905).

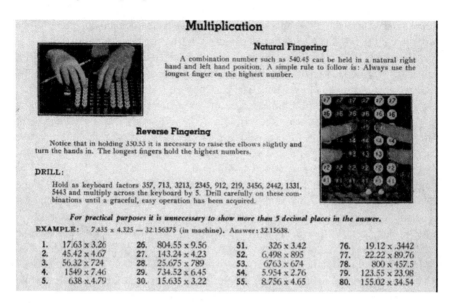

Fig. 6 Example of exercises in a textbook used by Comptometer Schools

At the same time, in the United States, two inventor–entrepreneurs, Dorr E. Felt (1862–1930) and William S. Burroughs (1857–1898) created new machines, one called Comptometer (Felt) and the other the Burroughs Adding Machine. Facing challenges in terms of improving the rate of entering numbers into the machine (input) and of the need to record the numbers that had been entered, they were successful in bringing innovative technological solutions which helped them not only to start a mass production of arithmometers but also helped keep the world-wide market under control until the mid-20th century (Campbell-Kelly et al. 2013). For instance, Felt found a way to introduce key-driven data entry, in a similar way to that of the typewriter. As described by Campbell-Kelly et al. (2013), when pressing the key with digits representing 7, 9, 2, 6, and 9 in each of the corresponding columns (the keys were organized in columns), the amount of $792.69 was added to the total. The authors mention that the skill needed for entering numbers at a fast pace was not an easy one and for this purpose, a number of Comptometer Schools were established to train operators. Figure 6 illustrates the drill-like exercises on multiplication from an instruction manual for operators.

In turn, Burroughs founded the American Arithmometer Company to build a machine which not only allowed rapid entry (as did the Comptometer) but also printed the entered numbers (second challenge mentioned above) (Campbell-Kelly et al. 2013).

In Russia, at the beginning of the 20th century there were arithmometers of two types, one produced by the Swedish inventor W. T. Odhner (1845–1905) who lived and worked in Russia and another one called *Brunsviga Rechnenmachine*.[8] When recognizing their utility for schools, Bradis (1929) mentions higher cost as the obstacle to their use. He argues, however, that being useful for calculations with multi-digit numbers they should be presented to school students. He argued that by means of a user's manual, everyone can learn how to use them within half an hour. According to the author, more time and effort were, however, needed to learn about the principles of their functioning (Bradis 1929). Regarding a possible use of arithmometers in teaching, Markushevich et al. (1951) mentioned the utility of the abacus (here, Russian abacus, or *schyoty*), slide rule, and particularly the arithmometer to construct graphs and solve equations.

In the United States, in their detailed account of the 200 years (1800–2000) of the development of tools for American teaching, Kidwell et al. (2008) devoted a chapter to calculators in which they analyzed "mechanical arithmetic for education." Namely, they noticed that in the early 20th century calculating machines became "commonplace in commerce, government, and science" (Kidwell et al. 2008, p. 246). The authors made reference to the work of Cortada (1993) who mentioned that business machine manufacturers were offering their own training classes, while also delivering their machines to bookkeeping classes at high schools and commercial colleges (Cortada 1993, cited in Kidwell et al. 2008). They also mentioned the suggestion made by Myers (1903) when describing a laboratory methodology for secondary schools. He included a calculation machine, along with a logarithmic slide rule, in his list of the equipment necessary for a mathematical laboratory.

Schlauch (1940) raised an important point about the use of calculating machines in performing computations while solving problems of school arithmetic bearing in mind the goal of (better) preparation of pupils for real life, where they would use the machines in retail business, banks and trust companies, statistics departments, as well as scientific laboratories and engineering offices where these devices were largely implemented. The author recalls an experiment he conducted 12 years before writing the text of the article (which means in about 1928) in which Monroe Calculating Machines were rented by the High School of Commerce in New York for its review course of business arithmetic with a goal to refresh students' skills in routine calculations and computations that the school graduates would likely have to accomplish once at work.

Unlike the usual lack of student enthusiasm for such types of tasks where the equipment was limited to paper and pencil, an immediate change in student attitude was noticed during the experiment, namely, students became fully engaged and reluctant to stop working at the end of the period (p. 35). In terms of the overall benefits of the experience, Schlauch (1940) enumerated an increased interest shown by students in performing fundamental operations rapidly and accurately; a greater speed of computations completed compared to those using pads and pencils;

---

[8]A calculating machine built based on Odhner's principles by *Grimme, Natalis & Co* (founded in 1871) in Germany (Braunschweig). The manufacture was led by Franz Trinks (1852–1931) who was also inventor of the Trinks-Arithmotyp, a key-driven mechanical calculator (Lenz 1924, p. 76).

checking of results became a habit in terms of the students' approach to calculations; students learned to determine the most efficient way of doing calculations
(traditionally or by machine); and finally, regarding the results, the experimental
class showed far greater improvement when using a machine in terms of their
computation speed and accuracy. As concrete examples of the students' gains, the
author mentions the discovery of number relations by the students when multiplying, for instance, 98 by 7. Students could make observations that the result can
be easily obtained if 100 is multiplied by 7, with the number 7 subsequently being
subtracted two times.

These findings, along with increased confidence in computations and a stimulated interest in problem solving using machines, led the author to conclude about
the usefulness of the calculating machines in teaching arithmetic especially in the
upper grades of junior high school and in senior high school (Schlauch 1940).

Concluding Remarks

From the point of view of mathematics education, the history of mechanical
computing devices does represent a "gold mine" for possible investigations and
discoveries which might have an unexpected influence in today's classrooms. First,
the long (more than 300 years) process of the invention of mechanical calculators
itself provides numerous examples of how practical needs to perform an increasing
number of calculations led to the development of computing devices, helping
humans to "lessen the labour of computations, to increase the speed with which
computations are made, and to guarantee the accuracy of the results of computations" (Schlauch 1940, p. 35). These objectives, being attained only in the first part
of the 20th century, attracted a great number of mathematicians and engineers who
spent years of their lives inventing, designing, and constructing a variety of
mechanical calculating machines that "could add and subtract automatically
through the interaction of gears and levers" (Kidwell et al. 2008, p. 245). When
making their attempts, they had to deal with a number of issues, mathematical and
technological in nature, which were related to some specific problems, such as how
to physically represent multi-digit numbers and to work with carried-numbers in
order to complete, for example, additions.

Moreover, these inventors had to find solutions to the problem of entering
numbers and keeping records of intermediate results. While transferring knowledge
and technological know-how from one generation of inventors to the next, and
consequently from one kind of device to another, certain technological and mathematical solutions were borrowed from anterior versions and used again, while
others provided further improvements, new insights, and eventually innovative
approaches. This could become a subject of study in today's classrooms, not only
from the historical perspective, but also from the perspective of thinking about
problems and solutions, learning about struggles and failures of the past to raise
innovators of the future.

Indeed, the idea of providing a mechanical aid for calculations was grounded not only in the quest for simplification of computational procedures, while handing the most complicated part of them to machines, but also in the challenges related to the complexity of the development and transmission of knowledge, in our case mathematical knowledge. Dealing with this complexity gave, on one hand, a boost for innovative ideas of computing devices, but, on the other hand, brought additional obstacles for their technical realization and implementation in everyday life. As an illustration of this complex relationship between the development of mathematics and technologies facilitating mathematical practices, we can mention the development of Indo-Arabic numerals and the system of recording and operating with numbers. This was considered an easier system to learn than calculations with an abacus, and gave rise to a new type of mathematics, according to Blikstein (2013). Blikstein also noticed that the appearance of paper-based algorithms, as a new kind of computing device, reflected a process of technological innovation with a "new set of societal needs, new technologies, new ways of using knowledge, and the recognition that a task previously monopolized by experts was potentially accessible to the masses." According to Poisard (2006), studying calculating instruments can contribute to a deeper understanding, by teachers and students, of the place-value system and the mechanism of the carried-number, which presented an obstacle to inventors of the first mechanical calculators in the 17th century.

But the educational value of the history of calculating machines goes beyond the issue of the mechanization (and future automatization) of basic computations. Being a process of invention other limitations existed, that is, a requirement for the correct technological tools, and the difficulty associated with multiplications and divisions compared to additions and subtractions. We showed in this chapter that solving these issues made possible the important progress in the 19th century which led to the eventual industrial production of very reliable and efficient computing devices (such as de Colmar's arithmometer, and its further improvements). We also saw that some pioneers–inventors, such as Leibniz and Babbage, while reflecting on computing devices not as mere technical aids for calculations but also as possible extensions of human thought, have raised more philosophical questions about a universal way of computing, not being limited to basic arithmetic operations but open to further (algebraic) generalizations. This path was supported by the development of the binary system of numeration, representing the basis of universal computing devices (as prototypes of modern computers), and by the development of more complex infrastructures of computational systems based on the idea of division of labor as well as using punch cards, keys, and printing devices for a better organization of the process of calculation in terms of input–output and working memory.

The final remarks are about the potential interest in the historical development of calculating machines for our present-day education in mathematics, related to the complexity of their use in different contexts in real life (in science, business, and other fields) and in teaching. The instrumental approach to understanding technological tools (including computing devices) as mediating instruments for human activities has been developed by Rabardel (1995), based on the ideas of Vygotsky

(1978), on how tools mediate learning. In the context of a mathematical task, this involves establishing a meaningful relationship between the artefact and the user who has to deal with a particular task and who sees the utility of the tool in terms of the realization of the task (Trouche 2005).

As a concrete example of the application of this approach to the context of today's classroom, related to the history of computing devices, we can cite a study conducted by Maschietto (2013), who used an arithmetical machine Zero+1, a kind of *pascaline*, at primary and secondary school levels, and studied its relationships with other instruments used by students to write numbers and make operations, such as a spike[9] abacus or calculator. One of the teaching experiments described in the paper was conducted with Grade 6 students who were asked to compare two instruments, namely, a spike abacus and a *pascaline*. The author reported that a discussion with the students was conducted about the type of task and the choice of instrument best suited to the task. For example, while working with small numbers, students found that the abacus was better suited for doing additions. However, when the numbers became bigger, like, for example, when doing the operation 237 +398, students perceived the advantage of using Zero+1, first of all, because the results of such an addition could be completed automatically.

The brief history of the invention of arithmetic machines presented in this chapter shows that it has not been an easy path for inventors to take. Inventors encountered many difficulties: the strength and durability of artefacts, affordances concerning the parts of the machine that already existed but had to be reshaped, the difficulty of mass production in a pre-industrial world, the cost of handmade manufacturing, and the resistance their inventions encountered in terms of acceptance by the community and by industry.

The inventors of these devices gave much of their time and devotion to the realization of arithmetic machines that did not always work (and often were complete failures). However, the sacred fire which animated them benefited others who took their ideas further on toward the modern era. As a result of this very complex development, from both mathematical and educational perspectives—the idea that the human brain can be replaced by a machine to do calculations gave birth to computers, and therefore, provided us with more complex affordances thus paving "the way for the sophisticated progeny that define much of modern society" (Raloff 1982, p. 171).[10]

While the educational implications of this lengthy process of the history of computing devices, and particularly mechanical calculators, are still not fully understood, the examples we analyzed here do illustrate its rich potential for further study in relation to the creation and transfer of mathematical knowledge to ingenious technological ideas (not always leading to success but yet fruitful in terms of their influence on future development).

---

[9]Another name used for a bead abacus.

[10]For more on modern development, see Martinovich's chapter in this book.

References

Ageron, P. 2016. L'arithmomètre de Thomas: sa réception dans les pays méditerranéens (1850–1915), son intérêt dans nos salles de classe. In *Proceedings of the 2016 ICME Satellite Meeting of the International Study Group on the Relations Between the History and Pedagogy of Mathematics*, ed. L. Radford, F. Furinghetti, and T. Hausberger, 655–670. Montpellier, France: IREM de Montpellier.

Anon. 1820. *Note sur la publication proposée par le gouvernement anglais des grandes tables logarithmiques et trigonométriques de M. de Prony*. Paris: Didot.

Babbage, C. 1832. *On the economy of machinery and manufactures*. London: Knight.

Babbage, C. 1961. On the division of mental labour. In *On the principles of and development of calculator*, ed. P. Morison and E. Morrison, 315–321. New York: Dover Publications Inc.

Bacon, F. 1623. *De dignitate & augmentis scientiarum*. London: J. Haviland.

Bacon, F. 1640. *Of the advancement and proficience of learning*. Oxford: L. Lichfield. [The first English translation of Bacon 1623.].

Baker, C. 1839. Infant schools. In *Central society of education*, vol. 3, 1–48. London: Taylor and Walton.

Beeson, M. J. 2004. The mechanization of mathematics. In *Alan Turing: Life and legacy of a great thinker*, ed. C. Teuscher, 77–134. Berlin-Heidelberg-New York: Springer.

Belanger, J., and D. Stein. 2005. Shadowy vision: Spanners in the mechanization of mathematics. *Historia Mathematica* 32 (1): 76–93.

Bloch, L. 2016. *Informatics in the light of some Leibniz's works*. Paper presented at XB2 Xenobiology Conference, Berlin.

Blikstein, P. 2013. Digital fabrication and "making" in education: The democratization of invention. In *FabLabs: Of machines, makers and inventors*, ed. J. Walter-Herrmann and C. Buching, 1–21. Bielefeld: Transcript Publishers.

Bradis, V.M. [Брадис В. М.]. 1929. Приближенные вычисления [Priblizhennye vychisleniya] (Approximative calculations). In *На путях к педагогическому самообразованию: На путях математики* [Na putyakh k pedagogicheskomu samoobrazovaniyu: Na putyakh matematiki] (Various approaches to the self-directed pedagogical studies: Ways to mathematics), ed. M.M. Rubinschtein [М.М. Рубинштейн], 87–116. Moscow: Mir.

Cajory, F. 1909. *A history of the logarithmic slide rule and allied instruments*. New York: The Engineering News Publishing Company & London: Archibald Constable & Co.

Campbell-Kelly, M., W. Aspray, N.L. Ensmenger, and J.R. Yost. 2013. *Computer: A history of the information machine*. Boulder: Westview Press.

Chazal, G. 2002. Les réseaux du sens: De l'informatique aux neurosciences. *Revue Philosophique de Louvain* 100 (1): 321–324.

Cole, J.R. 1995. *Pascal: The man and his two loves*. New York: New York University Press.

Cortada, J.W. 1993. *Before the computer: IBM, NCR, Burroughs, and Remington Rand and the industry they created, 1865–1956*. Princeton, NJ: Princeton University Press.

Cragon, H.G. 2000. *Computer architecture and implementation*. Cambridge: Cambridge University Press.

Davis, M. 2000. *Engines of logic: Mathematicians and the origin of the Computer*. New York: W. W. Norton & Company.

Diderot, D., and J.-B. le Rond d'Alembert (Eds.). 1751[-1772]. *Encyclopédie, ou dictionnaire raisonné des sciences, des arts et des métiers*. Paris: Briasson et al.

Durand-Richard, M.-J. 2010. Le regard français de Charles Babbage (1791–1871) sur le "déclin de la science en Angleterre." *Documents pour l'histoire des techniques* 19 (2): 287–304.

Freeth, T. 2002. The Antikythera Mechanism: 2. Is it Posidonius' Orrery? *Mediterranean Archaeology and Archaeometry* 2 (2): 45–58.

Freeth, T. 2014. Eclipse prediction on the ancient Greek astronomical calculating machine known as the Antikythera mechanism. *PLoS ONE* 9 (7): e103275.

Freeth, T., A. Jones, J.M. Steele, and Y. Bitsakis. 2008. Calendars with Olympiad display and eclipse prediction on the Antikythera mechanism. *Nature* 454 (7204): 614–617.

Glaser, A. 1981. *History of binary and other nondecimal numeration*. Los Angeles: Tomash Publishers.

Graf, K.-D. 1995. Promoting interdisciplinary and intercultural intentions through the history of informatics. In *Integrating information technology into education*, ed. D. Watson and D. Tinsley, 139–150. Dordrecht: Springer Science + Business Media. https://doi.org/10.1007/978-0-387-34842-1.

Green, C.D. 2005. Was Babbage's analytical engine intended to be a mechanical model of the mind? *History of Psychology* 8 (1): 35–45.

Halliwell, J.O. 1838. *A brief account of the life, writings, and inventions of Sir Samuel Morland, Master of Mechanics to Charles The Second*. Cambridge: Metcalfe & Palmer.

Horsburgh, E.M. 1914. *Modern instruments and methods of calculation: A handbook of the Napier tercentenary exhibition*. London: G. Bell & Sons.

Jackson, L.L. 1906. *The educational significance of sixteenth century arithmetic from the point of view of the present time*. New York: Teachers College, Columbia University.

Johnston, S. 1996. The identity of the mathematical practitioner in 16th-century England. In *Der 'mathematicus': Zur Entwicklung und Bedeutung einer neuen Berufsgruppe in der Zeit Gerhard Mercators (Duisburger Mercator-Studien)*, ed. I. Hantsche, 93–120. Bochum: Brockmeyer.

Johnston-Wilder, S., and D. Pimm. 2004. *Teaching secondary mathematics with ICT*. Maidenhead: Open University Press.

Jones, M. 2016. *Reckoning with matter: Calculating machines, innovation and thinking about thinking from Pascal to Babbage*. Chicago: Chicago University Press.

Kidwell, P.A., A. Ackerberg-Hastings, and D.L. Roberts. 2008. *Tools of American Mathematics Teaching, 1800–2000*. Baltimore: Johns Hopkins University Press.

Lardner, D. 1834. Babbage's calculating engine. *Edinburgh Review* 59: 264–327.

Leibniz, G.W. 1685/1929. Leibniz on his calculating machine (Translated from Latin by Mark Kormes). In *A Source Book in Mathematics*, ed. D.E. Smith, 173–181. New York: McGraw-Hill.

Leibniz [= Leibniz, G.W.] 1703 [1720]. Explication de l'arithmétique binaire qui se sert des seuls caractères 0 & 1; avec des remarques sur son utilité, & sur ce qu'elle donne le sens des anciennes figures Chinoises de Fohy. *Histoire de l'Académie royale, des sciences, avec les mémoires de mathématique & de physique*, [section "Mémoires",] 85–89.

Lenz, K. 1924. *Die Rechenmaschinen und das Maschinenrechnen*. Wiesbaden: Springer.

Lions, J.-L. 1991. De la machine à calculer de Pascal aux ordinateurs. *La Vie des sciences* 3: 221–240.

Lister, M., J. Dovey, S. Giddings, I. Grant, and K. Kelly. 2009. *New media: A critical introduction*. London and New York: Routledge.

Marguin, J. 1997. L'arithmomètre de Thomas n° 1398. *Bulletin de la Sabix* 18: 31–42.

Markushevich, A.I., A.Y. Hinchin, and P.S. Alexandrov [Маркушевич, А. И., А. Я. Хинчин, П. С. Александров]. 1951. *Энциклопедия элементарной математики* [Entsiklopediya elementarnoĭ matematiki] (Encyclopedia of elementary mathematics). Moscow & Leningrad: GTTL.

Maschietto, M. 2013. Systems of instruments for place value and arithmetical operations: An exploratory study with the Pascaline. *Education* 3 (4): 221–230.

Mayo, E., and C. Mayo. 1837. *Practical remarks on infant education, for the use of schools and private families*. London: Seeley and Co.

Menabrea, L.F. 1842. Notions sur la machine analytique de M. Charles Babbage. *Bibliothèque universelle de Genève* 41: 352–376.

Morar, F.-S. 2015. Reinventing machines: The transmission history of the Leibniz calculator. *British Society for the History of Science* 48 (1): 123–146.

Mosconi, J. 1983. Charles Babbage: vers une théorie du calcul mécanique. *Revue d'histoire des sciences* 36 (1): 69–107.

Myers, G.W. 1903. The laboratory method in the secondary school. *The School Review* 11 (9): 727–741.

Nepero, I. (Napier, J.). 1614. *Mirifici logarithmorum canonis descriptio*. Edinburgum: A. Hart.

Nepero, I. (Napier, J.). 1617. *Rabdologiae seu numerationis per virgulas libri duo: cum appendice de expeditissimo multiplicationis promptuario*. Edinburgum: A. Hart.

Nordhaus, W.D. 2007. Two centuries of productivity growth in computing. *The Journal of Economic History* 67 (1): 128–159.

Pascal, B. 1645/1998. Lettre dedicatoire a Monseigneur le Chancelier sure le sujet de la Machine nouvellement inventée par le Sieur B.P. pour faire toutes sortes d'operations d'Arithmetique, par un mouvement reglé, sans plume ny jettons, avec un advis necessaire à ceux qui auront la curiosité de voir ladite Machine, et de s'en servir. [n.p.] (Reprinted in B. Pascal, *Œuvres complètes. Tome 1*. Paris: Édition de Michel Le Guern.).

Poisard, C. 2006. The notion of carried-number, between the history of calculating instruments and arithmetic. *Proceedings of the Annual Conference of Mathematics Education Research Group of Australasia (MERGA)*, 2, pp. 416–423.

Project 2061. 1989. *Science for all Americans: A Project 2061 report on literacy goals in science, mathematics, and technology*. Washington, D.C.: American Association for the Advancement of Science.

Rabardel, P. 1995. *Les hommes et les technologies, une approche cognitive des instruments contemporains*. Paris: Armand Colin.

Raloff, J. 1982. On Beyond Babbage: The Rise of Automatic Digital Computers. *Science News* 121: 170–172.

Ratcliff, J.R. 2007. Samuel Morland and his calculating machines c.1666: The early career of a courtier–inventor in Restoration London. *British Society for the History of Science* 40 (2): 159–179.

Reuleaux, F. 1861. *Der Constructeur: Ein Handbuch zum Gebrauch beim Maschinen-Entwerfen*. Braunschweig: Friedrich Vieweg und Sohn.

Ryan, J.A. 1996. Leibniz' binary system and Shao Yong's "Yijing". *Philosophy East and West* 46 (1): 59–90.

Sawday, J. 2007. *Engines of the imagination: Renaissance culture and the rise of the machine*. Abingdon: Routledge.

Schlauch, W.S. 1940. The use of calculating machines in teaching arithmetic. *The Mathematics Teacher* 33 (1): 35–38.

Sonnenschein, W.S. 1879. Description of Sonnenschein's number-pictures, arithmometer and blackboard: their construction and use. London: W. Swan Sonnenschein.

Taton, R. 1963. Sur l'invention de la machine arithmétique. *Revue d'histoire des sciences et de leurs applications* 16 (2): 139–160.

Temam, D. 2009. La pascaline, la "machine qui relève du défaut de la mémoire." *Bibnum*. Retrieved from: http://bibnum.revues.org/548.

Tent, M.B.W. 2012. *Gottfried Wilhelm Leibniz: The polymath who brought us calculus*. Boca Raton, FL: CRC Press.

Trouche, L. 2005. Des artefacts aux instruments, une approche pour guider et intégrer les usages des outils de calcul dans l'enseignement des mathématiques. *Actes de l'université d'été de Saint-Flour*, pp. 265–290.

Vygotsky, L.S. 1978. *Mind in society: The development of higher psychological processes*. Cambridge, MA: Harvard University Press.

Walford, C. 1871. *The insurance cyclopaedia*. London: Layton.

Wiener, N. 1988/1950. *The human use of human beings: Cybernetics and society*. New York: Da Capo Press.

Viktor Freiman is a Full Professor of Mathematics Education at the Université de Moncton, Campus Moncton, Canada. His work, besides the history of mathematics education, focuses on the use of digital technology in teaching and learning, interdisciplinary learning, computational thinking, design thinking, creativity, mathematical problem solving, and giftedness. Since 2014, he is director of CompeTI.CA (Compétences en TIC en Atlantique) partnership network, studying a lifelong development of digital competences. He is co-editor of the Springer Book Series Mathematics Education in the Digital Era.

Xavier Robichaud is an Assistant Professor in Education at the Université de Moncton, Campus de Shippagan. He received his Ph.D. in Education from Université de Moncton and his Master's degree in Music from Université de Montréal. His research focuses on creativity, its development in children and the role of technology and culture in the creative process.

Computation Devices in Nineteenth-Century Mathematics Instruction in Europe

Gert Schubring

Abstract The main computational devices in use in schools in Europe during the nineteenth century were the logarithmic tables and the slide rule. This subject has not been researched upon before, and therefore, this chapter presents a first tentative to assess the development. The investigation has focussed on a number of European states, namely England, France, Germany, Italy and the Netherlands.

Keywords Logarithmic tables · Slide rule · England · France
Germany

Introduction

The use of computation devices has so far been very scarcely researched upon in studies on the history of mathematics teaching in nineteenth-century Europe. This chapter presents a first attempt to analyse the issue and to provide results for a number of specific European countries.

Two computation devices have been the principal instruments used in the teaching of arithmetic, algebra and trigonometry: logarithmic tables and slide rules. Although used less, slide rules have been more researched upon.

The standard types of logarithmic tables will be analysed together with the mathematical topics for which these tables had to be used by the students. The processes of simplification of these computation devices, originally developed for use by scientists and practitioners, upon their transposition to use in secondary schools, will be addressed.

Slide rules, originally also developed for professionals, began to enter schools during the nineteenth century, but at very different paces in the various countries. Also, this instrument became more perfected.

G. Schubring (✉)
Universidade Federal do Rio de Janeiro, Rio de Janeiro, Brazil
e-mail: gert.schubring@uni-bielefeld.de

© Springer Nature Switzerland AG 2018
A. Volkov and V. Freiman (eds.), *Computations and Computing Devices
in Mathematics Education Before the Advent of Electronic Calculators,*
Mathematics Education in the Digital Era 11, https://doi.org/10.1007/978-3-319-73396-8_14

The contribution will focus on Britain, France, Germany, Italy and the Netherlands. Not only differences between various school types within one country will be revealed but also significant differences between the studied countries.

The Logarithmic Tables

Logarithmic tables had already been established by the inventors of the logarithms. Henry Briggs (1561–1630) published in 1617 the first logarithmic tables. Due to the aim of providing utmost exactness in the results, his tables were established with 14 decimals. Exactness was required due to their use in astronomical computations. Thereafter, other applications emerged shortening computations for mathematical practitioners (Kauzner 1994, p. 219), and tables with a lesser number of decimals became produced. A new approach for exactness was initiated by Gaspard Riche de Prony (1755–1839), who, from 1792 onwards, contracted an enormous number of persons to calculate logarithmic tables with 29 decimals. This gigantic work was never achieved, but a shortened version became eventually published in 1891 (Grattan-Guinness 1990).

This approach to rigour, occasioned in the context of the French Revolution, became paralleled by another new application. The French Revolution initiated the first realizations of public education systems and significantly put a strong emphasis on the teaching of mathematics in their secondary schools—at least in general. Within these systems of education, logarithmic tables now began to enter mathematics instruction—with a twofold task: on the one hand, to simplify and facilitate any kind of calculation and, on the other hand, to prepare for use in some professional practice.

France

The use of logarithmic tables can be best assessed in France, thanks to its highly centralised educational structure. These tables were used practically in all school types as we will see, but with characteristic differences.

The first explicit use is documented as entrance requirement to higher education: for becoming admitted to the *École polytechnique* in Paris (EP), the leading higher education institution for the formation of engineers, both for military and civil services. In 1800, the programme for the entrance examinations to the prestigious school, besides listing the necessary content knowledge, demanded in an additional remark: "it is indispensable that the students are familiar [...] with the use of tables of logarithms and their application to trigonometry" (Belhoste 1995, p. 77). This requirement as previous knowledge is reiterated all over the nineteenth century. In 1828, for instance, the use of tables is even made the subject of an examination, for solving problems of elementary geometry:

An example of a problem to resolve a triangle question is proposed to each candidate, to verify that he is able to use the logarithmic tables. (my transl., G.S.)

Remarkably, it is added that these calculations have to be made with tables for seven decimals (ibid., p. 123) thus conceiving of a later use in professional practice, demanding a high degree of exactness.

Where should the candidates for the entrance examination have acquired this familiarity? The first curricula for the new secondary schools, the *lycées*, of 1802/1803, 1809, 1811 and 1814 were too succinct and too restricted to just indicating the textbooks as to mention methodical aspects of teaching. The political restoration had effected also a drastic reduction of mathematics teaching in the *lycées*, now renamed *collèges royaux*, in 1821. To remedy this situation at least to some extent, a "cours de mathématiques préparatoires" was introduced in 1826 for the two school years before the two concluding grades of "philosophie" in which mathematics teaching had been abolished in 1821. The curriculum of this stream was rather succinct again. To improve the quality of this teaching, methodical advices were given in 1829. There it was emphasised that the mathematics teacher had to clarify the construction of the tables and show how to use them (ibid., p. 125). In 1838, in the curriculum for all students of the colleges, in the *classe de seconde*, "theory and use of logarithms" was stipulated, which was supposed to imply the use of the tables (ibid., p. 152).

From 1843 on, however, one finds logarithms and their tables for a long time only indicated in the curricula for special teaching, now increasingly organised in separate courses, preparing for the entrance examinations of the *Grandes Écoles*. Thus, in 1843, for the "cours de mathématiques élémentaires" within the colleges, logarithms and the use of the tables should be taught. In the "cours de mathématiques spéciales", one understands trigonometry as a major field of applications: there, the construction, structure and use of logarithmic tables are one of the teaching subjects within trigonometry. Here, the first time, a particular author of tables is mentioned: Callet. In fact, François Callet's (1744–1798) *Tables portatives de logarithmes; contenant les logarithmes des nombres, despuis 1 jusqu'à 108,000; les logarithmes des sinus et tangentes ...* were a bestseller and the very often reprinted version of 1795, a five-figure table, might have been the tables also intended to be used in the earlier indications.

This trend to focus on the use of tables for specialised teaching became reinforced in 1852 when a bifurcation between classical colleges and a scientific school branch was established. In the first curriculum for this "enseignement scientifique" of 1852, the use of tables is amply mentioned there, within algebra (ibid., p. 283). And in 1853, for the *classe de mathématiques spéciales* within the *collèges*, logarithms are a major teaching subject within advanced algebra; again, Callet's tables are indicated (ibid., p. 304). The following year, extensive methodical advices were given for all teaching within the *lycées*. For their "section des sciences", there is a revealing discussion of the problems, shown by experience, whether students master effectively the use of the tables (ibid., p. 332). In particular, the number of figures in these tables is discussed: only five-figure tables should be used, and seven-figure tables should be absolutely banned from use in schools. Here, instead of Callet, the tables of the astronomer Joseph Lalande (1732–1807) are prescribed.[1] The aim of the use of the tables is to facilitate the calculations, and seven-figure tables do not serve for this aim in schools (ibid., p. 333). The extension of criticism of seven-figure tables shows that they must have really been applied so far—but one does not know, to what extent and where. Here, one finds for the first time the explicit demand to put the tables "into the hands of the students".

This structure is maintained over the following decades: the tables are used within the *enseignement scientifique*, in particular for trigonometry, their use is prescribed for the *classe de mathématiques élémentaires* within the *lycées*—while they are not mentioned in their classical sections. A real innovation is evidenced from 1880 when secondary school for girls became established. These schools adopted a modern type of curriculum. In their curriculum of 1882, logarithms are taught in the prelast year, but in a "cours facultative" of arithmetic—thus only as an optional subject (ibid., p. 470).

By the turn to the twentieth century, one observes a greater awareness for the quality of teaching and for methodology, the number of figures in the logarithmic tables as influencing the success of their use. The curricula of 1902, so important for the reform of secondary schools for boys in France (see Gispert and Schubring 2011), show this awareness. For the algebra provided in the "classe de troisième B", it is emphasised:

> As regards the logarithms, the main aim is to familiarise the students with the use of the table with four figures (or with five). If one limits the use to four-figure tables, as it seems to be very reasonable for this grade, one will familiarise the students also with the use of the antilogarithms. (ibid., p. 588; my transl., G.S.)

In the next grade, the "classe de seconde C et D", the aim is almost the same, namely to familiarise the students with the use of the tables. The number of figures is not mentioned. (ibid., p. 598) In the highest grade for mathematics, the "classe de mathématiques, sections A et B", one demands, within algebra, the use of five-figure tables—without any discussion (ibid., 607). In the somewhat revised curricula of 1905, achieving the final form of the reform process, one remarks a peculiar solution of this methodical question: instead of prescribing the same table for the entire schooldays, an "increasing" degree of exactness should be taught and learned throughout the respective school years. In the *classe de troisième,* it should be four-figure tables; in the *classe de seconde,* four- or five-figure tables; and eventually in the *classe de mathématiques,* five-figure tables (ibid., 661, 663 and 667).

It is important to emphasise that the teaching and use of logarithmic tables in France was not restricted to secondary schools, but that it was taught also in the primary school subsystem. One has to know that during the nineteenth century—

---

[1]Lalande's tables are another example of an almost timeless bestseller: it became reedited until the 1940s.

and even longer—primary schools were destined for the lower classes and that they constituted a separate system of schooling, without connection to the secondary schools for the upper classes—which had proper preparatory teaching. And logarithmic tables were in use for the upper part of the primary schools, the *écoles primaires supérieures* (EPS). Primary schools had become organised as part of the French public education system after the institution of the liberal monarchy in 1830. The programme of 1836 recommended for the EPS as logarithmic tables those by Callet and those by Prony (d'Enfert 2003, p. 90).[2] Teaching the use of the logarithmic tables was maintained there throughout the nineteenth century. The programme for the EPS of 1885—as well as those of 1893 and 1909—prescribed: "Use of logarithmic tables with 4 or 5 decimals" (ibid., p. 223, 263 and 324).

England

As it is well known, secondary schools in England have differentiated from the university system, organised there as the collegiate university, remarkably later than in Continental Europe. The few grammar schools of the nineteenth century were practising a curriculum dominated by classical languages. Mathematics played a minor or rather marginal role (Howson 2014). So far, there is no evidence of the use of logarithmic tables in this system of secondary schools. Given the strict focus of this kind of mathematics teaching on Euclid's geometry, there is no probability at all that calculations via logarithmic tables would fit into such a classically oriented teaching. One might assume that tables were used in military schools and in academies for training sailors, for the merchant and for the Royal Navy, but no research has been done on this aspect. An assessment of library catalogues for tables for school use resulted in tables published in 1908 by Frank Castle as the first instance—thus in the wake of the reform movement initiated by the struggles to replace the traditional domination of Euclid: Frank Castle, *Logarithmic and other tables for schools*. London 1908. As later editions show, Castle (1856–1928) had published five-figure tables.

In fact, the tables, which became very popular for school use in England in the first half of the twentieth century, are due to two authors highly active in the reform movement of the turn to this century: Charles Godfrey (1873–1924) and A. W. Siddons (see Fig. 1).

---

[2]Probably, this meant the following textbook: Gaspard Riche de Prony (1834), *Instruction élémentaire et pratique sur l'usage des tables de logarithmes*. Paris: Barbier.

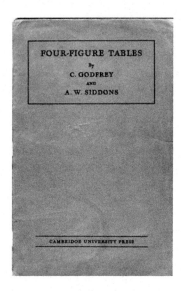

Fig. 1 Title page of the tables by Godfrey and Siddons

Italy

For the various states coexisting in Italy before its unification in 1860, there is so far no research but one can assume that logarithmic tables were in use. For instance, in the northern regions governed by Austria, one finds an Italian translation of tables published by Franz von Moĉnik, an influential author of textbooks in Austria, with translations into Italian: *Tavole logaritmiche-trigonometriche. Prima ed. italiana* (1864).

From 1860 on, the use of logarithmic tables was demanded in almost all curricula prescribed by the ministry, and this was so for the (classical) *licei* as well as for the *istituti tecnici*. Neither did the ministry prescribe any particular table of the various published tables nor did it pay attention to the number of decimals. In use were tables with seven and with five decimals, but it seems that 11-figure tables also have been used. Even Luigi Cremona (1830–1903), the staunch fighter for genuinely Italian textbooks (see Schubring 2004), translated Georg Vega's logarithmic tables, in the later edition by Bremiker (see below) into Italian—tables, which were reedited at least until the 1920s.

The Netherlands

The system of secondary schools in the Netherlands was also heavily determined by the confrontation of classical versus modern school forms and curricula. The so-called Latin schools dominated until the creation of the Hogere Burgerschoolen or HBS: higher modern schools in 1863 (see Smid 1997).

Before their reform in 1876, it seems that some school teachers did use loga-rithmic tables, and some did not. For instance, Jacob de Gelder (1765–1848), the activist within the Latin school system to promote mathematics, treated the use of logarithmic tables already in his *Allereerste Gronden der Cijferkunst* (1824) which was explicitly recommended by the government for Latin schools. Likewise, the textbook *Beginselen der Stelkunst* by Badon Ghijben and Strootman of 1838, which was in use both in some Latin schools as well as in some schools which can be considered as forerunners of the HBS, used the tables. But since there were no central examinations and hardly an official curriculum in those years, it is very well possible that some Latin schools did not teach logarithms at all.

On the other hand, it is evident that the use of logarithmic tables was quite common in the HBS right from their start in 1863. The tables themselves are not mentioned in the curricula, they require, however, the teaching of logarithms, or "the use of logarithms", but the "Algemeen Reglement voor de eindexamens" from 1870 states that the use of any books during these examinations is strictly forbid-den, "with exemption of logarithmic tables". So these tables were explicitly per-mitted, which only can mean that they were to be used during these examinations, and therefore were used in the forgoing teaching. Jan Versluys (1845–1920) already published in the 1870s logarithmic tables for use in schools, as did other authors. In the decennia afterwards, one finds discussions about the use of tables with 4, 5 or even 6 decimals. In 1937, it was officially decided that one should use tables with only 4 decimals. The logarithmic tables stayed in use until the curricular change of 1968, when they were finally abolished.

The modern form of the Dutch Gymnasium dates from 1876, and its curriculum for mathematics required concerning logarithms: in grade 3 to introduce logarithms, and in grade 4 to continue with logarithms. One can suppose that the use of logarithmic tables was part of the teaching of logarithms, just as in the HBS. The Gymnasium teachers participated in the discussions about the number of decimals to be used in the tables, and for instance, the official curriculum of 1919 demands "making calculations with logarithms", which makes sense only when one uses tables. Logarithmic tables for the gymnasia were also abolished in 1968.

Germany

The peculiar political situation of Germany does not allow obtaining general results about official regulations regarding the use of logarithmic tables. Since the end of the Napoleonic period, in 1815, there coexisted 39 sovereign political units on the German territory; and even since the formation of the *Deutsches Reich* in 1871, there were 25 states, each autonomous regarding its educational policy. The *Reich* was just a confederation. Moreover, Prussia—together with Austria a dominant state among this set of states—is distinguished, at least in the first half of the nineteenth century, by having established neo-humanist educational policy and this implied not only a high social status for its Gymnasium teachers, but also in particular to ascribe autonomy to the teachers: autonomy to determine their teaching method. Thus, the ministry consciously desisted in prescribing any teaching method (see Schubring 1991, pp. 180 sqq.).

On the other hand, it is evident that logarithmic tables were in constant use in secondary schools: evidence for this is given by the great number of logarithmic tables published throughout the nineteenth century. There existed also a special form of textbooks for mathematics teaching with logarithmic tables as an appendix.

A first change in the policy of the Prussian ministry occurred in 1861 when it recommended for the first time logarithmic tables: namely, the tables published by Carl Bremiker (1804–1877), a geodesist and astronomer.[3] In the 1850s, Bremiker dedicated himself to publish computational devices; he began in 1852 to publish a Latin version of logarithmic and trigonometric tables, with six decimals (Bremiker 1852)—in Latin, probably to be used in many countries. Then, he reedited in 1856 the famous tables of Georg Freiherr von Vega destined for astronomers and practitioners. And in 1857, he reedited the tables of August Leopold Crelle (1780–1855), which represented an original approach to renew the ideas of logarithms: to avoid multiplications and divisions (Crelle 1857). Eventually, in 1861, he published a version of the Latin tables for use in schools, still with six decimal places. But soon later, since 1874, he published the tables already with only four decimals. There were also editions with five decimals.

Soon emerged a competitor: since 1870, there were logarithmic tables, for use in schools and in practice, with five decimals, by Gauß. They proved to be a bestseller in Germany, reedited many times. In my school times, I used these tables, in an edition of 1958 (see Fig. 2). In fact, they were in use in secondary schools throughout Western Germany.

These tables present a revealing case for the relation between academic mathematics and school mathematics. The book gives as author "F. G. Gauß". The name of the famous mathematician is usually given as "C. F. Gauß". I happened to identify the author of the tables when I was searching—for the nineteenth century—for students of the *Rats-Gymnasium* in Bielefeld, the town of my university, who continued to study mathematics after having obtained the *Abitur* there in the nineteenth century. At first, I found a Friedrich Gauß, with *Abitur* of 1851: he studied mathematics and became mathematics teacher in Silesia—but published no table. Eventually, I found his brother Friedrich Gustav Gauß (1829–1915) who left the Gymnasium before the *Abitur*, continuing at a technical school. He became a highly influential geometer, in the service of Prussian ministries for topographic work, organising an immense cadastral land register in the Western provinces and later on in the Eastern provinces. Gauß succeeded not only in obtaining highest titles in the Prussian government (*Wirklicher Geheimer Rat* and *Exzellenz*), but also in academic respects: in 1899, the Strasbourg University conferred upon him the degree *doctor honoris causa* (Schubring 2008, p. 396). At a prominent place in the

---

[3]Zentralblatt für die gesamte Unterrichtsverwaltung in Preußen; 3(1861) 4, p. 220

Fig. 2 1958 edition of the logarithmic tables by F. G. Gauß

Fig. 3 Commemorative table for Friedrich Gustav Gauß in the Bielefeld city hall

Bielefeld city hall, there is a commemorative table for this eminent son of Bielefeld (Fig. 3).

Actually, it would have been not so far-fetched to relate the mathematician Gauß with logarithmic tables. It is characteristic for him that he combined research in pure

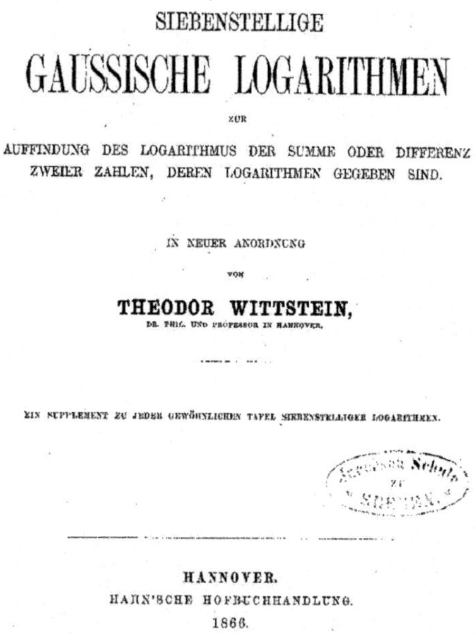

SIEBENSTELLIGE

GAUSSISCHE LOGARITHMEN

ZUR

AUFFINDUNG DES LOGARITHMUS DER SUMME ODER DIFFERENZ
ZWEIER ZAHLEN, DEREN LOGARITHMEN GEGEBEN SIND.

IN NEUER ANORDNUNG

VON

THEODOR WITTSTEIN,
DR. PHIL. UND PROFESSOR IN HANNOVER.

EIN SUPPLEMENT ZU JEDER GEWÖHNLICHEN TAFEL SIEBENSTELLIGER LOGARITHMEN.

HANNOVER.
HAHN'SCHE HOFBUCHHANDLUNG.
1866.

Fig. 4 Title page of the publication of the Gauß logarithms in book format 1866

mathematics and in applied mathematics and liked very much practicing geodesy and triangulation. His publications include topics on what he called higher geodesy. Even more, it is not so well known that he published, too, logarithmic tables— actually, of a special kind: tables for handy calculation of the logarithms of the sum or the difference of two numbers, which themselves are given only by their logarithms (Gauß 1812). Gauß himself explained the aim of these tables:

> The more widely the works of calculating astronomers become extended, the more important any facility becomes for them, which by itself could seem to be of minor importance. [...] The small table is not truly astronomical, but it will be particularly welcomed by the calculating astronomer [...]. The task, which this table should facilitate, is encountered at every moment in astronomical calculations; without the table one needs to search three times the logarithm – or, in any case, after an easy transformation, two times; but here the operations are reduced to just one. (Gauß 1828, p. 253; my transl., G.S.)

Originally published in an astronomical journal, Gauß' logarithms were reedited later on several times (see Fig. 4).

The preface of the 1958 edition of the Gauß tables (actually a reprint of the 1934 edition) relates of the development of didactical thinking regarding the use of the logarithmic tables. This development is called there "the elimination of unnecessary number ballast". In fact, the issue of the number of decimals has been amply discussed by mathematics educators in Germany. Walther Lietzmann, in his *Methodik* of 1916, the authoritative handbook for mathematics teachers, reported that it had not been so long time ago that tables with seven decimals were used at a large number of secondary schools. The conservative schools among them would have continued this practice, but fortunately, the Prussian ministry of education intervened (Lietzmann 1916, p. 242): in a decree of 1880, the Prussian ministry had declared tables with seven and with six decimals inappropriate and determined to use those with five decimals. And, according to Lietzmann, it had been Johann Traugott Müller, a mathematics teacher at the *Realgymnasium* in Wiesbaden, who had published in 1860 the first tables with only four decimals. But this remained an exception for quite a while.[4] Alois Höfler, the author of another influential *Methodik* handbook for teachers, mainly for Austria, pleaded for four-decimal tables as the best option—but added that Felix Klein had argued for three-decimal tables (Höfler 1910, p. 239). Lietzmann was the first to discuss the number of decimals in relation to the use of the slide rule: he mentioned the publication of handy tables with just three decimals, but he argued that such tables as "dispensable", since when restricting oneself to calculating with just three decimals one can use the slide rule with advantage (Lietzmann 1916, p. 244).

Thus, by the turn to the twentieth century, it became ever more acknowledged that, for use in schools, tables with four decimals are clearly sufficient; if the student has become well acquainted with the use of these tables, he will be easily capable to adapt himself to the use of tables with more decimals if he should need to use them in his professional life (Gauß 1958, p. iii). The Gauß tables were published since 1900 in a four-decimal edition; it became the major form since 1925, when the Prussian ministry had decreed this form.

The slide rule

The slide rule leads to the revealing result that this instrument developed in England was there enormously used by technicians and other professionals but never entered secondary schools, at least not during the nineteenth century. However, it became a favoured computing device in many types of schools in France. The history of the slide rule has been significantly more researched upon than that of logarithmic tables.

Edmund Gunter (1581–1626), who is supposed to have known Napier and who was professor of astronomy at Gresham College in London, had the idea that with a logarithmic scale and a pair of dividers one could find the sums and differences of logarithms of numbers and thereby multiply and divide with facility: the numbers should be indicated, scaled proportionally to their logarithms. First such computational devices were the carpenter's rules, produced since the 1680s (see Fig. 5):

> The carpenter's rule was made from two wooden one-foot rules that were held together at one end by a metal joint. Unfolded, one side became a simple two-foot measuring rule. The upper part of the other side of the twofold rule contained a groove that held a brass slide, with logarithmic scales on the upper and lower edges of both the slide and the adjacent parts of the groove. The inner edges and lower part of the carpenter's rule commonly were

---

[4]Actually, this was its second edition; the first one was published already in 1844.

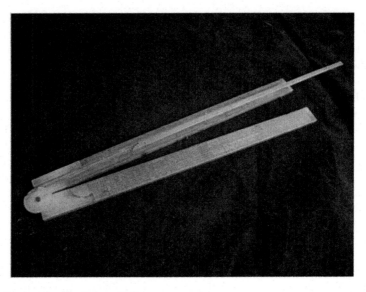

Fig. 5 Carpenter's rule of about 1830

marked with other scales that were useful to carpenters and spar makers. (Kidwell et al. 2008, p. 107).

The object, which should become the slide rule, was an instrument in use in Britain by practitioners and which was perfected by James Watt, about 1780, and known as Soho slide rule. It was the Soho type, which caused an enormous impact upon instrument making in France, when a French engineer and Egyptologist, Edmé-François Jomard (1777–1862), visited England in 1815 for examining both antiquities and the latest inventions in industrial practice. There, he purchased a Soho rule and presented it, together with a report to the *Société d'Encouragement*

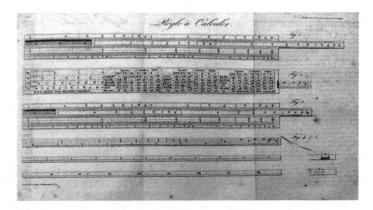

Fig. 6 Soho slide rule, as presented by Jomard in 1815 (Thomas 2016, p. 96)

pour l'Industrie Nationale, that same year (see Fig. 6). Thereafter, in France, the slide rule became ever more developed and sophisticated.

France

During the nineteenth century, France was leading—both in improving the slide rule and in its use: and uniquely in Europe not only in industry and practice, but also in teaching mathematics in schools.

The Instruments

On the side of instrument making, it is due to Amédée Mannheim (1831–1906), an engineer, to have developed a form of the slide rule, which became the model throughout. Thomas (2016) calls it the "modern form", developed by Mannheim as a young Polytechnicien. While one used to attribute the introduction of the cursor to Mannheim, this is not the case—the cursor had already been introduced by two new forms of slide rule in France in the 1820s, but the cursor obtained its characteristic form within the new construction by Mannheim (see Fig. 7)—clearly, the invention to install a cursor within the slide rule meant a decisive improvement, enabling to relate exactly the values on the various scales:

> The rule then used is of the type 'Soho' and is composed of three times the same scale: at the top of the rule, at the top and bottom of the slider; the fourth level, at the bottom of the rule, is a scale of squares, allowing in particular to obtain the square roots. Mannheim's idea is to modify this arrangement: he puts two scales of identical numbers below on the instrument, one on the rule and the other on the slider, and two scales of squares above, in the same manner. Thus the slide rule is, somehow, divided into two: in the vertical

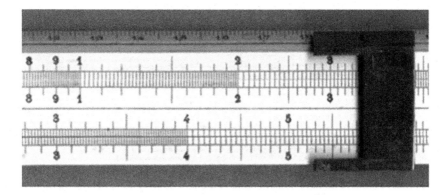

Fig. 7 Slide rule of Mannheim, with the cursor (Thomas 2016, p. 165)

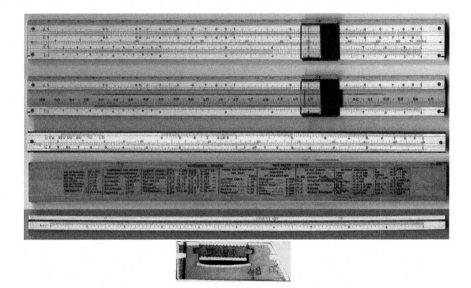

Fig. 8 Tavernier-Gravet slide rule, from about 1920; the figure below shows, in a cross section, how the cursor can move and how the sliding part is arranged

direction; this allows a better use of the limited space of the rule, and will give rise to several types of rules, all derived from that of Mannheim, greatly increasing the possibilities of the instrument. To facilitate the use of his 'modified slide rule', Mannheim thus generalises the use of the cursor. (Thomas 2016, p. 162; my transl., G.S.)

The Mannheim slide rule became the model in particular in Germany where from the 1870s one began to produce industrially slide rules (Thomas 2016, p. 160). Mannheim did not produce his slide rule—he had it produced, in a handicraft manner, by the instrument-maker Tavernier-Gravet in Paris. Tavernier-Gravet improved the instrument from the 1860s on (see Fig. 8). In particular, the cursor now embraced a transparent area, with the vertical line marked in it.

Use in Schools

Not only it is remarkable that slide rules were introduced for school use officially first in France, but it is even more remarkable that they were introduced there first for the primary schools. In a report of the ministry of education in 1847, it was emphasised:

we want that the slide rule permeats primary teaching – the slide rule being so popular in England, but so little known in France; it is by means of primary school that it should be given access to the workshops. (d'Enfert 2003, p. 121)

Actually, the assessment was not so correct: the slide rule had become meanwhile more developed in France. It is revealing, however, that the introduction in schools was justified by its later use in professional practice. One has to add that the teaching was not supposed to occur in the basic primary schools, but in the *écoles primaires supérieures* (EPS) (see above). This teaching was maintained, throughout the nineteenth century, in these higher primary schools. In 1893, the programme for the third year of the EPS demanded for its general section:

Use of the slide rule. (d'Enfert 2003, p. 263)

In 1909, for the same third year of the general section, it was demanded to explain logarithms by means of the slide rule. Moreover, it was added: "Study of the slide rule; realising arithmetical operations with entire numbers and with fractions". And for the second year of the "section industrielle", it was demanded regarding practical arithmetic: "Use of the slide rule as a means to control certain operations" (ibid., pp. 329–330). As operations in commercial practice had been indicated before: percentage, interest, discounts, money change.

While the teaching of the slide rule comprised the general section as well as special professional sections for the EPS, this teaching remained restricted for the science sections in secondary schools. The first mention occurred in 1852, for the newly founded *enseignement scientifique* (see above): "Use of the slide rule, restricted to multiplication and division" (Belhoste 1995, p. 283).

In 1853, also not within the teaching for all, but for the *classe des mathématiques spéciales* (see above), within an extended teaching of logarithms, the use of the slide rule had to be taught (Belhoste 1995, p. 304).

In 1854, the methodical instructions already mentioned above contained a revealing reflection about the aims of teaching the use of the slide rule:

The slide rule provides a quick and handy way to effect a great number of calculations with approximate results for which we do not need great accuracy. We have to make the students familiar with their use, regarding multiplication and division, as required by the programme. It will be sufficient to put slide rules of a low price in the hands of students, on which are given only those scales necessary for the proposed aim. (Belhoste 1995, pp. 333–334; my transl., G.S.)

In 1866, the mentioned programme for the *enseignement secondaire spécial* demanded, together with the use of the logarithmic tables, to teach the use of the slide rule (ibid., p. 438).

The use of the slide rule in the teaching for all students of the secondary schools became a topic only after the decisive curricular reforms of 1902, abolishing the inferior status of the modern school types in relation to the humanistic ones (see Gispert and Schubring 2011). In the important *Conférences Pédagogiques* of 1904 where the reform initiatives were discussed more broadly (see ibid., p. 78), the mathematician Émile Borel (1871–1956) made in his talk a revealing reflection on the role of computational devices in general; he even discussed the relation between our two topics, logarithmic tables and slide rules:

It appears to have only advantage to simplify as much as possible the material task of the student in the calculations by the use of auxiliary means. One will demand him to use as soon as possible the resources of logarithms; one can also teach him the use of the slide rule and even one can let him search their practice, permit to use tables of square roots and cube roots, sine tables, etc. There are, in Germany, collections of varied and simple numerical tables for use by students of secondary schools. I will not discuss the relative merits of these various procedures; for example, one may prefer the use of logarithms to four decimal places to the use of the slide rule, or vice versa; the essential is that the task of the calculator be simplified as much as possible, so that reaching without much trouble the result, the joy to have reached be not spoiled by the annoyance of too long a road. (Borel 1904, p. 434; my transl., G.S.)

Other European Countries

Thomas investigates in his book primarily the production of slide rules; their use in schools is discussed only for France, and thus, one learns a bit from him on the production outside of Britain and France, but nothing about their use in foreign schools. According to him, up until the middle of the nineteenth century, production of slide rules occurred only in Britain and in France (Thomas 2016, p. 193). From the 1860s, however, it were three German enterprises, which entered the market and became highly successful, in particular in providing slide rules for use in schools: Dennert & Pape, from 1872—better known by its later name Aristo; Nestler, from 1878; and Faber-Castell, from 1892. These companies transformed the former more artisanal production into industrial production.

As already told in the section on logarithms, one does not yet know for **Germany** whether there were formal requirements in the curricula of the various German states to teach the use of the slide rule. As Lietzmann reports, the first recommendation to use a slide rule had been published in 1872, by a mathematics teacher. This had no effect then, and impact was effected only by its republication in 1896. Lietzmann added as a comment that the change to using this device is due, on the one hand, to its widespread use in the practice and, on the other hand, to the reform in mathematics teaching: that school mathematics had approximated the mathematics of daily life. He ascribed the major impact for the introduction of the slide rule in schools to publications by C. H. Müller since 1897 and to a "precision school slide rule" developed by him (Lietzmann 1916, 244). In his book, Lietzmann even reproduced two school slide rules, a cheap one and a more sophisticated one (Fig. 9).

A use in schools since about the early twentieth century is confirmed by pertinent book publications. Firstly, there is a book on the theory and practice of the logarithmic slide rule, first published in 1911 and reedited until 1943 (Schrutka von Rechtenstamm 1911); since it was published in "Eigenverlag", one can assume that it was directed to practitioners. But a second book on the theory and practice of the

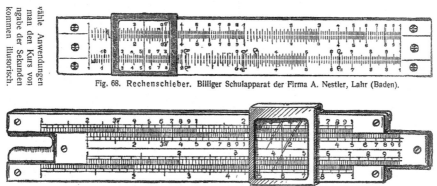

vähle Anwendungen
man den Kurs von
ngabe der Sekunden
kommen illusorisch.

Fig. 68. Rechenschieber. Billiger Schulapparat der Firma A. Nestler, Lahr (Baden).

Fig. 69. Rechenschieber. Ausführung der Firma F. E. Hertel & Co., Neu-Coswig-Dresden.

Fig. 9 German slide rules of about 1910

slide rule (Rohrberg 1916) was for the use of mathematics teachers, since it was published in the series "Mathematisch-physikalische Bibliothek", a highly popular series of booklets for mathematics teachers, in the interwar period.

I remember that during my school time, at a classically oriented West German Gymnasium by the end of the 1950s and the early 1960s, I regularly used a slide rule, of the type Aristo, for the mathematics lessons.

For **The Netherlands**, there is the surprising fact that teaching the use of the slide rule was introduced as late as 1968! Before, the slide rule was in use only in university physics courses. Being introduced so late, it became completely outdated within 10–20 years by all those electronic computational devices.[5]

Regarding **England**, like for logarithmic tables, there was no use throughout the nineteenth century in secondary schools, and probably for the same reasons. But even during the twentieth century, the slide rule did not enter schools. One knows that students would use the logarithmic tables by Godfrey and Siddons and by other authors and the teachers saw no need for another computing device.[6]

Regarding **Italy**, there were several textbooks since the middle of the nineteenth century, which document the use of the slide rule—*regolo calcolatore*—in schools. The best known among them is a book by Quintino Sella of 1859. Sella (1827–1883) had studied at the *École des Mines* in Paris and became thereafter professor at the engineering school in Turin. In 1860, he left higher education and turned to political activities. Between 1860 and 1870, he was several times Italian minister of finance (Thomas 2016, p. 199). It was probably due to his influence in the government that teaching the use of the slide rule was demanded in the programme of 1860 for the technical secondary schools. This was repeated in the next programme of 1864, but this one did not enter into practice. It seems that this prescription was an isolated fact and that thereafter the slide rule was not in use in secondary schools.

---

[5]Communication by Harm Jan Smid.

[6]Communication by Leo Rogers.

One can say in general that one made but rare use of computational devices in Italian schools throughout the nineteenth century.

Conclusion

We were able to give here only a first assessment, essentially based on official regulations available from ministerial curricula. What would be necessary to complement the analysis are reports by contemporaneous observers and in particular reports of persons remembering their school times and their mathematics learning. What became evident already now, however, are the considerable differences between the ways in which European countries realised the use of logarithms and the slide rule both in applications and in the classroom and the manners in which different cultures implemented computational practices into mathematics teaching.

Acknowledgements I am highly grateful for the assistance and information provided by Leo Rogers (England), Renaud d'Enfert (France), Marc Thomas (France), Roberto Scoth (Italy) and Harm Jan Smid (The Netherlands).

References

Badon Ghijben, Jacob, and H. Strootman. 1838. *Beginselen der Stelkunst*. Breda: Broese & Co.

Belhoste, Bruno. 1995. *Les sciences dans l'enseignement secondaire français. Tome 1: 1789–1914*. Paris: INRP.

Borel, Émile. 1904. Les exercices pratiques de mathématiques dans l'enseignement secondaire. Conférences Pédagogiques. *Revue générale des sciences pures et appliquées* 14: 431–440.

Bremiker, Carl. 1852. *Logarithmorum VI decimalium nova tabula Berolinensis et numerorum vulgarium ab 1 usque ad 100000 et functionum trigonometricarum ad decades minutorum secundorum*. Berlin: Nicolai.

Bremiker, Carl. 1869. *Logarithmisch-trigonometrische Tafeln mit sechs Dezimalstellen: mit Rücksicht auf den Schulgebrauch*. bearbeitet von C. Bremiker. Berlin: Nicolai.

Bremiker, Carl. 1874. *Tafeln vierstelliger Logarithmen*. Berlin: Weidmann.

Bremiker, Carl. 1877. *Tavole logaritmico-trigonometriche con cinque decimali; compilate dal dr. C. Bremiker*; edizione italiana eseguita per cura di L. Cremona. Milano: Hoepli.

Callet, François. 1795. *Tables portatives de logarithmes; contenant les logarithmes des nombres, depuis 1 jusqu'à 108,000; les logarithmes des sinus et tangentes*. Paris: Didot.

Castle, Frank. 1908. *Logarithmic and other tables for schools*. London: Nelson Thornes Ltd.

Crelle, August Leopold. 1857. *Rechentafeln welche alles Multipliciren und Dividiren mit Zahlen unter Tausend ganz ersparen, bei grösseren Zahlen aber die Rechnung erleichtern und sicherer machen*. Mit einem Vorworte von C. Bremiker. Berlin: Reimer.

d'Enfert, Renaud. 2003. *L'enseignement mathématique à l'école primaire: de la Révolution à nos jours. Textes officiels réunis et présentés par Renaud d'Enfert avec la collaboration d'Hélène Gispert et Josiane Hélavel, 1*. Paris: INRP.

de Gelder, Jacob. 1824. *Allereerste Gronden der Cijferkunst*. S'Grafenhage: vanCleef.

de Lalande, Joseph Jerôme. 1808. *Tables de logarithmes pour les nombres et pour les sinus*. Paris: Didot.

Gauß, Carl Friedrich. 1812. Tafel zur bequemern Berechnung des Logarithmen der Summe oder Differenz zweyer Größen, welche selbst nur durch ihre Logarithmen gegeben sind. *Monatliche Correspondenz zur Beförderung der Erd- und Himmels-Kunde*, Nov 1812, pp. 498–528.

Gauß, Carl Friedrich. 1828. Review: Table of logarithms of the natural numbers, from 1 to 108000, by Charles Babbage. London 1827. *Göttingische gelehrte Anzeigen*, 19. January 1828. (Reprint in: Gauß, *Werke*, vol. 3. Göttingen 1866, 253–254).

Gauß, Friedrich Gustav. 1870. *Fünfstellige vollständige logarithmische und trigonometrische Tafeln: zum Gebrauche für Schule und Praxis*. Berlin: Rauh.

Gauß, Friedrich Gustav. 1958. *Vierstellige logarithmische und trigonometrische Tafeln*. Große Schulausgabe, 171–180. Auflage. Stuttgart: K. Wittwer.

Gispert, Hélène, and Gert Schubring. 2011. Societal, structural and conceptual changes in mathematics teaching: reform processes in France and Germany over the twentieth century and the international dynamics. *Science in Context* 24 (1): 73–106.

Grattan-Guinness, Ivor. 1990. Work for the hairdressers: The production of de Prony's logarithmic and trigonometric tables. *Annals of the history of computing* 12 (3): 177–185.

Höfler, Alois. 1910. *Didaktik des mathematischen Unterrichts*. Leipzig: Teubner.

Howson, Geoffrey. 2014. England. In *Handbook on the history of mathematics education*, ed. Alexander Karp and Gert Schubring, 258–269. New York: Springer.

Kauzner, Wolfgang. 1994. Logarithms. In *Encyclopedia of the History and Philosophy of the Mathematical Sciences*, ed. Ivor Grattan-Guinness, 210–228. London: Routledge.

Kidwell, Peggy Aldrich, Amy Ackerberg-Hastings, and David Lindsay Roberts. 2008. The slide rule: Useful instruction for practical people. In *Tools of American Mathematics Teaching 1800–2000*, ed. Idem, 105–122. Washington: Smithsonian Institution.

Lietzmann, Walther. 1916. *Methodik des mathematischen Unterrichts. 2. Teil: Didaktik der einzelnen Unterrichtsgebiete*. Leipzig: Quelle & Meyer.

Müller, Johann Traugott. 1860. *Vierstellige Logarithmen der natürlichen Zahlen und Winkelfunctionen*. Halle.

Rohrberg, Albert. 1916. *Theorie und Praxis des Rechenschiebers*. Mathematisch-physikalische Bibliothek, Band 23. Leipzig: Teubner.

Schrutka von Rechtenstamm, Lothar. 1911. *Theorie und Praxis de logarithmischen Rechenschiebers*. Leipzig: Eigenverlag.

Schubring, Gert. 1991. *Die Entstehung des Mathematiklehrerberufs im 19. Jahrhun-dert. Studien und Materialien zum Prozeß der Professionalisierung in Preußen. 1810–1870*. Zweite, korrigierte und ergänzte Auflage. Weinheim: Deutscher Studien Verlag.

Schubring, Gert. 2004. Neues über Legendre in Italien. In *Mathematik im Fluss der Zeit*. Algorismus, Heft 44, ed. W. Hein, and P. Ullrich, 256–274. Augsburg: ERV Rauner.

Schubring, Gert. 2008. Gauss e a Tábua dos Logaritmos. *Revista Latinoamericana de Investigación en Matemática Educativa [relime]* 11 (3): 383–412.

Sella, Quintino. 1859. *Teorica e pratica del regolo calcolatore*. Torino: Stamperia reale.

Smid, Harm Jan. 1997. *Een onbekookte nieuwigheid? Invoering, omvang, inhoud en betekenis van het wiskundeonderwijs op de Franse en Latijnse scholen 1815–1863*. Delft: Delft Univ. Press.

Thomas, Marc. 2016. *La règle à calcul, instrument de l'ère industrielle: Le cas de la France*. Limoges: Presses Universitaires de Limoges.

von Močnik, Franz. 1864. *Tavole logaritmiche-trigonometriche. Prima ed. italiana*. Vienna: Gerold.

von Vega, Gottfried Freiherr. 1871. *Manuale logaritmico-trigonometrico. Nouva edizione a cura di C. Bremiker; tradotto in italiano per cura di L. Cremona*. Berlin: Weidmann & Roma: Loescher.

Wittstein, Theodor. 1866. *Siebenstellige Gaussische Logarithmen zur Auffindung des Logarithmus der Summe oder Differenz zweier Zahlen, deren Logarithmen gegeben sind: In neuer Anordnung*. Hannover: Hahn.

Gert Schubring is a Retired Member of the Institut für Didaktik der Mathematik, a research institute at Bielefeld University (Germany), and at present is a Visiting Professor at the Universidade Federal do Rio de Janeiro (Brazil). His research interests focus on the history of mathematics and the sciences in the eighteenth and nineteenth centuries and their systemic interrelation with social-cultural systems. One of his specializations is history of mathematics education. He has published several books, among which is *Conflicts between Generalization, Rigor and Intuition: Number Concepts Underlying the Development of Analysis in 17th–19th century France and Germany* (New York, 2005). He was Chief-Editor of the *International Journal for the History of Mathematics Education*, from 2006 to 2015.

Toward a History of the Teaching of Calculation in Russia

Alexander Karp

Abstract This chapter is devoted to the history of teaching calculation in Russia. It is a history in which political, pedagogical, and technological questions intersect, and much in it sounds quite contemporary and gives the mathematics educator much to think about. We analyze certain sections of Russian textbooks and approaches employed by them. Some characteristics specific to Russia are also discussed: the use of so-called number alphabets and the Russian abacus during certain periods.

Keywords Calculation · Educational goals · Russian abacus · Number alphabets
Mental calculation · Arithmetic textbooks

Introduction

The teaching of calculation has always occupied an important place in mathematics education, and Russian mathematics education is no exception. Calculation was seen both as a practical skill that it was difficult to do without in life, and as a means of intellectual development; not by accident one of the most famous paintings depicting a school in Russia, Nikolay Bogdanov-Belsky's (1868–1945) "Mental Calculation at S. A. Rachinsky's Country School," shows a class working out a complicated calculation—finding the value of the expression $\frac{10^2+11^2+12^2+13^2+14^2}{365}$. This process involved not merely calculating, but calculating mentally:[1] students were required to exhibit inventiveness and alertness, which were seen as being connected with the ability to calculate in general, since what was useful for quick and accurate calculation was not dogmatically following some memorized algorithm, but the ability to apply and even to discover special techniques that facilitated calculation. This was therefore the skill that schools had to teach.

---

[1]To solve the problem mentally one can note that $10^2 + 11^2 + 12^2 = 13^2 + 14^2 = 365$.

A. Karp (✉)
Teachers College, Columbia University, New York, NY, USA
e-mail: apk16@columbia.edu

© Springer Nature Switzerland AG 2018
A. Volkov and V. Freiman (eds.), *Computations and Computing Devices
in Mathematics Education Before the Advent of Electronic Calculators*,
Mathematics Education in the Digital Era 11, https://doi.org/10.1007/978-3-319-73396-8_15

The teaching of calculation thus lies at the intersection of several major lines of development in mathematics education. Operating with abstract numbers by itself presupposes a high level of development both in the ability to think and in the ability to use the language in which the calculations are carried out. Depman (1959, p. 111) cites a story by the writer Gennady Gor (1907–1981) about the difficulties experienced by a young man from a small Northern ethnic group in solving the simplest arithmetic problems involving addition: it is hard for him to add up "trees in general" without knowing whether they are tall or short. The teaching of basic arithmetic operations is connected both with psychology—which is why so many major scholars wrote about them (Menchinskaya 1955; Thorndike 1929)—and with the culture of a society. To make calculation easier, various counting devices have been invented, and here we enter into questions about the role of technology in mathematics education. The teaching of calculation was developed over decades and centuries, and both the techniques invented for calculating, as a part of elementary mathematics, and the techniques invented for teaching, as a part of pedagogy, merit detailed study. Finally, issues that concern hundreds of thousands and millions of people inevitably become points of conflict, and consequently the teaching of calculation, including its seemingly purely specialized details, have provoked ideological and political debates, which are of interest to those who are interested in the politics of mathematics education.

We cannot here describe all aspects of the teaching of calculation in Russia over the course of centuries—and not only for lack of room, but also because many of them have clearly not been sufficiently studied (in our view, this pertains above all to the didactic aspects, which have been analyzed, contrasted, and compared from a historical point of view only to a very small degree). The present chapter thus represents merely a kind of sketch of a description of the history of the teaching of calculation in Russia, in which many problems are deliberately left unexamined, while others are discussed only very briefly.

The discussion below will address the symbolism (language) of calculation, the simplest calculating devices (the Russian abacus), and above all, the didactic aspects of the teaching of calculation. But first, this paper will provide a brief history of the teaching of arithmetic in Russia, which, hopefully, will help readers to see the methodological and technological problems in social context.

A Brief History of the Teaching of Arithmetic in Russia

The art of calculation was known in Russia since ancient times. Every book on the history of science in Russia tells about the mathematical–chronological treatise of Kirik Novgorodets (Kirik of Novgorod), dated 1136 (Simonov 1980, 2007; Ivashova 2011; Polyakova 1997, 2002, 2010). The knowledge reflected in this work is considerable for its time, although it cannot be determined how widespread such knowledge was—more precisely, it cannot even be determined how many people had a basic ability to calculate. We possess no precise information about this.

Based on surviving birch bark manuscripts from Novgorod, researchers conclude that in mercantile Novgorod in the thirteenth–fourteenth centuries mathematical literacy was widespread, and they also point out that this was not the case in other parts of Russia even later (Polyakova 1997). At the Hundred Chapter Synod of 1551, which took place in Moscow with the participation of the Russian czar Ivan the Terrible, it was noted that "in olden days" there had been many schools in cities where calculation had been taught. The need to establish such schools anew was also mentioned at the Synod (Polyakova 1997). It remains unclear, however, to what extent this praise of the olden days corresponded to reality, and it is still less clear how far these educational plans came to being realized (not very far, it may be supposed).

The so-called Lithuanian Rus', that is, formerly Russian lands that had subsequently ended up being ruled not by the Mongols and their Muscovite representatives, but by Lithuanian princes (these regions roughly correspond to today's Ukraine and Belarus), found itself in somewhat more favorable circumstances during the pre-modern period (Karp 2014a). The influence of Polish Catholic and German Protestant cultures elicited a reciprocal striving to develop Russian Orthodox education as well. Consequently, these lands, which were subsequently annexed to the Muscovite state, often served as a source of literate people (and mathematically literate people also).

A significant leap forward was made during the era of Peter I (died in 1725). To reiterate, there is no reason to think, of course, that prior to this arithmetic was completely unknown—merchants simply could not have done without it. Much evidence of this fact has survived, including mathematical manuscripts (see Polyakova 1997). What was likely the first printed Russian book on mathematics —"Convenient Calculation; with which any man buying or selling can very conveniently find any number of any thing" (*Schitanie udobnoe, kotorym vsiakii chelovek kupuiuschchii ili prodaiushchii zelo udobno izyskati mozhet chislo vsiakiia veshchi*)—was published even before Peter, in 1682. This book consisted of a multiplication table up to 100×100 inclusively, and its very title indicates why and for whom it was published.

Nonetheless, the role and place of arithmetic at this time were quite insignificant. This can be seen by examining a book by Ilya (Elias) Kopievskii (about 1651–1714) from the early Petrine period—a time that was in some sense already transitional, when interest in the study of arithmetic was beginning to grow. Kopievskii's treatise (1699) was the first book on mathematics which went beyond providing tables of multiplication and was published in Russian, although it was printed not in Russia, but in Amsterdam. Kopievskii himself was officially considered a Pole, although in terms of his place of birth today he would have been considered Belarusian (Pekarskii 1862; Polovtsev 1903). After graduating from the Slutsk Protestant gymnasium and finding himself, due to a complicated set of circumstances, in Amsterdam, Kopievskii made use of Peter I's arrival in Amsterdam to secure support for several typographical projects. Among them was his *Arithmetic* (we omit the full title of this work, which was much longer, as was customary at the time). Kopievskii remarks that he wrote the book on instructions from the czar, but it is known that its publication was commissioned by certain Russian clerks (*prikazchiki*), the Filatyevs (Pekarskii 1862), who then rejected the book; one would like to explain this by the fact that they did not like the book (Russian mathematics educators do not like to mention its author among the esteemed founders of their discipline), but no evidence exists to support this supposition.

The book consists of 48 pages, 16 of which are devoted to arithmetic; the rest contain wise maxims and parables teaching "piety and pious ways to youths" (p. 2). Although the author remarks that he has explained everything "briefly, but clearly, candidly, and completely, so that even a small child will quickly understand it," to learn anything from his book appears to have been not easy. The book has five sections: numbers (we would say numeration), addition, subtraction, multiplication, and division. At times, definitions of sorts are given: "numeration is the intricate representation of any given number" (p. 3), but most of the book consists of descriptions and examples of algorithms for carrying out operations, for instance, subtraction: put 57 below 89; take 7 away from 9, leaving 2; likewise take 5 away from 8, leaving 3; thus, the answer will be 32. In all, seven examples of subtraction are provided; most are then checked for correctness (these checks are referred to as "tests" (*iskushenie*)). The text itself is not without misprints, but most importantly, all examples are purely numerical—there are no word problems (by contrast with the arithmetical manuscripts which circulated in Russia prior to the publication of Kopievskii's treatise, whose authors, however, also fail to explain why these problems should be solved using specifically one or another kind of operation). In other words, calculation is studied abstractly, without being applied to any kind of practical situation, and it is regarded (whether deliberately or simply because the author could not have done otherwise) as a discipline that needs to be accompanied by something more useful and engaging, such as parables, whose meaning Kopievskii always explains. Market demand or a royal commission for an arithmetic textbook did exist, but nothing existed that could satisfy such demand, and individuals capable of offering at least some kind of solution were not found at once.

Peter I's plans to modernize the army, build a fleet, and develop a manufacturing sector capable of competing with the West required reforms in education (Karp 2014b), which naturally affected the teaching of calculation as well as more advanced mathematical sciences. In 1701, *Mathematical and Navigational School* was opened in Moscow. In 1714, so-called arithmetic schools (*tsifirnye shkoly*) were established in provincial cities. Without discussing the extent to which Peter's plans were realized, we would note the importance of the recognition of the fact that certain categories of the population had to be required to obtain at least a basic education in mathematics. The importance of studying arithmetic (as well as more advanced mathematical disciplines) was also underscored by requirements at schools established then and later. Even the gentry who confined themselves to homeschooling found it necessary to teach their children calculation—the retired soldier hired to teach arithmetic to the protagonist of the famous eighteenth-century play "The Minor" ("Nedorosl") has the telling surname "Tsifirkin" (from *tsifra*, "numeral").

No less significant were the didactic reforms that were taking place. Polyakova (1997) rightly notes that pre-Petrine arithmetical manuscripts are in effect reference texts, which is why problems in them are grouped not in terms of methods for solving them, but in accordance with their application in real-life (commercial) situations. It now became evident that arithmetic had to be studied for application in many other situations as well, and consequently, it became necessary at least sometimes to study arithmetic as a subject, and not simply to memorize recipes for a few specific cases. In Russia, Western practices were broadly borrowed and Western teachers were invited, but very quickly Russian textbooks written on the basis of Western practices began to appear as well, of which the first that should be mentioned is the famous *Arithmetic* (1703) by Leonty Magnitsky (1669–1739) (see, for example, Freiman and Volkov 2015, and the bibliography there).

Polyakova (1997) argues that another distinctive didactic characteristic of pre-Petrine arithmetical manuscripts was the absence of explanations and proofs (these likely did not seem especially necessary to merchants). Now, after acquaintance with more developed Western mathematics, explanations became a focus of interest: the Russian academician Stepan Rumovsky (1734–1812) wrote that he endeavored to provide explanations for all rules in his textbook. On the other hand, Nikolay Kurganov (1725–1796), an author who became more popular and whose textbook *Arithmetic or Number Book* supplanted Magnitsky's *Arithmetic*, was of a different opinion and wrote: "the youth who is beginning his studies, due to the weakness of his reasoning, can benefit more from the use of books that contain rules alone, explained by means of examples and ingrained through repetition" (Depman 1959, p. 376).

The subsequent history of the teaching of arithmetic developed along the two lines mentioned above: the administrative–organizational and the specifically methodological. The first of these developments—more obvious and better studied —involved increases in the numbers of students who were taught arithmetic and in the gradual emergence of a more rigid description of required outcomes. The second consisted of changes in the understanding of the contents of what was taught and of the aims of teaching, and also of changes in the techniques and methods employed in such teaching. Of course, these two lines of developments are linked: increases in numbers of students and the transition to education on a mass scale usually necessitate a search for some kind of methodology that is capable of helping large numbers of teachers to instruct large numbers of schoolchildren, so that instruction might rely on something beyond the individual characteristics of teachers or students. Methodological changes (as well as administrative changes) will be mentioned very briefly in this section, but certain details of the methodological transformations and debates (far from all) will be discussed below.

Miliukov (1994) believed that Peter I's program of elementary education was a complete failure. Catherine II (reigned from 1762 until 1796), however, was able to establish a certain system of multilevel schools, which was subsequently supplemented with various departmental and professional, military, and other academies (Karp 2014b). At the same time, an enormous proportion of the population remained illiterate, as well as mathematically illiterate. The struggle to expand the number of people being educated may in fact be the main content of the two centuries following Peter I.

In 1802, the Ministry of Public Education was established, and sometime later (in 1826) Emperor Nicholas I wrote:

> Reviewing with especial attention the organization of the educational institutions in which Russian youth is educated to serve the state, I regret to observe that they lack that necessary and indispensable uniformity which must be the foundation of both child rearing and instruction. (*Perepiska...* 1826–1828, p. 1)

In accordance with the sovereign's wishes, educators began to introduce uniformity, both in teaching in general and in the teaching of calculation in particular. Textbooks were now put through a strict process of review and approval, and the Ministry began keeping a more watchful eye on their overall character and general contents as well. As a result, both in privileged schools (*gymnasia*) and in the far more widely accessible provincial schools, educators endeavored to teach arithmetic in a new way; these changes were connected first and foremost with the name of Fyodor Busse (1794–1859), author of textbooks and problem books in arithmetic (and other mathematical subjects) and handbooks for teachers (see, for example, Busse 1833).

The children of peasants—initially serfs and then even emancipated peasants—received practically no education.[2] Gradually, however, rural provincial and parish schools began to appear, in which arithmetic was taught. Among others, Leo Tolstoy or S. A. Rachinsky, whose name has already been mentioned, actively worked with such schools. The preparation of rural teachers, able and willing to teach in village schools (and to teach arithmetic also), was an important concern of pedagogy before the revolution of 1917.

Despite the rapid growth in the number of rural school in the decades prior to the revolution of 1917, illiteracy remained a problem. Probably the most optimistic estimate of the literacy rate is given by the former czarist minister of education Ignatiev (1933), who contends that by the time of the revolution, 56% of the population was literate and 91% of children attended schools (p. 663). The reality seems to have been far more lamentable, but even such estimates as these (and bearing in mind that one was considered literate even if one was able to read only by spelling out words, and that literacy did not mean mathematical literacy, and that a child might attend a school for only a brief period) demonstrate that quite large numbers of children were not taught basic arithmetic in school.

The revolution of 1917 radically changed everything, including the teaching of mathematics (Karp 2009, 2012). The old system of the organization of the school system was destroyed. It was replaced by unified labor schools—at primary and secondary levels. Arithmetic was supposed to be taught, and, in theory, to all children, at primary-level schools, but the organizational reform was accompanied by a methodological one: arithmetic as a separate subject was no longer taught;

---

[2] Serfdom remained in Russia much later than in other European countries. It was abolished only in 1861.

children had to learn to calculate as part of the so-called *complex* (*kompleksnyi*) approach, that is, in the context of studying various broad topics drawn from life. For example, they might study the topic "Postal System," which would involve discussions of geography (where do letters go?), arithmetic (calculate the postage for a letter or parcel), the Russian language (write the address), and many other subjects. To study rules and algorithms in this way was awkward (Karp 2012), but this was not regarded as an objective of prime importance.

Stalin's counter-reforms brought traditional teaching back into schools, with the difference that universal education (at least at the elementary level) now had the entire might of the state behind it. Ivashova (2011) identifies three stages in the development of mathematics in Soviet elementary schools—from 1932 to 1969, from 1969 to 1990, and since the 1990s. The watershed moments, according to this view, are the educational reform resolution passed by the Central Committee of the Communist Party in 1931, the introduction of new curricula at elementary schools in 1969, and finally the collapse of the Soviet Union as the nominal date for a certain democratization of society.

The didactic–methodological (in the broad sense of the term) changes in the teaching of calculation that occurred since 1990 were rather substantial, but probably still more substantial were the technological changes, thanks to which certain, and indeed many, skills that had been diligently developed previously became unnecessary and the aims of the teaching of calculation shifted somewhat.

The Symbolism of Calculation: Number Alphabets

In this section and the one that follows, we will discuss not, of course, all possible aspects, but only a few that are characteristic specifically of Russia. A very important distinctly Russian feature of this type was the use of number alphabets during certain periods (see, for example, Simonov 1974). Ancient Russian numeration was based on the Cyrillic alphabet: letters were used as numerical symbols. Its prototype was the Ionian alphabetic numeration system, which was later used by the Byzantines. The ancient Russian symbolism, however, has its own distinctive attributes (Simonov 1977).

This system is non-positional; includes special symbols, for example, for thousands; and does not allow for fractional notation. We would find it extremely inconvenient to use, but possibly Depman (1959) is right to note that the inconvenience of the Ionian system for us is due to the fact that we do not attempt to master it completely in order to use it for calculation, but do no more than merely familiarize ourselves with it.

There is, however, a different view (Maistrov 1974), which holds that this system was never used for calculation at all, neither in ancient Russia nor by the Greeks. Maistrov notes that "we have no evidence of any operations being carried out using this system [in ancient Russia]" (p. 48). As for the Greek evidence, it is called into question by Maistrov. "Moreover," he writes, "there are a number of considerations demonstrating that there in fact could not have been any calculation using this system" (p. 48). Among such considerations, he cites the inconvenience of expressing certain numbers, as well as the fact that words rather than symbols were used to denote fractions.

In Maistrov's view (shared, for example, by Polyakova 1997), calculations in Greece and Ancient Russia were made on an abacus, and only the result was written down. Consequently, no convenient system for making written calculations was needed: what was needed was only a convenient notation system. To a certain extent, it may be said that today we often operate under similar conditions, understanding the notation of a number, including the notation of the results of operations, but not caring too much about how these operations are carried out—a calculator knows what it is doing.

What is clear at any rate is that the alphabetic method of number notation endured for a very long time—the "Convenient Calculation" of 1682 referred to above employed precisely this type of numeration, which was replaced only in the eighteenth century, and not entirely. The well-known statement by a character in Pushkin's story "History of the Village of Goryukhino"—that he "could not learn Slavic numbers"—cited, for example, by Depman (1959, p. 58), shows not that Slavic numbers were unknown during the first third of the nineteenth century, but on the contrary, that they were then still considered something that an educated person needed to know.

The Technology of Calculation: The Russian Abacus

This brings us to the topic of the abacus, which was for centuries the main means of calculation in Russia. Pre-revolutionary schools used the abacus until the end of their existence (Evtushevsky 1871; Talalay 1903), nor was it forgotten in Soviet textbooks (Popova 1933).

During the period 1820–1830, the abacus was improved to make it convenient not only for addition and subtraction, but also multiplication and division, and this new abacus was quite widely promoted (Tikhomirov 1830). We have archival records from the middle of the nineteenth century of reviews of a more recent textbook devoted to the subject—a review process was required before a textbook could be used in schools on a comparatively wide scale (O rassmotrenii 1852). It is noteworthy that the aforementioned Busse, reviewing such a book by the teacher Lesnevskii, remarks that the book could be used only when specifically such abaci are employed as the ones discussed by its author (p. 11). One of the versions of Lesnevskii's book, however, was published (Lesnevskii 1852), which indicates that it was used along with other texts, oriented toward a different device.

Without attempting to present the history of the Russian abacus in any detail, we will nonetheless say a few words about it, relying first and foremost on the thoroughgoing study by Spasskii (1952). In the spirit of the early 1950s, Spasskii painstakingly refutes theories of the foreign origins of the Russian abacus (Chinese,

Golden Horde, Armenian, and Turkish), insisting that it was an original product of the Russian national genius. To refute the established theory of the Chinese and Mongol origins of the abacus, Spasskii draws attention to the fact that the Russian abacus is decadic, while the Chinese abacus is in essence pentadic (p. 419). He likewise remarks that the modern abacus likely appeared comparatively late, since, for example, in Western regions of the former ancient Russian world it was unknown.

Holding the view discussed above that during certain periods calculations were performed not on paper, but using a device, the abacus, Spasskii rightly notes that even with the development of calculations on paper, the Russian abacus did not disappear. Manuscripts from the seventeenth century devoted to the art of calculation, including numerical calculation (i.e., calculation that used Arabic numbers rather than Slavic numeration), contain descriptions of various techniques and devices for calculating. It is unclear, however, whether all of these were actually used in practice, or whether some were simply copied to the list from a foreign textbook. Western travelers described several calculating devices used in Russia that were unfamiliar to them. One of them wrote explicitly in his journal: "They calculate and represent numbers in a different manner from other people" (Spasskii 1952, p. 356). Since this comment concerned representation as well as calculation, one might ask whether the differences lay less in the nature of the calculating device and more in the numeration system. On the other hand, one copy of an old manuscript contains an addendum indicating that in one specific case of using a device for calculation, the German rather than the Russian manner of calculation ought to be employed, in other words, not the kind ordinarily used in Russia, but a different kind, likely known from Western textbooks. Based on numerous records and surviving abaci, Spasskii (1952) shows that likely both in the seventeenth century and even more so earlier, different devices were employed, and that in general a certain evolution of the abacus occurred, which was to a certain extent connected with the specific nature of the problems that people engaged in calculation had to solve, and thus ultimately with social-economic development.

Today, it is difficult to share the pathos displayed by Spasskii in refuting theories of the foreign origins of the abacus. The fact that the Russian abacus has its distinctive features is evident, but no less evident is the fact that other countries also had their calculating devices, and it is unlikely that these instruments were invented without knowledge of what was happening in other places. What is remarkable is rather the incredible popularity of the abacus in Russia, which endured even after such instruments had become far less popular in Western Europe. Spasskii (1952) quotes various authors, who claim that they can carry out calculations using an abacus much faster than other people can do on paper, and this can easily be believed; it is difficult to imagine, however, that such virtuosity can be achieved on a mass scale more easily than the ability to perform calculations on paper. It is difficult to explain the conservative nature of Russian society in this regard merely by arguing that the Russian abacus was superior to all others. There is evidence (Pchelko 1940, pp. 30–31) that many graduates of elementary schools in the mid-nineteenth century learned nothing by being taught how to calculate in school, but continued to calculate as they had at home—mentally or using an abacus—since the methodological improvements that had already taken place, as noted above, reached many schools very slowly, while at home children were taught to use an abacus in "the easy way." Pchelko (1940) points out that classroom subtlety in general was considered useless in practice, since no peasant would carry paper and pencil with him. The factors behind the abacus's popularity—whether it had to do with how arithmetic was taught, or with the general low literacy rate of the population, which ruled out writing, or with the high cost or unavailability of paper—merit a separate discussion.

Certain Examples of the Teaching of Calculation

Going over certain problems from Magnitsky's *Arithmetic* and the solutions given there, Pchelko (1940) concludes:

> The problems were not given to students to solve on their own. The schoolchild had to know how to use a given rule to solve only those problems which he had solved many times, or at least problems similar to those which he had solved earlier. (p. 15)

Algorithms for calculating were also not explained in any way. Students thus learned that certain problems were solved using, say, multiplication, while multiplication was carried out according to certain rules. In this way, the skilled practitioner could obtain the answer he needed without much mental effort. It may be argued that one of the fundamental ways in which the development of mathematics education changed course in the centuries after Magnitsky consisted in an effort to show students the meaning of the operations they performed and to develop their independent ability to carry them out. Below we will examine certain examples from different textbooks which demonstrate how teaching changed.

Addition and Subtraction of Integers in Kurganov's "Number Book"

Kurganov's textbooks, as has already been noted, gradually pushed out Magnitsky's *Arithmetic*. Depman (1959, p. 376) argues that they are already considerably closer to modern textbooks. Nonetheless, the explanation of addition in them is quite similar to what is found in Magnitsky (and, indeed, in Kopievskii). After stating that addition "gives rules on the basis of which may be found a number equal to two or many uniform numbers" (Kurganov 1791, p. 10), the author goes on to say that "if the given numbers are simple [that is, single-digit], their sum is quite handily found, as in the example: 2 and 3 make 5."

After this, he must explain about numbers that are less simple. First, a certain rule is given, about which it is immediately said that "this may be shown more handily and clearly with an example." The example shows the addition of three numbers at once—two four-digit numbers and one three-digit number—in a column, but it is accompanied by separately written out sums, which must be found in the course of the calculation. Kurganov then offers four more numerical examples of column addition of multidigit numbers.

After this, finally, several word problems are given "in order to show what kinds of problems or questions may be solved by means of addition" (p. 11). Among them, for example, is the following:

> During the destruction of the city of Troy, 880000 people were killed on the Greek side and 686000 on the Trojan side. Determine how many people were killed in all.

The answer is found by means of column addition of the given numbers without further commentary. There are four such examples in all and nothing more is said about the topic of addition.

Subtraction is covered in the same manner. First, we are told that "subtraction is a method by which from one given number, another may be taken away, and the remainder may be determined" (p. 15). Then the subtraction of simple numbers (i.e., one-digit numbers) is discussed. After this, quite a number of examples are given, although not all possible situations are presented. Then, Kurganov examines a complicated example of column subtraction (a four-digit number from a four-digit number), and then separately talks about cases in which one must borrow from the next order of magnitude—a rule algorithm and an example are given. Five more examples of subtraction follow.

The following section states that subtraction may be checked by addition. This is not explained, except by the word "of course"—that "of course" the greater of the given numbers must be obtained after addition if the subtraction was carried out correctly. Everything again concludes with examples of problems that are solved using subtraction.

The Addition of Integers in F. Busse's Textbook

The textbook by Busse (1833, first edition 1830) opens with a detailed, many page discussion of what a unit is, and how numbers are obtained, written, and pronounced. Once one knows this, the author notes, one can go on to operations involving numbers, the simplest of which is addition. He begins his explanation with the following problem: how much must be paid for two books, if 5 rubles were paid for the first and 3 rubles were paid for the second? Then, the terms "addends" and "sum" are introduced, and the "plus sign" is presented. A single-digit addition table is given. This is followed by sections devoted, respectively, to the addition of one-digit and two-digit numbers, two-digit numbers, three-digit numbers, and multidigit numbers, and finally any numbers. It is pointed out that, for example, in adding 25 and 9, we add 2 tens, 5 ones, and 9 more ones, and therefore, we obtain 2 tens and 14 ones, and so on. Column addition notation is introduced. In order to clarify the addition of several three-digit numbers, a word problem is given that requires students to determine how many years have passed from the founding of the Russian state to the present moment. These sections contain no other word problems, which appear only much later, when the text addresses nominative numbers and operations involving them.

The Addition of Integers in V. Agloblin's Textbook

The textbook by Agloblin (1846), a teacher of mathematics at the Moscow Institute for the Nobility, was not especially popular and did not introduce any special changes into teaching. This is just what makes it interesting for our purposes: we may suppose that it shows how ordinary teachers taught. The textbook is not very different from Kurganov's when discussing addition. The material is presented in a manner much less systematically gradual and sequential than it is in Busse's textbook, but instead, it contains many verbal discussions. Concerning addition, we are told that this is an "operation used to combine several given numbers into a single number." From this, the author concludes that addition is used to determine the number of units of different orders of magnitude contained in several given numbers. Then, an example is used to calculate how many units of each order of magnitude are contained in a certain sum—it is determined that the sum has 16 ones, 17 tens, and so on. Then, the result is simplified. Then, the addition of the same numbers is written down in the form of a column. To conclude, the author offers two other examples of column addition. The whole section is far shorter than it is in Busse's text.

The textbook contains no word problems. Much later, after covering all the operations, the author begins discussing operations with nominative numbers, but here too very few problems are given (if we consider them problems at all, as they are given in simplest form, e.g., there are so many pounds and so many more, determine how many there are in all). Finally, there is a section entitled: "Using the Preceding Methods of Calculation to Solve Problems Encountered in Society." This section, however, contains analyses of far more complex problems than ones reducible to addition (starting with proportions, etc.). It may, of course, be imagined that the author, a working teacher, simply did not consider it necessary to add certain problems, believing that other working teachers would add them themselves if they deemed it necessary to do so.

The Addition of Integers in the Textbook by A. F. Malinin and K. P. Burenin

Arithmetic (or *Manual on Arithmetic*) by A. F. Malinin (1835–1888) and K. P. Burenin (1836–1882) went through numerous editions and in a certain sense became a classic (Prudnikov 1956). The difference between this text and the ones discussed above can be seen already in the fact that the authors begin by discussing the concept of an operation in general; offering examples, they show situations in which "out of two given numbers, a new number must be composed" (we quote from Malinin and Burenin's edition of 1897, pp. 14–15).

The investigation of addition begins with a problem about the number of apples in three baskets (the number of apples in each basket is indicated). After explaining that one could simply count all the apples one by one, the authors ask whether there is a more convenient way to find the answer. In a natural fashion, the following definition arises: addition is an operation which may be used to compose a number containing as many units as were contained in all of the given numbers. (Note that the text is written in such a way that the word "operation" (*deistvie*) is understood in this case as a technique or method.)

Subsequently, in the customary order, the addition of single-digit numbers is discussed, and then the addition of multidigit numbers—in a way quite similar to Agloblin's textbook. But by contrast with that textbook, the authors also make remarks about the order in which numbers are added, and recommend that addition be carried out in the order that is most convenient. And most importantly, they then offer examples of problems that may be solved using addition, along with a certain generalization to the effect that those problems are solved using addition in which one number must be increased by the number of units contained in another, or when a number must be found that is equal to several numbers taken together. The section concludes with questions about the material covered, which make it possible to check how well the student has grasped what has been said. Among these questions, for example, are the following: *What is that which we call addition? How is addition carried out in those cases in which the number of terms to be added is very large? Which problems are solved using addition?* As a supplement to the textbook, A. F. Malinin and K. P. Burenin wrote a problem book, so it was assumed that material for practicing skills would be drawn from there.

The Addition of Integers in N. S. Popova's Textbook

N. S. Popova's (1885–1975) textbook arrived in schools immediately after the education reform resolutions passed by the CPSU's Central Committee in 1931 (actually, her textbooks—coauthored with I. N. Kavun—had been widely used in schools even before this, although the new edition differed from them). The textbook for first grade (references below will be to the edition Popova 1937a) differs strikingly from the textbooks mentioned thus far, if only by the abundance of illustrations in it. Addition appears for the first time at the very beginning precisely on an illustration, in the context of becoming acquainted with the ten digits; the illustration shows two white chickens and one black chicken, and it is accompanied by the notation: $1 + 2 = \cdots$ (p. 5). Later comes a section entitled: "Adding and taking away by ones"; this is followed by "Adding and taking away by twos," and so on, up to adding and taking away by nines. In each case, many examples and word problems are given. Then addition and subtraction up to 20 are introduced —at first without going over 10, and then with going over 10. Multiplication and division are covered in the same fashion, followed by all operations with answers up to 100. This is the point at which the first-grade textbook ends.

In second grade, students cover calculation up to 1000. Addition of multidigit numbers is introduced only in textbooks for the third and fourth grades (Popova 1937b). These contain almost no illustrations. After discussing an addition problem (by simply stating how it is solved), the author demonstrates column addition, accompanying it with brief comments. Note that Popova's textbooks were supplemented by problem books, which contained many varied assignments, both for improving technical skills and deeper problems.

Toward a Methodological Analysis of Changes in School Textbooks

Naturally, textbooks that have appeared since Popova's textbook have not followed her in every respect. On the contrary, many new methodological ideas have emerged, including ideas about the teaching of calculation. In today's textbooks, the sections devoted to this topic contain elements of algebra, ties to measurements, elementary set theory, and much else (Ivashova (2011) describes certain contemporary textbook sets). Nonetheless, even stopping at textbooks from over a half-century ago, we can identify certain trajectories in the changes that occurred.

The most important of these, in my view, is connected with changes in the goals of the teaching of calculation. Pchelko (1940) rightly sees here a transition from a purely practical (*material*) approach to a developmental (*formal*) approach, associated in pedagogy with the name of Pestalozzi. While for Kurganov, as for those who preceded him, it was important above all to teach students how to solve certain practical problems without thinking too deeply about why they were solved specifically in this way, subsequently, the understanding of the process and the independence of the students became increasingly important.

Fyodor Busse wrote that nothing could be left without a thorough explanation (Pchelko 1940, p. 21), but this leaves open the question of how such an explanation is supposed to be given. In eighteenth-century textbooks, explanations are practically equated with verbal explanations, which were additionally formulated in ways that we would today consider extremely imprecise and sloppy. Better words could usually be found, and such improvements did continue being made (Agloblin's textbook has advantages over Kurganov's), but this development was accompanied by a conspicuous decrease in the relative share of verbal material. Of course, such a decrease was connected in part with the fact that textbooks began to be aimed at increasingly younger students, who were learning to read at the same time as they were learning to calculate. But there was more to the matter than just this. Views of the role of problems in education were gradually changing. Semyon

Shokhor-Trotskii (1853–1923) was perhaps more articulate than others in voicing the view that problems were not only an aim, but also a means of education (Shokhor-Trotskii 1886), but similar ideas had been expressed before him as well (although opinions about the role of problems were by no means always identical— see Pchelko 1940). And the number of problems offered to students for solving on their own visibly grew.

Ivashova (2011) quotes Kavun and Popova (1934, p. 10), who state that the theory behind calculation in their textbooks remained "unspoken, not set out in precise language" (p. 54). This was indeed the case, and students followed the models they were given without being offered a precise theoretical formulation (likely impossible at a certain stage in any event). The very organization of the material, however, its structure and sequence, gave students an opportunity to grasp the logic behind what was being done and said. Consequently, another important transformation, which took place over an extended period of time, consisted in the increasing gradualness and sequentiality of instruction. Although we might suppose, of course, that teachers using Kurganov's and even Agloblin's textbooks added problems and examples of their own, nonetheless, the fact that these textbooks so quickly introduce the addition of multidigit numbers remains surprising.

Gradualness and suitability to the age of the students were already mentioned by Busse (Pchelko 1940, p. 21), but nonetheless the fact that it was necessary to spend such a long time studying topics that were obvious to an educated adult provoked resistance. This resistance was expressed in particular in discussions surrounding the so-called Grube method or the *method of studying numbers* (Karp 2006; Pchelko 1940). The method of the German pedagogue August Wilhelm Grube (1816–1884) entailed the so-called *monographic study of numbers*, in which each number was investigated from all angles and very thoroughly, while operations were discussed within the framework of this investigation. The method spread around the world and had its adherents in Russia as well, including the author of what was probably the most popular problem book in arithmetic during the period 1870–1890, V. A. Evtushevskii (1836–1888). The method was criticized from various perspectives, in particular by supporters of the so-called *method of studying operations*, who saw operations as the correct foundation for the course in arithmetic. It is noteworthy, however, that many critics, including Leo Tolstoy (1874), criticized Grubbe's method specifically for its tedious gradualness and tendency to explain what children already knew anyway (for example, thanks to the popular Russian game *babki*—a version of knucklebones—as Tolstoy wrote). This, for example, is how Rachinskii (1902) assessed Grubbe's approach:

> This technique, which may be necessary when one is dealing with five-year-old children (or with idiots), smacks of extreme artificiality when one is dealing with children twice that age, who already know how to calculate beyond 100, who already have a practical understanding of the decimal system, because they know how to count kopecks, ten-kopeck coins, and rubles. (p. 72)

Evtushevskii (1874), in response to Tolstoy's criticism, pointed to the results of a study conducted by the Moscow Committee on Literacy, which showed that schoolchildren who were taught using Tolstoy's preferred method had difficulty solving the following problem: "10 blackbirds were sitting in a tree; 3 flew away; how many were left?" "Apparently," Evtushevskii noted wryly, "these students had never played *babki*" (p. 52).

The monographic investigation of numbers was indeed probably often carried out in an excessively exhaustive manner, although children were taught how to calculate—if not always, then very often—starting not at age ten, as Rachinsky made it, but earlier. We cannot overlook the fact that despite all the criticisms directed at this method, its influence can be felt even in the textbook by Popova, cited above. In actual teaching, some degree of compromise triumphs, which takes into account different, including seemingly contradictory, points of view.

The idea of gradualness and suitability to the distinctive characteristics of different ages leads to the idea of concentric teaching, which can be clearly seen in Popova's textbooks, but which to some extent is present already in Agloblin (the fact that focusing on abstract numbers first and on nominative numbers second is probably not very helpful is a separate issue). Pchelko (1940) connects this idea also with the influence of Pestalozzi, to whom he also attributes the emphasis on visual demonstration. Indeed, this last concept, promoted by Pestalozzi, undoubtedly had an impact on the teaching of calculation also, manifested most obviously in the appearance of figures and illustrations in textbooks, but also in the use of various models and visual aids, which became increasingly popular in Russian schools over the course of the nineteenth and twentieth centuries.

Mental Calculation in Russian Schools

Another idea that goes back to Pestalozzi had a special fate in the teaching of arithmetic in Russia: the idea of mental calculation. By all appearances, as was noted above, written, "schoolroom" calculation was not particularly popular outside the walls of the school. Mental calculation turned out to be at once that which was needed in life and that which gave proof of inventiveness and alertness. In addition, mental calculation is a skill that lends itself easily and conveniently to competitions, which is also a concern not of least importance when dealing with children. Rachinskii (1902), whom we have already cited numerous times, wrote that children could become absorbed in competitions in written calculation too, surrounding him with their slates, on which they made their calculations:

> Sergey Alexandrovich, division! – Give me division by hundreds! – Give me division by ones! – Give me division by millions! – Give me division by thousands! And the solutions were presented so quickly that I barely had time to write down the problems. (p. 63)

But mental calculation was even more successful. Rachinsky regarded the carrying out of mental calculations as a manifestation of many different abilities:

In addition to memory, other abilities are obviously involved here as well: imagination, which vividly shows them the composition of a number from the initial factors and their combination; the ability to connect the outward appearance of a number with this composition. (p. 71)

As proof, he tells how one of his students found the product $84 \cdot 84$ instantaneously. To do so, the student noticed that $84 = 7 \cdot 12$, and therefore, $84 \cdot 84 = 7 \cdot 7 \cdot 12 \cdot 12 = 49 \cdot 144 = 50 \cdot 144 - 144$, which is not hard to calculate in one's head.

Such virtuosity was not achieved by everyone, of course, and by no means everyone considered it necessary. Nonetheless, recognition of the importance of mental calculation was widespread. The famous mathematics educator Fyodor Yegorov (1845–1915) wrote that "Mental calculation is mostly calculation based on free reasoning" (quoted in Pchelko 1940, p. 185), and he went on to say that "Practice at mental calculation develops in children the skill of choosing techniques that are best suited for carrying out an operation in a given case" (p. 186).

Consequently, there arose a tradition of giving students numerous oral problems and training not only their memory, but also their skill at finding the most sensible method of calculation. Popova's (1937b) textbook contained a special section which discussed, for example, effective procedures for mentally multiplying by 5 or carrying out relatively complicated mental subtractions by breaking up one operation into several ones, and so on. The problem book by Popova and Pchelko (1941) contains a large number of oral problems that involve choosing the optimal order of operations. Among them, for example, are such altogether simple but useful calculation assignments as the following:

$$(1000000\text{-}9) + (10000\text{-}9991) \quad (\text{p. 14})$$

or

$$25 \cdot 37 \cdot 4 \quad (\text{p. 117})$$

Popularity was (and continues to be) enjoyed not only by "special" problems, in which numbers and operations were deliberately chosen with a view to showing, for example, the importance of using the properties of operations (commutativity, associativity, distributivity), but also by quite arbitrary ones, as when a teacher gives a sequence of assignments by pointing to numbers on a chart and indicating which operations are to be performed. Shokhor-Trotskii (1886) considered such a process important both in educational and psychological respects.

Conclusion

To reiterate yet again: our survey is incomplete—many important textbooks have not even been mentioned, and in discussing those that have been mentioned, we have touched on only a few aspects. Russian mathematics education was developed in interaction with foreign mathematics education, and this interaction was by no means limited to the influence of Pestalozzi and Grube, discussed above. It would be interesting to compare the processes, the tendencies, and the difficulties encountered by elementary mathematics education in different countries (the history of the study of arithmetic in different countries is discussed by Bjarnadóttir (2014), but such comparisons must be continued and deepened). And many other aspects of the material examined above require additional study.

Nonetheless, much may be seen even from what has been said here. The first point of interest is what may be called the durability of existing problems. Just as centuries ago, we again find ourselves in a situation in which algorithms for written calculation are probably not used by anyone in practice; and while a peasant, who really did not carry pencil and paper with him, also did not carry an abacus, calculators now are in fact carried by everyone. Just as before, we face the question of explanations: does anything about calculations need to be explained, and if yes, what exactly? Just as before, we find it important to contrast thoughtlessness and conscious understanding in the carrying out of operations. And just as before, we also face the issue of the relation to practical applications: the importance of solving real-world problems was also recognized in earlier times, and in earlier times many of the "real-world problems" assigned in schools, and their solutions, were stupefyingly artificial and pointless—just as they are, alas, now. Even the idea of involving children in a search for their own methods and algorithms for carrying out calculations, which seems so modern, is not at all new.

One conclusion that may be drawn from this seems obvious: although, of course, future elementary school teachers of mathematics cannot be told about all the upheavals in the development of their subject, some acquaintance with its history is indispensable for them. The conviction with which most future teachers leave college—that in earlier times there was some kind of so-called traditional education, which is always regarded as horrible, but that now, maybe five years ago, maybe ten years ago, genuine and innovative education has begun—hardly helps them in their actual pedagogical work, not infrequently leading them into absurdities which were familiar to people already in the eighteenth century.

The changes that have taken place in education in general and in the teaching of calculation in particular are, of course, vast and due not to technological achievements alone. In many countries (although far from all), what may be called the political problem of offering all children the opportunity to acquire some kind of basic education in calculation has been solved. But the qualitative problem— teaching all children calculation well—remains unsolved even in quite wealthy and prosperous countries.

As often happens, from this it is sometimes concluded that there must be no need to teach calculation at all. Meanwhile, elementary calculation remains what it has been for centuries: an example of a substantive, visually clear, obviously useful, and at the same time not difficult section of mathematics. Therefore, let us conclude this chapter with the words of the Russian mathematics educator Alexander Goldenberg (1837–1902):

Conscious assimilation of calculation techniques; thoughtful carrying out of arithmetical operations to solve problems, even simple ones; confidence in the means that always unmistakably lead to the goal; accurate assessment of these means; and finally unfailing trust in them—all of these, in our firmest conviction, are precious aspects of the teaching of the art of calculation to children. (Goldenberg 1940, p. 104)

References

Agloblin, V. 1846. *Arifmetika* [Arithmetic]. Moscow: Tipoigrafiya Nikolaya Stepanova.

Bjarnadóttir, K. 2014. History of teaching arithmetic. In *Handbook on the history of mathematics education*, ed. Alexander Karp, and Gert Schubring, 431–458. New York: Springer.

Busse, F. 1833. *Rukovodstvo k arifmetike dlya upotrebleniya v uezdnykh uchilischakh Rossiiskoy Imperii* [A Manual in Arithmetic to be used in the District Schools of Russian Empire]. Moscow: Universitetskaya tipografiya.

Depman, I.Ya. 1959. *Istoriya arifmetiki* [History of Arithmetic]. Moscow: Gosudarstvennoe uchebno-pedagogicheskoe izdatel'stvo.

Evtushevsky, V.A. 1871. *Arifmeticheskie schety* [The Arithmetical Abacus]. St. Petersburg: Typografiya Golovina.

Evtushevsky, V.A. 1874. *Otvet na stat'iu grafa L. Tolstogo "O narodnom obrazovanil"* [Response to Count L. Tolstoy's Article, "On the People's Education"]. St. Petersburg: Tipografiya Bezobrazova.

Freiman, V., and A. Volkov. 2015. Didactical innovations of L. F. Magnitskiĭ: setting up a research agenda. *International Journal for the History of Mathematics Education* 10 (1): 1–24.

Goldenberg, A. 1940. Obrazovatel'noe znachenie obucheniya detey proizvodstvu arifmeticheskikh deistvii [The Educational Significance of Teaching Children to Carry Out Arithmetical Operations]. In *Khrestomatiya po metodike nachal'noi arifmetiki*, ed. A.S. Pchelko, 103–105. Moscow: Gos. Uchebno-pedagogicheskoe izdatel'stvo Narkomprosa RSFSR.

Ignatiev, P. 1933. Education. In *Russia, U.S.S.R.: A complete handbook*, ed. P. Malevsky-Malevich, 651–676. New York: William Farquhar Payson.

Ivashova, O. 2011. The history and the present state of elementary mathematical education in Russia. In *Russian mathematics education: Programs and practices*, ed. Alexander Karp, and Bruce R. Vogeli, 37–80. Hackensack, NJ: World Scientific.

Karp, A. 2006. "Universal responsiveness" or "splendid isolation"? Episodes from the history of mathematics education in Russia. *Paedagogica Historica* 42 (4–5): 615–628.

Karp, A. 2009. Back to the future: The conservative reform of mathematics education in the Soviet Union during the 1930s–1940s. *International Journal for the History of Mathematics Education* 4 (1): 65–80.

Karp, A. 2012. Soviet mathematics education between 1918 and 1931: A time of radical reforms. *ZDM/International Mathematics Education* 44: 551–561.

Karp, A. 2014a. On mathematics education in the Orthodox Europe (Chap. 7). In *Handbook on the history of mathematics education*, ed. Alexander Karp, and Gert Schubring, 143–151. New York: Springer.

Karp, A. 2014b. Mathematics education in Russia (Chap. 15). In *Handbook on the history of mathematics education*, ed. Alexander Karp, and Gert Schubring, 303–322. New York: Springer.

Kavun, I.N., and N.S. Popova. 1934. *Metodika prepodavaniya arifmetiki. Dlya uchitelei nachal'noi shkoly i studentov pedagogicheskikh tekhnikumov* [The Methodology of the Teaching of Arithmetic. For Elementary School Teachers and Students at Pedagogical Institutes]. Moscow-Leningrad: Uchpedgiz.

Kopievskii, I. 1699. *Kratkoe i poleznoe rukovedenie vo arifmetiku, ili v obuchenie i poznanie vsyakogo scheta, v sochtenii vsyakikh veschey* [A Brief and Useful Manual in Arithmetic, or

Teaching and Learning Calculation of All Kinds, in the Counting of All Kinds of Things].
Amsterdam: Tessing.

Kurganov, N.G. 1791. *Arifmetika ili chislovnik* [Arithmetic or Number Book]. St. Petersburg:
Imperatorskaya Akademiya nauk.

Lesnevskii, I.V. 1852. *Arifmetika i prilozhenie pravil ee k vykladkam na schetakh* [Arithmetic and
the Application of Its Rules to Calculations on the Abacus]. Novgorod.

Magnitskii, L.F. 1703. *Arifmetika, sirech' nauka chislitel'naya, s raznykh dialektov na slovenskii
yazyk perevedenaya, i vo edino sobrana, i na dve knigi razdelena* [Arithmetic, or learning of
numbering, translated into Slavic language from different dialects, assembled and divided into
two books]. Moscow.

Maistrov, L.E. 1974. Rol' alfavitnykh system numeratsii [The Role of Alphabetic Systems of
Numeration]. In *Istoriko-matematicheskie issledovaniya*, vol. 19, ed. A.P. Yushkevich, 39–49.
Moscow: Nauka.

Malinin, A., and K. Burenin. 1897. *Arifmetika* [Arithmetic]. Moscow: Nasledniki brat'ev Salaevykh.

Menchinskaya, N.A. 1955. *Psikhologiya obucheniya arifmetike* [The Psychology of Teaching
Arithmetic]. Moscow: Uchpedgiz.

Miliukov, P.N. 1994. *Ocherki po istorii russkoy kul'tury* [*Outlines of history of Russian culture*],
vol. 2. Moscow: Progress-Kul'tura.

O rassmotrenii knigi Lesnevskogo "Arifmetika na schetakh" [On the Review of Lesnevskii's Book
"Arithmetic on the Abacus"]. 1852. Central Historical Archive of St. Petersburg, f. 439, op. 1,
d. 3161.

Pchelko A.S. 1940. *Khrestomatiya po metodike nachal'noi arifmetiki* [Anthology on the
Methodology of Elementary Arithmetic]. Moscow: Gos. Uchebno-pedagogicheskoe izdatel'stvo Narkomprosa RSFSR.

Pekarskii, P.P. 1862. *Nauka i literatura v Rossii pri Petre Velikom* [Science and literature in Russia
of the time of Peter the Great]. St. Petersburg: Obshchestvennaya Pol'za.

Perepiska ob uchrezhdenii Komiteta po ustroistvu uchebnykh zavedenii [Correspondence on the
Establishment of the Committee on the Organization of Educational Instititutions], (1826–
1828). Russian State Historical Archive, f. 738, op. 1, d. 1.

Polovtsev, A.A. (ed.). 1903. *Russkii biograficheskii slovar'* [Russian Biographical Dictionary],
vol. 9. St. Petersburg: Tipografiya glavnogo upravleniya udelov.

Polyakova, T.S. 1997. *Istoriya otechestvennogo shkol'nogo matematicheskogo obrazovaniya* [The
history of Russian school mathematical education]. Book 1. Rostov na Donu: RGPU.

Polyakova, T.S. 2002. *Istoriya matematicheskogo obrazovaniya v Rossii* [The history of
mathematical education in Russia]. Moscow: Moscow State University Publishers.

Polyakova, T.S. 2010. Mathematics education in Russia before the 1917 Revolution. In *Russian
mathematics education: History and world significance*, ed. Alexander Karp and Bruce R.
Vogeli, 1–42. London, New Jersey, Singapore: World Scientific.

Popova, N.S. 1933. *Uchebnik arifmetiki dlya nachal'noi shkoly. Chast' vtoraya. Vtoroi god
obucheniya* [Textbook in Arithmetic for Elementary Schools. Part Two. Second Year
Instruction]. Moscow-Leningrad: Uchpedgiz.

Popova, N.S. 1937a. *Uchebnik arifmetiki dlya nachal'noi shkoly. Chast' pervaya. Pervyi god
obucheniya* [Textbook in Arithmetic for Elementary Schools. Part One. First Year Instruction].
Leningrad: Uchpedgiz.

Popova, N.S. 1937b. *Uchebnik arifmetiki dlya nachal'noi shkoly. Chast' tretya. Tretii i chetvertyi
god obucheniya* [Textbook in Arithmetic for Elementary Schools. Part Three. Third and Fourth
Year Instruction]. Moscow-Leningrad: Uchpedgiz.

Popova, N.S., and A.S. Pchelko, 1941. *Sbornik arifmeticheskikh zadach i uprazhnenii dlya
nachal'noi shkoly. Chast' chetvertaya* [Collection of Problems and Exercises in Arithmetic for
Elementary Schools. Part Four]. Moscow-Leningrad: Uchpedgiz.

Prudnikov, V.E. 1956. *Russkie pedagogi matematiki XVIII-XIX vekov* [Russian Mathematics
Educators of the 18th–19th centuries]. Moscow: Uchpedgiz.

Rachinskii, S.A. 1902. *Sel'skaya shkola: sbornik statey* [Rural Schools: Collection of Articles]. St. Petersburg: Sinodal'naya tipografiya.

Shokhor-Trotskii, S.I. 1886. *Metodika arifmetiki* [Methodology of Arithmetic]. Moscow: Uchpedgiz.

Simonov, R.A. 1974. "Tsifrovye alfavity" i sostoyanie gramotnosti v Drevney Rusi ["Number Alphabets" and the State of Literacy in Old Rus']. *Matematika v shkole* 1: 80–82.

Simonov, R.A. 1977. O formirovanii drevnerusskoy numeratsii [On the Formation of Ancient Russian Numeration]. In *Istoriko-matematicheskie issledovaniya*, vol. 22, ed. A.P. Yushkevich, 237–241. Moscow: Nauka.

Simonov, R.A. 1980. *Kirik Novgorodets, uchenyi XII veka* [Kirik of Novgorod, scholar of the 12th century]. Moscow: Nauka.

Simonov, R.A. 2007. *Matematicheskaya i kalendarno-astronomicheskaya mysl' Drevney Rusi* [Mathematical and Astronomical-Calendrical Thought in Old Rus']. Moscow: Nauka.

Spasskii, I.G. 1952. Proiskhozhdenie i istoriya russkikh schetov [The Origin and History of the Russian Abacus]. In *Istoriko-matematicheskie issledovaniya*, vol. 5, ed. G.F. Rybkin, and A. P. Yushkevich, 269–420. Moscow: Gosudarstvennoe izdatel'stvo tekhniko-teoreticheskoy literatury.

Talalay, A.A. 1903. *Noveishie tablitsy dlia bystrogo vychisleniya. Umnozhenie i delenie na schetakh ili na pis'me vsekh chisel tselykh i drobnykh* [New Tables for Rapid Calculation. Multiplication and Division of All Integers and Fractions on an Abacus or on Paper]. St. Petersburg: Tipografiya Bysselya.

Thorndike, E.L. 1929. *Psychology of arithmetic*. New York: Macmillan.

Tikhomirov, P. 1830. *Arifmetika na schetakh ili legchaishii sposob proizvodit' vse arifmeticheskie deistviya na schetakh, usovershenstvovannykh general-maiorom g. Svobodskim* [Arithmetic on the Abacus, or the Easiest Method for Carrying Out All Arithmetical Operations on the Abacus, as Perfected by Major General Svobodskoi]. St. Petersburg.

Tolstoy, L. 1874. O narodnom obrasovanii [On the People's Education]. *Collected Works, 90 volumes* (1928–58), vol. 17, 71–132. Moscow.

Alexander Karp is a professor of mathematics education at Teachers College, Columbia University, NY, USA. He received his Ph.D. in mathematics education from Herzen Pedagogical University in St. Petersburg, Russia, and also holds a degree from the same university in history and education. Currently, his scholarly interests span several areas, including the history of mathematics education, gifted education, mathematics teacher education, and mathematical problem solving. He served as the managing editor of the *International Journal for the History of Mathematics Education* and is the author or editor of over one hundred publications, including over twenty books among which are *Handbook on the History of Mathematics Education* (co-edited with Gert Schubring), *Russian Mathematics Education* (co-edited with Bruce Vogeli), *Mathematics in Middle and Secondary School: A Problem Solving Approach* (with Nicholas Wasserman).

Part V
Conclusion

The Unsettling Playfulness of Computing

Jean-François Maheux

Abstract Calculation algorithms are fated to be supplanted, and the art of using them replaced by different sets of skills, and eventually: machines. Yet, using and creating algorithms is still mathematically relevant and in fact contributes to the evolution of mathematics itself. In school and everyday life, however, calculating is often seen as a mindless, boring activity. Algorithms are indeed designed to require as less thinking as possible. I suggest that taking a playful attitude towards algorithms in school is essential so that they can be appreciated in full. Playing with algorithms allow them to really become objects of interest, and be offered as one particular way of experiencing mathematics.

Keywords Play · Algorithms · Epistemology · Perfection–imperfection

A Frolicsome Start

When was the last time you did longhand division? Would you be able to extract a square or a cubic root by hand? How about finding a value for *sin* (37°) to a precision of 5 decimals? For many, computing, or the act of calculating, seems a mindless, automated activity. And it is probably true to say that not many among us use calculation algorithms on a daily basis, unless we are teaching them! Nevertheless, there seems a general agreement that children should learn to add and make by-hand divisions in school. Some will argue that these skills are useful for solving interesting or important problems (i.e. to create a rectangle of a given area) and that children should become proficient in obtaining results without having to reflect too much on the reasons why 6384×215 makes 1,372,560. But in the face of today's technological developments, it becomes increasingly difficult to justify the effort and the time spent on handwritten calculations. Pre-installed on everyone's phones, calculators have established themselves in our pockets and will soon

J.-F. Maheux (✉)
Université du Québec à Montréal, Montréal, QC, Canada
e-mail: jfmaheux@mail.com

© Springer Nature Switzerland AG 2018
A. Volkov and V. Freiman (eds.), *Computations and Computing Devices in Mathematics Education Before the Advent of Electronic Calculators*, Mathematics Education in the Digital Era 11, https://doi.org/10.1007/978-3-319-73396-8_16

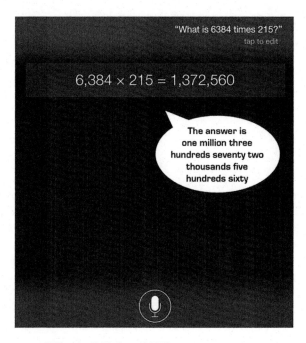

Fig. 1 Siri's answer to "What is 6384 times 215?"

be very good at interpreting natural language problems in terms of mathematical operations to perform (Fig. 1).

On the other hand, it is still interesting that children develop "computational fluency" (Kilpatrick et al. 2001), which some associate with forms of "conceptual understanding" that encourage mental flexibility (Carpenter 1986; Baroody and Dowker 2003). This is where one might calculate 25×99 by thinking of an equivalent expression such as $25 \times 100 - 25$ instead of mechanically using a general algorithm. In a similar vein, we easily agree that exercising mental calculation enriches one's problem-solving "toolbox" (Butlen and Peizard 2000). But when do the strategies for calculating become serious objects of attention *in themselves* in mathematics education? Very often in schools, the variety of means by which a calculation can be done are considered only as pathways to mastering so-called standard algorithms (e.g. Maddel 1985). There are, of course, nuances to this view (e.g. Heege 1983; Thompson 1994). But it is still a common idea that calculation strategies and algorithms are means to an end, where obtaining (accurate, quick) results is what matters most. In that sense, we are generally happy in a mathematics classroom to privilege calculators over algorithms, thus "reliev[ing] students of cumbersome computation, allowing them to concentrate on more meaningful mathematical activities" (e.g. Ballheim 1999, p. 4).

There may be good reason to embrace this logic. As Plunkett (1979) explains, our written algorithms were selected in great part precisely *because* they are standardized, contracted, mechanisms. Any reflection on the historical development of algorithms should certainly account for the fact that algorithms are a kind of instrument designed with utility in mind. According to Leontiev's (1979) analysis, tools are destined to become invisible, unattended and further mechanized. This is precisely what Babbage did for calculation algorithms when he designed the difference machine in 1830 (Babbage 1864/2011). But we must realize that this goes in a direction opposite to what we find in schools, where, as many argue, algorithms should be learned "with meaning" in order to develop students' number sense and understanding of our numerical system (e.g. McIntosh et al. 1992). An interesting tension arises here. In this context, computing algorithms are, in a sense, *diverted* from their original, intended use. Algorithms were not introduced in school curricula to promote "conceptual understanding", even if some want to use them with that intention. Just like a pen makes a fair letter opener, and if I have one handy, I may use it as such—but this is certainly not why I carry a pen.

I am under the impression that a utilitarian take on mathematics education is still the mainstay keeping algorithms in schools despite their increasing obsolescence. As such, they will be at risk of disappearing from schools when their obvious futility finally overcomes the prevalent resistance to such change. The case of the root extraction algorithm illustrates this: long a component of many mathematics curricula, it was eventually abandoned, despite its pedagogical potential (e.g. Askey 1995). As Hilton (1980) has put it: "Long division and square root extraction, by paper-and-pencil algorithms, are dull and dreary activities and quite unnecessary today" (p. 248).

On the other hand, calculation algorithms are a very important part of the history of mathematics (e.g. Chabert et al. 1994), and we have been asking for some time how we might bring this history into the classroom (e.g. Katz 2000) without referring students back to stale textbook images of Pythagoras in a peplum. We wonder, for instance, how we could introduce the study of the history of mathematics not merely as an *aid*, but also as an *aim* in itself (e.g. Jankvist 2007). In that respect, calculation algorithms could offer us something interesting. For one, some are *already* integrated into many curricula. Inhabiting the mid-point between utility and futility, they could also ease a transition from an emphasis on technical relevance (i.e. the position that algorithms are helpful tools and that an understanding of their history helps us teach/learn them) to the status of interesting historical artefacts that also create occasions to do mathematics.

In this chapter, I offer a short overview of the history of calculation algorithms in general, to trace the natural evolution of that domain of mathematics, and to survey its enduring presence in mathematical research. I also present a reading of that history emphasizing the *playful* aspects of computing, as observable through some historical examples. This then leads me to reflect on the nature of algorithm-based calculation, and the kind of mathematical experiences generated by such computation. With this approach, I hope to recall that algorithms are mathematical objects interesting in themselves, especially from a historical perspective. Situating this contention within the context of mathematical activity is important for me. As I see it, this allows us to move away from the questions traditionally associated with teaching and learning mathematics (e.g. asking what students learn, how much they know, how to get them to do or to understand this or that in such or such a way), which I profoundly object to on epistemological, ontological and ethical grounds (see, e.g., Maheux 2010). An alternative for me consists of thinking about mathematics education in terms of offering students occasions to engage in various kinds of mathematical activities and experience mathematical ideas in diversified ways. I thus finish by briefly discussing how we can look to engaging with calculation algorithms (with students) for no reason other than *doing mathematics.*

Computing: An Art in Decay?

Crunching numbers has not always been an easy, merely technical, mindless task. Through history, number systems have often made difficult what appear today to be the simplest operations, such as multiplying or dividing integers or rational numbers. For example, The Rhind Papyrus, a document considered to be a teaching aid for scribes, demonstrates the means by which ancient Egyptians obtained sums, products, differences and quotients. Mastery of these systems was not at everybody's fingertips and was accessible only through the study of numerous examples of how to operate with these numbers (something also evidenced by the study of Sumerian clay tablets). Abacus-based multiplication is also quite complex and challenging, which also explains why there are many types of abacus, and many ways to use them.

The art of the abacus only began to be seriously challenged in the Middle Ages. It was then that handwritten computations became more widespread, thanks to the refinement and popularization of a revolutionary invention slowly making its way into Europe: paper. Yet the technological replacement of the abacus did not move swiftly, despite the numerous advantages (both from mathematical and practical perspectives) of paper algorithms. On paper, one faces none of the numerical limitations of the abacus, and it is possible to go back and check every partial result—or even interrupt calculations and easily return to them later. The risk of mis-manipulation is minimized, and successive computations are simultaneously available. On paper the demands in terms of memorization and organization are much less pressing. But traditions have deep roots. A famous woodcut by Gregor Reisch (circa 1503), in his *Margarita Philosophica* encyclopaedia, allegorizes this revolution (Fig. 2). Using written calculations, Boethius beats Pythagoras and his reckoning tables (a board-style abacus) in a mathematical competition overseen by a prim Typus Arithmetica.

A reflection of those changing times, this picture also mirrors the modern-day battle between advocates of the teaching of "traditional" by-hand algorithms, and those in favour of replacing them with calculator use (e.g. Ruthven 2009). If mathematicians and astronomers of the sixteenth century adopted the new way of calculating relatively quickly, reckoning tables were still used for business or accounting, and often seen as a (more) reliable way to verify by-hand calculations. The art of computing on a reckoning table was actually taught until the eighteenth century, which might very well speak to something deep at play concerning the

Fig. 2 Arithmetica supervising a competition between written and abacus calculations

nature of mathematics and mathematical activity, and its relationship with technology. From a cultural-historical perspective, tools and their usage are embodied ideas, concepts, or ways of doing and knowing strongly connected with the activity in which they are embedded (e.g. Rabardel 1995). Calculating with a reckoning table or an abacus involves moving and grouping tokens: gestures that are strongly evocative of what we do when we count objects (money, bags of flour, chips, etc.). This embodiment is not as easily found when working with numbers and addition or multiplication tables, for example. Mathematicians, however, are much more at ease with the idea of thinking about numbers and operations elastically. Euclid's algorithm to find the greatest common divisor of two numbers that are not relatively prime (book 7, proposition 2) is developed in terms of segments of relative length, a way of thinking still visible in the work of Descartes, Newton and so on.

Letting go of traditional calculation aids was easier for mathematicians in part because for them, algorithms are also objects of interest in themselves. Euclid's books, as we know, are primarily about proofs. Showing a result is great, but knowing how, when and why it can be obtained is even more central to his and others' mathematical activity, something true even in the development of computing techniques. Some 20 years after the publication of Reisch's woodcut, mathematicians were still trying to calculate solutions for specific third-degree equations. Saying this, we must acknowledge that there are no clear distinctions between what we call algorithms, formulas and methods. If a calculation algorithm is an organized set of operations one must perform in order to associate some numerical inputs with one or more resulting numerical values, "zeroing" a function (i.e. calculating its roots) is not fundamentally different than multiplying. The quadratic formula is a compact way of presenting an algorithm: the step-by-step series of basic operations performed on the coefficients of a second-degree polynomial to calculate its zeros. Similarly, a common multiplication algorithm can be expressed in the form of a formula, of maybe more conveniently a set of formulas, for example multiplying two 2-digit numbers: $a_1a_2 \times b_1b_2 = 100 (a_1b_1) + 10(a_1b_2 + a_2b_1) + a_2b_2$. Although this can also be generalized in some ways to n-digit numbers, we are closer here to the idea of a "method". Take, for example, the bisection method used to calculate roots; i.e. using a positive image and a negative image f(a) and f(b), find the mid-point c between a and b, and f(c), and repeat with either a and f(a) or b and f(b) depending on the sign of f(c), until you get a "close enough" answer. In its structure and process, the recursive but technically simple step-by-step procedure is not all that different from the longhand division algorithm. Developing such algorithms, formulas or method has always been a significant mathematical challenge. Having developed techniques allowing them to find numerical values for various cases, Tartaglia and Fiore famously dared one another to solve a number of specific cubic equations. And in his *Ars Magna* of 1545, Cardano finally published a general solution to third-degree polynomial equation in the form of an explanation elaborated through case examples.

These instances illustrate something vital to the history of computing: time and again, algorithms have been at the heart of many crucial mathematical inventions or discoveries. Around 1250, Iranian mathematician al-Tūsī not only explained how to use a sand table, or give a method to extract roots, but also presented a thorough discussion of the *theory* of cubic equation (Rashed 1997). A very similar approach was taken some 400 years later by Vieta, who created, out of the swarm of early algebraic procedures imagined by al-Khwārizmī, the first real symbolic algebra (e.g. Klein 1992). In a sense, the algorithms Vieta's approach seeks to explain are rendered obsolete by his more general approach. This movement of replacement is a natural one. Tools and theories support and enrich one another, even if they sometimes look in completely opposite directions. The problem of actually solving simultaneous equations representing a given situation (some algorithms do require a high level of mathematical proficiency) and that of understanding how, and of proving why the method works, certainly serve two different purposes and appeal to an audience in two very distinctive ways. But we clearly observe both preoccupations in mathematicians' work, and in their interactions with one another. Newton's 1685 methods for calculating roots through a sequence of polynomials improved Vieta's implementation, but also produced a new understanding of functions and roots. Raphson's reprises of Newton's work (in 1690) both simplified the algorithm and redefined it in terms of successive approximation. Similarly, Simpson's 1740 take on the method both revisited it, in terms of calculus and what would become a theory of iterations, and generalized the method. In this sense, calculation algorithms are fated to be supplanted, and the art of using them replaced by different sets of skills. Another great example can be found in the work of Napier, whose mathematical career was essentially devoted to the simplification of calculations. Along the way, he created the immensely powerful idea of logarithms and designed different tools to facilitate their use (his famous Napier's bones), together with numerous other ingenious inventions.

Moreover, we need to add to this movement the inevitable progression of automated computations. While Pascal's calculator of 1640 could add or subtract two numbers, Babbage's Difference Engine, in 1830, was designed to compute logarithmic tables by quickly articulating basic operations in complex ways. Another two hundred years later, Mauchly and Eckert's ENIAC machine (in 1940) could add or subtract ten-digit numbers thousands of times in a second and thus quickly realize multiplications of larger numbers (more than 300 per second), do divisions, extract square roots, and so on (Wilkes 1956). In the process, the art of calculation increasingly turned into the science of devising efficient machines in order to rapidly obtain reliable results. The science of computing became that of transforming complex numerical problems into a series of simple steps (including loops, branches and subroutines). That is to say, what it meant to work on a computation problem completely changed. Thus, while William Shanks spent years calculating and checking 180 decimals for π and produced incorrect figures because of a little mistake,[1] we now have phones capable of accurately finding thousands of decimals in no time. Computing reigns today in a way incommensurable with what it meant before. In essence: The King is dead, long live The King!

Much more could be said about the technological replacement of by-hand calculations by machines. One particularly interesting observation on the matter was made by Daston (1994) in her study of calculations during and after the French Revolution. She recalls the ambitious project of Gaspard Riche de Prony, who organized a group of about 90 "human calculators" placed under the supervision of 8 foremen preparing instructions and worksheets for them based on equations devised by a few mathematicians, including Legendre and Carnot (Grier 2013). This massive enterprise, described by de Prony as the "largest and most imposing monument to calculation ever executed or even conceived" (de Prony 1824, p. 3), aimed at calculating logarithmic and trigonometric tables from 1 to 200,000 with a precision from 14 to 29 decimals. But beside its scale, what was really unique about this project (completed within 2 years!) was the participation of workers who were by no means experts in mathematics, but mostly laid off servants and hairdressers only able to add and subtract numbers. In Daston's analysis, this moment in history coincided with a crucial shift in the history of calculation: it marked the point after which computing began to be considered mundane and unremarkable. Until then, the art of calculation had been associated with high intelligence and morality and was seen as an intellectual activity elevated above manual work. But after de Prony's demonstration, this began to change. Calculating was becoming a form of blue-collar labour at the bottom of the economic hierarchy imagined by Adam Smith, whose *Inquiry into the Nature and Causes of the Wealth of Nations* provided de Prony with the source inspiration for his project. In his turn, Babbage fell into

---

[1]He actually dedicated 20 years of his life computing a total of 707 digits without noticing the mistake he made in the 528th digit.

step with these notions, inspired, as he put it, to mechanize de Prony's approach (Babbage 1864/2011).

But then again, the automation of by-hand calculation was not unaccompanied by other mathematical developments. For one, the figures produced by Prony's or Babbage's calculators still needed to be checked. This necessitated the creation of a procedure comparing adjacent values, in order to identify potential miscalculations. In the same way, the tradition of "sanity tests" (e.g. comparisons to an approximated value, casting out nines) gave rise to the search for ways to program machines to check their results efficiently. More and more, the art of calculation became the search for and study of algorithms, not merely "solutions", and this growing interest in algorithms promoted the exponential growth of new mathematical ideas. Napier's 1617 publication *Rabdologiae,* for example, in which he discusses a variety of computation aids including his celebrated bones, contains a thorough presentation of binary arithmetic (using 0 and 1), including all the necessary procedures for the five basic operations, and for the conversion of numbers between base 10 and base 2. Around 1650, Brouncker calculated decimals of π using continued fractions. More than 300 years later, in 1960, Anatolii Alexeevich Karatsuba developed a new multiplication algorithm (yes! an algorithm to calculate something like 12345×12345) which was *improved* a few years later by Andrei Toom (1963) and then again by Stephen Cook in 1966. And as of today (2016), finding the fastest algorithm for multiplication of two n-digit numbers is still an open mathematical problem!

In this we see how algorithms are results in themselves. In his well-known 1900 address on important problems in mathematics, Hilbert asked for an algorithm to determine whether certain Diophantine equations are solvable in rational integers (Problem 10), mentioning also the Riemann hypothesis (Problem 8). Speaking back to Hilbert in 1970, Yuri Matiyasevich completed a proof showing that such an algorithm cannot exist. As for the zeros of the zeta function, mathematicians undertook the epic task of calculating as many as they could, in hopes of finding one that would contradict the conjecture. New algorithms were devised and computerized: in 1968, 3,500,000 roots were calculated; by 1986, we hit 1,500,000,001 roots; and then over 10,000,000,000,000 using an algorithm created in 1988.

It might then be that the dullness of calculating really only exists at the surface level. Certainly, not many people nowadays spend time calculating by hand, but this is the kind of computation still generally taught in schools, even at the secondary level (e.g. Sewell 1981; Hart 1981). In 2007, French mathematician Gilles Dowek introduced his book on *The Metamorphosis of Computation* stating:

> Algorithms such as those used to perform addition and multiplications are perceived as elementary part of mathematics, and calculating seems a boring, un-creative activity. [...] Making people dream, get excited with computation is certainly a challenge... (pp. 9–10, my translation)

Getting people thrilled about calculation might be a challenge, but not an unachievable one! Through his book, Dowek manages to show us how calculating might actually be the key to the next major revolution in/of mathematics, replacing axioms and axiomatic thinking altogether with algorithms. That is to say, the actual act of calculating might itself be re-gaining some terrain, becoming, moreover, more highly regarded in relation to the kind of problems it poses, and the new mathematical ideas and theories that can be developed around it. Not all advocates of empirical mathematics go this far, but many stress the importance of the rehabilitation of computation to mathematics, and embrace the new possibilities it promises (Tymoczko 1998). Quasi-empiricism in mathematics invites us to think in terms of *computational truth,* in which accumulated evidence, obtained by calculations, stands in its own right. We might think here of computer proof such as that of the four colour theorem (un-checkable by a human mathematician, because it would be simply too long and complex (e.g. Franklin 1987)), but not only that. Experimental mathematics considers a large number of calculations as "relevant" as a logical proof can be, in relation to the pursuit of mathematical knowledge (e.g. Borwein and Bailey 2003; Borwein et al. 2004). The Goldbach conjecture or the Riemann hypothesis mentioned above bring to light a crucial fact: enormous empirical evidence has been found for which some sort of "truth" must be granted (e.g. Chaitin 2004). But this sort of truth will have to embrace the possible existence of counterexamples. This is what happened with Euler's sum of powers conjecture (a generalization of Fermat's Last Theorem).[2]

So all this said, even though computations in the way of longhand division do seem to come to an end, the movement towards discovery is part of the evolving nature of (doing) mathematics and is far from exhaustion. Computing, as part of mathematical practices and other human activities, participates in a continual process of adaptation. Traces of the past are still present everywhere. And this is not necessarily a bad thing! Sharp's EL-8048 *soroban* calculators (Fig. 3) are still available for purchase through eBay; the French École Polytechnique mentions that it is permitted, in its entry tests, to use slide rulers when the calculator is not allowed (many institutions specify whether both are permitted or not); and children still learn longhand division in elementary school. Calculating has always been evolving, and its evolution clearly depends on the tools available (e.g. Abdeljaouad 2005): the human body, pebbles and sticks, sandboxes, clay tablets, pen and paper, mathematical tables… and computers. More so, these evolutions are grounded in the usefulness of computations in all areas of life. In this sense, it is only natural to let go of previous ways of calculating in favour of newer, faster, easier and more reliable means of obtaining arithmetical results. On the other hand, old ways of (doing) mathematics can still be interesting from a mathematical perspective. They represent an important part of our history and illustrate cultural diversity, but also correspond to a particular kind of mathematical experience.

---

[2]Another powerful example concerns the inferiority of the prime counting function versus the logarithmic integral. A large number of confirming observations were made before the first counterexamples were identified, hence the interest in continuing calculations.

Fig. 3 A Sharp EL-8048 *soroban* abacus calculator for sale on eBay

In a short but inspiring comment, Morley (1932) reacts to the opposition between techniques and concepts, positing that the latter should be given priority in mathematics education. His argument does not rest on the utilitarian idea that children need to know how to calculate, nor on their potential interest in terms of conceptual understanding. He rather stresses the importance of algorithm in mathematics: "we seem to have lost any vision of the magnitude of the achievement of their construction" (p. 50). He then suggests presenting students with the problem of constructing algorithms, encouraging them to experience "the personal feeling of power and satisfaction that possession of an algorithm gives" (idem). The few examples he mentions of what this might look like are quite convincing to me. More importantly, his general orientation strongly conveys what appears to me a crucial shift: consider algorithms in themselves, he argues. See them as interesting mathematical creations and embrace their exploration as an end, rather than as a means to something else.

Playful Computations

Mathematicians seem to love making use of common words in a playful manner. In mathematics, "perfect", "elegant", "trivial", "well-behaved", "deficient" and "solid" carry meanings very different from their everyday connotations. Although generally considered chiefly in regard to its utility, and surely in relation to its "unreasonable effectiveness" in the organization of our lives and in the sciences, the pursuit of mathematics is also, and quite importantly, playful. As Whitton (1998) notes, play has an important role in all branches of mathematics and contributes to the evolution of mathematics itself. Here, I am talking about not only mathematical puzzles, but also the creation and development of new mathematical concepts. Perforce inventive, many mathematicians in their daily work simply try to solve essentially mathematical problems, engaged in a form of mathematical research that, Holton et al. (2001) argues, *is* a form of play.

Moreover, chances are this is not a new phenomenon. It is indeed easy to find examples of apparently "recreational" mathematics in very early traces of mathematical activity. For example, the 3600-year-old Egyptian Rhind Papyrus presents a calculation riddle: "Seven houses have seven cats that each eats seven mice that each eats seven grains of barley, that each would have produced seven hekat of grain: How many things are we talking about?". We also know of Chinese magic squares dating back to 2200 BC, and in Ancient Greece, mathematicians enjoyed challenging one another with arithmetic puzzles, among which was the famous cattle problem attributed to Archimedes. Analysing mathematical texts from that era, Netz (2009) discusses the mathematical "style" of various documents, characterized by the presence of what he calls "extravagant calculations". The best-known example of such a text is probably Archimedes' *Sand-Reckoner*, in which he investigates the maximum number of grains of sand it would take to fill the cosmos! This tradition of playful mathematics is also attested in Arab mathematics of the ninth–fifteenth centuries (e.g. Djebbar 1997). Work in number theory often involves "frivolous" considerations, which might, of course, turn out to be useful. But Mersenne primes were not really created first for their potential applications. The useful and playful aspects of calculating can also come hand in hand. Fibonacci's 1202 *Book of Calculation* demonstrates the practical use of Hindu-Arabic numbers for everyday applications, but also speaks about perfect numbers, and presents the rabbit-counting algorithm from which the Fibonacci sequence derives! A strong sense of the glorification of mathematics and mathematical activity "for itself" comes through all this. Beside the customary ideas of the usefulness of calculating, reckoning can also be done "for the fun of it", or for the pleasure of figuring out the result of this or that, or for the sake of adding a number to a list of numbers, or for the joy of bringing forth unsuspected patterns, and so on. Why is it that this playful disinterestedness seems nowadays to vanish into the background of a utilitarian discourse which emphasizes over and over again the (eventually) pragmatic outcomes of such mathematical work? Even a book like Körner's (1996) *The Pleasures of Counting* reflects this tendency, hasty to justify computational pleasure by connecting it with practical applications.

What a contrast with the playful art of basic arithmetic observable in Renaissance manuscripts, where the elegantly arranged division algorithm (Fig. 4) reminds us of the pleasure that can come from doing or encountering a beautiful thing, even if not the easiest one to use! Some scholars have made a wonderful job collecting various basic algorithms of that sort (e.g. Cerquetti-Aberkane and Rodriguez 2002), in which the potentially playful side of calculating can be seen.

Of course, these algorithms "work", they can be useful, but there is also something *more* to them, something enjoyable beyond the mere reach of an arithmetical solution. The aesthetic pleasure of constructing the symmetrical figure

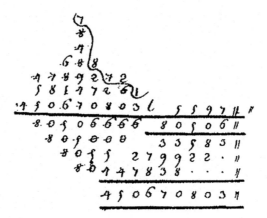

Fig. 4 A division algorithm and its verification from a Renaissance manuscript: 450670803 divided by 80506 yields a quotient of 5597 and a remainder of 78721

is one aspect of this kind of gratification. Another, I would argue, can be found in the simple act of achieving something (variably) difficult. Algorithmic play excites the kind of challenging pleasure essential to gaming. The gigantic project designed and carried on by de Prony shows this very well. The level of precision of the table he calculated went far beyond the requirements of any practical application, as de Prony said himself. These tables have to be seen as work of art, the celebration of craftsmanship, like a cathedral erected in tribute to the triumph of reason:

> Prony's tables, whose symbolic intensity increased in inverse proportion to their utility. Like the temples and torches of the Festival of Reason, the tables became fetishes of rationalism, admired for their most irrational qualities. The very extravagance of their precision, their unwieldy bulk, their uniqueness as unpublished manuscripts, and, above all, the remarkable methods by which they had been assembled – these were the aspects of the tables that turned them into monuments, as opposed to tools, in the history of calculation. (Daston 1994, p. 184)

I can hardly move on to the next section without a word about the many calculation games that are now spreading through the "mobile" platforms, some of which are incredibly popular. There are numerous games, like the well-liked *2048*, in which players have to compose numbers together to create given values; *Number ninjas,* where the four basic operations are at play; or one of 2014s all-star games, *Slice Fraction*, which engages players to operate on ratios (Fig. 5). Here, as in many mathematical games, the pleasure of computing blends with common characteristics of games identified by Shechtman and Knudsen (2009) or Huszár and Rolínek (2014) and many others. We can see elements of freedom and uncertainty as players spontaneously engage with and generate possibilities within a holistic, ruled context which is variously rewarding, permits risk-taking, and is pleasing to watch and interact with. In brief, they offer good "gameplay" undergirded by mathematical operations. In most mathematics-related games, however, it is not evident how the mathematical activity itself is a source of pleasure, instead of being simply part of a game that is enjoyable for other reasons.

Fig. 5 Screenshots from the games *2048, Number Ninjas* and *Slice Fraction*

Yet the playful side of mathematics is not what first comes to mind when we think of doing divisions and multiplications. Neyland (2001, 2010) asserts that mathematics "education, at its best, is comical" (2010, p. 85), rather than focused on so-called important facts and enduring truths. He makes a good case. Pointing to the dominance of the scientific management model imposed on mathematics education (an agenda strongly advanced by some political forces) to explain the prevalent situation, he warns us that this drive towards efficiency, certainty, order, etc., is taking away the *spirit* of both mathematics and education. A spirit that permeates the history of algorithms, and the non-utilitarian motives behind them.

Unsettling

For Davis (1996), the importance of play has not been overlooked within mathematics education, although "it might be argued to have been undervalued" (p. 211). Nuancing this position, Holton et al. (2001) respond that "play has been overlooked for many older students in our schools and universities" (p. 414). I think it might also easily be argued that although forms of play are clearly present in mathematics education, the playful face of mathematics is not made explicit and drawn on as often as it could, except perhaps in early childhood education (e.g. Tucker 2014). Children, of course, learn a lot through play. Ginsburg (2006), for instance, explains that everyday mathematics is naturally part of children's play and that young students also spontaneously play with the mathematics they experience in school. Moreover, all sorts of correlations can be found between children's play and mathematics: constructing models, extending patterns, making conjectures, organizing information, developing methods and identifying relationships are all mathematical actions "consonant with children's behaviors during play" (Whitton 1998, p. 474). This in part explains why mathematical thinking easily emerges through play (e.g. Van Oers 2010). Pitri (2001) makes similar observations, discussing how artistic play (in the form of imitation, exploration, experimentation or construction) contributed to her students' reasoning and problem-solving. To explain this, she refers to Bretherton's (1984) work on children's play and reasoning abilities.

Curiously enough, these scholars often stress the ways in which mathematic play is "useful" and contributes to the development of essential capabilities, something underwritten by Shechtman and Knudsen (2009) when they write "bringing play-fulness to mathematics is by no means intended to undermine its seriousness" (p. 10). In contrast, Eisner's (1990) analysis of the notion of play points out that it is generally not regarded as a part of the "serious business of schooling" (p. 43). To return to Neyland's observations about the scientific management model imposed on mathematics education, the modern push towards efficiency, certainly propelled by mathematics, may need to be interrogated (e.g. Maheux 2011). In that spirit, I would like to argue here in favour of an "authentically playful" mathematical activity—not one whose worth depends on achieving external goals such as the development of number sense or problem-posing abilities as part of some specific or meta curriculum. Play for play!

The word "play" traces back to the pseudo-Indo-European root *dlegh*—"to engage oneself", from which we derive words like the German *pflegen* "take care of, cultivate", and the Middle Dutch *pleyen* "to rejoice, be glad". The meaning, "to take part in a game", only originates from circa 1200 and has only been seen as an opposite of "work" since the fourteenth century. That is, play originally meant participation in something joyful, pleasurable. How might computing be engaged with as a source of joy? Scholarly investigations of the notion of play are numerous, and I content myself here to invoke some of the ideas often encountered in that literature. In his analysis of play in children's development, Vygotsky (1966) opposes the interpretation of play as pure pleasure, arguing that play can be frus-trating at times. Rather, he insists on the importance of the *imaginary*, when we do "as if", and of the *rules* players make up, accept or challenge. At the heart of play, Vygotsky explains, is the idea of liberating oneself from the contingence of "real-word" actions, rules and consequences, which evolve in a separate realm governed by its own set of actions, rules and consequences—something many might recognize as a key feature of mathematics itself!

Computing can be useful. It can be a "real-life" activity; it can powerfully serve many very important purposes. Something serious, something real. But it is not limited by the constraints of utility: it never has and never will be. Engaging in playful computations, I argue, is first and foremost a question of attitude. What are we going to play up? What will we play down? Because moving away from the obligations of "real-life" reckoning can also mean playing with the rules. That said, I use the word "play" here only for lack of a better one. English language does not easily allow us to talk about doing something "just for the sake of doing it". In French, the adverb "gratuitement" means just that: doing something without *ex-pecting* anything in return, without any external motivation or goal. Though kin to the English adjective "gratuitous", *gratuitement* carries none of its negative con-notations. In the French, it also signifies "pointless" and "useless", but again in no bad way. Rather, it conveys a feeling of spontaneity and generosity, of freedom— the idea of doing things not out of obligation, but "just like that", because we want to, or find pleasure in it, or think it is important or interesting. But that importance or interest has a texture very different from the practical, utilitarian imperative;

otherwise, we don't say the thing is done *gratuitement*! It can only serve a broad, fuzzy or temporal end, such as having a great time, or making good to someone, or cultivating oneself, or advancing humanity. It is in this spirit that I am trying to think and talk about calculation algorithms, and deconstruct some common ideas about them.

Even from a "purely" mathematical perspective, we can ask: "What is calculating?" and find rich possibilities. One way to see computing is in association with discrete thinking. Calculation concerns the making of relations between elements of some groups, while algorithms are ways to find, or see, or maybe even define those relations. In that sense, computing is reliable; it is checkable, precise and permanent, but also very contextual. Indeed, we don't multiply complex numbers the same way we do integers, and the meaning we can give to those operations and the various steps involved can also be quite different. But what makes calculations so compelling probably has more to do with how stable and regular they are in their defined domain. We assume that by the time we finish our calculations, whatever we calculated is still in the same state and that by doing it again we will get the same results, no matter how many times we try. But then again, this assumption can be challenged. I can imagine multiplying two random numbers (e.g. rand[0,1], numbers between zero and one). Each calculation will give me a different number in predictable range. The predictability of that range can also be rendered problematic with some mathematical sophistication: in algorithmic information theory, we can think of algorithms for calculating numbers whose results cannot be predicted, but only observed after the fact.

Computing can also be a way of understanding something. Some numbers somehow only exist through their algorithmic definition: for example, quasiperfect numbers are *theoretical* numbers for which the sum of all divisors of n is equal to $2n + 1$. We have an algorithm (a formula) to calculate them, but we still don't know if a single one exists or not. Another mind-puzzling case is that of Chaitin omega number. Omega is a real number representing the probability that a randomly constructed program will halt. But this number is not computable, which means that there cannot be an algorithm allowing us to find its digits (even if it is an arithmetical number in the sense of first-order Peano arithmetic!). There is also something quite fascinating in the nature of approximative algorithms. Using a Maclaurin or a cumbersome Taylor series expansion to calculate the sine of an angle gives both the powerful feeling of an unstoppable convergence and the hopeless feeling of an infinite process. Continued fractions also provide us with a step-by-step process to calculate the same value in a very different way. Contrasting both provides wonderful context, allowing us to *get a feel* for what those convergence and imprecisions are about. Newton's method also plays this estimation game, though in a different way. For example, when an analytical expression for the derivative is too complex to obtain or evaluate, we can approximate it using the slope of a line going through two nearby points on the function, which in turn slows down the convergence. Using the method also requires a good general understanding of the function for which a value is to be computed: the algorithm does not converge for all functions. We also know that large errors in the initial estimate can lead to non-convergence of the algorithm, and there are cases for which very small variations of the initial value can lead to very different (correct) roots to be calculated. These and other similarly strange phenomena take place around algorithms. They draw a picture of algorithms very different to what commonly comes to mind for those less familiar with them. Why is this richness so rarely brought forth when people talk about mathematics?

In practice, most people learn about maths in school, but what they are offered there is a simplified version of mathematics which privileges the measurable performance of technique expositions and concept recognitions over frolicking and complexity (see also Neyland 2004). I believe an extant epistemological substructure also helps to explain why the playful side of mathematics and reckoning is often effaced. Mathematics is commonly considered a flawless, orderly, stable-yet-evolving field of investigation, to which students are introduced. Russell (Russel 1967) powerfully expressed this, saying that mathematics "possesses not only truth, but supreme beauty, [it is] sublimely pure, and capable of a stern perfection" (p. 162). This perfection Kant (Kant 1934/1781) describes as inherent and objectively clear, thus perpetuating the Platonic conceptualization of mathematics as part of the perfection of the heavens, a realm of timeless truths (Plato's *Republic*) in which earthly objects are mere shadows. In such an account, mathematics is generally taken as the pinnacle of *perfection*: it is certain, clear, universal and so on. This image of mathematics is still present on the contemporary scene (Shapiro 2000), but also very salient in mathematics education—and not only for defenders of a "frontal" teaching approach. Ernest (1994, 2014) presents a good example of this with his emphasis on rationality as a means to recover/reproduce mathematics' exactness and orderliness in the classroom. Naive versions of this image of mathematics seem to be very strongly anchored outside the circle of the philosophers and historian of mathematics. Furinghetti (1993), for example, sees this view of mathematics prevalent throughout popular culture, particularly in literature, movies, television, etc. Ideas of mathematic certainty and exactitude are also often encountered in the discourse of teachers and students and seem to correlate with how they engage in teaching and learning activities (e.g. Lerman 1990; Thompson 1984; Young-Loveridge et al. 2006).

At the beginning of the twentieth century, however, mathematicians faced what was called a "foundational crisis". Suddenly confronted with the impossibility of establishing perfectly indubitable and coherent bases to define mathematical objects and assertions (e.g. Aspray and Kitcher 1988), mathematicians realized that mathematics is often done based on intuitions that cannot be trusted. They eventually found that some mathematical statements (e.g. every even integer greater than 2 is equal to the sum of two primes) cannot be proven to be true or false and that it is impossible to know which ones are not. This is Gödel's famous incompleteness theorem. Also, important oversights and a great number of errors were identified in the work of otherwise great mathematicians, as reported by Davis (1972), who pushes one step forward: "[...] arithmetic operations are enveloped in a smog of uncertainty. The sum 12345 + 54321 is not 66666. It is not a number. It is a probability distribution of possible answers in which 66666 is the odds-on favorite"

(p. 171). Davis then explains how mathematics (and, thus, algorithms and their results) is inherently uncertain. It requires creating, recognizing, reproducing and concatenating symbols, which always implies the possibility of inaccuracy. As he does this, Davis shows us that the actual nature of what we take as a "basic" operation is still open to interpretation. Of course, we know the logical necessity for 12345 + 54321 to be 66666; that was even proven in the most un-debatable way (e.g. Dedekind-Peano axioms). But we can also look at 12345 + 54321 very differently. First, there is always the possibility of changing the rules (What if we are in base 6?) or giving or taking a different meaning to or from some symbols ("+" is a concatenation operator in a JavaScript). We can also take a pragmatic perspective and consider the empirical evidence of alternative results for such sums in the work of young students learning addition. And what about taking into account, as Davis (1972) suggests, the result of a faulty computer programme? Whether or not such instances should count in/as mathematics is open to discussion: it is a philosophical position, but one that touches the very nature of mathematics.

Taken together, such observations about the nature of mathematics and mathematical activity are elements of what I have begun to call the *imperfection* of (doing) mathematics. This imperfection is not only a necessary evil, in the way Wiener (1915) phrased it when he wrote: "The average mathematician neither knows, nor, I grieve to say, cares, what a number is [...] but he can [still] make advances in it, and discover mathematical laws previously unknown" (p. 569). It is also incredibly powerful, allowing new mathematics to be created outside what would otherwise be predetermined limits. Yes, (doing) mathematics intrinsically implies making mistakes, working in disorder, accepting uncertainty, going along with ambiguity, and so on (e.g. Ormell 1992), but these are essential conditions for innovation in concepts and methods (e.g. Lakatos 1976; Châtelet 2000; Davis 2006). Playing with something perfect, useful, certain or absolute seems somehow out of order. There is no room to play in the serious business of (doing) mathematics, if the rules that ultimately govern it are already and permanently fixed in Nature, by God, or in some other realm. The imperfect, uncertain, relative and useless, on the contrary, provide good material for playing. They seem to offer a hospitable environment for changing rules and meaning, trying out things, exploring for the sake of it, or simply enjoying the experience of doing this or that without bowing to external expectations.

Mathematics education, however, still generally appears unaware of this possibility, and school curricula usually reflect the common view of mathematics as a certain, unequivocal, well-organized network of ideas, our understanding of which develops steadily, and on a very rational basis. That is, the imperfect nature of (doing) mathematics is not yet part of mathematics education epistemologies, which might in part explain why it seems so difficult to imagine mathematics, or calculation, being done by student in an authentically playful way. On the other hand, both the history and the nature of calculating and algorithms clearly evidence how "imperfect" they are in some respects. There is also something essential to mathematics beside the seriousness of the "basic needs" they help fulfil—regardless of the problems they also create (e.g. Emmer 1998)—and the impeccable architecture to which they contribute—regardless of how shaky the foundations might be (e.g. Kline 1982). The unsettling playfulness of computing participates in this "essentialness". I realize that I am saying this while questioning altogether the way we think about mathematics, mathematical activity *and* mathematics education, and without much elaboration on the latter: I do so elsewhere (e.g. Maheux and Proulx 2018). Let us at least agree that questioning is a healthy practice and that algorithms, their history and role in mathematical practices and in today's schools provide rich grounds for such questioning.

Some Educational Suggestions, Because Why Not?

I mentioned earlier how Netz's (2009) analysis of ancient Hellenistic mathematical texts shows that Greek mathematicians were in the habit of engaging in extravagant calculation (he uses the expression: "a carnival of calculations"). In those texts, we see them not only digressing from their intended proofs with pages of unreadable computations, but also presenting one another obviously inaccurate figures to work with:

> Aristarchus emphasizes the tentative nature of the calculation by choosing values for the simple observations that are just obviously too pat. He gives the angle Moon-Sun-Earth at the exact moment of dichotomy of the moon as 1/30 a right angle – while the correct value is far lower, about nine minutes. More troubling (since this is much easier to measure) is the very false value provided for the seen angle of the moon, 1/45 a right angle (the true value is one-fourth that; anyone who can measure anything in the sky should be able to see the difference). (p. 29)

Why is that? Because just like Archimedes' calculation of grains of sand, Aristarchus' results did not have to be right, or even close to real-world contingence. Calculation was an *art* (here, I think of the word *logistike,* or "art of calculation") and could be playful. Which does not mean that it could not eventually turn out to be useful, or less imaginary (e.g. by refinement of the initial values). But that was not its primary purpose, obviously.

Learning from that lesson, I propose that we think of engaging students in much more playful mathematics, even when it comes to computing. Calculations such as "How many apples has Benny", or more fantastic versions thereof, where apples are traded for lightsabers or baby dragons, are meaningless. We have more than enough of them already. I look here to reckonings that are *mathematically* intriguing and challenging, and rooted in an essentially mathematical ethos rather than existing in the service of something else. I want to insist here on my resistance to having such mathematical experiences justified or quantified by so-called educational goals, like the development of "number sense" or "conceptual understanding". These are for me improbable institutional constructs. Though good for the management of mathematics education, they have little to do with actual mathematical activity. They follow on from an industrial, modern approach to (mathematics) education which I personally find important to challenge on every occasion.

Mathematics is full of potentially playful occasions to compute. How about exploring digital arithmetic: the many ways of counting and calculating on the fingers or other body parts? Along the way, there might be, of course, something to encounter about the neatness of our decimal system. Why not be a bit wilder and play cryptography? From the simplest substitution scheme à la Caesar to the RSA algorithm, opportunities abound to make great play of a variety of mathematical ideas and processes. Feeling a bit riskier? Why not try to get as many decimals for pi as we can, using various formulas—not because we need them, but to see how well we can do, and to experience something about convergence. How about writing a programme that does the same thing, or a sketchier one that might look for the roots of a polynomial of various degrees? Perhaps the most subversive among us will make a play for the incommensurable pleasure of the Fermi estimate, with which one might attempt to answer questions such as "How many hours have children spent computing playlessly, and often uselessly, in school since the beginning of mandatory education?", or "How many trees are wasted each year for tissue paper to dry the tears of students who have only learned how to hate mathematics, and not how to have fun with it?". It is obvious to me how, in such cases, the problem posed by the operationalization of calculation in technological tools simply vanishes. Computing here connects with a mathematical activity that is completely beyond what computers do, since it is about *playing* with numbers, exploring, experimenting, getting a sense of some phenomena, and so on. Even better, technologies can clearly play a constructive role in those games, by helping with technicalities that are *not always* relevant to the issue, by quickly extrapolating what is done at a finer grain to produce a "big picture", and by feeding this picture back into the minute details of operations, one way or another.

Introducing students to old mathematical texts is often touted as a good way of including history in mathematics classrooms. Tzanakis et al. (2002) also suggest, for example, re-enacting parts of the history. Examining, using or designing algorithms could also play a part in reformulating students' in-class interactions with mathematics. There is something really wonderful about the way algorithms devised for simple operations compare across culture and history. Inventing new ones might be a stimulating challenge! One could even try to expand such algorithms to deal with situations beyond what they are normally used for. This is neatly exampled in Proulx and Beisiegel (2009), in which they explore Euclidian division with negative integers. What about digging into the old tricks of nine casting and proof by 7? It is said that the mathematician ibn al-Bannā (1256–1321) proposed an extension making it possible to check the result of multiplication and division of fractions (Abdeljaouad and Oaks 2012)!

I will not so endanger myself as to conjecture *what* students might learn, engaging with mathematics in such a playful manner, but something mathematically interesting could certainly take place. I have mentioned Morley's (1932) short communication about children creating algorithms; one of the most powerful observations he makes is that an operation is transformed through algorithmization. It comes to signify something different and is regarded as a *new* operation altogether. Mathematical ideas live through their representations and the ways in which we use them. Algorithms embody some of these ideas, allowing us to experience them first-hand. Historically speaking, we can also recognize this notion in the work of mathematicians who developed richer conceptualizations of mathematical concepts through their algorithmization, and were able in turn to create new algorithms reflecting those refinements. Providing students with occasions to experience mathematical ideas, and in various ways, is for me what mathematics education is all about. Algorithms can encourage such encounters, for mathematicians and students alike. As for me, that is all I ask.

References

Abdeljaouad, M. 2005. *Les arithmétiques arabes: 9e-15e siècles*. Tunis: Ibn Zeidoun.
Abdeljaouad, M., and J. Oaks. 2012. De la découverte d'al-Lubâb fî sharh 'a'mâl al-hisâb d'al-Hawârî al-Misrâtî. *Actes du 11e Colloque Maghrébin sur l'Histoire des mathématiques Arabes*, Alger-Kooba: École Normale Supérieure.
Askey, R. 1995. Cube root algorithms. *Mathematics in School* 24 (2): 42–43.
Babbage, C. 1864/2011. *Passages from the Life of a Philosopher*. Cambridge: Cambridge University Press.
Ballheim, C. 1999. How our readers feel about calculators. In *Mathematics education dialogues*, ed. Z. Usiskin, 4–5. Reston, VA: NCTM.
Baroody, A.J., and A. Dowker (eds.). 2003. *The development of arithmetic concepts and skills: Constructive adaptive expertise*. Mahwah, NJ: Lawrence Erlbaum Associates.
Borwein, J., and D. Bailey. 2003. *Mathematics by experiment: plausible reasoning in the 21st century*. Natick, MA: A K Peters.
Borwein, J., D. Bailey, and R. Girgensohn. 2004. *Experimentation in mathematics: computational paths to discovery*. Natick, MA: A K Peters.
Bretherton, I. 1984. *Symbolic play: The development of social understanding*. New York: Academic Press.
Butlen, D., and M. Peizard. 2000. Calcul mental et résolution de problèmes numériques au début du collège. *Repères IREM* 41: 5–24.
Carpenter, T.P. 1986. Conceptual knowledge as a foundation for procedural knowledge. In J. Hiebert (Ed.), *Conceptual and procedural knowledge: The case of mathematics*, 113–132. Hillsdale, NJ: Lawrence Erlbaum Associates, Inc.
Cerquetti-Aberkane, F., and A. Rodriguez. 2002. *Faire des mathématiques avec des images et des manuscrits historiques du cours moyen au collège*. Champigny-sur-Marne, France: CRDP de l'académie de Créteil.
Chabert, J.-L. (dir.), E. Barbin, M. Guillemot, A. Michel-Pajus, J. Borowczyk, A. Djebbar, J.-C. Martzloff. 1994. *Histoire d'algorithmes: du caillou à la puce*. Paris: Belin.
Chaitin, G.J. 2004. Thoughts on the Riemann hypothesis. *The Mathematical Intelligencer* 26 (1): 4–7.
Châtelet, G. 2000. *Figuring space: Philosophy, mathematics, and physics*. Dordrecht & Boston: Kluwer.
Daston, L. 1994. Enlightenment calculations. *Critical Inquiry* 21 (1): 182–202.
Davis, B. 1996. *Teaching mathematics: Toward a sound alternative*. New York: Garland Publishing Inc.
Davis, P.J. 1972. Fidelity in mathematical discourse: Is one and one really two? *American Mathematical Monthly*, 252–263.

Davis, P.J. 2006. *Mathematics & common sense: A case of creative tension.* Natick, MA: A K Peters/Boca Raton, FL: CRS Press.

de Prony, G.R. 1824. *Notice sur les grandes tables logarithmiques et trigonometriques: adaptées au nouveau système métrique décimal.* Paris: Didot.

Djebbar, A. 1997. *Matériaux pour l'étude des problèmes récréatifs de la tradition mathématique arabe* (IXe-XVe siècles). Nantes: Université d'été de Nantes.

Dowek, G. 2007. *Les métamorphoses du calcul: une étonnante histoire de mathématiques.* Paris: Le Pommier.

Eisner, E.W. 1990. The role of art and play in children's cognitive development. In *Children's play and learning: Perspectives and policy implications,* ed. E. Klugman and S. Smilansky, 43–56. New York: Teachers College Press.

Emmer, M. 1998. The mathematics of war. *ZDM* 30 (3): 74–77.

Ernest, P. 1994. *Mathematics, education, and philosophy: An international perspective.* London & Washington, D.C.: The Falmer Press.

Ernest, P. 2014. Certainty in mathematics: Is there a problem? *Philosophy of Mathematics Education Journal* 28: 1–22.

Franklin, J. 1987. Non-deductive logic in mathematics. *British Journal for the Philosophy of Science* 38 (1): 1–18.

Furinghetti, F. 1993. Images of mathematics outside the community of mathematicians: Evidence and explanations. *For the Learning of Mathematics* 13 (2): 33–38.

Ginsburg, H.P. 2006. *Mathematical play and playful mathematics: A guide for early education.* In *Play = Learning: How play motivates and enhances children's cognitive and social-emotional growth,* ed. D. Singer et al., 145–165. Oxford: Oxford University Press.

Grier, D.A. 2013. *When computers were human.* Princeton, NJ: Princeton University Press.

Hart, K.M. (ed.). 1981. *Children's understanding of mathematics.* London: John Murray.

Hilton, P. 1980. Math anxiety: Some suggested causes and cures: Part 2. *Two-Year College Mathematics Journal* 11 (4): 246–251.

Heege, H.T. 1983. The multiplication algorithm: An integrated approach. *For the Learning of Mathematics* 3 (3): 29–34.

Holton, D., A. Ahmed, H. Williams, and C. Hill. 2001. On the importance of mathematical play. *International Journal of Mathematical Education in Science and Technology* 32 (3): 401–415.

Huszár, K., and M. Rolínek. 2014. *Playful Math—An introduction to mathematical games.* Sommer campus am IST Austria. Online at: https://repository.ist.ac.at/312/1/Playful_Math.pdf.

Jankvist, U.T. 2007. Empirical research in the field of using history in mathematics education. *Nordic Studies in Mathematics Education* 12 (3): 83–105.

Kant, I. 1934/1781. *Critique of Pure Reason.* London: Macmillan.

Katz, V.J. (Ed.). 2000. *Using history to teach mathematics: An international perspective.* Washington, D.C.: The Mathematical Association of America.

Kilpatrick, J., J. Swafford, and B. Findell. 2001. The strands of mathematical proficiency. In J. Kilpatrick, J. Swafford, and B. Findell. (eds.), *Adding it up: Helping children learn mathematics* (pp. 115–155). Washington, D.C.: National Academies Press.

Kitcher, P., and W. Aspray (Eds.). 1988. *History and philosophy of modern mathematics.* Minneapolis: University of Minnesota.

Klein, J. 1992. *Greek mathematical thought and the origin of algebra.* New York, NY: Dover Publications.

Kline, M. 1982. *Mathematics: The loss of certainty.* New York, NY: Oxford University Press.

Körner, T.W. 1996. *The pleasures of counting.* Cambridge: Cambridge University Press.

Lakatos, I. 1976. *Proofs and refutations: The logic of mathematical discovery.* Cambridge etc.: Cambridge University Press.

Leontiev [Leontyev], A.N. 1979. The problem of activity in psychology. In *The concept of activity in Soviet psychology,* ed. J.V. Wertsch, 37–71. Armonk, NY: Sharpe.

Lerman, S. 1990. Alternative perspectives of the nature of mathematics and their influence on the teaching of mathematics. *British Educational Research Journal* 16 (1): 53–61.

Madell, R. 1985. Childrens' natural processes. *Arithmetic Teacher* 32: 20–22.

Maheux, J.F. 2010. *How do we know? An epistemological journey in the day-to-day, moment to-moment, of researching teaching and learning in mathematics education.* Doctoral thesis, Victoria, BC: University of Victoria.

Maheux, J.F. 2011. Epistemological issues to educational use of technology: The case of calculators in elementary mathematics. In L. Gómez Chova, D. Martí Belenguer, A. López Martínez (Eds.), *3rd International Conference on Education and New Learning Technologies, July 4th-6th, 2011. EDULEARN11 Proceedings CD.* Barcelona, Spain: International Association of Technology, Education and Development, 495–505.

Maheux, J.F., and J. Proulx. 2018. Mathematics education (research) liberated from teaching and learning: Towards (the future of) doing mathematics. *The Mathematics Enthusiast* 15 (1): 78–99.

McIntosh, A., B.J. Reys, and R.E. Reys. 1992. A proposed framework for examining basic number sense. *For the Learning of Mathematics* 12: 2–8.

Morley, A. 1932. Teaching and learning algorithms. *For the Learning of Mathematics* 2 (2): 50–51.

Netz, R. 2009. *Ludic proof: Greek mathematics and the Alexandrian aesthetic.* Cambridge: Cambridge University Press.

Neyland, J. 2001. *An ethical critique of technocratic mathematics education: Towards an ethical philosophy of mathematics education.* Doctoral dissertation, Victoria University of Wellington.

Neyland, J. 2004. Toward a postmodern ethics of mathematics education. In *Mathematics education within the postmodern,* ed. M. Walshaw, 55–73. Greenwich, CT: Information Age Publishing.

Neyland, J. 2010. *Rediscovering the spirit of education after scientific management.* Rotterdam & Boston & Taipei: Sense Publishers.

Ormell, C. 1992. *New thinking about the nature of mathematics.* Geelong, Vic.: Deakin University, Centre for Studies in Mathematics, Science and Environmental Education.

Pitri, E. 2001. The role of artistic play in problem solving. *Art Education* 54 (3): 46–51.

Plunkett, S. 1979. Decomposition and all that rot. *Mathematics in Schools* 8 (3): 2–5.

Proulx, J., and M. Beisiegel. 2009. Mathematical curiosities about division of integers. *The Mathematics Enthusiast* 6 (3): 411–422.

Rabardel, P. 1995. *Les hommes et les technologies; approche cognitive des instruments contemporains.* Paris: Armand Colin.

Rashed, R. 1997. *Histoire des sciences arabes,* Paris: Seuil.

Russel, B. 1967. *The Autobiography of Bertrand Russell: 1872–1914 (volume 1).* London: George Allen & Unwin Ltd.

Ruthven, K. 2009. Towards a calculator-aware number curriculum. *Mediterranean Journal of Mathematics Education* 8 (1): 111–124.

Sewell, B. 1981. *Use of mathematics by adults in daily life.* Leicester: Advisory Council for Adult and Continuing Education.

Shapiro, S. 2000. *Thinking about mathematics.* Oxford: Oxford University Press.

Shechtman, N., and J. Knudsen. 2009. Bringing out the playful side of mathematics: Using methods from improvisational theater in professional development for urban middle school math teachers. *Play and Performance: Play and Culture Studies* 11: 105–134.

Thompson, A.G. 1984. The relationship of teachers' conceptions of mathematics and mathematics teaching to instructional practice. *Educational Studies in Mathematics* 15 (2): 105–127.

Thompson, I. 1994. Young children's idiosyncratic written algorithms for addition. *Educational Studies in Mathematics* 26 (4): 323–345.

Tucker, K. 2014. *Mathematics through play in the early years.* London: SAGE.

Tymoczko, T. 1998. *New directions in the philosophy of mathematics.* Princeton, NJ: Princeton University Press.

Tzanakis, C., A. Arcavi, C.C. de Sa, M. Isoda, C.K. Lit, M. Niss, et al. 2002. Integrating history of mathematics in the classroom: An analytic survey. In *History in mathematics education: The ICMI Study,* eds. J. Fauvel and J. Van Maanen, 201–240. New York etc.: Kluwer Academic Publishers.

Van Oers, B. 2010. Emergent mathematical thinking in the context of play. *Educational Studies in Mathematics* 74 (1): 23–37.

Vygotsky, L.S. 1966. Igra i ee rol' v umstvennom razvitii rebenka (Play and its role in the mental development of the child; in Russian). *Voprosy psikhologii, 12*(6), 62–76. [English translation by N. Veresov and M. Barrs is published in 2016 in International Research in Early Childhood Education 7 (2): 3–25 and available online at https://files.eric.ed.gov/fulltext/EJ1138861.pdf.]

Weeks, C. 2008. Interview with Evelyne Barbin. *HPM Newsletter* 67: 1–4.

Whitton, S. 1998. The playful ways of mathematicians' work. In *Play from birth to twelve and beyond: Contexts, perspectives, and meanings*, ed. D.P Fromberg and D. Bergen, 473–481. New York and London: Garland Publishing.

Wiener, N. 1915. Is mathematical certainty absolute? *The Journal of Philosophy, Psychology and Scientific Methods*, 568–574.

Wilkes, M.V. 1956. *Automatic digital computers*. New York: Wiley.

Young-Loveridge, J., M. Taylor, S. Sharma, and N. Hāwera. 2006. Students' perspectives on the nature of mathematics. In P. Grootenboer, R. Zevenbergen & M. Chinnappan (Eds.), *Identities, Cultures and Learning Spaces*. (Proceedings of the 29th Annual Conference of the Mathematics Education Research Group of Australasia). Adelaide, SA: MERGA, vol. 2, pp. 583–590. Available online at https://researchcommons.waikato.ac.nz/handle/10289/2129.

Jean-François Maheux is a Professor of Mathematics Education in the Mathematics Department of the Université du Québec à Montréal. His work focuses on mathematical activity from an epistemological (both historical and philosophical) perspective, which often hinges on conceptualizing tools. He also explores phenomenological and deconstructive research writing within the field. His Ph.D. dissertation, completed in 2010, is entitled "How Do We Know? An Epistemological Journey in the Day-to-day, Moment-to-Moment of Researching, Teaching and Learning in Mathematics Education". Professor Maheux regularly publishes in the journal *For the Learning of Mathematics* (FLM) of which he is an associated editor. One of his recent multifarious pieces was published in the *Mathematics Enthusiast Journal* with the title "Mathematics Education (Research) Liberated from Teaching and Learning: Towards (the future of) doing mathematics".

Calculating Aids in Mathematics Education Before the Advent of Electronic Calculators: Didactical and Technological Prospects

Dragana Martinovic

Abstract This chapter presents a synopsis of ideas, movements, and reforms that influenced education in the twentieth century before the electronic calculators started being routinely used in mathematics classrooms around the world. This was an era of rapid industrialization which required efficient skilling of the population; the new approaches to how mathematics is taught and learnt were influenced by the succession of psychological theories and technological innovations. Along the way, some didactical and technological approaches were abandoned, while others kept reappearing, although altered. The readers are introduced to selected technological innovations of the period (e.g., teaching machines, thinking machines, and calculating/computing machines), along with the questions, expectations, and disappointments they gave rise to.

Keywords Teaching machines · Thinking machines · Computing machines Aspects of design and use in schools · Place in education · Psychological theories

Although calculators are considered a twentieth-century phenomenon, their mechanical predecessors existed as far back as mid-seventeenth century (e.g., Pascal's calculator), building on the much earlier ideas of calculating aids (e.g., abacus), which are explored elsewhere in this book. The first prototype of a handheld electronic calculator appeared in 1967 when Texas Instruments released their Cal Tech calculator with four arithmetic functions and a print option. But it was not until the late 1970s/early 1980s that electronic calculators reached a wider audience including, ultimately, students.

Historically, the first half of the twentieth century was a time of intensified industrialization along with the popularization of science. Industry needed skilled workers who could do more than manual work, which led to the need for more accessible education. This was also the time when the works of Sigmund Freud, Albert Einstein, and Alan Turing popularized psychiatry, science, and mathematics.

D. Martinovic (✉)
University of Windsor, Windsor, ON, Canada
e-mail: dragana@uwindsor.ca

© Springer Nature Switzerland AG 2018
A. Volkov and V. Freiman (eds.), *Computations and Computing Devices in Mathematics Education Before the Advent of Electronic Calculators*,
Mathematics Education in the Digital Era 11, https://doi.org/10.1007/978-3-319-73396-8_17

By mid-century, the rapid social change and political turmoil of a world coming out of a world war and going into the Cold War brought about dreams of intelligent machines and an expectation that they could be used for both intellectual and manual tasks. Many of these ideas had been conceived of much earlier, but now they could be revived and brought to fruition.

This chapter explores the technological and didactical prospects of various educational aids before the electronic calculators started being routinely used in mathematics classrooms around the world. It opens with an overview of ideas, movements, and reforms that influenced education in the twentieth century, followed by the initiatives to use machines to solve some of the problems in education, such as: (a) dealing with repetitive teaching and learning tasks involving drill and practice (by means of teaching machines), (b) providing one-to-one training in a specific domain of expertise (by means of thinking machines), and (c) performing complicated or long calculations accurately (by means of calculating/computing machines, see Fig. 1). While the sources for this chapter are international, the chapter focuses on the North American experience.

The distinction between these three types of educational aids is not clear-cut; some educational intelligent systems may have applied ideas of programmed instruction to evaluate student knowledge in a particular domain, or some calculators may have been equipped with the intelligent decision making. However, most of these aids were rule-based and users were required to take standard steps to receive feedback or be accurately assessed. In other words, the ideas of one period were carried forward (i.e., rule-based approach), built upon and transformed (e.g., with new technologies), and sometimes replaced as the new teaching approaches emerged (e.g., as cognitivism became more popular than behaviorism).

Didactical Prospects

The first half of the twentieth century saw several reforms in mathematics education in the USA. These reforms were guided by certain dominant theories about the psychology of learning (e.g., Edward Thorndike's theories in the early 1900s,

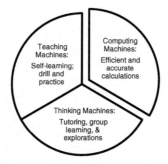

Fig. 1 Tools that affected mathematics education in the twentieth century and their main educational functions

William A. Brownell's in the 1940s, and B. F. Skinner's in the 1960s) and by movements in modern mathematics established by the 1950s (e.g., modern algebra and set theory, modern analysis and non-Euclidian geometries; Grouws and Cebulla 2000). Not only curriculum content but also teaching approaches were influenced by these reforms. Internationally, the theories of, for example, John Dewey (1859–1952; an American philosopher, psychologist, and educational reformer), Jean Piaget (1896–1980; a Swiss developmental psychologist) and Lev Vygotsky (1896–1934; a Soviet psychologist) created opportunities for mathematics educators like Guy Brousseau (1933– ; a French mathematician and mathematics educator) to make significant contributions to the field.

Influence of Psychological Theories

In the 1989 NCTM Yearbook, Lambdin Kroll made a connection between influential psychology of learning theories of the first half of the twentieth century and the elementary school mathematics curriculum in the USA of the same and later periods. She described how the mathematics curriculum went through several phases between the 1920s and the 1970s: (a) the "drill-and-practice phase" (1920–30, influenced by behaviorists such as Thorndike), followed by (b) the "meaningful arithmetic" phase (1930–50, influenced by Gestalt psychologists), and finally (c) the "new math" phase (1960–70, influenced by Bruner and Piaget).

Influence of Behaviorists

During the "*drill-and-practice*" *phase* (Lambdin Kroll 1989), pedagogy was inspired by Thorndike's theory of stimuli response (S-R bond theory), which conditioned the learner to properly respond to stimuli presented in the form of exercises grouped to exemplify a particular rule (a bond). For example, there were seven rules in a two-column addition, such as "learning to keep one's place in the column as one adds" and "learning to add a seen to a thought-of number" (Lambdin Kroll 1989, p. 201). As a consequence, (a) the mathematics curriculum became a collection of disconnected bits of knowledge, (b) learners obtained the skill of doing mathematics through small, easy steps, and (c) learners were expected to demonstrate mastery of the prescribed procedures from which any deviation was discouraged. Since Thorndike demonstrated in several studies that learning mathematics did not readily transfer to improved reasoning in other domains, this "threatened the very existence of mathematics in schools" (Grouws and Cebulla 2000, p. 213), in terms of its utility.

Drill and practice continued to be part of schooling beyond the 1930s, but stopped being solely emphasized in the curriculum. Instead, it became an incentive to design technological aids which would allow students to practice under minimal supervision and even on their own. The idea that learning through small and manageable steps could maximize success and minimize errors was at the core of programmed learning (e.g., Skinner 1965; see the section on teaching machines), which was considered most suitable for learning subjects that could be broken into small units, such as foreign languages or mathematics (Benjamin 1988). As part of the *UNESCO Educational Media Conference Recommendations*, Lumsdaine (1962) highlighted the importance of programmed instruction:

> Special emphasis should be placed on developing the potential of individual programmed-instruction methods, whose unique advantages and relatively recent development provide a major addition to our educational resources in the struggle against ignorance. (p. 340)

This report summarized the principal recommendations made by participants at the UNESCO conference in Paris on new methods and techniques in education, such as using films, individual programmed instruction, and television as teaching tools. These "[n]ew methods and techniques in education were thus seen as a great opportunity in all fields of education–in school, out of school, among adults, and for the fundamental education of illiterates" (UNESCO; as cited in Lumsdaine 1962, p. 338).

In the mid-1960s, Skinner's programmed instruction seemed to hold more excitement and promise than any other previous innovation in the whole history of education (Kulik et al. 1980a). Programmed instruction was guaranteed to help learners reach their potential faster and more fully, along with removing frustration and boredom from the classroom. But, when researchers carried out experimental and comparison studies to determine whether this form of instruction contributed positively to the cognitive and affective outcomes of students, their findings were discouraging. Nonetheless, even if the studies did not confirm the superiority of programmed instruction in terms of student achievement, they did report savings in instruction time and cost.

Consequently, the drill-and-practice programs continued to grow. At that time it was unrealistic that schools would have means to afford and utilize computers, simply put, they were too large and expensive to purchase and maintain (e.g., Blikstein 2013). Computers were thus situated in central locations, from which they served different customers, including selected schools. Suppes and Jerman (1969) describe how in 1965, in one elementary school in California, Grade 4–6 students received daily drills in arithmetic as the school had installed terminals which were connected through telephone lines to a computer at Stanford University. The lessons, reviews, and tests were organized into units, each of which emphasized a particular skill or concept (e.g., addition, fractions, and inequalities) presented at different levels of difficulty. In the 1967–68 school year alone, almost 4,000 Grades 1–6 students from 31 schools completed nearly 300,000 drill-and-practice arithmetic lessons from the Stanford program. After several years, the connectivity expanded to 15–90 teletype terminals being installed in each of the five states, so that in 1969, "approximately 2,500 students [were] receiving daily lessons in arithmetic and algebra" (p. 23).

Influence of Gestalt Psychologists

The *"meaningful arithmetic" phase* (Lambdin Kroll 1989) that followed between 1930 and 1950 was a pushback on the part of educators and theorists who believed in incidental learning and in learning related to real-life applications. During this phase, the importance of developing both concepts and skills in the context of real-world applications found support in the work of Gestalt psychologists such as Kurt Koffka, Max Wertheimer, and Wolfgang Köhler (Trafton and Suydam 1975). Trafton and Suydam reflected on this era by saying,

> In the 1930s and 1940s when the effects of programs that focused only on mastering isolated skills in a rote manner were clear, emphasis was placed on careful initial work with [arithmetic] operations, on the use of models and manipulative materials, on the rationalization of computational procedures and on the relationship to the real world. From this twenty-year effort, much was learned about developing a better perspective for teaching computation that has relevance for curriculum workers today. (p. 530)

Gestalt psychologists argued (a) that learning happens through making connections, (b) that "the course of mental development is from a broad, vague, and indefinite total to the particular and precise detail," and (c) that "the properties of parts are functions of the whole or total system in which they are imbedded" (Hartmann 1966, pp. 657–8). When implemented in a mathematics curriculum, this meant that teachers needed to provide students with conditions suitable for discovery and with activities rich with stimuli. It was suggested that this kind of instruction could track the development of a mathematical concept, rather than diligently adhering to the material presented in the curriculum. Consequently, teachers needed to give students more time to develop understanding of mathematical concepts.

William A. Brownell (1895–1977), an American educational psychologist, was a proponent of such an approach (Grouws and Cebulla 2000). He challenged Thorndike by stating that when arithmetic is meaningfully taught (i.e., when children understand why algorithms work and how concepts connect), it provides "a sound foundation for increasing a student's ability to transfer what was learned to new situations" (p. 217). Brownell's "meaning theory" formed the groundwork for the "modern mathematics movement" and generated special attention in the 1980s for the notion of "teaching for understanding."

Indeed, education in the 1930s and 1940s was considered predominantly as preparation for work, thus having social utility. At that time calculating machines were used in "business, industry and finance as well as in research and statistics" (Schlauch 1940, p. 35), which prompted Schlauch to ask, "Should the pupils be taught to perform the computations arising in the solution of problems in arithmetic by the use of machines, as they would in a similar situation in real life?" (p. 35). He argued that since the cost of a calculating machine (i.e., Monroe Calculating Machine Company's "Educator") was similar to the cost of a good typewriter, high schools should purchase enough of them so that each student could learn to operate one under the direction of a teacher.

This was truly an era of embracing various technologies as teaching aids and looking for evidence that they contribute to classroom instruction. Examples of such technologies are: (a) *radio*—Woody (1935) claimed the superiority of radio schools over control schools in an experiment involving Grades 2 and 3 students, where in radio school, an 18-week investigation consisted of having two radio lessons per week accompanied by drill sheets and a test every sixth lesson; (b) *lantern slides*—Woody (1935) described an experiment involving Grades 2 and 3 students, which did not find conclusive evidence that favored the presentation of number combinations by means of lantern slides, compared to the blackboard; (c) *various visual aids*—geometrical solids and surfaces, photographs of geometric shapes found in nature and industry; (d) *diagnostic testing devices*—a Dictaphone record was used to find an error in student/peer computations and correct them; and (e) *multisensory aids*—Fehr (1947) suggested that these bring a sense of reality to a subject without the subject losing its abstract aspects and thus the class could be turned into a mathematical laboratory.

The idea of a mathematics laboratory continued to live on throughout the twentieth century, despite shifts in the curriculum and ideology. For Ruth Irene Hoffman, a professor of mathematics at the University of Denver, a mathematics laboratory was "a state of mind," but also like

> a physical plant equipped with such material objects as calculators, overhead and opaque projector, filmstrips, movies, tape-recorder, measuring devices, geoboards, solids, graph board, tachistoscope, construction devices, etc. Since a student learns by doing, the lab is designed to give him the objects with which he can do and learn. (1968, p. 88)

In it, much like in a science laboratory, students could make discoveries through experiments, questioning, and problem-solving, while the teacher facilitated the activities that were both relevant to the students and connected to real life. With these tools, students could create flowcharts of the algorithms as they moved from a real-life problem (e.g., starting a car) to a corresponding mathematical problem. Hoffman also mentioned that teaching multisensory aids were used differently than in the more conventional classroom, for example:

> [a] dictating machine with earphones is used, sometimes to provide explanations to those with reading difficulties and sometimes for drill in basic addition and multiplication facts, for commentaries on filmstrips, for related activities. The overhead projector is used for the usual purposes of presenting prepared material, but is also used in conjunction with a copying machine to allow students to present their own flow-charts, graphs, or solutions of problems to the rest of the class. Other equipment includes calculators, scales, slide rules, standard mathematical equipment and models, and much equipment for individual student use. (1968, p. 90)

Kline (1973), who was one of the major opponents of the "new math" curriculum which arose in the 60s, promoted turning mathematics classrooms into laboratories, where students could use "clever and helpful devices" which would help them make intuitive leaps in understanding mathematics phenomena, before they move toward more rigorous analysis and proofs.

Influence of Mathematicians and Cognitive Psychologists

The "*new math*" *phase* (Lambdin Kroll 1989) that followed between 1960 and 1970 emphasized that mathematics education must be based on learning the fundamental principles of the mathematics discipline (e.g., that "elementary ideas of algebra depend upon the fundamentals of the commutative, distributive, and associative laws"; Bruner 1960, p. 19). This phase was famous for its emphasis on mathematics structures, such as sets and number theory.

Inspired by the ideas featured in a series of monographs published by a group of mathematicians under the common pen name of N. Bourbaki, Botts and Pikaart (1961) underlined that "all of mathematics can be viewed as stemming from a single axiomatic system called axiomatic set theory" (p. 504). Accordingly, the new curriculum accentuated set theory, the precise language of mathematics, symbolic manipulation, and "an understanding of the formal, logical foundations of mathematics" (Grouws and Cebulla 2000, p. 218). For Botts and Pikaart (1961), the axiomatic approach of the "new math" was illuminating and applicable even to "familiar arithmetic manipulations of elementary school" (p. 498).

Jerome Bruner, a psychologist and Harvard University professor, had ideas that fueled the new math movement, such as that, "Fourth-grade children can play absorbing games governed by the principles of topology and set theory, even discovering new 'moves' or theorems [..., although] they cannot put these ideas into formal language or manipulate them as grownups can" (Bruner 1960, p. 13). He promoted discovery-based learning and a "spiral" curriculum through which any idea could be presented to any child at an age-appropriate time: first at a more basic level for the younger child, and then later revisited on a higher, more complex and sophisticated level for an older one.

In his 1960 presentation at the National Council of Teachers of Mathematics, Bruner asserted:

> I cannot help but feel that we are on the threshold of a renaissance in education in America… With the active attitude that an emphasis on discovery can stimulate, with greater emphasis (or fewer restraints) on intuition in our students, and with a courteous and ingenious effort to translate organizing ideas into the available thought forms of our students, we are in a position to construct curricula that have continuity and depth and that carry their own reward in giving a sense of increasing mastery over powerful ideas and concepts that are worth knowing, not because they are interesting in a trivial sense but because they give the ultimate delight of making the world more predictable and less complex. (2007, p. 55)

As it often happens, some of Bruner's ideas were misinterpreted or could not be properly applied in practice, which stimulated opposition to the reform. The elementary school teachers lacked support and training in formal mathematics, and Bruner had to repeatedly explain that discovery learning is not haphazard, but has to be goal-driven and to reach an agreed-upon set of norms with respect to what is to be discovered.

While there were many critics of the "new math" as well as debates about which should be taught first—abstractions or concrete applications—Evans (1965) did not see an irreconcilable conflict between the ideas of Skinner and Bruner and those of the Gestalt psychologists. In fact, he found a common thread in Skinner's and Bruner's shared belief that with proper instruction, children could be exposed to mathematical concepts at an earlier age. Therefore, Evans looked for studies that used programmed instruction to teach advanced mathematical topics to students several grade levels below the age when this topic was usually taught. He found examples of teaching "algebraic structures to first graders..., geometry to second graders..., and equations and inequalities to bright fifth and sixth graders..." (p. 404). Other examples included bright fifth graders gaining deductive reasoning skills through mastering "specially prepared materials in mathematical logic" and first graders being taught "set concepts, including union, difference of sets, set variable, and the empty set and zero" (p. 405).

Evans also found common ground between programmed instruction and "discovery learning" proposed by the Gestalt psychologists. Using the example of learners self-discovering generalizations from a sequence of prearranged exemplars, he concluded that "proponents of 'errorless' [programmed] and 'discovery' [trial-and-error] learning, although conceptually far apart, are empirically close together" (p. 416). That it was indeed possible to combine these two paradigms, showed O'Shea (1979) with his "quadratic tutor". This Intelligent Tutoring System (ITS; see more in the section about thinking machines) was designed to adapt to student's response and to improve its teaching strategy. The tutor, which itself was based on a set of if-then rules—and thus prescriptive—taught solving quadratic equations by the discovery method. The student was given a problem of the type $x^2 + c = bx$ and asked to guess two values for x, for the given b and c. While the initial examples would be relatively easy to solve (e.g., when $c = 2$, $b = 3$, then $x_1 = 1$, $x_2 = 2$), in the next round of prompts, the tutor would increase their difficulty, until it became convinced that the student has indeed discovered the targeted rule (e.g., one of the Viète's rules, namely that $x_1 + x_2 = b$ and $x_1 * x_2 = c$). The tutor was adaptable, thus providing easier examples if the student was not "getting it," and reinforcing the rule after noting that the student started providing correct answers.

Morris Kline, a mathematician and mathematics educator, whom we introduced earlier as opponent of the "new math," wrote that "the new curricula ignore pedagogical difficulties; they ignore the lack of appeal which mathematics might have as a set (the word is occasionally useful) of rigorous deductive structures; and they ignore the broader and more significant values of mathematics in science and other areas of our culture" (1966, p. 323). Kline called for teaching mathematics as a liberal arts subject, so that students can get ideas about it and its role in the culture and the society. Beyond the arithmetic that is daily needed, the curriculum should contain mathematics related to physical situations and let the students develop their understanding constructively (as opposed to deductively). In 1973, he published an influential book against the "new math," titled, *Why Johnny Can't Add: The Failure of the New Mathematics*.

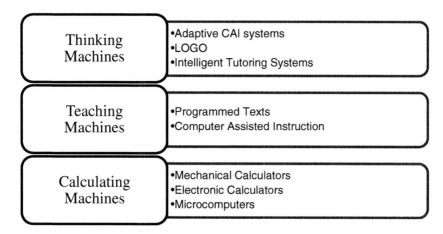

Fig. 2 Main technological ideas that influenced the twentieth-century education

Criticized as making school mathematics "too abstract, confusing, and impractical" (Grouws and Cebulla 2000, p. 219), the "new math" phase ended in the 1970s. It was followed by a new phase that was described as "a decade of direct instruction, behavioral objectives, and an increase in the use of standardized achievement tests" (p. 219). The general public had asked for a back-to-basics approach because they felt that teachers (especially in elementary school) had underestimated the importance of drill in learning mathematics. However, the term "basic" was a challenging one, as it was suggested that "it must change as society and technology change" (p. 220).

Throughout the twentieth century, novel technologies were implemented in schools (see Fig. 2), and more time and thought were dedicated to their potential effects on mathematics learning and teaching. The idea of mechanizing education was not new. After all, it seemed natural that the tools and processes used in large-scale manufacturing could also be used in education so that more people could efficiently obtain basic schooling or gain job-related skills. More details about the technological and design aspects of the teaching, calculating, and thinking machines in the era are given in the next section.

Technological Prospects: Aspects of Design

Teaching machines dealt with repetitive teaching and learning tasks involving drill and practice, calculating machines helped students to efficiently and accurately perform complicated or long calculations, while thinking machines assisted in a self-learning and a small group learning, tutoring, and experimentation. However, the technological changes had a much faster turnaround than changes these tools brought about in education practices.

Teaching Machines: Aspects of Design and Use in Schools

Various scholars have argued that using the term "teaching machine" is wrong and confusing; the words "machine" and "mechanism" did not always refer to a physical assembly of parts that work together; rather, they referred to the fact that teaching machines consisted of highly structured items and rules for using them. This inappropriate use of the terminology was a probable cause of the resentment among teachers (Tobin 1968) since they felt that such a mechanistic approach deprived education of humanism. Some teaching machines mentioned in this section were designed to support learning through practice, some were designed for testing, and some were designed for both.

The first US patent for a device for teaching arithmetic was issued to George Altman in 1897 (Räsänen 2015). Controlled by the learner, the device could pose addition, subtraction, multiplication, and division questions and would also inform the learner if the answers he or she provided were correct. In his patent documentation, Altman described one task of addition:

> The child is requested to add "3" and "2." [The child] mentally calculates that "3+2 = 5," and to ascertain if the calculation is correct, [the child] presses on the spot of the innermost [addition] circle on the result-sheet containing the number "5," so that the push-pin belonging to said circle will be depressed and a signal given, which indicates that the child is right. If any other spot on said circle is pressed on, no response whatever will be made. (Altman 1897, para. 90, 95)

In 1913, Herbert Austin Aikins (1867–1946), a psychologist and a professor at Western Reserve University in Cleveland, patented an educational aid that could be used to teach or test knowledge in "arithmetic, reading, spelling, foreign languages, history, geography, literature, or any other subject in which questions can be asked in such a way as to demand a definite form of words, letters, or symbols" (Aikins 1913, p. 1). This device consisted of a wooden case into which wooden blocks (which Aikins referred to as cards) were fitted. Aikins's device presented the learner with questions, provided a means for responding, and indicated the correctness of the response:

> The problem represented may be the sum or product of two numbers [which] may be... tested by inserting, in proper order, selected cards bearing individual numbers... [For] the determination of the product of '11x12', which being 132, the cards bearing the numerals in question are the only ones formed so as to fit, and in this order, the notches concealed within the case. (Aikins 1913, p. 2)

In the 1920s, Sidney Pressey (1888–1979), an educational psychology professor at Ohio State University, developed the Automatic Teacher, a machine suitable for teaching and testing students in different subjects (Benjamin 1988). In testing mode, the machine would record if the answer was right or wrong, then move on to the next question, while in teaching mode, it would not allow the learner to move on before he or she provided the correct answer. This machine looked like a manual typewriter with a window that showed a multiple-choice question with four or five possible answers. The student had to press one of the keys that (presumably) corresponded to the correct answer. In testing mode, when the student pressed a key, the machine recorded the answer on a counter and revealed the next question. After the test was completed, the score on the counter would inform the teacher of the number of correctly answered questions.

Pressey's machine never made a breakthrough in the classroom even though he continued to improve it until 1932 when he finally abandoned the project; it seems that the public was not ready for his product (Benjamin 1988). His motivation for creating the Automatic Teacher was not to eliminate teachers, but to relieve them from the drudgery of standardized testing so that they could focus on doing more enjoyable and creative work with their students. Also, when the machine was used in teaching mode, the students could practice at their own pace. In his design, Pressey incorporated ideas from educational psychology, expounding upon the benefits of the Automatic Teacher in: (a) providing instant feedback to the learner when the machine was in teaching mode; (b) intensifying the learner's recall of facts by providing a series of related questions; and (c) providing extrinsic motivation for further practice: The machine could be set to reward the student with a candy after a number of correct answers was received.

Contrary to his predecessors, B. F. Skinner (1904–1990), a psychologist and professor at Harvard University, was very successful in promoting his teaching machines. It seems that after the World War II and boost of economy, education in the USA was readier to accept novel ideas, such as using audiovisual tools and educational television (Benjamin 1988). It also allowed Skinner to make a breakthrough with his ideas about teaching machines and programmed learning. Skinner considered that the purpose of teaching was to expedite learning and that the purpose of mechanizing education was to expedite teaching. He distinguished between a machine and its program: The machine provided the instrumentation and the environment for a program, but it was the program itself that was fundamental for learning (Skinner 1965). From a perspective of a behaviorist, such as Skinner, the objective was to provide student with a series of carefully arranged learning sequence (i.e., the program), which will result in student demonstrating achieved goals (presumably in a testing situation).

In Skinner's interpretation, both the programmer (i.e., a designer of the program), who knows "what was to be taught and could prepare an appropriate series of contingencies," and the teacher, "who knows the student, has followed his progress, and can adapt available machines and materials to his needs" (Skinner 1965, p. 8), used the teaching machine to expedite student learning. Teaching machines could be programmed the same for the whole class or could be customized for a specific student. In both cases, the machine could help the student learn through a carefully designed sequence of small steps, each of which presented a unit of information. After receiving the student's response, the machine would provide immediate feedback (e.g., true, not true), thereby reinforcing the correct answer. Skinner was clear that for student, "the test is not given to measure what he has learned, but to show him what he has not learned and thus induce him to listen and read more carefully in the future" (p. 14).

After Skinner proposed a distinction between a machine and its program, it was soon realized by scholars that a physical machine was not necessary because programmed instruction in and of itself could achieve the teaching goals just as well. Consequently, one could replicate the function of a teaching machine in a textbook (see Table 1) because "anything which uses response-dependent information to change the learner's performance is a teaching machine (whether it is a book or a computer-based system)" (Stolurow and Davis 1965, p. 169).

According to Benjamin (1988), by "1962 there were 65 different teaching machines on the market …, and nearly 200 private companies were producing teaching machines, programmed texts, or both" (p. 708). These machines were priced from $20 to several thousands of dollars, and only within two years, 100,000 of the $20 machines were sold (Benjamin 1988).

Although the computers in the 1980s could be used for different educational purposes (e.g., to instruct, calculate, perform various tasks), they truly represented a pinnacle of a long and rich history of calculating devices. These machines existed for centuries in different forms and found their place in education before the twentieth century, but the era of industrialization and especially The World War II (Larrivee 1958a, b) intensified the need for aids which could perform complicated and long calculations both accurately and effortlessly.

Computing Machines: Aspects of Design and Use in Schools

In an article written for mathematics teachers, Larrivee (1958b) distinguishes between analog and digital computing devices; while digital devices help people to count (e.g., abacus, adding and calculating machines, and high-speed electronic digital computers), the analog devices help them to measure (e.g., graphs, charts, nomograms, slide rules, differential analyzers, and planimeters). The well-known examples include *abacus* (Chinese *suanpan*: used for adding, subtracting, multiplying, dividing, and taking the square root of values), *soroban* (the Japanese version of the Chinese abacus), *Napier's bones* (seventeenth-century Scotland: a set of numbering rods that simplified multiplication, division, and extraction of roots), and the *slide rule* (movable logarithmic scale) (Spencer 1968). These aids are still used, although their pervasiveness has diminished with the advent of modern electronic calculators.

Horton (1937), a statistician of Institute of Educational Research at Columbia University, described two types of digital computing machines used at that time in schools: *key-driven machines* (e.g., calculators or comptometers) and *non-key-driven machines*. In Horton's estimate, about 200 hours of practice or a 4- to 6-month course was needed for operators to learn how to properly use a key-driven machine. These machines had buttons with digits in columns, each column representing a place for the digit (i.e., units, tens, etc.) and could be used to add, subtract (add the 9-complement), multiply (use repeated addition), divide (use repeated subtraction), and find a square root. Since they were to be used efficiently in the workplace, these machines were designed for quick blind-typing: Each key had a large digit on it for addition and its 9-complement as a small digit for subtraction. Here are examples of addition and subtraction done on a comptometer (see Table 1):

Assuming it is desired to multiply 158 by 49: put one's fingers over 1 on the hundreds column, 5 in the tens column, and 8 in the units column. […] These three fingers should be simultaneously depressed nine times, and the result is that we have multiplied 158 by 9. Then, without disturbing the relative positions of the fingers, shift them one place to the left, and depress them four times, the effect of which is to multiply 158 by 40. […] If we take the Comptometer with the register set at zero, and press the key 7, the figure 7 shows in the register. If we want to subtract 7 (thus reducing the register again to zero) we work from the small figures of the keys, choosing the key with a small figure one below the number we desire to subtract [i.e., 6, since 6 = 7-1]. Accordingly, in this case, we select the key with the small figure 6, i.e., the key with the large figure 3 [i.e., 3 is a 9-complement of 6, as 9-3 = 6]. Depressing this key, when we already have 7 on the register, would cause the register to show 10; but, by slightly pushing forward with the finger or thumb of the left hand a little projecting lever, we prevent the machine from carrying, so that, while the units column is altered to zero, there is no 1 carried forward to the tens column. (Dicksee 1916–1918, pp. 54–56, as cited on http://www.officemuseum.com/calculating_machines_key_driven.htm)

Table 1 Examples of subtraction as addition to 9-complement as described in Dicksee and Horton[a]

Examples	Result with subtraction	Use a 9-complement of 1 less than subtrahend	Use addition	Result without carryforward
Dicksee (1916–18)	7 −7 0	9 −(7−1) 3	7 +3 10	7 +3 0
Horton (1937)	100 −25 75	99 −(25−1) 75	100 +75 175	100 +75 75

[a]Such a method of subtraction in its binary form (as the 1's complement and the 2's complement) has been implemented in the architecture of most computers

The non-key-driven machines (Horton 1937) were much easier to operate and did not require special training because they used operations in a straightforward form (without complements). Enthusiastically, Horton proposed that calculating machines should be "the standard equipment of any progressive school to be used not only in the office practice classes, but wherever mathematics is taught" (p. 276), and even at the graduate level, especially for extensive statistics applications. "There is hardly a position in any capacity where the ability to use one or more of the various types of calculating machines would not be advantageous; in some cases, it is absolutely necessary" (p. 271), she wrote. This article published by the National Council of Teachers of Mathematics in their journal for mathematics teachers brought to teachers' attention different types of calculating machines as well as the necessity to equip students with skills for using them. In the meantime, schools were using a spectrum of computing devices (e.g., Larrivee 1958b) and some more sophisticated calculating machines especially suitable to prepare students to work on the job-related clerical, billing, and bookkeeping tasks.

Felix Klein (1849–1925), a German mathematician and mathematics educator, who is regarded as a precursor in the use of technology in schools (e.g., he created a mechanical calculating machine for children; Borba and Bartolini Bussi 2008), believed that calculating machines with their speed and reliability would eventually make logarithmic tables redundant. Once the issue of their high cost was resolved, he wrote, "a new phase of numerical calculation will be inaugurated" (Klein 1932, p. 174). That was indeed well predicted. Through the technological advances in the later twentieth century, which brought about the invention and propagation of microcomputers, teachers could make available to their students multifunctional devices, such that could calculate, instruct, and much more; teaching machines became computer-controlled and subsequently influenced by the ideas coming from artificial intelligence (AI) research. The time had arrived when thinking machines could be made to help people do their intellectual work or study.

Thinking Machines: Aspects of Design and Use in Schools

In his book, *Thinking Machines*, Irving Adler (1913–2012), a mathematician and prolific writer, stated that "we have entered the age of 'thinking' machines and automation" (1961, p. 9). Although the author made a parallel between humans and thinking machines (i.e., computers), he did not claim that the process of "thinking" is the same for both. According to Adler, there are two types of human thinking: one that is creative and unpredictable, and the other that is routine; only the latter could be associated with what the thinking machines could do. Albeit the brain consists of nerve cells while the computer of that era consisted of vacuum tubes or transistors, Adler found multiple parallels between the two: Both could receive input and produce output, and both have a memory and a control unit. However, the brain had more capacity and was faster than any computer of that time. To Adler, these comparisons were important because "...the more we develop computers, the better we shall understand them. And the better we understand our computers, the better we shall understand ourselves" (Adler 1961, p. 184).

The AI projects of the mid-twentieth century strived "to fully duplicate the human capacities of thought and language on a digital computer" (Winograd 1991, p. 201). Researchers approached this goal with two objectives in mind: one to explain in mechanical terms how the human mind functions (which aligned with Adler's ideas), and the other to create intelligent (thinking) machines that could serve some purpose (i.e., apply expertise in the domain). These, so-called, expert systems used a set of basic facts as their domain of knowledge and a set of "if–then" rules of logical and heuristic reasoning to function.

According to Winograd, the AI researchers' ambitious goal was based on two flawed premises: (a) that the human mind has a mechanistic and deterministic nature, and (b) that the mind's functioning can be modeled in a manageable number of steps. Winograd wrote that "the optimistic claims for artificial intelligence have far outstripped the achievements, both in the theoretical enterprise of cognitive modeling and in the practical application of expert systems" (1991, p. 208).

Since its early projects, AI has taken several different and diverging routes and has established itself as a viable discipline. Of specific interest in this chapter is the work done in providing computer-assisted instruction (CAI), intelligent tutoring systems (ITS), and LOGO, a learning environment and a programming language rooted in AI.

Computer-Assisted Instruction. The early educational computer programs were designed for recall of facts (e.g., arithmetic and vocabulary, such as in Uhr 1969). These were essentially automated flash card systems, which provided student with a simple question (e.g., calculation) and, after receiving the student's response, could present the results of student's performance on the task. However, these early programs could not change their actions based on the student response and thus required to have stored all correct and likely incorrect responses at interaction points with the student. The adaptation capability was necessary for the computer program to perform intelligently (Shute and Psotka 1994), to the extent that the term "adaptive" was sometimes used interchangeably with the term "intelligent" system (Nwana 1990).

In his review of how computers could be used to assist in the educational process, Hansen (1966) described characteristics of an adaptive model for tutoring designed by Stolurow. In this model, adaptability was related to having a number of different teaching programs for a given subject, and using the student's existing performance, ability, or personality measures to make decisions on what teaching program to offer to the student on the fly. Using various performance measures (i.e., speed or accuracy of response), the system would build a model of the student's learning. The idea of self-learning, which permeated the teaching machines era, was also a driving force of the tutoring programs which had an intelligent component to them.

Intelligent Tutoring Systems. According to Nwana (1990), intelligent tutoring systems (ITS) emerged from the Skinnerian programs of the 1950s. These computer programs were designed to provide individualized instruction and feedback, along with knowledge of the domain. For example, *WEST* (designed by Brown and

Burton in 1978) provided tutoring in basic arithmetic skills; *LMS* (designed by Sleeman and Smith in 1981) did tutoring in basic algebra; and *GEOMETRY Tutor* (designed by Anderson, Boyle, and Yost in 1985) helped with high school geometry proofs. The early ITSs were designed to have knowledge of: (a) the domain (domain expert model; what the student needs to know), (b) the learner (student model; what the student already knows), and (c) teaching strategies (what unit of instruction should be taught next, and how it should be presented). In effect, a properly designed ITS was supposed to compare its student model with its domain expert model to identify the best way to alter its tutoring strategy.

According to Spiers (1996), ITSs could use one of two strategies: (a) one aimed at diagnosing and repairing the student's misconceptions about a given topic (i.e., the performance-driven approach) and (b) one where students had to follow the teaching strategy regardless who they are and whether they understood the material or not (i.e., the tutor model-tracing approach). Both strategies, however, had drawbacks. In the first approach, where all diagnostics were performance-driven, it was impossible to design a system that could fix a student's misconceptions, since there would always be systematic errors and bugs that could not be easily diagnosed (VanLehn 1990). The tutor model-tracing approach, on the other hand, meant imposing a method or methods that might be very different from the student's. The tutor would seemingly know the best approach, but this was not always the case and could consequently lead to the necessity for overly elaborate explanations to the envisioned typical student (Nwana 1993).

Shute and Psotka (1994) describe a challenge of creating "a computerized instructional system to help second graders learn double-digit addition" (p. 5). When adding multiple-digit numbers, students can make various errors—because they do not understand the place value (i.e., do not carry or carry even when it is not necessary), or because they cannot add single digits accurately, or are careless:

> If student A answered the following two problems as: $22 + 39 = 61$, and $46 + 37 = 83$, you'd surmise (with a fair amount of confidence) that A understood, and could successfully apply, the "carrying procedure." But consider some other responses. Student B answers the same problems with 51 and 73 [fails to carry a one to the ten's column], student C answers with 161 and 203 [incorrectly adding one's column result to the ten's column], and student D answers with 61 and 85 [mistakenly adding $6 + 7 = 15$]. (p. 5)

An intelligent system would recognize different sources of these errors and remediate accordingly. Shute and Psotka suggested that ITSs should do that by taking a mid-way approach "between too much and too little learner control" (p. 28), and by being more responsive to the student shifting needs.

Close to the end of the twentieth century, Shute and Psotka cautioned the ITS community that: "There are actually very few ITS in place in schools, yet they exist in abundance in research laboratories. We need to move on" (1994, p. 50). As the ITSs continued to improve, 15 years later, the well-known Carnegie Mellon University researchers in this domain, Aleven et al. (2009) stated that with the evolvement of relatively simple authoring tools for customization of ITSs without use of programming, the ITSs will become widespread.

LOGO. Compared to the ITSs which were created by different research groups around the world and for students of different levels, Seymour Papert's LOGO system was created for young children by the research group at Massachusetts Institute of Technology (MIT) in 1967. Papert had an idea that children should learn to think like mathematicians rather than simply learn about mathematics. He knew Jean Piaget and accepted Piaget's challenge to create a constructivist environment for learning mathematics. By manipulating and directing the moves of a turtle figure on a computer, children could develop algorithmic, or what Papert referred to as, computational thinking (Papert 1980).

McCorduck (1979) visited the LOGO environment at the MIT which at that time consisted of computer terminals of which one was connected to a TV. The child could control the movements of the white dot (the "turtle") on the screen by instructing it where to move (and possibly to leave a trace behind). Through programming the turtle's moves, the child could draw, decompose, combine, multiply, and transform geometric figures. The LOGO environment also contained a device that could produce music based on the child's commands as well as a simple pictogram keyboard for young children who could not yet read and write. McCorduck explained:

> The name Turtle Geometry comes from the fact that in addition to the cathode ray tubes, or TV screens, there are small mechanisms at the LOGO project with humped backs that rather resemble turtles. They crawl over the bare tile floor, attached by those haphazard wires to the computer and [are] manipulate[d] by a child at the terminal. The turtles have pens in their innards, and with a command PENDOWN, can execute on paper (or on the floor) any design that the child commands. PENUP signals the end. (1979, p. 298)

To control the turtle's movements on the screen or on the floor, the child would count the steps, calculate the distances, and consider the angles that the turtle should make to reach the goal or to follow a pattern. LOGO did not provide tools for drill in arithmetic, but it provided

> [A] context in which there is a real need for these processes and in which children must clearly conceptualize which operation they should apply. For example, first-grade children determined the correct length for the bottom line of their drawing by adding the lengths of the three horizontal lines that they constructed at the top of the tower: 20+30+20=70 (Clements 1983–84). (Clements and Sarama 1997, p. 15).

The examples presented in this section describe thinking machines that were designed to teach students mathematics (e.g., CAI and ITS) and to help children develop thinking skills (e.g., LOGO). While both the CAI and ITS utilized self-learning, LOGO was envisioned for a social learning, preferably in small groups and under teacher's guidance.

Dreams, Criticisms, and Disagreements

This chapter has portrayed some of technological innovations used in mathematics education during the twentieth century. As new sciences (e.g., computer science, cognitive science) and new technologies emerged, mathematics educators became excited about the prospects they held for the subject. Discussions around drill versus meaning, purpose of school mathematics, and place of computers in education continued to fuel discussions among educators even in this day and age.

Is There an Irreconcilable Dichotomy Between Drill and Meaning?

As early as in 1935, Brownell emphasized that "[a]rithmetic is best viewed as a system of quantitative thinking. [...] If one is to be successful in quantitative thinking one needs a fund of meanings, not a myriad of 'automatic responses.' [...] Drill does not develop meanings. Repetition does not lead to understandings." (Brownell 1935, p. 10; cited in Weaver and Suydam 1972, p. 10). Furthermore, he wrote, "[u]ntil teachers are differently selected and differently trained, it is fruitless to expect them adequately to teach children arithmetic through incidental experience" (Brownell 1935, p. 18; cited in Weaver and Suydam 1972, p. 11).

To replace the existing drill-based teaching, Brownell proposed the "meaning theory," which

> ...conceives of arithmetic as a closely knit system of understandable ideas, principles, and processes. According to this theory, the test of learning is not mere mechanical facility in "figuring." The true test is an intelligent grasp upon number relations and the ability to deal with arithmetical situations with proper comprehension of their mathematical as well as their practical significance. (Brownell 1935, p. 19; cited in Weaver and Suydam 1972, p. 11)

Overall, the period 1935–1960 saw recurring discussions that stressed (a) the rejection of the mechanical approach to learning, with its principal concern for skill mastery, and (b) the acceptance of the Gestalt psychology approach, with its focus on learning as a growth process. In addition, there was general agreement that the development of skill should come after the development of meaning and understanding (Weaver and Suydam 1972).

However, not everybody agreed with Brownell's ideas or how they were applied; in 1945, even Brownell himself criticized distortion of his ideas in practice. Swain (1960), for example, cautioned that in 25 years since its inception, "[m]eaningful learning [became] for [educational leaders and writers on arithmetic] a catchall phrase useful for justifying whatever pedagogical procedure they may advocate [p. 272]" (as cited in Weaver and Suydam 1972, p. 14). Also, the studies were showing that results of programmed and conventional instructions are comparable in their effects on mathematics learning—as summarized in a meta-analysis of 89 studies in elementary and secondary schools (Hartley 1977).

The fact that programmed instruction was less effective than computer-assisted instruction (i.e., drill-and-practice systems or computer-based tutoring systems with each student working at his/her own rate; Hartley 1977) and that even Skinner (1986) thought that the small computer is "the ideal hardware for Programmed Instruction" (p. 110) and encouraged integration of the two. Moreover, as Trafton and Suydam (1975) argued, even in a curriculum oriented toward discovery and understanding, there was still a need for drill and practice; nor did drill have to be boring and based only on "flash cards or endless paper-and-pencil exercises" (p. 533). On the contrary, they wrote:

> The proliferating use of hand-held calculators adds a new consideration to the elementary school mathematics curriculum. Calculators may decrease the need for drill and practice at more advanced algorithmic levels, but they will not eliminate it. Children need a base on which to build mathematical ideas, and computational facts and skills help to form this base. With careful planning, the calculator can serve to strengthen the base; it should be used to facilitate the development of mathematical ideas and algorithmic learning and not merely as a crutch that the child must carry with him. (Trafton and Suydam 1975, p. 534)

All the same, the discussion about the place of drill and practice in mathematics education continues even today. Should mathematics be taught as a liberal arts subject as Morris Kline suggested in 1966?

What Is the Purpose of Teaching Mathematics in School?

The Second Report of the NCTM Commission on Post-War Plans (1945) stated:

> We must conceive of arithmetic as having both a mathematical and a social aim. The fundamental reason for teaching arithmetic is represented in the social aim. No one can argue convincingly for an arithmetic which is sterile and functionless. If arithmetic does not contribute to more effective living, it has no place in the elementary curriculum. To achieve the social aim of arithmetic, children must be led to see its worth and usefulness. (p. 200)

After reflecting on the US elementary and high school mathematics curriculum in the twentieth century and the language that permeated media of the time, Schoenfeld (2001) argued that "throughout the century, the rhetoric has been changed, but that (with the exception of the implementation of the 'New Math,' which represented a major curricular shift) changes in practice were not nearly as dramatic as the rhetoric that surrounded them" (p. 244). According to Schoenfeld, the practice was such that throughout most of the twentieth century, the US elementary school mathematics was arithmetic-oriented and calculation-heavy; high school mathematics in mid-century did not build on what students learned in elementary school; and high school mathematics was mostly considered to be preparation for work. At the same time, the "new math" was rejected by parents and teachers, mostly because it was impenetrable to them, and consequently, it was replaced in the 1970s by the "back-to-basics" movement. This conundrum was "the result of a fundamental conflict of paradigms" (Ellis and Berry III 2005, p. 7), such as learning mathematics through drill and practice at the start of the century, followed by unleashing ideas about utility of mathematics (i.e., steering students toward vocational courses), creating the pretentious "new math" program, and then falling back to the shallowness of the back-to-basics movement.

On a more positive note, Schoenfeld (2001) presented mathematics education in the last quarter of the century as a unification of mind (cognitive revolution, artificial intelligence, and expert systems based on the human problem-solving processes), mechanism (information processing, knowledge organization), and culture (situated cognition). On a less positive note, it is questionable to what extent these powerful theoretical advancements reformed the school practice in equally powerful ways. Is it possible for the formal learning environments to unify different paradigms, as Evans suggested in 1965?

What Is the Place of Calculating/Computing Machines in Mathematics Education?

In his seminal book, *Elementary Mathematics from an Advanced Standpoint: Arithmetic-Algebra-Analysis*, Klein (1932) concluded the chapter in which he described the principles of operation of calculating machines:

> [...] with the wish that the calculating machine, in view of its great importance, may become known in wider circles than is now the case. Above all, every teacher of mathematics should become familiar with it, and it ought to be possible to have it demonstrated in secondary instruction. (Klein 1932, p. 22)

Similar to Pressey, Skinner, and other inventors of teaching machines who were hoping for "an 'industrial revolution' in education" (as cited in Benjamin 1988, p. 707), for Klein, calculating machines were relieving mathematicians from the "mechanical work of numerical calculation [and] error" (1932, p. 21). In the late 1950s, "farsighted educators began dreaming about a computer age in higher education" (Kulik et al. 1980b, p. 525). The ideas of Pressey, Skinner, Klein, and others were repeated in envisioning computers as infinitely patient tutors, scrupulous examiners, and tireless schedulers of instruction; both teachers and students in these imagined classrooms would have a more pleasant and productive work experience.

Similar sentiments about revolutionary changes in education existed among the LOGO community. Since its induction, LOGO provided a vision of changes that computers could make in the society (McCorduck 1979) and Papert was acknowledged for creating "a revolutionary programming language, the first designed expressly for use by children, at a time when computers used to fill entire rooms and were impossibly complicated" (Blikstein 2013, para. 6).

Kieren (1977), a Canadian professor of mathematics education, anticipated that computing technology would profoundly affect mathematics education. He suggested that algorithm design and coding might be key activities in one's formal development of mathematics concepts and that computational power of technology allows one to progress toward advanced mathematics through exploration, which were exactly the ideas that motivated Papert to create LOGO.

However, the evidence of benefits from using these technologies was lacking; Beck (1960) reported that "[we] believe the calculator [is] valuable as a teaching tool in the intermediate grades, but at the present time, we are not certain how much it improves learning, when it should be introduced, or what methods and materials are most effective" (p. 103). A decade later, Cech (1970) could not confirm that the use of electric-printing desk calculators (with the four whole number operations) improves the attitude toward mathematics of low-achieving ninth-grade general mathematics students, although the students did compute better with calculators than without them.

Even though results from various studies of the LOGO system and its impact on education have been decidedly inconclusive (Voogt et al. 2015), Papert's work prompted ongoing discussions on the importance of students' computational thinking skills, such as planning and debugging, designing and executing algorithms (Burkhardt 1985), and learning mathematics "by direct programming rather than by indirect instruction" (Nwana 1990, p. 251).

As the pendulum of reforms turned teachers from the "'false promises' of 'modern mathematics'" (Shult 1981, p. 181) toward the back-to-basics ideas, the students were expected to master basic mathematics skills before they start relying on calculators. For Shult, "one of the most complex, controversial issues in mathematics education [in the 1980s was] the use of hand-held calculators in schools" (p. 181).

In contrast, ITSs have remained of interest to educators over decades because, among else, they could be adapted to changing trends in mathematics education. Although they originated from the ideas of drill and practice and were criticized for being relatively good at passing on knowledge, but poor at teaching how to use the knowledge (Milech et al. 1993), some ITSs embraced a constructivist approach to education. While the student acts with both a goal and a plan for reaching that goal in mind, the ITS behaves as a more knowledgeable peer, providing supports within the student's zones of proximal development (Vygotsky 1978).

The debate has continued, stimulating research efforts to establish the advantages in using calculators and computing devices in mathematics classes. Can calculators truly strengthen the basic skills, without becoming a habitual crutch for mathematics learners; can they help students develop mathematical ideas and algorithmic learning, as Trafton and Suydam suggested in 1975?

Conclusions: The Confluence of Machines and Mathematics Education

This chapter has presented the main technological and pedagogical ideas in mathematics education of the first 80 years or so of the twentieth century and has shown the use of machines in mathematics education as both cyclical (i.e., ideas keep recurring) and transforming. The intent was to walk the readers through the technological innovations of the period, along with the questions, hopes, and disillusions they gave rise to. The history of the International Congress on Mathematics Instruction, as reported by Furinghetti et al. (2008), shows how in relatively short time the discourse changed from discussing using concrete materials in mathematics teaching (i.e., games, worksheets, films, manipulatives, and overhead projectors; ICME-1, 1969) to adopting Papert's LOGO Turtle and TVs (ICME-2, 1972).

This was an era when various technologies were promoted as teaching aids, such as films, radio, television, lantern slides, visual and multisensory aids, and mathematical laboratory (Hoffman 1968). While new educational methods were developed, employing individual programmed instruction, discovery-based learning, and real-life mathematics applications, education was gradually abandoning the model of transmitted knowledge in favor of a knowledge growth model. Also computers, once when they became miniaturized and affordable, started leaving research centers to find their place in schools. Through their ever-increasing capacity, they could potentially serve different functions—teaching, calculating, and thinking—of the machines described in this chapter.

However, Benjamin's (1988) words were prophetic when he, after reflecting on the history of teaching machines, reminded us that "if past behavior is a predictor of future behavior, then it seems unlikely that computers or any other teaching machines will play more than a supporting role in the classroom" (p. 711). Throughout most of the twentieth century, formal education system remained slow to change and continued to lag behind technological advances in society as a whole. Of the three types of machines discussed here, calculating machines were most successful in their spread. During the last decades of the century, personal calculators became a common companion at work and in school and are nowadays even available on mobile phones. Yet, the history tells us that the more intelligent and helpful technology becomes, there is a growing fear that its users may become "de-skilled" in functions they use technology for (Hoffman and Militello 2009). We see that in the persistence of the "back-to-basics" ideas (Grouws and Cebulla 2000; Schoenfeld 2001) and questions around when to use calculators in mathematics classes (Shult 1981).

Mathematics educators still wonder: "Do new technologies transform old ones or erase them?" (Borba and Bartolini Bussi 2008, p. 298). Transformation can be seen in the creation of virtual manipulatives that are based on physical ones. Regardless that, for example, a Tower of Hanoi in its virtual form asks for the same algorithmic solution as its physical counterpart, the process of using the manipulative virtually, on the computer, requires different sets of skills and engages the student's mind differently. On the other hand, some tools, such as compass and abacus, are disappearing from the classrooms because of the availability of dynamic geometry software and calculators; Felix Klein's prediction that calculating machines will make logarithmic tables redundant was justified. The question that Borba and Bartolini Bussi ask is the question that the readers of this book may ponder over too: "Does mathematics remain untouched as different technologies are used?" (2008, p. 298). But also, how are the learning and teaching of mathematics affected?

Using the historical lens and revisiting some "old" ideas, hopes, and discussions is both revealing and inspirational. We saw how during the mid-twentieth century, modern mathematics had a very strong influence on mathematics education, but the mathematics and education communities were each divided in their view on the "new math" era. In the meantime, mathematics education continued to evolve as "a field of study and practice distinct from mathematics" (Kilpatrick 2008, p. 25). As mathematics education continued to be part of the heated debates, its history shows that development of computational skills, in particular, and learning and teaching of mathematics, in general, remained of interest for various entrepreneurially minded individuals, psychologists, mathematicians, didacticians, and computer scientists.

References

Adler, I. 1961. *Thinking machines: A layman's introduction to logic, Boolean algebra, and computers*. New York, NY: The John Day Company.

Aikins, H. A. 1913. *Educational appliance*. Retrieved from https://www.google.com/patents/US1050327.

Aleven, V., B.M. McLaren, J. Sewall, and K.R. Koedinger. 2009. A new paradigm for intelligent tutoring systems: Example-tracing tutors. *International Journal of Artificial Intelligence in Education* 19 (2): 105–154.

Altman, G. G. 1897. *Apparatus for teaching arithmetic*. Retrieved from http://www.google.com/patents/US588371.

Beck, L.L. 1960. A report on the use of calculators. *The Arithmetic Teacher* 7 (2): 103.

Benjamin, L.T. 1988. A history of teaching machines. *American Psychologist* 43 (9): 703–712.

Blikstein, P. 2013. *Seymour Papert's legacy: Thinking about learning, and learning about thinking*. Available at https://tltl.stanford.edu/content/seymour-papert-s-legacy-thinking-about-learning-and-learning-about-thinking.

Borba, M., and M. Bartolini Bussi. (2008). Resources and technology throughout the history of ICMI (working groups—reports). In *The first century of the international commission on mathematical instruction (1908–2008). Reflecting and shaping the world of mathematics education*, ed. by M. Menghini, F. Furinghetti, L. Giacardi, and F. Arzarello, 289–300. Roma: Istituto della Enciclopedia Italiana.

Botts, T., and L. Pikaart. 1961. Mathematics from the modern viewpoint. *The Mathematics Teacher* 54 (7): 498–504.

Brownell, W.A. 1935. Psychological considerations in the learning and the teaching of arithmetic. *The National Council of teachers of mathematics. The Tenth Yearbook. The Teaching of Arithmetic*, 1–31. New York, NY: Teachers College, Columbia University.

Bruner, J.S. 1960. *The process of education*. Cambridge, MA: Harvard University Press.

Bruner, J.S. 2007. On learning mathematics. *The Mathematics Teacher* 100: 48–55 (Paper presented before The National Council of Teachers of Mathematics, Salt Lake City, Utah, August 1960). Reprinted from *The Mathematics Teacher* 53: 610–619.

Burkhardt, H. 1985. Computer aware curricula: Ideas and realisation. In *The Influence of computers and informatics on mathematics and its teaching, proceedings from a symposium held in Strasbourg, France in March 1985*, ed. by R. F. Churchhouse et al., 147–155. Cambridge etc: Cambridge University Press. http://dx.doi.org/10.1017/CBO9781139013482.015.

Cech, J.P. 1970. *The effect the use of desk calculators has on attitude and achievement in ninth-grade general mathematics classes*. Unpublished thesis at School of Education, Indiana University, Bloomington.

Clements, D.H., and J. Sarama. (1997). Research on LOGO: A decade of progress. In *Logo: A retrospective*, ed. by D. Maddux Cleborne, and D. LaMont Johnson, 9–46. Binghamton, NJ: The Haworth Press, Inc.

Commission on Post-War Plans of the NCTM. 1945, May. Second report of the commission of post-war plans. *The Mathematics Teacher* 38: 195–221.

Ellis, M.W., and R.Q. Berry III. 2005. The paradigm shift in mathematics education: Explanations and implications of reforming conceptions of teaching and learning. *The Mathematics Educator* 15 (1), 7–17.

Evans, J.L. 1965. Programming in mathematics and logic. In *Teaching machines and programmed learning II, data and directions*, ed. R. Glaser, 371–440. Washington, D.C.: National Education Association.

Fehr, H. 1947. The place of multisensory aids in the teacher training program. *The Mathematics Teacher* 40: 212–216.

Furinghetti, F., M. Menghini, F. Arzarello, and L. Giacardi. 2008. ICMI Renaissance: The emergence of new issues in mathematics education. In *The first century of the international commission on mathematical instruction (1908–2008). Reflecting and Shaping the World of Mathematics Education*, ed. by M. Menghini, F. Furinghetti, L. Giacardi, and F. Arzarello, 131–147. Roma: Istituto della Enciclopedia Italiana.

Grouws, D. A., and K. L. Cebulla. 2000. Elementary and middle school mathematics at the crossroads. In *American education: Yesterday, today and tomorrow, Part II*, ed. by T. L. Good, 209–255. Chicago, IL: University of Chicago Press.

Hansen, D.N. 1966. Computer assistance with the educational process. *Review of Educational Research* 36 (5): 588–603.

Hartley, S. S. 1977. *Meta-analysis of the effects of individually paced instruction in mathematics*. Unpublished doctoral dissertation, University of Colorado.

Hartmann, G.F. 1966. Gestalt psychology and mathematical insight. *The Mathematics Teacher* 59 (7): 656–661.

Hoffman, R.I. 1968. The slow learner—Changing his view of mathematics. *NASSP Bulletin* 52 (87): 86–97.

Hoffman, R.R., and L.G. Militello. 2009. *Perspectives on cognitive task analysis: historical origins and modern communities of practice*. New York, NY: Psychology Press.

Horton, E.M. 1937. Calculating machines and the mathematics teacher. *The Mathematics Teacher* 30 (6): 271–276.

Kieren, T.E. 1977. Mathematics education research in Canada: A prospective view. In *Educating teachers of mathematics: The Universities' responsibility. Proceedings of the Canadian Mathematics Education Study Group (CMESG)*, ed. by A.J. Coleman, W.C. Higginson, and D.H. Wheeler, 2–21. Kingston, ON: Queen's University.

Kilpatrick, J. 2008. The development of mathematics education as an academic field. In *The first century of the International Commission on Mathematical Instruction (1908–2008). Reflecting and shaping the world of Mathematics education*, 25–39. Roma: Istituto della Enciclopedia Italiana.

Klein, F. 1932. *Elementary mathematics from an advanced standpoint: Arithmetic-algebra-analysis*. Translated from the third German edition by E. R. Hedrick and C. A. Noble. London, England: Macmillan and Co.

Kline, M. 1966. A proposal for the high school mathematics curriculum. *The Mathematics Teacher* 59 (4): 322–330.

Kline, M. 1973. *Why Johnny can't add: The failure of the new math*. New York: St. Martin's Press.

Kulik, J.A., P.A. Cohen, and B.J. Ebeling. 1980a. Effectiveness of programmed instruction in higher education: A meta-analysis of findings. *Educational Evaluation and Policy Analysis* 2 (6): 51–64.

Kulik, J.A., C.-L.C. Kulik, and P.A. Cohen. 1980b. Effectiveness of computer-based college teaching: A meta-analysis of findings. *Review of Educational Research* 50(4): 525–544.

Lambdin Kroll, D. 1989. Connections between psychological learning theories and the elementary mathematics curriculum. In *New Directions for Elementary School Mathematics: 1989 Yearbook*, ed. P.R. Trafton, 199–211. Reston, Virginia: National Council of Teachers of Mathematics.

Larrivee, J.A. 1958a. A history of computers I. *The Mathematics Teacher* 51 (6): 469–473.

Larrivee, J.A. 1958b. A history of computers II. *The Mathematics Teacher* 51 (7): 541–544.

Lumsdaine, A.A. 1962. UNESCO educational media conference recommendations. *Audio Visual Communication Review* 10 (6): 338–342.

McCorduck, P. 1979. *Machines who think: A personal inquiry into the history and prospects of Artificial Intelligence*. San Francisco, CA: W. H. Freeman and Company.

Milech, D.K., G.R. Kirsner, and B. Waters. 1993. Applications of psychology to computer-based tutoring systems. *International Journal of Human-Computers Interaction* 5 (1): 23–40.

Nwana, H.S. 1990. Intelligent tutoring systems: An overview. *Artificial Intelligence Review* 4: 251–277.

Nwana, H.S. 1993. An approach to developing intelligent tutors in mathematics. *Computers & Education* 20 (1): 27–43.

O'Shea, T. 1979. A self-improving quadratic tutor. *International Journal of Man-Machine Studies* 11 (1): 97–124.

Papert, S. 1980. *Mindstorms: Children, computers, and powerful ideas*. New York, NY: Basic Books.

Räsänen, P. 2015. Computer assisted interventions on basic number skills. In *The Oxford handbook on numerical cognition*, ed. by R. Coohen Kadosh and A. Dowker, 745–766. Oxford, UK: Oxford University Press.

Schlauch, W.S. 1940. The use of calculating machines in teaching arithmetic. *The Mathematics Teacher* 33 (1): 35–38.

Schoenfeld, A.H. 2001. Mathematics education in the twentieth century. In *Education across the century: The centennial volume. Part I*, ed. by L. Corno, 239–278. Chicago, IL: National Society for the Study of Education.

Shult, D. 1981. Calculator use in schools. In *Calculators in the classroom: With applications for elementary and middle school teachers*, ed. D. Moursund, 181–183. New York: Wiley.

Shute, V.J., and J. Psotka. 1994. *Intelligent tutoring systems: Past, present and future*. Interim Technical Paper for Period April 1M3–April M4. Air Force Materiel Command Brooks Air Force Base, Texas. Retrieved from www.dtic.mil/dtic/tr/fulltext/u2/a280011.pdf.

Skinner, B.F. 1965. Reflections on a decade of teaching machines. In *Teaching machines and programed learning II, data and directions*, ed. R. Glaser, 4–20. Washington, D.C.: National Education Association.

Skinner, B.F. 1986. Programmed instruction revisited. *Phi Delta Kappan* 68 (2): 103–110.

Spencer, D.D. 1968. Computers: Their past, present, and future. *The Mathematics Teacher* 61 (1): 65–75.

Spiers, G.F. 1996. *An analogical reasoning based mathematics tutoring system*. Unpublished doctoral dissertation, Computing Department Lancaster University, Lancaster, Great Britain.

Stolurow, L.M., and D. Davis. 1965. Teaching machines and computer-based systems. In *Teaching machines and programed learning II, data and directions*, ed. R. Glaser, 162–212. Washington, D.C.: National Education Association.

Suppes, P., and M. Jerman. 1969. Computer assisted instruction at Stanford. *Educational Technology* 9: 22–24.

Tobin, M.J. 1968. *Teaching machines in programmed instruction*. A contribution to the UNESCO seminar on Programmed Instruction, Varna, Bulgaria, August 1968.

Trafton, P.R., and M.N. Suydam. 1975. Computational skills: A point of view. *The Arithmetic Teacher* 22 (7): 528–537.

Uhr, L. 1969. Teaching machine programs that generate problems as a function of interaction with students. In *ACM '69: Proceedings of the 1969 24th national conference*, 125–134.

VanLehn, K. 1990. *Mind Bugs: The origins of procedural misconceptions*. Cambridge, MA & London, UK: The MIT Press, A Bradford Book.

Voogt, J., P. Fisser, J. Good, P. Mishra, and A. Yadav. 2015. Computational thinking in compulsory education: Towards an agenda for research and practice. *Education and Information Technologies* 20: 715–728.

Vygotsky, L. 1978. *Mind in society*. Boston, MA: Harvard University Press.

Weaver, F.J., and M.N. Suydam. 1972. *Meaningful instruction in mathematics education*. ERIC Information Analysis Center for Science, Mathematics, and Environmental Education, Columbus, Ohio.

Winograd, T. 1991. Thinking machines: Can there be? Are we? In *The boundaries of humanity: Humans, animals, machines*, ed. J.J. Sheehan, and M. Sosna, 198–223. Berkeley, CA: University of California Press.

Woody, C. 1935. Arithmetic. *Review of Educational Research* 5 (1): 14–30.

Dragana Martinovic is a Professor of Mathematics Education at University of Windsor, ON, Canada, where she leads the Human Development Technologies Research Group. Dragana is a Fellow of the Fields Institute for Research in Mathematical Sciences and a Co-Director of the Fields Centre for Mathematics Education. In her research, she explores ways in which technology can improve teaching and learning of mathematics. She is dedicated to providing opportunities for all involved in education to collaboratively work toward increasing student engagement, student success, and love for mathematics.

Index of Technical Terms

© Springer Nature Switzerland AG 2018
A. Volkov and V. Freiman (eds.), *Computations and Computing Devices
in Mathematics Education Before the Advent of Electronic Calculators*,
Mathematics Education in the Digital Era 11, https://doi.org/10.1007/978-3-319-73396-8

Index of Individuals

© Springer Nature Switzerland AG 2018
A. Volkov and V. Freiman (eds.), *Computations and Computing Devices
in Mathematics Education Before the Advent of Electronic Calculators*,
Mathematics Education in the Digital Era 11, https://doi.org/10.1007/978-3-319-73396-8

Index of Primary Materials

© Springer Nature Switzerland AG 2018
A. Volkov and V. Freiman (eds.), *Computations and Computing Devices in Mathematics Education Before the Advent of Electronic Calculators*, Mathematics Education in the Digital Era 11, https://doi.org/10.1007/978-3-319-73396-8

Printed by Printforce, the Netherlands